FOUR BILLION YEARS AND COUNTING

Canada's Geological Heritage

Robert Fensome, Graham Williams, Aïcha Achab, John Clague,
David Corrigan, Jim Monger, and Godfrey Nowlan (editors)

Copyright 2014: Canadian Federation of Earth Sciences

All rights reserved. No part of this book may be reproduced, stored in a retrieval system or transmitted in any form or by any means without the prior written permission from the co-publishers, or, in the case of photocopying or other reprographic copying, permission from Access Copyright, 1 Yonge Street, Suite 1900, Toronto, Ontario M5E 1E5.

Co-publishers: Nimbus Publishing and the Canadian Federation of Earth Sciences

Website: www.EarthSciencesCanada.com /4by (selected illustrations available for download)

Editorial Board: Robert Fensome (chair), Graham Williams, Aïcha Achab, John Clague, David Corrigan, Jim Monger, and Godfrey Nowlan

Publication Committee: Jennifer Bates (chair), Sandra Barr, Thomas Clark, Robert Fensome, Elisabeth Kosters, and Godfrey Nowlan

Technical Editor: Kate Kennedy; additional technical editing by Clare Goulet

Special Advisors to the Editorial Board: Richard Grieve, Andrew MacRae, Raymond Price, Christy Vodden, and Chris Yorath

French Edition Committee: Aïcha Achab and Thomas Clark

French Editorial Review Committee: Aïcha Achab, Thomas Clark, Mathilde Renaud, and Marc Bélanger.

Fund Raising Committee: Graham Williams (chair), Aïcha Achab, Peter Bobrowsky, Peter Dimmell, Michael Enachescu, Linda Ham, Michel Jebrak, Judi Lentin, David Lentz, and Grant Wach

IYPE Fund Raising Drive: J. D. (Jim) Reimer, John Boyd, and Godfrey Nowlan

Graphics Coordinator: Bill MacMillan; additional graphics by Richard Franklin and Philip O'Regan; and paleogeographic maps by Ron Blakey.

Layout and Design: Neil Meister, Meisterworks Graphic Design

Review Committee: Jennifer Bates, Sandy McCracken, and Robert Fensome

Website: Godfrey Nowlan, Skinnyfish Media, and Fleiva Media

Cover: Artwork by Christopher Hoyt

Library and Archives Canada Cataloguing in Publication

Four billion years and counting : Canada's geological heritage / Robert Fensome, Graham Williams, Aïcha Achab, John Clague, David Corrigan, Jim Monger, and Godfrey Nowlan (editors) ; Canadian Federation of Earth Sciences.

Includes bibliographical references and index.
ISBN 978-1-55109-996-5 (pbk.)

1. Geology—Canada—History. 2. Geology—Canada—Pictorial works. I. Fensome, Robert Allan, 1951– , editor II. Canadian Federation of Earth Sciences.

QE13.C3F68 2014   557.1   C2014-903173-4

Frontispiece:
The Rocky Mountains may be Canada's best-known landscape. Part of the range is a UNESCO World Heritage Site. Although they do not reach the heights of the Coast and St. Elias Mountains, the Canadian Rockies attract and inspire many visitors. And their striking geology has motivated many budding geologists. This is a view of the Kananaskis Country of Alberta. RON GARNETT / AIRSCAPES.CA.

Nimbus Publishing acknowledges the financial support for its publishing activities from the Government of Canada through the Canada Book Fund (CBF) and from the Province of Nova Scotia through Film & Creative Industries Nova Scotia. We are pleased to work in partnership with Film & Creative Industries Nova Scotia to develop and promote our creative industries for the benefit of all Nova Scotians.

Major Supporters

Geological Survey of Canada (Natural Resources Canada)

Canadian National Committee for the International Year of Planet Earth
(this committee adopted the book as one of its major projects and approached
sponsors including Nexen Inc. and the chapter supporters)

Nexen Energy ULC

Canadian Geological Foundation

Ministère des Ressources naturelles du Québec

Individual Chapter Supporters
(see list on next page)

Chapter Supporters

Chapter 1 (On the Rocks): John Maher

Chapter 2 (Dance of the Continents): James D. Reimer

Chapter 3 (It's About Time): Roger Smith, Suncor Energy (retired)

Chapter 4 (Fossils and the Bush of Life): Ian Young

Chapter 5 (From Stardust to Continents): Ken Gillis, Hudbay Minerals Inc.

Chapter 6 (Laying the Foundations): The Mining Association of Nova Scotia

Chapter 7 (Southern Sojourn): John Boyd, RPS Boyd PetroSearch

Chapter 8 (Crossing the Equator): Michael Rose, founder,
Berkley Petroleum, Duvernay Oil and Tourmaline Oil

Chapter 9 (Pangea Breaks Up): Dale Leckie, Nexen Energy ULC

Chapter 10 (Final Approach): John 't Hart, Talisman Energy (retired)

Chapter 11 (The Ice Age): Stephen Marston

Chapter 12 (Forging a Nation): Robert Horn

Chapter 13 (Having the Energy): Clayton Riddell, founder, Paramount Resources Ltd.

Chapter 14 (Building Canada): Institut national de la recherche scientifique, Québec

Chapter 15 (Water: A Clear Necessity): Canadian Chapter of the I. A. H.,
in memory of Canada's groundwater pioneers

Chapter 16 (At the Beach): Stephan Benediktson, Taylor Hill Exploration Ltd.

Chapter 17 (On Dangerous Ground): Darcy Marud

Chapter 18 (Environmental Challenges): Canadian Geoscience Education Network,
in memory of Ward Neale

Chapter 19 (Toxins in the Rocks): Tamaratt Teaching Chair in Geoscience,
University of Calgary

Chapter 20 (Canada's Geological Heritage): Scott Jobin-Bevans, Caracle Creek
International Consulting Inc., in memory of Douglas Robert Bowie (1942–2008)

The Atlantic Geoscience Society provided funding for the acquisition
of additional illustrative materials.

TABLE OF CONTENTS

Introduction 1

PART 1: FOUNDATIONS

Chapter 1: On the Rocks 4
Graham Williams, Aïcha Achab, Robert Fensome, Catherine Hickson, Alain Leclair, Andrew MacRae, Andrew Miall, Alan Morgan, Godfrey Nowlan, and Brian Pratt

Box 1: Crystal Clear 19
Graham Williams, Pierrette Tremblay, and Robert Fensome

Box 2: Getting to the Core 22
Raymond Price, Jim Monger, and Graham Williams

Chapter 2: Dance of the Continents 24
Brendan Murphy, Jim Monger, Jean Bédard, Robert Fensome, Andrew Hynes, Michel Malo, Andrew Miall, Raymond Price, and Graham Williams

Chapter 3: It's About Time 38
Andrew Okulitch, Robert Fensome, John Gosse, Jim Monger, Léopold Nadeau, Godfrey Nowlan, and Graham Williams

Box 3: Mapping, Probing, and Sensing 45
Godfrey Nowlan, John Clague, Ron Clowes, Sonya Dehler, Gordon Fader, Robert Fensome, Wayne Goodfellow, Roger Macqueen, Raymond Price, and Graham Williams

Chapter 4: Fossils and the Bush of Life 51
Godfrey Nowlan, Aïcha Achab, Robert Fensome, Brian Pratt, and Graham Williams

Box 4: Spheres of Influence 65
John Clague, Raymond Price, and Graham Williams

Box 5: The Lay of the Land 69
Robert Fensome, David Corrigan, Paul Fraser, Fran Haidl, Jim Monger, and Graham Williams

PART 2: THE EVOLUTION OF CANADA

Chapter 5: From Stardust to Continents: before 2,500 million years ago 74
Wouter Bleeker, David Corrigan, Andrew Knoll, John Percival, Robert Fensome, Brian Pratt, and Graham Williams

Chapter 6: Laying the Foundations: Canada 2,500 to 750 million years ago 85
David Corrigan, Andrew Knoll, Rob Rainbird, Bruce Ryan, Thomas Clark, Robert Fensome, and Graham Williams

Chapter 7: Southern Sojourn: Canada 750 to 444 million years ago 99
Keith Dewing, Sandra Barr, Ronald Blakey, Fran Haidl, Christopher Harrison, Denis Lavoie, Jim Monger, Guy Narbonne, Bruce Sanford, Cees van Staal, Aïcha Achab, Doug Boyce, Robert Fensome, Patricia Gensel, Ian Knight, Michel Malo, Michael Melchin, Godfrey Nowlan, Raymond Price, Rob Rainbird, David Rudkin, Graham Williams, and Graham Young

Box 6: Walcott's Legacy: the Burgess Shale 122
Jean-Bernard Caron and David Rudkin

Box 7: The Green Revolution 126
Howard Falcon-Lang, Robert Fensome, Patricia Gensel, and Graham Williams

Chapter 8: Crossing the Equator: Canada 444 to 251 million years ago 130
Christopher Harrison, Sandra Barr, Ronald Blakey, Martin Gibling, Fran Haidl, Denis Lavoie, Jim Monger, Cees van Staal, Wayne Bamber, Doug Boyce, John Calder, Maurice Colpron, Keith Dewing, Robert Fensome, Patricia Gensel, Ian Knight, Michel Malo, Michael Melchin, Jim Monger, Pierre Jutras, JoAnne Nelson, Godfrey Nowlan, Raymond Price, David Rudkin, Bruce Sanford, Cees van Staal, Graham Williams, Graham Young, and John-Paul Zonneveld

Chapter 9: Pangea Breaks Up and Mountains Rise: Canada 251 to 65.5 million years ago 160
Jim Monger, Ronald Blakey, David Eberth, Christopher Harrison, Maurice Colpron, Fabrice Cordey, Sonya Dehler, Carol Evenchick, Robert Fensome, Andrew MacRae, JoAnne Nelson, Godfrey Nowlan, Paul Olsen, Terry Poulton, Raymond Price, Arthur Sweet, François Therrien, Hans Wielens, Graham Williams, Chris Yorath, Darla Zelenitsky, and John-Paul Zonneveld

Box 8: Growing Fur 189
John Storer, Natalia Rybczynski, Graham Williams, and Robert Fensome

Chapter 10: Final Approach: Canada 65.5 million years ago to today 192
Robert Fensome, Ronald Blakey, John Clague, Christopher Harrison, Jim Monger, John Storer, Graham Williams, Sonya Dehler, Alejandra Duk-Rodkin, David Eberth, Catherine Hickson, Dale Leckie, Andrew MacRae, Rolf Mathewes, Walter Nassichuk, Raymond Price, Natalia Rybczynski, Hans Wielens, Marie-Claude Williamson, and Chris Yorath

Chapter 11: The Ice Age 216
Lynda Dredge, Alwynne Beaudoin, John Clague, Alan Morgan, Gilbert Prichonnet, Robert Fensome, and Graham Williams

Box 9: Blowing Hot and Cold 240
John Clague and Martine Savard

PART 3: WEALTH AND HEALTH

Chapter 12: Forging a Nation 246
Wayne Goodfellow and Benoît Dubé

Chapter 13: Having the Energy 261
Martin Fowler, Hans Wielens, Graham Williams, Charles Jefferson, Lesley Chorlton, Fran Haidl, Denis Lavoie, Jim Monger, Reuben Murphy, Godfrey Nowlan, and Grant Wach

Chapter 14: Building Canada 278
Dixon Edwards, Lawson Dickson, Robert Fensome, Robert Ledoux, and Randall Miller

Box 10: Québec City's Extraordinary Geological Legacy 293
Robert Ledoux, Pascale Côté, Dixon Edwards, and Robert Fensome

Chapter 15: Water: a Clear Necessity 298
Stephen Grasby

Chapter 16: At the Beach 307
Philip Hill, Donald Forbes, and Bernard Long

Chapter 17: On Dangerous Ground 320
John Clague and Andrée Blais-Stevens

Box 11: Extraterrestrial Visitors 333
Ann Therriault, Robert Fensome, and Graham Williams

Chapter 18: Environmental Challenges 338
Nick Eyles and Michael Lazorek

Chapter 19: Toxins in the Rocks 346
Patricia Rasmussen, David Gardner, and Michel Camus

Chapter 20: Canada's Geological Heritage 354
Robert Fensome and Graham Williams

Box 12: Milestones 374
Robert Fensome and Graham Williams

INTRODUCTION

The vastness of Canada is nowhere better appreciated than in the Prairie Provinces. This aerial view is of the South Saskatchewan River near Saskatoon, Saskatchewan. RON GARNETT / AIRSCAPES.CA.

Canada is vast, with an area of almost 10 million square kilometres and a coastline of over 202,000 kilometres—the longest of any nation. These are well-celebrated facts. Less celebrated, but perhaps worthy of greater wonder, is Canada's trove of geological treasures. Canada's geological story spans 4 billion years or more, from Earth's oldest-known rocks to those being created today on the sea floor off British Columbia. Surprises abound: few people know, for example, that Canada once had the largest lake ever known, or that its rocks have yielded the oldest evidence for sexual reproduction, the remains of some of the first animals, and the bones of the earliest-known reptile. In recent decades, geologists have come to the amazing conclusions that some 1.9 billion (or 1,900 million) years ago, parts of Saskatchewan and Manitoba were separated by a huge ocean; that 500 million years ago, parts of Nova Scotia were attached to what is now Africa; that about 450 million years ago, Revelstoke in British Columbia's interior was located near the edge of the continental shelf;

that a mere 45 million years ago, a dawn-redwood forest thrived on Axel Heiberg Island, then as now in the high Arctic; and that 10,000 years ago (barely a blink of the eye in geological terms), the site on which Ottawa now stands was largely submerged by the sea. Such evidence from Canadian rocks and fossils demonstrates that our planet's surface is constantly changing—and has been doing so for a long time.

Four Billion Years and Counting explores the exciting history revealed by Canada's geology and how the natural resources, which are the foundation of Canada's wealth and development, were formed.

The book is divided into three sections:

Foundations explores the basic concepts of topics such as the rock cycle, plate tectonics, geological time, and the fossil record. Geology-savvy readers may opt to skip this section and leap directly into the second section.

Evolution of Canada takes us from the Big Bang, about 13.75 billion years ago, to the present. We reveal how Canada's modern landscape has geology at its foundations, and we investigate the country's spectacular fossil record and how it contributes to our understanding of the evolution of life.

In *Wealth and Health*, we look at contemporary economic and social issues that are rooted in Canada's geological past and its rocks. We delve into the origins and exploration of natural resources, the whys and wherefores of natural hazards, and environmental and health issues that relate to geology.

The writing of *Four Billion Years and Counting* has been a team effort of Canada's geoscience community, involving more than a hundred contributors, who have provided up-to-date information and insights from their fields of expertise. Please join us in the exhilarating voyage through time that is Canada's geological heritage.

An early amphibian navigates its way around submerged stems of the horsetail *Calamites* in this scene based on fossil finds from rocks about 310 million years old in the Maritime Provinces. STEPHEN GREB.

PART 1 · FOUNDATIONS

The collision of tectonic plates causes the Earth's crust to be uplifted on a regional scale. As the surface rises it is sculpted by erosion into mountains and valleys, as at North Howser Tower in the Kootenays of southeastern British Columbia. JOHN SCURLOCK.

1 • ON THE ROCKS

CHAPTER SUPPORTED BY A DONATION FROM JOHN MAHER

This view, from Skihist Provincial Park, British Columbia, shows the Thompson River flowing through a precipitous canyon before joining the Fraser River at Lytton. The scene shows how the rock cycle can have an impact on everyday life. WALTER LANZ

Niagara

As they hunted caribou, mammoth, and mastodon over 10,000 years ago, early Paleo-Indian explorers would surely have witnessed the falls of the Niagara River. Although the identity of the first European to see Niagara Falls is unknown, we do know that Samuel de Champlain heard about them from the Iroquois in 1604 and that Louis Hennepin, a member of the La Salle Expedition of 1678, was the first European to write of them in detail. Today, three great falls plunge over the edge at Niagara. Almost 800 metres wide and over 50 metres high, the mainly Canadian Horseshoe Falls is the most impressive. The American Falls and Bridal Veil Falls, both within the United States, are smaller and lower. The idea that the retreating falls had created the Niagara Gorge was first proposed by three brothers, Andrew, Benjamin, and Joseph Ellicott, in 1790. They estimated that the lip had receded by about 6 metres over 30 years. By extrapolating back in time and along the length of the cut, the brothers concluded that Niagara Gorge was about 55,000 years old, considered an unbelievable (and blasphemous) span of time by most of their contemporaries.

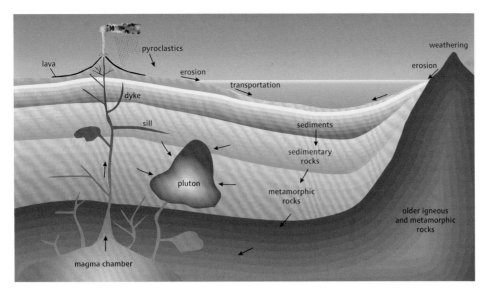

The rock cycle: an eternal process of recycling rocks through erosion, deposition, subsidence, melting, cooling, uplift, and other processes. Igneous rocks started out as molten material that solidified either at the surface as volcanic (extrusive) rocks or below the surface as intrusive rocks in structures such as plutons and dykes. Erosion of igneous and other rocks produces sediments that may become sedimentary rocks. The effects of heat and pressure on all pre-existing rocks may change (metamorphose) them to metamorphic rocks and might eventually melt them to produce molten magma—and the cycle starts again. FROM COLMAN-SADD AND SCOTT (1994). USED WITH PERMISSION OF THE AUTHORS, THE GEOLOGICAL ASSOCIATION OF CANADA, AND THE GEOLOGICAL SURVEY OF NEWFOUNDLAND AND LABRADOR.

Visiting Niagara Falls in 1841, Victorian geological superstar Charles Lyell enthusiastically wrote that "The sun was shining full upon them—no building in view—nothing but the green wood, the falling water, and the white foam." Lyell estimated that the river had gradually excavated the great gorge by slowly eroding the Falls back over 11 kilometres in about 35,000 years. That this process continues today is clear from the occurrence of major rock falls. Lyell also noticed a gap in the rocky walls of the Whirlpool section of Niagara Gorge, perhaps reflecting a previous course of the river. Later studies confirmed the existence of this route, St. David's Buried Valley, which led downstream from the Whirlpool prior to the last advance of the glaciers that once cloaked much of Canada.

Thus the story of Niagara Falls and Gorge involves glaciation and erosion. But these are widespread processes, so what else would explain this famous landmark? The main reason for the existence of the Falls (and for the Niagara Escarpment of which the Falls are part) is the presence of a hard limestone that forms a cap over which the Niagara River plummets at the Falls on its relentless journey from Lake Erie to Lake Ontario. Beneath the caprock are less resistant shales. These softer rocks are rapidly eroded by swirling water, which undercuts the overlying caprock, causing chunks of it to collapse. Today we know that the limestone and shale, now gently tilted westward, were deposited in a tropical continental sea some 425 million years ago (Chapter 8).

Steeped in legend, tradition, and hoopla, Niagara Falls is one of those places where vacationers come face-to-face with geology, and where we can start exploring one of the most fundamental of natural systems—the rock cycle. It's easy to see how Niagara Falls has eroded its lip and cut back its gorge over recent millennia. The boulders at the base of the falls resulted from this erosive process. It's not hard to imagine that finer eroded particles, such as sand and mud, have been swept away in the torrent toward Lake Ontario and ultimately the sea. Wherever they eventually settle and accumulate, these sedimentary particles may build up and become buried to such depths that they are compressed into sedimentary rocks, similar to those exposed in the eroding rock wall behind the Falls. So we begin to get a sense of a cycle. The full rock cycle, though, involves more than just erosion, sediments, and sedimentary rocks, as we will see in this chapter.

Rocks, which are made up of minerals (Box 1), are of three types: igneous, sedimentary, and metamorphic; these might be thought of as fire rocks, wind-and-water rocks, and changed rocks, respectively. The rock cycle connects all three types to each other and involves processes such as weathering, erosion, transportation, crushing, and melting.

A view of the Niagara River's Horseshoe Falls from the Canadian side. The boulders to the left at the base of the falls testify to the erosive power of the River as it plummets over the precipice. KEITH VAUGHAN.

Ash cloud from the 2010 eruption of Eyjafjallajökull volcano in Iceland, looking northeast. Fine ash from this volcano travelled in the atmosphere as far as northwestern Europe, causing disruptions to flights and the closure of major airports. ALAN MORGAN.

Hot Rocks

The Earth formed from cosmic dust particles that coalesced and melted over 4,500 million years ago, meaning that the earliest rocks formed on Earth—derived from that primordial melt but now lost—must have been igneous. Igneous rocks form from the cooling and crystallization of molten material called magma, which exists within the Earth at temperatures of 700 to 1,300°C. As magma cools, mineral crystals separate out; generally, the faster the cooling, the smaller the crystals. Magma that reaches the surface is called lava. Under surface conditions, lava cools rapidly to form fine-grained volcanic rocks that have crystals usually too small to be seen by the naked eye. Because volcanic rocks form when magma is extruded from the Earth's interior, we call them extrusive. Extrusive rocks can form from lava flows or explosively produced particles of lava such as volcanic ash and bombs. Rocks derived from such fragments are called pyroclastic rocks, or just pyroclastics. Occasionally, magma cools so quickly that it doesn't crystallize at all, instead producing volcanic glass, also known as obsidian.

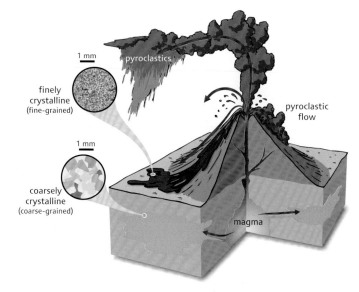

A volcano, partly cut away, to show how magma reaches the surface and how it is then distributed as lava and pyroclastics.

Columnar basalt near Mount Garibaldi, British Columbia. Mount Garibaldi is one of the volcanoes of the Cascade Range, which extends from southwestern British Columbia to northern California. PAUL ADAM.

A boulder of 200-million-year-old basalt from Newport Landing, Nova Scotia. When molten, the lava gave off bubbles of gas, leaving vesicles once the rock had solidified, and these were later filled by crystals (white) precipitated from mineral-rich water percolating through the rock. ROB FENSOME.

A block of rhyolite from Campbellton, New Brunswick. As the original lava cooled and set about 410 million years ago, bands of different composition were stretched out into layers, like multicoloured toffee. HEINZ WIELE, COURTESY OF THE ATLANTIC GEOSCIENCE SOCIETY.

Pillow lava from near the Lillooet Glacier, southwestern British Columbia. JOHN CLAGUE.

A volcanic rock's colour commonly gives clues about the minerals it contains. Dark-coloured igneous rocks get their colour from dark minerals such as pyroxene and olivine, which are both rich in magnesium and iron. The first two letters of magnesium, along with the chemical symbol for iron (Fe), provide our term for these dark igneous rocks and their molten precursor: mafic. Because they have low silica (silicon dioxide) content, mafic magmas and lavas are runny or, more technically, are said to have low viscosity. Rocks that are composed entirely of mafic minerals, and thus have very low silica content, are called ultramafic.

Basalt, a mafic volcanic rock that makes up vast areas of ocean floor, is the most abundant rock on the planet's surface. Being runny in its molten state, basalt tends to flood large areas rather than to build classic cone-shaped volcanoes. Sometimes basalt forms what are known as shield volcanoes, as in Hawaii, so named because of their shape. Basaltic lavas can be extruded from elongate fissures as well as from circular vents. When basalt flows stop moving before they solidify, cooling and consequent contraction produces a polygonal pattern of cracks, resulting in a columnar structure. Underwater, basaltic lava commonly forms pillows, as cold water rapidly chills the outside of a blob (pillow) of fluid lava. Within the chilled margin, the lava is still molten, and this breaks through the thin outer crust of the pillow to form another pillow, a process that is repeated until all the lava cools.

Light-coloured (or felsic) volcanic rocks, such as rhyolite and dacite, have large amounts of the generally light-coloured mineral feldspar and a high silica content. (The term felsic comes from feldspar and from Si, the chemical symbol for silicon.) This makes their parent lavas more viscous (less runny, or more like molasses) than mafic lavas.

Andesite is a type of fine-grained volcanic rock, so named because it is common in the Andes Mountains of South America. In silica content and viscosity it is intermediate, falling between basalt and rhyolite, and may be explosively shot out of volcanoes or form slow-moving lava flows. Felsic and intermediate magmas tend to extrude from central vents and craters. They produce classic cone-shaped structures, known as stratovolcanoes and formed from a mix of flows and pyroclastics. Perhaps the most famous stratovolcano is Mount Fuji in Japan. Very viscous lava can plug the volcano's vent, blocking the magma's escape route; eroded remnants of such plugs are called necks. Gases and pressure may build up within a volcano, especially one whose outlet is plugged, leading to enormous explosions in which hot gases, ash, and volcanic bombs are hurled thousands of metres into the air, or cascade down the volcano's flanks as deadly avalanches known as pyroclastic flows. All the explosive activity, steep slopes, and rapid changes associated with volcanoes cause a lot of breakage; imagine keeping your china cabinet on the flanks of an active volcano. This disruption commonly results in a type of rock known as volcanic breccia (pronounced "brechia"), which consists of larger angular fragments within a much finer-grained matrix of ash or mud.

Plumbing the Depths

In the plumbing system beneath a volcano, magma may cool and crystallize in place to form intrusive rocks, so called because they have been intruded into pre-existing rocks. Because they cool underground and hence more slowly, intrusive igneous rocks are generally coarser grained than volcanic rocks. Crystal size can vary, depending in part on the depth at which the magma cools. Some types of minerals crystallize earlier than others, so the rock may consist of large crystals (phenocrysts) in a fine-grained matrix of other minerals, a bit like cherries in a cake; such a rock is called a porphyry.

Bodies of magma commonly solidify several kilometres underground as large masses called plutons or batholiths, the latter term usually denoting the largest such bodies. An example of an intrusive rock commonly found in plutons and batholiths is granite, usually formed by cooling and crystallization over thousands to millions of years. Most of the Coast Mountains, from Vancouver to southwestern Yukon, are underlain by granitic plutons that cooled and crystallized between 170 and 45 million years ago.

Granitic magmas are usually produced by partial melting of rocks at depths of 25 to 100 kilometres. As temperature rises to 600 to 900°C, silica-rich minerals tend to melt, but iron- and magnesium-rich minerals do not. Where present, water lowers the temperature at which rocks melt and so accelerates the process. Blobs of melt separate from the host rock and collect to form granitic magma. This magma then migrates toward the surface, either because the melt is less dense than the surrounding rock, or because movements within the crust squeeze it upward like toothpaste. As the melt rises, a process called stoping occurs: the

This polished specimen of granite from southern New Brunswick is part of a monument to Quebec historian François-Xavier Garneau near the House of Assembly in Québec City. ROB FENSOME.

Gabbro, about 420 million years old and probably from New Brunswick, was used for gravestones of *Titanic* victims in Fairview Lawn Cemetery, Halifax, Nova Scotia. ROB FENSOME.

Remnant of what was originally a block of sedimentary rock enclosed by granite of the South Mountain Batholith, a large intrusion dating from 375 million years ago. Such inclusions of surrounding rock within an igneous intrusion are known as xenoliths, and many are visible in the granite around Peggys Cove Lighthouse, Nova Scotia. ROB FENSOME.

This 590-million-year-old dyke near Powassen, Ontario, cuts through older, lighter-coloured rock. WOUTER BLEEKER.

Aerial view of resistant, mafic sills, about 120 million years old, within more easily eroded, dark-grey marine shales, over 200 million years old, in the Blue Mountains, northwest Ellesmere Island, Nunavut. ASHTON EMBRY.

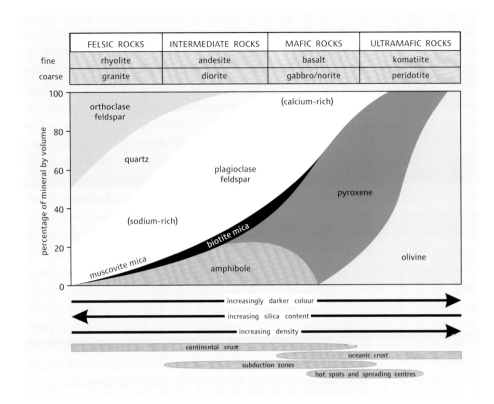

Different igneous rock types have different mineral compositions, different physical properties such as grain size and colour, and different distributions. Explanations for the features represented by the four bars at the base of the diagram are provided in Box 2 and Chapter 2. ADAPTED FROM VARIOUS SOURCES WITH INPUT FORM BARRIE CLARKE.

Cool Rocks

Rocks are broken down by freezing and thawing, by the action of chemicals in rain and groundwater, and by the activities of animals and plants. This in-place breakdown of rocks is called weathering. Once rocks have been weathered, fragments are eroded away by physical forces such as gravity, wind, rain, currents, waves, and ice. Eventually the fragments accumulate as sediment, layers of clay, silt, sand, and gravel at the bottom of rivers, lakes, or seas (as fluvial, lacustrine, and marine deposits respectively). When buried for long intervals of time, sediments are generally transformed into sedimentary rocks. This change results from processes such as compaction and cementation—the cement derived from the precipitation of materials such as silica and calcite (calcium carbonate) from groundwater. The entire conversion process from sediment to sedimentary rock is called diagenesis. Through diagenesis, gravel becomes conglomerate (or breccia if the fragments are angular rather than rounded), sand becomes sandstone, and silt and clay become siltstone, claystone, or (if easily split along bedding surfaces) shale. In the field, siltstone, claystone, and shale can be difficult to tell apart and so can be collectively called mudstone. We refer to all fragmental deposits as clastic sediments, or clastics. Sedimentary rocks usually occur in layers called beds, collectively termed strata. Stratigraphy is the study of these bedded, stratified sedimentary rocks.

melt surrounds and isolates blocks of the host rock, which break off and fall into the molten mass, eventually melting to become part of the magma. In this way, the magma may absorb large volumes of the host rock.

Sometimes magma moves upwards along fractures and solidifies, forming sheets of rock called sills if they are sandwiched between other rock layers, or dykes if not. Dykes and sills commonly occur in swarms, sometimes in giant radiating systems such as the Mackenzie and Franklin dyke swarms in the Northwest Territories and Nunavut. Igneous rocks in sills and dykes tend to have a grain size intermediate between that of volcanic rocks and that of plutons and batholiths.

Many names exist for igneous rocks based on grain size and composition, a few of which are shown in the illustration above. Wherever possible in this book we use general names such as mafic and felsic or, for example, granitic for rocks that are similar to granite but not precisely a granite in the technical sense. We also occasionally use the word crystalline. Geologists use this term to describe rocks composed of tightly interlocking crystals, and thus it can be applied to most igneous rocks, and also most of the metamorphic rocks that we describe below. Crystalline rocks contrast with granular rocks, usually sedimentary, such as sandstones.

Clastic sediments and sedimentary rocks vary in composition. Clay is composed of tiny, flat crystals of so-called

Tree roots disrupt a resistant sandstone ledge, thus contributing to its weathering and erosion, near Chutes-de-la-Chaudière, Charny, Quebec. ROB FENSOME.

Erosion in the badlands of Dinosaur Provincial Park, near Brooks, Alberta. Weathering breaks down the rocks into sediment, which is moved downhill by gravity and water. The sediment accumulates in small rills as shown here, but may eventually end up in rivers feeding into lakes, seas, or oceans. JOHN WILLIAM WEBB.

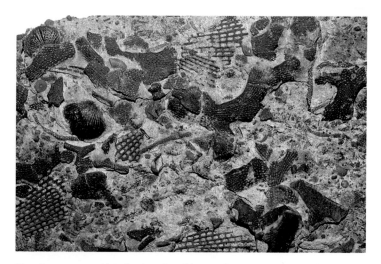

Many limestones are rich in fossils, such as this 390-million-year-old example from Hungry Hollow, near Arkona, Ontario. The fossils in this sample are mostly bryozoa, small organisms that build net-like colonies. ALAN MORGAN.

A plume of sediment entering Hudson Bay from Witchekat Creek, Wapusk National Park of Canada, Manitoba. N. ROSING, COPYRIGHT PARKS CANADA.

Erosion has created these "hoodoos" in sedimentary rocks in the badlands near Drumheller, Alberta. The layering, or bedding, was continuous before erosion sculpted the landscape. The lower brown marine shales are about 75 million years old; the white sandstone above is a slightly younger fluvial deposit. THE ROYAL TYRRELL MUSEUM OF PALAEONTOLOGY.

Granular sediments saturated with water, such as this beach sand at Lawrencetown, Nova Scotia, can liquefy. That explains why these students have sunk into the sand. MARTIN GIBLING.

clay minerals that originate from the chemical breakdown of minerals such as feldspar and the dark-coloured minerals of igneous rocks, especially mafic ones. Clay minerals give clay its distinctive properties. Grain size is also important in classifying clastics. Sand is composed of grains larger than silt but smaller than gravel. Beach sand is often composed of quartz grains, a testament to that mineral's durability. Quartz sand grains commonly have a surface veneer (or staining) of iron oxide, which explains the yellow to orange colour of many beaches—pure quartz sand is white.

In contrast to physically accumulated clastic deposits, some sedimentary rocks form through chemical or biological processes. Chemical sedimentary rocks include carbonates—principally limestone (made usually from calcite) and dolostone (made mainly from the calcium-magnesium carbonate mineral dolomite). Limestones and dolostones can be precipitated directly from warm seawater. Calcium carbonate is one of nature's great building materials, used by many organisms to construct shells. In warmer water, where life thrives, calcium carbonate may be precipitated and act as a cement, binding together shells to form limestone beds. Such conditions are conducive to the growth of reefs—buildups of invertebrate shells and skeletons, precipitates from micro-organisms, and carbonate mud. Reefs can be enormous structures, such as the modern Great Barrier Reef, which stretches for over 2,000 kilometres off northeastern Australia. Carbonates produce some of Canada's most spectacular scenery, including the towering cliffs of the southern Canadian Rocky Mountains (from now on called just the Rocky Mountains, or Rockies). Some sedimentary rocks are a mixture of carbonates and clastics; in these cases we can use the term calcareous, as in, for example, calcareous shales. Calcareous mudstone is sometimes referred to as marl.

Carbonates are the most common chemical sedimentary rocks but, when bodies of salt water dry up under hot arid conditions, they evaporate and leave a succession of rocks known as evaporites. Gypsum (calcium sulphate) is precipitated first, followed by halite (or rock salt—sodium chloride), and then potash (mainly potassium-bearing minerals). Canada has bounteous deposits of evaporites, including the thick salt sequences beneath the Goderich and Caledonia areas of southern Ontario and the subsurface potash deposits of Saskatchewan.

Going with the Flow

The layering, distribution, textures, and structures associated with clastic sedimentary rocks can tell us a lot about their origin. Useful sedimentary structures include ripples and their larger-scale cousins, dunes. Ripples and dunes may be symmetrical when formed by currents that swish back and forth, as on a beach washed by waves, or they may be asymmetrical when deposited by currents of wind or water moving in one direction, as in a river. If a sedimentary rock has ripples or dunes preserved within it, we can deduce something about the ancient environment (or paleoenvironment) in which it formed and perhaps even the direction of currents that prevailed at the time. If we dig through a modern ripple or dune, we may see that the bedding is at an angle to the horizontal. This effect is created as the ripple or dune migrates

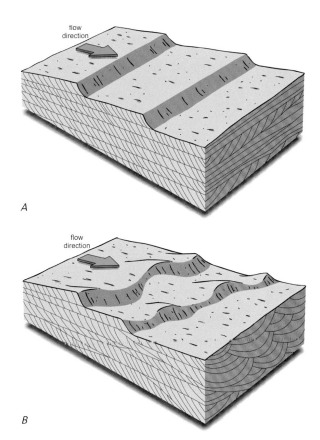

A

B

The flow of water and air over loose sediment results in the formation of ripples and dunes on the sediment surface. Factors such as current speed, direction, and variability, as well as the size of sediment particles (grain size) result in the formation of different styles of ripples and dunes. This diagram illustrates two such styles and the respective styles of cross-bedding seen in cross-section. ADAPTED FROM VARIOUS SOURCES BY LYNN DAFOE.

Modern dunes underwater and on the banks of the Saint John River, New Brunswick. RON GARNETT / AIRSCAPES.CA.

Cross-bedding in 200-million-year-old sandstones (redbeds) of eolian (wind-blown rather than subaqueous) origin, Wasson Bluff near Parrsboro, Nova Scotia. ROB FENSOME.

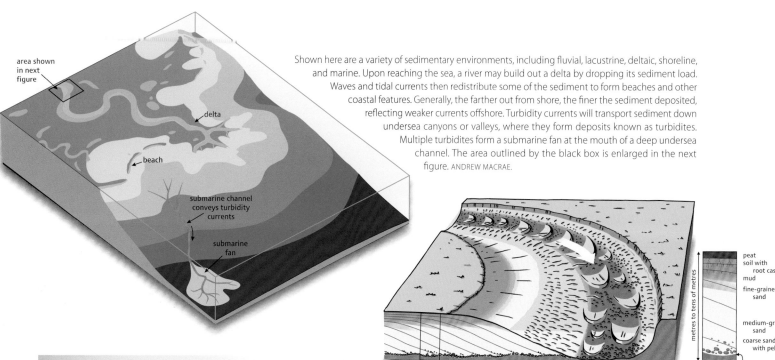

Shown here are a variety of sedimentary environments, including fluvial, lacustrine, deltaic, shoreline, and marine. Upon reaching the sea, a river may build out a delta by dropping its sediment load. Waves and tidal currents then redistribute some of the sediment to form beaches and other coastal features. Generally, the farther out from shore, the finer the sediment deposited, reflecting weaker currents offshore. Turbidity currents will transport sediment down undersea canyons or valleys, where they form deposits known as turbidites. Multiple turbidites form a submarine fan at the mouth of a deep undersea channel. The area outlined by the black box is enlarged in the next figure. ANDREW MACRAE.

An example of what happens at a river bend. The main diagram shows the bend with a cutaway view facing the reader. Momentum carries the main river current around the outside of the bend, towards the right side of the drawing. The strength of the current there erodes the river bank, and only coarser material settles to the bottom; the stronger current also leads to the development of large dunes on the channel floor. On the inside bend of the channel, the weaker current deposits finer sand and mud, building out an arc-shaped platform called a point bar. Because of undercutting on the outside bend and deposition on the point bar, the channel migrates, leaving a sediment record of shifting environments and processes. Eventually vegetation and soil may develop on the point bar. The column at right gives an idea of what a vertical section through the sediments would look like at the location shown by the box in the main diagram. ANDREW MACRAE.

Ripples on tidal flats, Five Islands, Nova Scotia. ROB FENSOME.

Ripples in about 325-million-year-old sedimentary rock from West Bay, near Parrsboro, Nova Scotia. ROB FENSOME.

and sediment accumulates in layers parallel to the structure's sloping surfaces, usually the ones facing down-current. Such angled bedding is called cross-bedding.

Mudcracks are often preserved as sedimentary structures. They form under conditions of alternate wetting and drying, such as on river flood plains or in the intertidal zone. Tool marks, dents, scratches, and grooves on sediment surfaces record where objects were dragged along the bottom. Flute marks (teardrop-shaped scours on a sediment surface) echo the presence and direction of strong eroding currents. Channels (grooves or gullies in a former surface) are often recognizable as a body of sediment with a U- or lens-shaped cross-section.

Sedimentary textures can also tell us about past environments. An example is graded bedding, in which grain size changes uniformly, usually decreasing upward within a single bed. This upward decrease is caused by the weakening of water currents transporting the sediment. Initially, in a strong current, only coarser material falls to the bottom, but as the current wanes, the material being dumped is increasingly finer grained. On sandy beaches, the constant wave action tends to keep clay-sized particles in suspension while depositing sand; consequently, coarser material tends to be deposited closer to shore and finer material farther offshore. Similarly, a river deposits coarser sediments on the outside of a bend, where the current is faster, and mud on the inside of the same bend, where the current is slower.

Mudcracks in modern muds, western Cape Breton Island, Nova Scotia. MARTIN GIBLING.

Mudcracks in 350-million-year-old mudstone, Clarke Head, near Parrsboro, Nova Scotia. ROB FENSOME.

A 300-million-year-old channel filled with layers of sand, now preserved as sandstone, cuts into older sedimentary rock near the village of Jacquet River, New Brunswick. ROB FENSOME.

This sequence of sedimentary rocks at L'Islet-sur-Mer, Quebec, deposited about 505 million years ago, have been steeply tilted, so imagine the original sea floor coming vertically out of the ground. The grey, regularly bedded rocks to the left represent turbidites, and the blockier, less well-bedded rocks in the foreground are sandstone and conglomerate that originated as sediment in a deep-sea channel. ROB FENSOME

Deltas form where rivers enter the sea or a lake and deposit their sediment load, as here at Peyto Lake in Banff National Park of Canada, Alberta. RON GARNETT / AIRSCAPES.CA.

CHAPTER 1 · ON THE ROCKS

Coarse-grained rocks such as conglomerates are most commonly associated with deposition in environments with relatively steep gradients or strong currents.

In deeper water, sediments are deposited from turbidity currents and debris flows. Turbidity currents are avalanche-like flows of sediment-laden water, triggered by events such as the Grand Banks Earthquake in 1929 (Chapter 17). The turbidity current set off by that event hurtled down the continental slope at up to about 65 kilometres per hour, snapping transatlantic cables in its path. It turns out that turbidity currents have been responsible for many distinctive sedimentary rocks, which we call turbidites. Examples of ancient turbidites, several hundred million years old, can be seen in exposures along the south shore of the St. Lawrence River at L'Islet-sur-Mer, Quebec. Debris flows, like turbidity currents, may be triggered by earthquakes, but they involve largely unsorted pebbles, cobbles, and boulders floating in a mud-silt-sand mix that tumbled together down a relatively steep slope into deeper water. A good example, formed of limestone blocks, can be seen preserved in the rocks at Cow Head in western Newfoundland.

Low cliffs of peat, Escuminac Point, New Brunswick. ROB FENSOME.

Coal and Soil

Coal is a sedimentary rock formed from compressed ancient plant material, microscopic to tree-sized, which resisted decay because it accumulated under anoxic (oxygen-reduced) conditions in wetlands, initially as peat. In the modern tropics, peat accumulates at the electrifying rate of 2 to 4 millimetres per year. As peat is buried, with increasing time and temperature it becomes compressed and most of its water is squeezed out. Over millions of years, peat hardens to form different grades of coal— lignite, sub-bituminous coal, bituminous coal, and anthracite, successively. A layer, or seam, of coal may represent only one tenth of the thickness of the original peat. So, a coal seam that's a metre thick was originally 5 to 10 metres of peat that took about 2,500 years to accumulate.

Folded strata, including coal, at Canmore, Alberta. ALAN MORGAN.

Soils are derived partly from weathering processes and partly from the remains and activities of organisms. They played a vital role in the evolution of life on land; before plants came ashore, land surfaces would have looked similar to the rocky, dusty barrens of Mars. Lichens (symbiotic organisms that are part algae and part fungi) may have existed on land millions of years before the first true plants made landfall and still play a role in starting the breakdown of bare rock surfaces. But true plants have had a much greater impact. Plants not only contribute to soil formation materially but also have roots that help to break up rocks and change drainage patterns by anchoring surface material.

Once soil began to form, plants and animals were able to invade the land and evolve there. Microscopic organisms such as bacteria, algae, and fungi provide the nutrients that sustain plant growth and convert dead organic matter to humus, which makes up to 90 percent of some soils. Burrowing or tunnelling animals, especially worms and ants, play a key role by mixing mineral and organic matter and by facilitating drainage.

Soils are classified according to their texture, chemical composition, and layering. These three attributes reflect biological activity, climate, the nature of the parent material, and local topography. Though some soils are unlayered, most have several layers (horizons) as shown in the accompanying diagram. As well as being important for agriculture, soils are a vital reservoir for carbon dioxide, a greenhouse gas (see Box 9). Soils preserved in the fossil record are called paleosols. Geologically

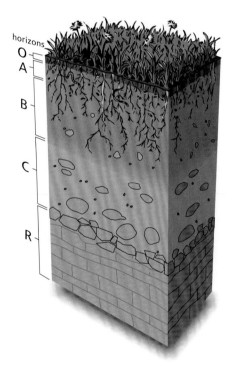

A soil profile. At the top is the O horizon, consisting of not-yet-decomposed organic matter, mainly plant litter. The A horizon is the topsoil, generally darker than horizons lower down because it still contains much organic matter. The B horizon is known as the subsoil and consists mainly of minerals and clay; it also includes many plant roots and organisms that churn the soil. The C horizon is made up mainly of chunks of rock and smaller rock particles. The R horizon is the bedrock on which the soil forms; the nature of the bedrock influences the chemical characteristics of the soil. ADAPTED FROM VARIOUS SOURCES.

An ancient calcareous paleosol, known as calcrete, near St. Martins, New Brunswick. MARTIN GIBLING.

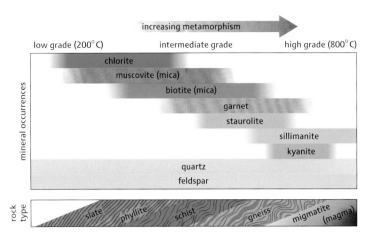

Minerals and rocks that form at increasing grades of metamorphism. Laboratory studies show that different minerals form under different conditions of pressure and temperature. The bottom row shows the succession of rock types formed as metamorphism progresses, assuming that the original rock was a mudstone. ADAPTED FROM VARIOUS SOURCES.

recent fossil soils look like modern soils, just buried, whereas older paleosols are generally rock-like and have a characteristic knobbly and disrupted appearance.

In a Pressure Cooker

Metamorphic rocks are produced when igneous, sedimentary, or even previously existing metamorphic rocks are subjected to high temperatures and/or pressures. Metamorphism involves major changes in mineral types and textures, the resulting metamorphic rock being determined in large part by the composition of the original rock. Clay-rich rock can be changed to slate and then to schist; quartz-rich sandstone is metamorphosed to quartzite; limestone converts to marble. Clay-rich rocks can also become argillite, a uniform fine-grained, low-grade metamorphic rock that doesn't split like slate. In naming metamorphic rocks, geologists sometimes simply add the prefix meta. Thus, a metasandstone is simply a metamorphosed sandstone and a metavolcanic rock is a metamorphosed volcanic rock.

Rocks can be weakly to strongly metamorphosed, depending on the extent of changes wrought by increased temperature and pressure. Slate and metasandstone are low-grade metamorphic rocks formed by weak metamorphism, and such rocks usually retain some vestige of their original sedimentary features, such as bedding. Often these rocks develop cleavage: an alignment of minerals at right angles to the direction of pressure or compression. Micas—silicates with broad, flat (platy) crystals—are especially liable to show such alignment, so rocks with lots of fine-grained mica, such as slates, split generally along cleavage planes rather than along the original bedding surfaces. Schist has lots of mica, and gneiss (pronounced as in "have a gneiss day") tends to have little mica; both are coarsely crystalline, high-grade metamorphic rocks. They are usually so altered that it is difficult to tell what the original rock was. The crystals in gneiss tend to become layered in discrete bands of different composition, which reflect concentrations of dark and light minerals; the bands are not related to the original bedding.

Cleavage in this 480-million-year-old slate from Blue Rocks, Nova Scotia, is vertical and more or less perpendicular to the original sedimentary bedding (the light and dark grey banding). Because sedimentary layers are still discernible, this slate is considered a low-grade metamorphic rock. MARTIN GIBLING.

This roadside block near Pont Rouge, Quebec, illustrates the banded nature of gneiss. ROB FENSOME.

Migmatite formed deep in the crust by the partial melting of grey gneiss and later injection of pink granitic melt. The grey gneiss formed about 3,000 million years ago, and the granitic melt was injected about 400 million years later. This beautiful rock is exposed on mainland Nunavut. LEOPOLD NADEAU.

Colourful marble consisting largely of orange calcite and a green pyroxene from a quarry near Bancroft, Ontario. The rock was metamorphosed from limestone about 1,100 million years ago. GRAHAM WILSON.

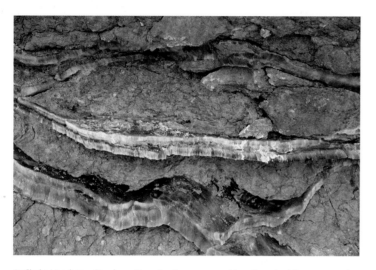

At Clarke Head, near Parrsboro, Nova Scotia, orange and translucent white gypsum veins have been injected into older (about 350 million-year-old) mudstone. ROB FENSOME.

At lower metamorphic grades, platy crystals of chlorite and mica are common. As higher metamorphic grades are reached, minerals such as garnet, staurolite, and sillimanite may form. Such high-grade metamorphic rocks form at depths of 15 to 25 kilometres within the crust. If pressure (usually the result of deep burial) is a major factor during metamorphism, minerals such as kyanite and glaucophane may grow. The blue colour of glaucophane gives rise to the name blueschist, a rock formed under conditions of low temperature and high pressure.

Igneous rocks also show interesting metamorphic changes. When basalt is metamorphosed at low pressures and temperatures, some of its constituent minerals convert to the green minerals chlorite, actinolite, and epidote, producing a type of rock called greenstone or greenschist. At higher metamorphic grades, greenstone becomes amphibolite, a dark green to black rock made up of interlocking amphibole crystals.

Under extreme conditions, especially of rising temperature and if groundwater is present, metamorphic rocks may begin to melt as the rock cycle comes full circle. The felsic components of rocks melt first, well before the mafic components. Such partly melted rocks are called migmatites. Felsic melt may collect to form granitic magma.

Groundwater, commonly rich in dissolved minerals, may thus affect chemical reactions during metamorphism. However, mineral-rich fluids are widespread in rocks and have an impact beyond metamorphic processes. When hot, groundwater is termed hydrothermal. Fluids seep through pores and fractures, and the minerals in them can precipitate, for example, replacing organic material to produce petrified fossils (Chapter 4), or helping glue sedimentary rocks during diagenesis. Substances dissolved in hydrothermal fluids can be precipitated in cracks and pores within rocks to form pockets and veins of minerals—a process responsible for many commercially viable deposits, as we will see in Chapter 12.

Bending and Breaking, Squeezing and Stretching

Sediments are usually deposited in horizontal layers that eventually form horizontal sedimentary strata. But all rocks can be squeezed (compressed) or stretched (extended), generally in directions parallel to the Earth's surface. Compression causes rocks to buckle, producing folds, and both compression and extension cause rocks to break, producing joints and faults. Folding can produce upward bulges called anticlines, or troughs called synclines. Such structures can be on any scale, from microscopic to regional. Extreme folding sometimes turns rocks upside down.

Where rocks fracture, but no relative movement occurs between the two sides, the fracture is called a joint. Joints are common in rocks that were once deeply buried and are now at the surface, the cracks caused by the removal of the heavy load above them. When one side of a fracture moves relative to the other, we

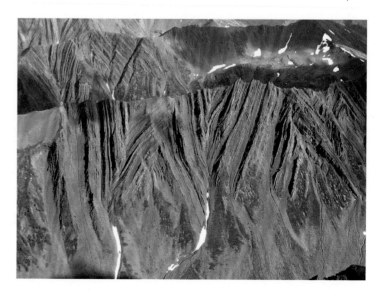

Tight folds in 150-million-year-old sedimentary rocks in the Skeena Mountains, British Columbia. MARGO MCMECHAN.

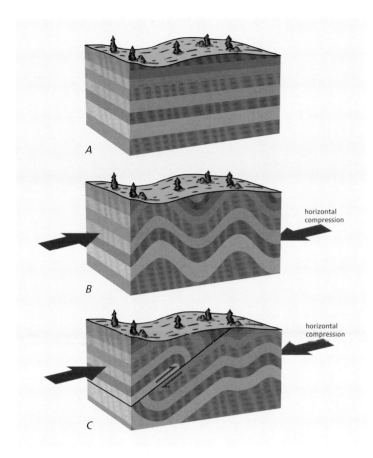

A. Sedimentary strata before deformation. *B.* The same rocks after horizontal compression rumples them into folds, forming anticlines and synclines. *C.* Continuing pressure may form thrust faults (low-angle reverse faults—see also the next figure) that push layers of rock on top of other layers.

have a fault. Normal faults occur in rocks that are being stretched; as a response to this extension, a fault forms and one block slips downward relative to the other. In fractures formed under compression, one block is forced up relative to its neighbour, forming either a reverse fault or a thrust fault, depending on the angle that a fracture makes with the bedding (see diagram below left).

In strike-slip faults, blocks slide past each other horizontally, like two trains going in opposite directions on parallel tracks. A good example is British Columbia's north-south trending Fraser Fault between the towns of Lytton and Hope (Chapter 10). There, rocks on the west side have moved 140 kilometres north relative to those on the east side. This movement has created a zone of easily eroded fractured rocks that have been selectively worn away by the Fraser River to form a deep canyon. Strike-slip faults come in two flavours: right-lateral and left-lateral. To understand the difference, imagine standing on one side of a strike-slip fault: if movement of the opposite side is to the right (think clockwise), it is right-lateral; and if the movement is to the left (counterclockwise), it is left lateral.

Fractured rocks such as those created by the Fraser Fault are called fault rocks, and are loosely classified as metamorphic rocks. Fault rocks that form close to the Earth's surface can consist of ground-up rock flour. The grinding action of some faults can form a fault breccia, a jumbled, chaotic mixture of angular fragments embedded in a matrix of rock flour.

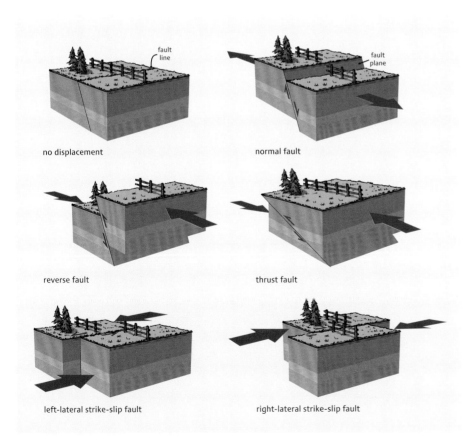

Different kinds of faults. Check the location of the fence to see which way fault blocks are moving.

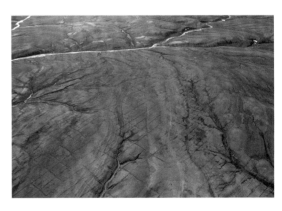

Rectangular arrangement of joints in sedimentary rocks on Ellef Ringnes Island, Nunavut. CAROL EVENCHICK.

Normal faults in sedimentary rocks, Prince Patrick Island, Northwest Territories. CHRISTOPHER HARRISON.

In this cliff near Parrsboro, Nova Scotia, the steeply tilted grey rocks at far right are about 350 million years old and have been thrust up relative to younger (around 200-million-year-old) grey volcanic and reddish sedimentary rocks to the left. The older and younger strata are separated by light-green, finely crushed material produced by grinding as the fault moved the two sets of rocks past each other. ROB FENSOME.

Ice-smoothed surface showing folds in dark amphibolite and pink gneiss on the west side of Franklin Island, Georgian Bay, Ontario. Such metamorphic rocks form in the crust beneath mountain ranges, perhaps as deep as 40 kilometres. That these rocks are now exposed at the surface implies that the mountains have eroded away and the crust in the region has been uplifted. CHRISTOPHER HARRISON.

This view of the Wernecke Mountains in Yukon reveals an unconformity between older, steeply tilted, brown sedimentary rocks below and younger, gently tilted, light-coloured carbonates above. The brown rocks were deposited before 1,710 million years ago in the Wernecke Basin (Chapter 6). R. T. BELL, REPRODUCED WITH THE PERMISSION OF NATURAL RESOURCES CANADA 2013, COURTESY OF THE GEOLOGICAL SURVEY OF CANADA.

Bobbing Up and Down

In our discussion of metamorphic rocks, we found that some minerals provide clues about the temperatures and pressures under which a rock was formed. For example, the mineral sillimanite forms at depths of about 20 kilometres. From similar observations, we know that metamorphic rocks exposed around Parry Sound on Georgian Bay, Ontario, were originally at depths of over 40 kilometres about 1,000 million years ago. These rocks are now exposed at the surface because of later uplift and erosion of overlying rocks. Evidence for vertical downward movement comes from the coastline around Joggins, Nova Scotia. There, a sequence of sedimentary rock 4 to 5 kilometres thick was deposited in rivers and flood plains over an interval of 2 to 3 million years. Since flood plains usually maintain an elevation reasonably close to sea level, this means that the Earth's crust locally dropped down, or subsided, about 4 to 5 kilometres while the sediments were accumulating to fill the space. Such subsidence over such a short time is strikingly fast, geologically speaking.

Uplift and subsidence are important in reshaping the Earth. As rocks are uplifted and deformed, the mountains or uplands so formed are constantly being eroded. Eventually, the forces producing the uplift will relax, and high land will be worn away. New sediment will be deposited on the roots of old mountains. When the new sediments compact to become sedimentary rocks, they will differ from the older rocks because of both their different nature and their lack of deformation. Since the rocks of the two ages do not conform to each other, we call the surface that divides them an unconformity.

We tend to think of rocks as static objects that provide us with ground to stand upon. But if we know how to read them, rocks open a window into our planet's history. The first person to understand the rock cycle was the Scot James Hutton, a member of the "Scottish Enlightenment" of the eighteenth century. Upon recognizing the seemingly never-ending story told in the rocks, Hutton proclaimed that the Earth showed no vestige of a beginning, no prospect of an end. How amazed and delighted Hutton would have been by another kind of geological system, a system that is the focus of the next chapter, a system unravelled only during the second half of the twentieth century—plate tectonics.

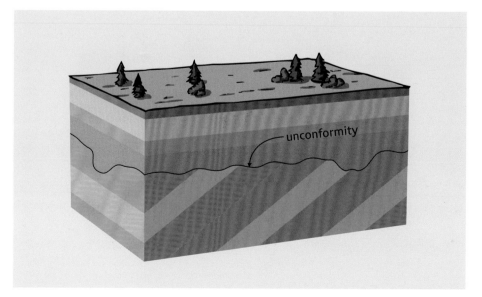

An unconformity (irregular line) separates a lower suite of tilted strata from an upper suite of horizontal strata. Unconformities represent ancient erosion surfaces, commonly subaerial, and preserve relics of former landscapes such as channels.

BOX 1 • CRYSTAL CLEAR

A Diversity of Minerals

A mineral is an element or compound that is normally crystalline and that has formed as a result of geological processes. Some minerals are composed of just one element. Examples of single-element minerals include native copper and native gold (native being the geological term indicating a pure natural state). The minerals diamond and graphite are both made solely of carbon—but what a difference: designer jewellery or pencil lead! Minerals such as diamond and graphite that have identical compositions but different crystal forms are called polymorphs. Most of the 4,000 known minerals are compounds (made up of two or more elements). Some rocks are made up of a single mineral, but most are composed of two or more.

Silicates are the most abundant minerals on Earth. They have a core framework of oxygen and silicon atoms arranged in a tetrahedral structure. Other elements are commonly attached to the tetrahedron, combined either with silicon or oxygen atoms. Among silicates, quartz (silicon dioxide or silica) is the most familiar mineral, occurring typically as clear glassy or milky white crystals. Other important groups of silicate minerals are feldspars, garnets, micas (muscovite and biotite), olivines, amphiboles (including hornblende), and pyroxenes (including augite). Feldspars make up about 50 percent of the Earth's crust and come in several varieties depending on chemical composition. Two common varieties are plagioclase, which is usually white to grey and ranges in composition from sodium-rich to calcium-rich, and orthoclase, which is potassium-rich and commonly pink.

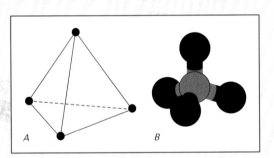

Silicates are based on a tetrahedral molecular arrangement. *A* shows the basic shape of a tetrahedron. In silicates, a silicon atom sits in the middle surrounded by four oxygen atoms, representing the four corners, as shown in *B*.

The carbonates form another large group of minerals. Their molecular structure is based on a triangle formed by three oxygen atoms, with a carbon atom in the centre. Calcite, made of calcium carbonate, is the main component of limestone and the most abundant carbonate mineral. Calcite and aragonite are polymorphs, aragonite being a less stable form of calcium carbonate, although it forms some shells. Dolomite, a calcium-magnesium carbonate mineral, is another example, as is siderite, an iron-carbonate mineral.

Oxides contain oxygen and one or more other elements. Examples are iron oxides and aluminum oxides, the latter including the mineral corundum, of which ruby and sapphire are varieties. Iron oxide minerals such as hematite, goethite, and magnetite are extensively mined as a source of iron for making steel. Oxides form only in oxygen-rich environments.

Evaporite rocks are made up of evaporite minerals, which include halides and sulphates. Halides contain one of the halogen elements—fluorine, chlorine, iodine, or bromine—combined with certain metallic elements such as sodium, potassium, copper, or silver. Halite is the most abundant and most mined halide mineral; we consume it as table salt, and it is also known as rock salt. Sulphates contain four oxygen atoms surrounding a central sulphur atom plus one other element in a tetrahedral structure. One of the best-known sulphates is gypsum, found in manufactured products such as drywall plasterboard, bread, and toothpaste.

Some of the most commercially valuable minerals are the sulphides. These are composed of metallic elements, such as zinc and lead, combined with sulphur. Sulphide minerals include pyrite (iron sulphide), chalcopyrite (copper-iron sulphide), galena (lead sulphide), and sphalerite (zinc sulphide).

Native copper, Cape d'Or, Nova Scotia. HEINZ WIELE, COURTESY OF THE ATLANTIC GEOSCIENCE SOCIETY; SPECIMEN COURTESY OF THE NOVA SCOTIA MUSEUM.

Native gold from Timmins, Ontario.

A one-carat diamond from the Chidliak project, at the southern tip of Baffin Island, Nunavut. The diamond was mined from a kimberlite pipe (Chapter 12) about 150 million years old. PEREGRINE DIAMONDS LIMITED.

Halite from a potash mine at Rocanville, Saskatchewan. HELEN TYSON, FROM THE COLLECTION OF ROD AND HELEN TYSON.

Quartz crystals, Salmon River Gold District, Nova Scotia. HEINZ WIELE, COURTESY OF THE ATLANTIC GEOSCIENCE SOCIETY; SPECIMEN COURTESY OF THE NOVA SCOTIA MUSEUM.

Smoky quartz from Nova Scotia. HEINZ WIELE, COURTESY OF THE ATLANTIC GEOSCIENCE SOCIETY; SPECIMEN COURTESY OF THE FUNDY GEOLOGICAL MUSEUM.

Amethyst from Thunder Bay, Ontario. HELEN TYSON, FROM THE COLLECTION OF ROD AND HELEN TYSON.

Multicoloured agate surrounds a core of purple amethyst in this example from around the Bay of Fundy in Nova Scotia. HEINZ WIELE, COURTESY OF THE ATLANTIC GEOSCIENCE SOCIETY; SPECIMEN COURTESY OF THE FUNDY GEOLOGICAL MUSEUM.

Labradorite, a variety of feldspar, probably from Nain, Labrador. ALAN MORGAN.

Garnet from the Jeffrey Mine, Quebec. HELEN TYSON, FROM THE COLLECTION OF ROD AND HELEN TYSON.

Crystals of biotite (black) and apatite (green) southwest of Bancroft, Ontario. GRAHAM WILSON.

Hornblende, the common rock-forming variety of amphibole, from Bear Lake, west of Bancroft, Ontario. HELEN TYSON, FROM THE COLLECTION OF ROD AND HELEN TYSON.

Crystals of calcite or dolomite from Wawa, Ontario. HELEN TYSON, FROM THE COLLECTION OF ROD AND HELEN TYSON.

Crystals of aragonite from Thetford, Quebec. HELEN TYSON, FROM THE COLLECTION OF ROD AND HELEN TYSON.

Satin spar, a variety of gypsum, from Nova Scotia. HEINZ WIELE, COURTESY OF THE ATLANTIC GEOSCIENCE SOCIETY; SPECIMEN COURTESY OF THE FUNDY GEOLOGICAL MUSEUM.

Selenite, a variety of gypsum, from Winnipeg, Manitoba. HELEN TYSON, FROM THE COLLECTION OF ROD AND HELEN TYSON.

Goethite, an iron oxide mineral, from Bridgeville, Nova Scotia. HEINZ WIELE, COURTESY OF THE ATLANTIC GEOSCIENCE SOCIETY; SPECIMEN COURTESY OF THE NOVA SCOTIA MUSEUM.

Pyrite from the former Nanisivik Mine, northern Baffin Island, Nunavut. HELEN TYSON, FROM THE COLLECTION OF ROD AND HELEN TYSON.

Chalcopyrite from Vancouver Island, British Columbia. HELEN TYSON, FROM THE COLLECTION OF ROD AND HELEN TYSON.

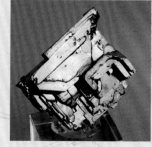
Galena from Watson Lake, Yukon. HELEN TYSON, FROM THE COLLECTION OF ROD AND HELEN TYSON.

Mineral Properties

A diagnostic property of most minerals is their crystal shape, a reflection of their atomic structure. Seven major categories, or crystal systems, are recognized, of which the easiest to visualize is the cubic system. Minerals with cubic crystals include pyrite and halite. Slightly more complex is the trigonal system, represented by calcite. A trigonal crystal looks like a cube that has been squashed into a rhombic shape. Quartz crystals are in the hexagonal system, in which the peaks of crystals have six faces.

Thin section of a granitic rock from Nova Scotia's South Mountain Batholith (Chapter 8) viewed under a microscope. The photo to the left is in normal light and reveals blue tourmaline in a matrix of mica and quartz. The photo to the right was taken with polarized light. The area of view is 5 millimetres across. Thin sections are primary tools for determining the mineral composition of rocks and consist of paper-thin polished slices. BARRIE CLARKE.

The surest way to identify a mineral is by chemical analysis, but that's not easy to do, even for geologists. A cheaper and simpler way, if not quite as precise, is to cut and polish a rock to produce a thin slice. We can then examine this wafer of rock by mounting it on a glass slide under a microscope and transmitting light through it or reflecting light off it. Under transmitted light and polarizing lenses, minerals reveal startlingly vivid and beautiful colours and patterns that help us to identify them.

For hand specimens we must rely on differences such as colour, lustre, hardness, and crystal shape. Colour can be misleading because it is so variable in some minerals, even though their crystals are always the same shape. For example, quartz can be colourless and clear like glass—in fact, we do make glass from quartz. But minor impurities of other elements produce a dazzling array of hues. Amethyst, one of the most beautiful and treasured varieties of quartz, gets its deep violet colour from iron impurities within the crystals. The delicate pink tint of rose quartz comes from traces of titanium, iron, or manganese in the crystal. And the grey to brown tint of smoky quartz results from the natural gamma radiation within the crystals. Perhaps the most eye-catching variety is the multi-hued agate, with its different coloured bands of minute quartz crystals. In contrast to quartz, some minerals are consistently of one colour—for example, the copper carbonate minerals malachite and azurite are invariably green and blue, respectively.

Minerals have other features or properties that help us to distinguish them. One is streak, the colour of the powder made from the mineral. To determine a mineral's streak, we scratch it across the surface of an unglazed porcelain tile, known as a streak plate. The colour of the powder streak produced is always the same for a particular mineral. Lustre is the way in which the surface of a mineral reflects light. Some minerals, such as pyrite (fool's gold), have a metallic lustre. Other minerals have a vitreous lustre (glassy like quartz), a brilliant lustre (gems like diamonds), or an earthy lustre (dull like hematite).

One of the most useful properties for identifying a mineral is its hardness. This is expressed on the Mohs Scale—named for the German mineralogist Frederick Mohs—of 1 (softest) to 10 (hardest). In order of increasing hardness, 1 is talc, 2 is gypsum, 3 is calcite, 4 is fluorite, 5 is apatite, 6 is orthoclase feldspar, 7 is quartz, 8 is topaz, 9 is corundum, and 10 is diamond. We can determine relative hardness by dragging a crystal of one mineral over the surface of another and observing which of the two scratches the other. Thus, calcite will scratch gypsum but not quartz. It is also useful to know that a fingernail has a hardness of 2.5 so will scratch gypsum but not calcite or quartz. A copper cent coin is 3, a steel knife is 5, glass is 5.5, porcelain is 7, and emery cloth is 8. If you have a diamond, rest assured that nothing will scratch it. Hardness reflects the molecular structure of a mineral rather than its composition, as witnessed by the fact that graphite has a hardness of 1 and diamond 10. Graphite is the most stable form of carbon under normal temperatures and pressures, but its molecular structure involves sheets of carbon atoms held together weakly. In contrast, the carbon atoms in diamonds have a tetrahedral arrangement that is almost unbreakable.

Minerals hold endless fascination for rockhounds and jewelry aficionados alike, and it is easy to be attracted by their beauty. In their precious form, they have proved to be the foundation of nations. As components of ores, minerals are the life-blood of many a Canadian mining community. And as the building blocks of rocks, minerals are the basis of much of the subject matter of this book.

Malachite (green) and azurite (blue) from Whitehorse, Yukon. HELEN TYSON, FROM THE COLLECTION OF ROD AND HELEN TYSON.

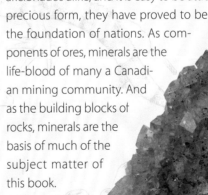

Fluorite from Daniel's Harbour, Newfoundland. HELEN TYSON, FROM THE COLLECTION OF ROD AND HELEN TYSON.

BOX 2 • GETTING TO THE CORE

The Solar System, one of about two billion such systems in the Milky Way Galaxy, has four large, outer, gaseous planets—Jupiter, Saturn, Uranus, and outermost Neptune—and four smaller, inner, rocky or terrestrial planets—Mercury, Venus, Earth, and Mars. Earth is the largest rocky planet, with a radius of 6,378 kilometres at the Equator and 6,357 kilometres at the poles. We are familiar with the Earth's surface, but what lies between our feet and our planet's centre? Some mines, such as the Creighton Mine at Sudbury, Ontario, are over 2 kilometres deep; but that still leaves a long way to go. Drilling has gone deeper: a borehole on the Kola Peninsula in northwestern Russia reached a depth of about 12.3 kilometres in 1992. The target depth was 15 kilometres, but at 12.3 kilometres the temperature was already 180°C instead of an expected 100°C. At that rate, temperatures at the target depth would be 300°C, far too hot for the monitoring equipment. So we can only dig or drill to relatively shallow depths.

A few of the Earth's deeper secrets have been revealed by tectonic uplift and erosion. In certain places, these processes have laid bare rocks once buried more than 40 kilometres below the surface, as along the shores of Georgian Bay on Lake Huron. And some volcanic vents contain rock fragments (and diamonds) that formed at depths of at least 150 kilometres. Yet even that is still more than 6,000 kilometres from the centre of the Earth. Meteorites (Box 11) also provide clues. Some are chunks of broken planets that formed around the time of the Solar System's birth 4,557 million years ago but didn't survive.

To acquire more detailed knowledge of the Earth's deep interior, we need to use remote sensing techniques. One of the most successful is analysis of seismic waves (Box 3) generated naturally during earthquakes or artificially by mechanical means. The speed of the waves varies depending on the composition, temperature, pressure, and state (solid or liquid) of the rocks they are passing through. Although most waves return to the surface, where they can be recorded by seismometers around the globe, they arrive at variable times after the initial shock, depending on the distance of the recording instrument from the earthquake source and what they've passed through. These different travel times are used to interpret the nature and depth of the rocks traversed by the seismic waves, a technique something like that used by doctors during an ultrasound examination, allowing them to "see" the unborn baby in its mother's womb.

So what are the results? Analysis of seismic waves shows that the Earth's interior is like a series of nested balls. At the centre is a solid inner core, a sphere with a radius of 1,210 kilometres. Largely through analogy with meteorites, it is thought that the inner core consists of a dense iron alloy, but at a sizzling 5,000 to 6,000°C and a pressure of over three million times that of the atmosphere at the Earth's surface. Surrounding this inner core is a fluid, metallic outer core. The motion of this 2,260-kilometre-thick spherical shell of liquid iron is responsible for generating the Earth's magnetic field, which is vital to our life-sustaining environment as it shields the Earth from destructive solar winds.

Surrounding the core is the mantle, which is about 2,855 kilometres thick and makes up more than 80 percent of the planet's volume. The mantle consists of mafic and ultramafic rocks such as peridotite. Mantle rocks undergo changes (called phase changes) in mineral structure but not chemical composition, with increasing temperature and pressure. To better understand phase changes, think of graphite and diamond as two different phases in the mineral representation of carbon. One important phase change, defining the boundary between the upper and lower mantle, occurs at a depth of about 660 kilometres.

Even though the mantle is effectively solid, it may contain extremely small amounts of melt. All of the mantle except its uppermost part is hot enough to deform, albeit at rates typically less than 10 centimetres per year, by a process known as creep. Creep can be thought of as exceedingly slow flow; glass can creep in a similar way over centuries. Heat is continuously generated in the mantle by decay of small amounts of radioactive isotopes of uranium, thorium, and potassium. Cooler, denser rock sinks, and hotter, more buoyant rock wells up and spreads laterally, setting up thermal convection that transfers heat from the Earth's interior to its surface, where heat is lost mostly by conduction at the sea floor, eventually escaping to space. Thermal convection is the engine that drives the creep of mantle rock.

The Earth's crust is a thin, cold, rigid skin 5 to 60 kilometres thick that surrounds the mantle, from which it differs by general composition. The crust is richer in silica and aluminum (and thus more felsic) than the more iron- and magnesium-rich mantle. The crust comes in two flavours: oceanic and continental. Oceanic crust is made mainly of extrusive and mafic rock, is mostly between 5 and 10 kilometres thick, and is present at or near the Earth's surface for less than 200 million years before being recycled back into the mantle (Chapter 2). Continental crust is usually between 30 and 60 kilometres thick and consists mainly of felsic to intermediate rocks that are much less dense than basalt. Their lower density ensures that continents are high-standing relative to ocean floors and keeps continental crust from being recycled into the mantle—with the result that continental crust is mostly much older than the oldest oceanic crust. Continental crust is thickest beneath mountains that are still building, as below the modern Himalayas and Andes, where it is at least 60 kilometres thick. The height of these ranges reflects the buoyant support of mountain roots, which—just as a boat displaces water—displace an equivalent mass of the underlying higher-density mantle.

Distinction of the crust from the mantle is based on composition. But rock strength provides another way of distinguishing layers of the Earth's interior. The uppermost part of the mantle is made from relatively cool, stiff, strong rock firmly attached to the bottom of the crust. Mechanically, the rigid uppermost mantle and the crust behave as a unit, called the lithosphere. Below the lithosphere, temperature increases, partial melting occurs, and so rocks are physically weaker. This weak zone in the upper mantle is known as the asthenosphere,

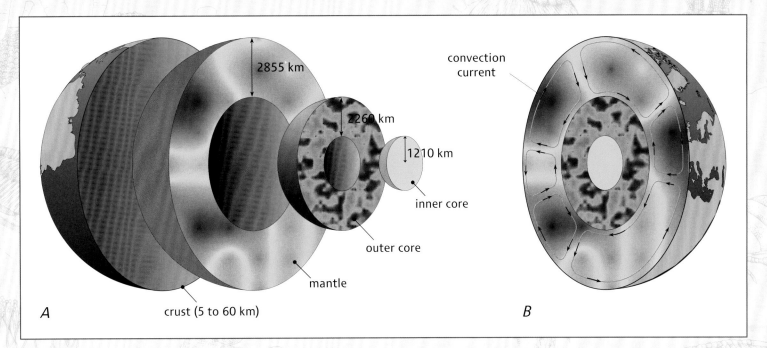

Structure of the Earth. *A* shows approximate radius measurements of the different layers from crust to inner core. *B* shows the possible distribution of convection currents within the mantle. FROM COLMAN-SADD AND SCOTT (1994); USED WITH PERMISSION OF THE AUTHORS, GAC, AND THE GEOLOGICAL SURVEY OF NEWFOUNDLAND AND LABRADOR.

and here rocks flow more readily than in the rest of the mantle. The base of the asthenosphere is poorly defined but, from observations of seismic waves, we know it to be about 250 kilometres down.

How do we know the asthenosphere can flow? In Chapter 1 we discovered the ups and downs of the Earth's surface. Such vertical movements provide the answer. The mobile nature of at least part of the planet's interior was recognized about ninety years ago from observations of regions such as Hudson Bay and the Baltic Sea, once depressed by the weight of ice that accumulated during the last glaciation (Chapter 11). The ice sheet was 2 to 3 kilometres thick and very heavy. Where the weight of an ice sheet pushes down on the lithosphere, the underlying asthenosphere is displaced. The land or sea floor directly beneath the ice sinks up to several hundred metres, and the land beyond the edge of the ice sheet bows up slightly. After the ice melts, the Earth's surface gradually recovers to its original levels. The process of vertical uplift and subsidence of the lithosphere in response to the addition and removal of material at the surface is known as isostasy. Old, cold lithosphere such as that of the Canadian Shield is still rebounding from the most recent glacial retreat. In contrast, younger lithosphere, such as that under Canada's western mountains, has already fully rebounded.

The lithosphere therefore "floats" on the underlying asthenosphere much as an iceberg floats in the ocean. This is why the dense basaltic ocean floors are at an average depth of about 4.5 kilometres below sea level, and the less dense continents stand at an average elevation of about 500 metres above sea level. Moreover, as we will see in Chapter 2, the lithosphere—the rigid outer shell of the Earth—is broken into tectonic plates, with the asthenosphere acting like a lubricating oil that allows these plates to move.

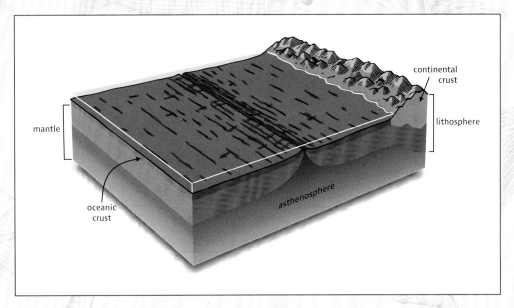

Structure of the outer layers of the Earth in terms of both rock composition (crust versus mantle) and rock strength (lithosphere versus asthenosphere). The latter relationship is the most important when considering plate tectonics (Chapter 2) because the outer, stiff lithosphere, consisting of rigid crust and outermost mantle, forms plates that float on the more plastic asthenosphere.

2 • DANCE OF THE CONTINENTS

CHAPTER SUPPORTED BY A DONATION FROM JAMES D. REIMER

Without tectonic forces within the Earth's lithosphere we would not have magnificent mountain ranges such as the Rockies. This is a view of the Kananaskis Country of southwestern Alberta. RON GARNETT / AIRSCAPES.CA.

On the Move

In 1596, following the discovery of the New World and the creation of the earliest reasonably accurate global maps, the Antwerp-born cartographer Abraham Ortelius raised the possibility that continents had not always been in their present positions. And in 1620, the English scientist Francis Bacon noted that the coastlines of South America and Africa would fit together snugly if the Atlantic were closed. Such ideas stayed in limbo until the early twentieth century, when American geologist Frank Taylor and German meteorologist Alfred Wegener independently suggested that continents had drifted apart. In 1910, Taylor reasoned that mountain ranges such as the Himalayas and the Cordillera resulted from the movement of continents. Wegener began to publish his ideas in 1912. In a book published in German in 1915 (and in English in 1924 as *The Origin of Continents and Oceans*), he argued for the former existence of a single landmass or supercontinent, which he named Pangea. Further, he implied that the modern Atlantic Ocean originated when Pangea broke up. One of Wegener's most persuasive arguments involved the distribution of fossils of the freshwater reptile *Mesosaurus*, the land-dwelling reptile *Lystrosaurus*, and the fern-like plant *Glossopteris*. All three organisms lived about 250 million years ago in Africa, Antarctica, Australia, South America, and India, but nowhere else. Wegener concluded that if the continents were fixed in position, the distribution of these extinct life forms would make no sense, and that the group of now mostly southern continents must have once been joined together. He made other persuasive arguments, involving the distribution of ancient glacial deposits now found near the Equator, and the resemblance of geological structures along the eastern coast of South America to those in southwestern Africa.

The photo to the left shows an outcrop of rocks, about 200 million years old, from Parrsboro, Nova Scotia. The photo to the right shows rocks of identical age from northwestern Morocco. Both successions consist of lower red sediments with some lighter-coloured layers toward the top, overlain by basalt (brown in Nova Scotia, grey-green in Morocco). These two images show how strikingly similar rocks from opposite sides of the Atlantic can be. PAUL OLSEN.

Although Wegener's ideas were enthusiastically embraced by some, many geologists did not accept the concept of continental drift, mainly because there was no known mechanism that would allow continents to plough through oceans. To explain the unlikely fossil distributions, the doubters proposed convenient land bridges thousands of kilometres long that would allow organisms to migrate between continents.

Magnetic Proof

Proof of continental drift came in the 1950s thanks to paleomagnetism, the study of the Earth's ancient magnetism as preserved in rocks. The characteristics of the Earth's magnetic field (Box 2) are closely related to our planet's spin axis, even though the magnetic poles are not identical in position to the planet's rotational poles. One way to think about the Earth's magnetic field is to recall the school science experiment: when a piece of paper is placed over a bar magnet and iron filings sprinkled over the paper, they settle in a pattern with alignments inclined steeply inwards at the ends, or poles, of the magnet and parallel to its length near the middle, with gradations in between. In a similar fashion, inclination of the Earth's magnetic field varies from an angle of 90 degrees to the surface at the magnetic poles to 0 degrees at the Equator.

This inclination can be preserved in rocks. When, for example, a lava flow containing iron-bearing minerals cools below 550°C, some of these minerals (such as magnetite) become magnetized parallel to the Earth's magnetic field. Magnetic grains in sediment may also align themselves parallel to the Earth's magnetic field as they are deposited. This alignment is inclined at lesser or steeper angles to the surface, depending on latitude. To determine the latitude at which a rock formed, the inclination of the "frozen-in" (remanent) magnetism is measured (corrected for any deformation the rocks may have undergone since they formed). The angle between the remanent magnetism and the horizontal tells us the latitude at which the rock formed.

As well as the inclination of the ancient magnetic field, the remanent magnetism also records the direction in which the magnetic poles lay at the time the rocks formed. In 1954, such measurements were made by Cambridge geophysicist Keith Runcorn and his graduate students. They determined the position of ancient magnetic poles (paleopoles) for different times in the geological past as

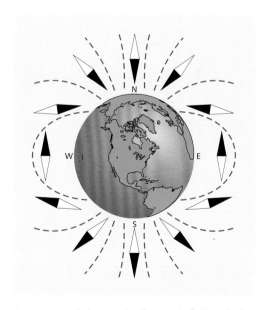

A compass works because in all magnetic fields, including that of the Earth, electrons flow from one magnetic pole to the other, causing the needle to point toward the magnetic north. The Earth's magnetic field also shows varying degrees of inclination to the ground depending on latitude. As shown here, a free-standing compass needle will point vertically at the magnetic poles and parallel to the ground at the Equator, with gradations in between. An analysis of a rock's magnetic signature, frozen-in at the time of formation, thus indicates the approximate latitude at which the rock originated.

indicated by European rocks. They found that the paleopoles from rocks of different ages plotted at different positions along a line that they called a polar-wandering curve. They then obtained results for other continents and discovered that, although the locations of individual paleopoles and curves were different for each continent, the different curves had essentially the same shape. Since at any one time the Earth can have only one spin axis and thus one set of magnetic poles (north and south), the existence of several polar-wandering curves is impossible. Thus the curves cannot represent true polar wandering. However, if the continents were brought together as envisaged by Wegener, a bit like assembling a gigantic jigsaw puzzle, the polar-wandering curves came together. This was convincing proof that the continents were wandering, not the poles. But how can continents move?

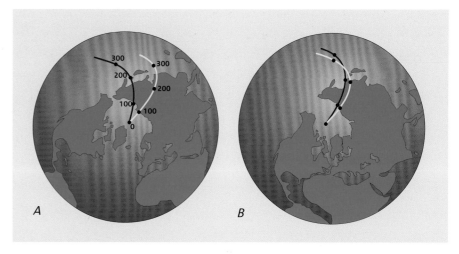

A shows apparent polar-wandering curves for two continents, plotted on a present-day map. The curves were determined by measuring magnetic inclinations from different-aged rocks in each continent. The numbers refer to ages in millions of years. Because only one magnetic north pole can exist at any one time, the fact that two curves result from plotting such data is strong evidence for continental drift. When the curves are drawn together and the continents "pulled" with them, as in *B*, we can reconstruct what the Earth's geography looked like before the Atlantic Ocean opened. ADAPTED FROM VARIOUS SOURCES.

Revelations from the Deep

During the laying of submarine telephone cables in the mid-nineteenth century, a north-south trending ridge had been discovered in the middle of the Atlantic. Oceanographic expeditions in the twentieth century revealed that this Mid-Atlantic Ridge is part of an interconnected oceanic-ridge system that winds its way across some 65,000 kilometres of ocean floor. This system of ridges has an average depth of only 2 kilometres below sea level and thus stands some 2 kilometres above the average ocean floor. Parts of this ridge system, such as the Mid-Atlantic Ridge, are in the middle of an ocean. Others converge with and locally intersect the margins of continents. Equally significant was the discovery of narrow continuous trenches up to 11.5 kilometres deep, which encircle much of the Pacific Ocean.

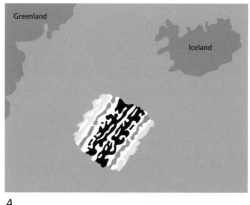

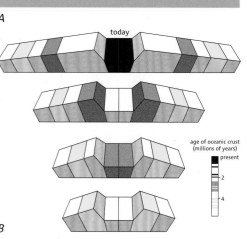

The *JOIDES Resolution*, which was built in Halifax, Nova Scotia, departs Honolulu on May 9, 2009. The *JOIDES Resolution* is one of several ships dedicated to exploring the geology of the oceans. WILLIAM CRAWFORD, COURTESY OF THE INTEGRATED OCEAN DRILLING PROGRAM US IMPLEMENTING ORGANIZATION.

A shows magnetic anomalies on the Mid-Atlantic Ridge southwest of Iceland. *B* shows an idealized series of cross sections representing stages in the development of the pattern of anomalies shown in *A*. ADAPTED FROM JANE RUSSELL IN KIOUS AND TILLING (1996), WITH PERMISSION FROM THE US DEPARTMENT OF THE INTERIOR (USGS).

Knowledge of the sediments and rocks beneath the ocean floor accelerated dramatically in the mid-1960s, when scientific drilling in deep water became technically feasible and was carried out by the Deep Sea Drilling Project. As part of this and later projects, over 1,300 core holes have been drilled in water depths of up to about 2,500 metres and thousands of metres of cored rock have been recovered and analyzed. From such material, we now know that all ocean floors are underlain by basalt, generally covered by a veneer of sediment. We have also learned that with increasing distance from oceanic ridges, the ocean floors become progressively older and deeper. An enormous surprise was that the entire ocean floor is no older than about 180 million years, and most of it is less than 100 million years old.

More surprises followed through the use of magnetometers, instruments that measure the strength of the Earth's magnetic field at any one location. Magnetometers towed behind a ship in the northeastern Pacific Ocean in 1961 revealed a distinctive linear or striped pattern. The stripes represented alternating bands of sea floor with high and low magnetic intensity. Similar magnetic stripes showed up across the Mid-Atlantic Ridge south of Iceland. These bands are not only parallel to the ridge, but show a striking mirror-like symmetry on either side of it. We now know that most ocean floors show such distinctive magnetic patterns, in complete contrast to the complex and irregular magnetic patterns on continental lithosphere. How could scientists explain both the young age and the magnetic stripes of the ocean floors?

British geologist Arthur Holmes, an early proponent of continental drift, had suggested in the 1930s that ocean floors spread as the result of convection currents circulating in the underlying mantle. In 1962, American geologist Harry Hess proposed that oceanic ridges lie directly above diverging upwelling currents of such cells. According to Hess, the currents spread laterally at the base of the lithosphere, pulling apart sections of oceanic lithosphere on each side of a ridge as they do so. This allows molten basaltic lava to well up near the crests of the oceanic ridges. As dykes of mafic magma are continuously intruded parallel to the ridge crest, they cool, solidify, and are in turn intruded and shoved aside by younger dykes, in a symmetrical pattern. Could this conveyor-belt-like process of sea-floor spreading that takes place at oceanic ridges (or spreading ridges) be the continental-drift mechanism that had eluded Wegener?

In 1963, Lawrence Morley of the Geological Survey of Canada and, independently, two Englishmen, Fred Vine and Drummond Matthews, reasoned that those stripes of rock with high magnetic intensity cooled when the Earth had normal polarity—that is, when the Earth's "North Magnetic Pole" was in the north, as it is today. Because their remanent magnetic signal reinforces the signal produced by the present magnetic field, it gives high magnetic intensity readings. In contrast, stripes of rock with low magnetic intensity cooled when the Earth had reverse polarity—when the "North Magnetic Pole" was in the south. At such times, the ancient remanent magnetic signal subtracts from the present magnetic field, resulting in a lower intensity. In the past 100 million years, our planet's magnetic field has flipped poles, or reversed polarity, about 200 times, the latest being 780,000 years ago. Reversals ranging in duration from 30,000 to tens of millions of years have now been widely recorded in the Earth's rocks.

A Disappearing Act

Over time, the amount of new oceanic lithosphere added at spreading ridges is staggering. In the past 20 million years, a strip of new ocean floor up to 3,000 kilometres wide has been generated in the Pacific Ocean, and another about 1,000 kilometres wide in the Atlantic Ocean. Over the past 90 million years, sea-floor spreading has added an average of 3.4 square kilometres of new oceanic lithosphere each year. No oceanic lithosphere older than about 180 million years is preserved beneath modern oceans, although older ocean-floor remnants are embedded in many mountain belts. As no evidence exists to suggest that the Earth has expanded like a giant balloon, equal amounts of oceanic lithosphere must have been created and consumed. But where has the consumed portion gone?

We find the answers around the Pacific Ocean. There, oceanic trenches lie on the Pacific side of curved chains of volcanoes known as magmatic arcs, which collectively form the well-known Ring of Fire. Magmatic arcs come in two varieties, island arcs and continental arcs. The Mariana Islands in the western Pacific are an example of an island arc. The Cascade Range, involving a chain of volcanoes extending from southwestern British Columbia to northwestern California, is an example of a continental arc. Big

Ring Mountain, British Columbia, a subglacial volcano known as a tuya (see Chapter 11) that was active sometime during the past 2.5 million years. This mountain is one of the volcanoes forming the Cascade Range, a continental magmatic arc above the Cascadia Subduction Zone. STEVE GORDEY.

This 375-million-year-old granite at Peggys Cove, Nova Scotia, is associated with the closing of the long-lost Rheic Ocean (Chapter 8). ROB FENSOME.

earthquakes are also common all around the Pacific. How are magmatic arcs, big earthquakes, and oceanic trenches related to one another? And how are they related to the riddle of expanding ocean floors on a non-expanding Earth?

In 1928, geophysicist Kiyoo Wadati discovered that earthquakes near Japan originate along an inclined surface that slopes westward at an angle of about 60° beneath the Japanese Island Arc. This surface can be traced from the bottom of the trench, which lies about 300 kilometres east of Japan, down into the mantle to a depth of about 700 kilometres. Subsequently, it was found that similar earthquake-defined surfaces occur all around the Pacific, and all slope down from trenches, generally toward the nearest continent. To explain these relationships, researchers concluded that the earthquakes resulted from downward movement of ocean floor on enormous reverse faults. This explanation conformed with the ideas of Holmes and Hess that the trenches are indeed where oceanic lithosphere descends and disappears into the mantle. The process is called subduction and the inclined upper boundary of the descending lithosphere is a subduction zone (see figure on page 33).

Transforming Our Ideas

Yet another seminal discovery was made in the early 1960s, this one by Canadian John Tuzo Wilson. Researchers had noticed that in many places the linear magnetic stripes and the spreading ridges on the ocean floors are truncated and offset by enormous linear fractures, or faults. Between the truncated ridges, these fractures are seismically active, with earthquakes caused by rocks on either side of the fracture grinding against one another as they move in opposite directions. Although the fracture may continue beyond the offset spreading ridges, there the ocean floor on both sides of the fracture moves in the same direction, because these segments of the fault are inactive. As ocean-floor mapping progressed, the active section of these faults was seen to connect with, and so become "transformed" into, spreading ridges or subduction zones. Some of these faults link discrete segments of spreading ridges, others link a spreading ridge to a subduction zone, and still others link two subduction zones. Realizing that these faults were a new and special kind of fracture, Tuzo Wilson named them transform faults. In contrast to the situation at spreading ridges and subduction zones, lithosphere is neither created nor destroyed at transform faults.

The vast majority of transform faults occur within oceans. But along the western margin of North America, transform faults alternate with subduction zones. To the south is the notorious San Andreas Fault, which was the locus of the San Francisco Earthquake of 1906 and several later large earthquakes. This extensive transform fault separates Pacific lithosphere from North American lithosphere, with the former moving northward relative to the latter. At the current rate of motion, eastern San Francisco will be a suburb of western Los Angeles in 11 million years. Offshore, between northwestern California and southwestern British Columbia, is the Cascadia Subduction Zone; north of this is another transform fault, the mostly submerged Queen Charlotte-Fairweather Fault. At its northern end, the Queen Charlotte-Fairweather Fault connects with the Aleutian Subduction Zone, which extends westward for about 4,000 kilometres across the northern Pacific Ocean.

The composite nature of the western boundary of the North American Plate, showing plates and plate boundaries. ADAPTED FROM VARIOUS SOURCES.

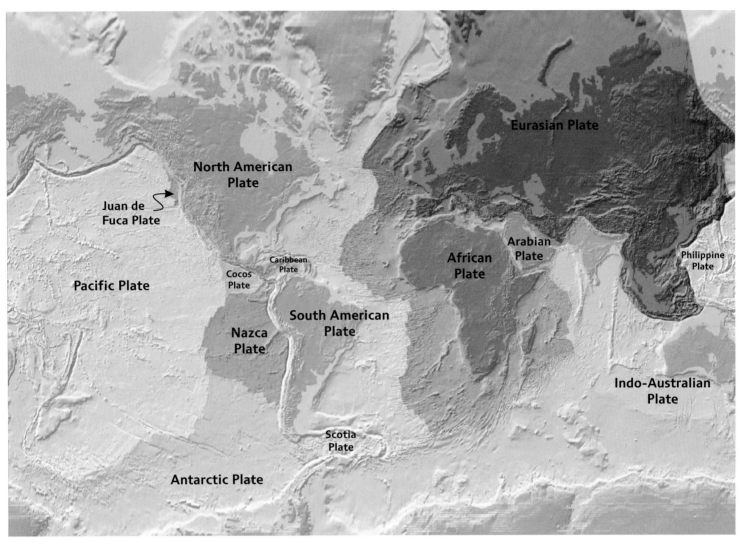

The present global distribution of tectonic plates. WALTER ROEST.

A Grand Synthesis

In the mid-1960s, researchers realized that the global network of spreading ridges, subduction zones, and transform faults defined the boundaries of several rigid lithospheric plates that covered the Earth's surface. They established that plates are constantly moving relative to one another, and changing continuously as they grow and diverge along spreading ridges, as they converge and are consumed in subduction zones, and as they slip past each other along transform faults. This relatively simple concept has become geology's great unifying theory, known as plate tectonics. Plate tectonics provides a compelling explanation of how oceans and continents have evolved over time. The lithospheric plates are fragments of the planet's thin outer shell, generally 50 to 100 kilometres thick, that "float" on the underlying asthenosphere.

A major problem that delayed early general acceptance of continental drift was the absence of a plausible driving mechanism. Wegener's idea of continents ploughing across ocean floor seemed far-fetched. We now know that the energy for driving the plates comes from the Earth's internal heat. Radioactive decay of unstable chemical elements within the mantle warms the planet's interior and provides enough heat to cause mantle convection. So Arthur Holmes was on the right track back in the 1930s when he proposed convection currents as the drivers of global tectonics. Now we realize that the pattern of mantle convection is controlled by rigid plates that are the constantly moving outermost parts of convection cells.

A few plates are made entirely of oceanic lithosphere, such as the Nazca Plate off western South America. The huge Pacific Plate is entirely oceanic except for the sliver of California west of the San Andreas Fault. Most large plates contain both continental and oceanic lithosphere. The North American Plate is one of these: its western part, including all of Canada and its eastern continental shelf, is composed of continental lithosphere, and its eastern part from the continent's eastern continental slope edge to the Mid-Atlantic Ridge, is made of oceanic lithosphere. Within each plate, the lithosphere is mostly rigid, although plate margins carrying magmatic arcs above subduction zones are hot and relatively soft and deformable.

In the past few years, global-positioning systems (GPS) using orbiting satellites have allowed us to measure plate movements

Plate tectonics explains the distribution of major physical features on the Earth's surface, including Canada's highest mountains in the St. Elias Range of northern British Columbia and Yukon. In the distance is Mount Logan, Canada's highest peak. W. LYNCH, COPYRIGHT PARKS CANADA.

Rift basins form when continents begin to break up. The basins become filled with non-marine sediments such as the redbeds in this cliff at Cape Blomidon in Nova Scotia. The overlying darker layers are of basalt similar to that underlying the ocean floors. BOB TAYLOR.

Away from plate boundaries, the interiors of continents are dominated by flat to gently rolling landscapes, as here near Foremost, Alberta. ROB FENSOME.

directly. GPS measurements show that modern plates can attain speeds of up to 20 centimetres per year, but more commonly they move at less than 10 centimetres per year, or about the rate that a fingernail grows. This seemingly slow rate translates into enormous distances over millions of years. Ten centimetres per year is equivalent to 10,000 kilometres in 100 million years.

Clues to the location and shape of past oceans and continents come from rocks, as we will see later. By using such clues, we can learn something about ancient plate activity and the evolution of pieces of lithosphere, such as the one underlying Canada.

Oceans Come and Oceans Go

The birth of a new ocean starts with the breakup of a large continental landmass, perhaps arched up from below by upwelling mantle. Doming causes the regional continental lithosphere to stretch, thin, crack, and form rift valleys such as today's East African Rift System. At first, fluvial and lacustrine deposits accumulate in the rift valleys. Thinning of the lithosphere causes a drop in pressure in the underlying mantle, which consequently partly melts. The resulting magma rises toward the surface and can be extruded as basaltic lava flows among the rift valley sediments, as in the 200-million-year-old deposits around the Bay of Fundy in the Maritimes. With continued thinning, continental lithosphere splits completely and oceanic lithosphere begins to form between the two newly created continents as they drift apart. Thus, rifting evolves into drifting—two terms to remember because they are used a lot by geologists. Widespread deposition of deep-marine sediments in a former rift basin generally indicates the onset of sea-floor spreading and drifting.

During sea-floor spreading, some mafic magma travels up vertical fractures (feeder dykes) to the surface, where it cools to form pillow lavas on the ocean floor. Some of the magma remains near the base of the crust in large chambers, where it cools slowly and forms bodies of the mafic rock known as gabbro, the coarse-grained equivalent of basalt. Other magma solidifies within the feeder dykes, which are progressively split apart as new basaltic magma flows into the ridge axis. As the solidified basalt cools below 550°C, it preserves a continuous record—the magnetic stripes—of the normal and reverse polarities of the Earth's magnetic field. What remains in the mantle after the mafic melt has been removed are ultramafic rocks such as peridotite (Box 2).

The total volume of the spreading ridges is enormous, and variations in this volume over millions of years have caused global sea levels to rise and fall by as much as 100 metres. At times, the seas have flooded (transgressed) over large areas of continental interiors; at other times they have retreated (regressed). Today, seas lap over the edges of continents to form variably wide but generally modest continental shelves, which are simply the flooded margins of continents. So, off today's east coast of North America, the boundary between the continental shelf and slope on the one hand, and the deep ocean basin on the other, lies within the North American Plate. Although an echo of the original rifted margin, the boundary between the continental slope and the ocean basin is no longer a plate boundary. Such within-plate continental margins are relatively stable tectonically and are consequently known as passive margins.

As oceanic lithosphere moves farther from a spreading ridge, it cools and becomes denser, eventually sinking into the underlying mantle at a subduction zone. The depth in the mantle that can be reached by subducting lithosphere has been imaged by seismic tomography, a technique that is somewhat like taking a CT-scan of the Earth. Results show that over the past 100 million years or so, Pacific lithosphere has subducted beneath western North America and sunk slowly to the mantle-core boundary, with its tip now about 2,900 kilometres beneath the eastern seaboard of the continent.

How subduction zones create magmatic arcs was yet another insight of the 1960s. When oceanic lithosphere descends into the mantle, it expels superheated water upon reaching depths of 100 to 250 kilometres. (Superheated water is above 100°C, but remains a liquid because it is under high pressure.) The released water rises into the lithosphere of the overriding plate, where its molecules significantly weaken the crystal structures of the minerals present, lowering their melting temperatures. Consequently, vast volumes of magma are created and a magmatic arc is born. If a subduction zone is below oceanic lithosphere, as under the Mariana and Tonga islands in the western Pacific, the arc magmas tend to be mafic. Where the subduction zone is overridden by silica-rich continental lithosphere, the rising arc magma tends to be felsic to intermediate, as along the present-day western margins of the Americas.

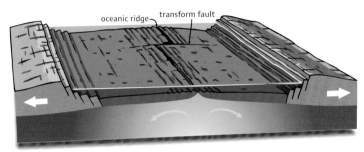
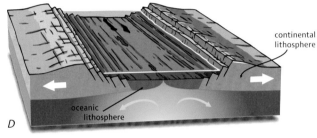
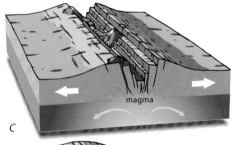
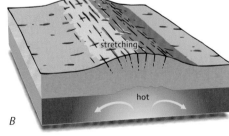
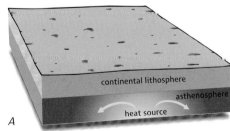

Stages in the breakup of a continent to form a new ocean. *A* shows undisturbed continental lithosphere overlying asthenosphere. Energy sources in the mantle heat, stretch, and weaken the continental lithosphere, sometimes causing doming (*B*). Eventually the continental lithosphere rifts and volcanic rocks are extruded (*C*). The rifted continental lithosphere may split and the two sides drift apart, creating two continents and leaving a small ocean basin in between, as in Baffin Bay and the Red Sea (*D*). Continuing separation leads to formation of a large ocean basin such as the modern Atlantic Ocean (*E*).

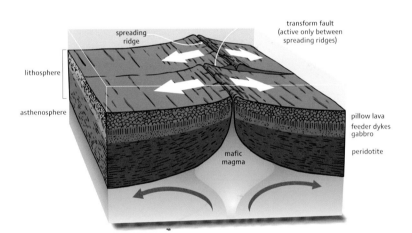

The structure beneath an oceanic spreading ridge.

Part of the Mid-Atlantic Ridge outcrops in Iceland, as shown here near Thingvellir, where the rift at the centre of the ridge is clearly visible. This marks the boundary between the North American and Eurasian plates. CHRISTOPHER HARRISON.

CHAPTER 2 · DANCE OF THE CONTINENTS

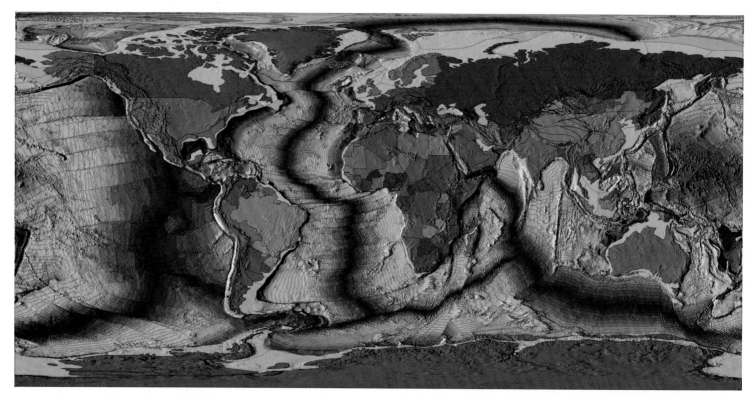

Time of formation of ocean floor has been revealed through paleomagnetism and radiometric dating (Chapter 3) and is shown on the map by the colour pattern. Red through yellow represents ocean floors formed between 70 million years ago and the present, and pale green through dark blue represents ocean floors created between about 180 and 70 million years ago. The growth rate of Pacific Ocean floor is about three times that of the Atlantic, as reflected by the wider colour bands in the former. The symmetrical spreading in the Atlantic and Indian oceans is in marked contrast with the asymmetrical pattern in the Pacific Ocean, a result of the Americas overriding huge areas of the eastern Pacific floor. The light tan colour represents continental shelves and continents are in various brown to pastel shades that indicate countries. CHRISTOPHER SCOTESE.

Mountains of Convergence

Mountain belts form some of the Earth's most inspiring landscapes. The most famous are products of converging and colliding tectonic plates. Some of the most climactic collisions occur when subducting plates carry continents. For instance, about 50 million years ago, an ancient ocean called Tethys, which lay between central Asia and India, became completely subducted. India, carried on the same plate as Tethys, arrived in the ocean's wake. Continental lithosphere resists subduction because it is too buoyant to sink, and so India was pushed into and beneath what is now Tibet, causing the mighty Himalayas and the Tibetan Plateau to rise.

Collisions also occur when continental and oceanic plates converge so rapidly that the oceanic plate cannot subduct fast enough. During such a collision, the weak arc lithosphere in the overriding continental plate takes up the strain and is squeezed

Northward view of the Rocky Mountains of southern Alberta, showing imbricated thrust sheets tilted upward to the east (Chapter 9). These mountains are perhaps the best known example of a thrust-and-fold belt. ALAN MORGAN.

Ancient volcanic rocks underlie the Swallowtail Light on Grand Manan Island, New Brunswick. Although the exact age of these rocks is unknown, they are older (perhaps much older) than about 400 million years and formed on a microcontinent called Ganderia, which originated on the southern margin of the Iapetus Ocean but eventually fused with the ancient continent of Laurentia to the north when the ocean closed (chapters 7 and 8). ROB FENSOME.

and thickened. One dramatic expression of uplift resulting from such a collision has been the change in the course of the Amazon River, which from 100 to about 15 million years ago drained into the Pacific Ocean. Collision between the Pacific and South American plates raised the Andes about 15 million years ago, reversing the flow of the Amazon so that it now drains into the Atlantic. We will explore another such collision, important to the story of the Canadian Cordillera, in Chapter 9.

An important feature associated with some plate collisions are thrust-and-fold belts. As a result of the collision, thick sequences of rocks are thrust faulted and folded toward the continental interior. The Rocky Mountains are the best-known example of this process; another, more ancient example can be found in the Appalachians of southeastern Quebec.

All modern oceans contain areas where the lithosphere is thicker than regular oceanic lithosphere. These areas include island arcs, oceanic plateaus perhaps bearing atolls, and isolated fragments of continental lithosphere such as present-day Madagascar. If subduction continues and these within-ocean features are swept toward the continent, they will ultimately collide with it. Because high-standing islands or plateaus are more buoyant than regular oceanic lithosphere, they will be scraped off the subducting plate and will stick, or accrete, to the overriding plate rather than be subducted. Many mountain belts contain remnants of such former within-ocean features; such remnants are called terranes (a term not to be confused with terrain, which denotes topography). Terranes thus have a variety of origins: they may be continental fragments (microcontinents); former island arcs; or former pieces of thickened oceanic lithosphere such as Hawaii may become if it is accreted to a continent. Many terranes are a mixture of these elements. The convergence and collision of terranes with continental margins commonly leads to the rise of mountains.

Remnants of former deep oceanic lithosphere can be preserved within an ancient mountain system. Such remnants are known as ophiolite suites or ophiolites, and examples can be seen in a belt from central Newfoundland to southern Quebec, along what was the edge of a long-vanished ocean called Iapetus

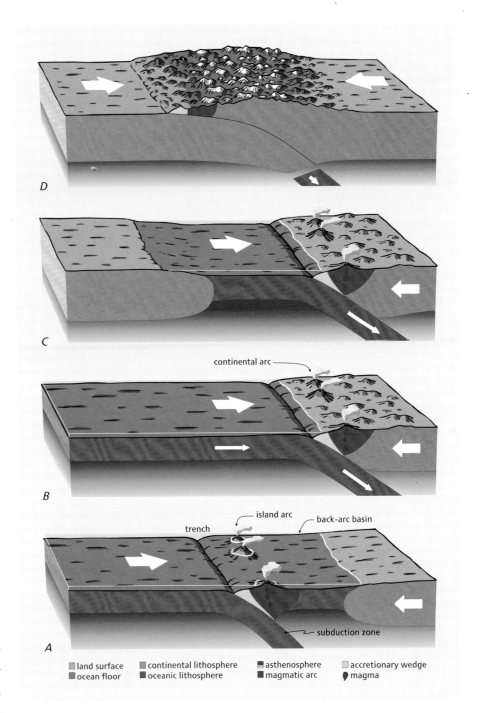

Different kinds of convergent plate margins. When the lithosphere disappears at trenches by subduction into the mantle, chains of volcanoes and bodies of magma are generated in magmatic arcs in the overriding plate. The arrows indicate relative motion of the plates. *A* shows a setting such as that in the present western Pacific. *B* shows a setting such as today's eastern Pacific margin. *C* shows a similar situation to *B*, but with another continent approaching. *D* shows a continuation of this process—the two continents have collided and one is being thrust beneath the other, a process that will eventually stall, shutting down subduction and causing changes in plate interactions and boundaries.

(Chapter 10). Here, during an ancient episode of convergence, a fragment of oceanic lithosphere was lifted "out of the water" by the buoyant continental lithosphere after the continental fragment failed to subduct. The belt was part of the Appalachians, a mountain range that formed through subduction, the collision of terranes, and ultimately the unyielding clash of continents between 500 and 300 million years ago. The Appalachians are now worn down by erosion to a shadow of their former glory.

Anatomy of the still-active Cordilleran Orogen in Canada. *A* shows the distribution of: accretionary wedges; microcontinents, which commonly incorporate remnants of island magmatic arcs and granites. *B* shows the distribution of the various major microcontinent/island arc terranes in the Cordillera, names and details of which will be introduced in later chapters. ADAPTED FROM MONGER AND BERG (1987), COURTESY OF THE US GEOLOGICAL SURVEY.

Processes similar to those of the modern world have been deforming rocks and building mountains for at least the past 2,500 million years. Much of the flat or rolling Canadian Shield is made of the roots of ancient mountain belts that formed around 2,000 million years ago. Just over 1,000 million years ago, a continent that included most of today's Canadian Shield collided with another continent to raise mountains as impressive as the modern Himalayas. The worn-down roots of these ancient mountains form the Grenville belt that extends southwestward from southeastern Canada (well exposed for example in the Laurentian Highlands of southern Quebec) beneath the American Great Plains to Mexico.

To emphasize the common origin of ancient and modern belts of rocks deformed by plate convergence and collision, whether or not mountainous today, geologists use the term orogenic belt or orogen. Hence, the Grenville belt is more correctly called the Grenville Orogen. A convergent or collisional event within an orogen is known as an orogeny.

Hot Spots and Rift Mountains

Mountains are not only produced along convergent plate boundaries. Some are associated with divergent plate boundaries. Earlier in the chapter, we learned that the rifting of large continents may be accompanied by bulging of the lithosphere. This bulging can produce uplands known as rift-shoulder highlands or mountains. The Baffin Moun-

Ultramafic mantle rocks, part of a suite of rocks known as ophiolites, are exposed in Gros Morne National Park of Canada, Newfoundland. These rocks were thrust up about 480 million years ago (Chapter 7). The brown colour and lack of vegetation are characteristic of outcrops of ultramafic rocks. MICHAEL BURZYNSKI.

The almost vertical structures in this cliff are the solidified remains of feeder dykes that supplied magma to spreading ridges. Such dykes are preserved in ophiolite suites. This example is from Gros Morne National Park of Canada, Newfoundland. JEAN BÉDARD.

Glaciers fill a valley in mountains on eastern Baffin Island. These rift-shoulder mountains were formed by uplift during the initial opening of the Labrador Sea. ALAN MORGAN.

tains on Baffin Island and the Torngat Mountains in Labrador, up to about 2,000 and 1,650 metres high, respectively, are rift-shoulder mountains that arose about 60 to 30 million years ago through rifting that ultimately led to the formation of Baffin Bay.

Some mountains show no relation to plate boundaries. Indeed, the tallest mountain on Earth is the island of Hawaii, located within the Pacific Plate. (Mount Everest has the highest elevation above sea level, but is not the tallest mountain from base to summit.) Hawaii is a giant basaltic shield volcano that rises some 9,000 metres from the surrounding ocean floor. It is being generated at a hot spot, the term given to a magmatic centre whose origin seems to be independent of plate tectonics. Volcanoes of the Hawaiian Islands become younger from northwest to southeast and are extinct except for the southeasternmost islands of Maui and Hawaii. In 1963, Tuzo Wilson suggested that Hawaiian volcanism originates from a relatively fixed source deep within the mantle, above which a narrow, buoyant column of material, called a mantle plume, rises to produce hot-spot volcanic activity at the surface. As the Pacific Plate moved northwestward, a succession of volcanic islands sprang up above the plume head and became extinct when the plate moved farther northwest. Some hot-spot traces are found in continental lithosphere,

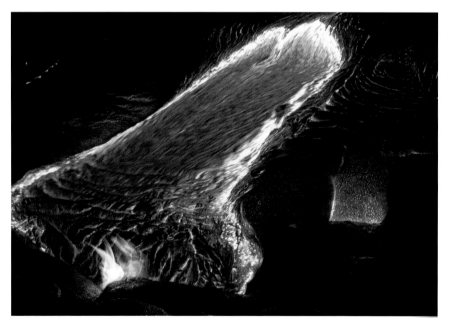

Basaltic lava flow on Kilauea volcano, Hawaii, perhaps the world's most famous hot spot. This photo was taken about 4 metres from the lava. LASZLO PODOR.

such as the still-active example now underlying Yellowstone in northwest Wyoming. Another example, no longer active, might be represented by the Monteregian Hills of southern Quebec.

Making and Filling Holes

Sediments accumulate in sedimentary basins, often simply referred to as basins, depressions on the Earth's surface that are commonly related to plate tectonics. Once basins have formed,

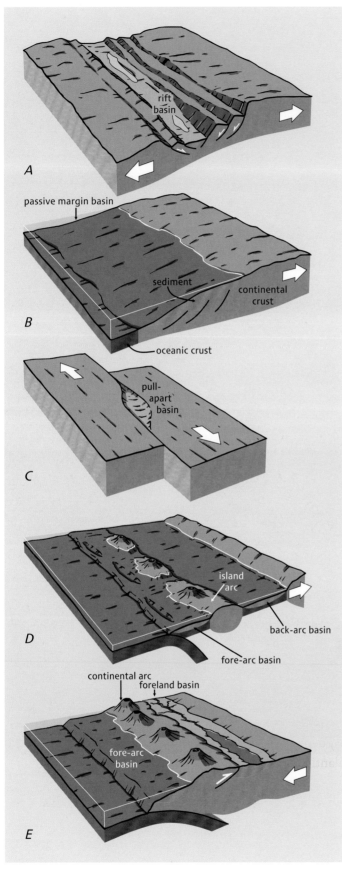

The relationships between sedimentary basins and plate tectonics. Extensional plate tectonic settings with rift basins formed during rifting (A) and passive margin basins formed during drifting (B). Pull-apart basins, which are never large, can occur in transform tectonic settings (C). Convergent-plate settings may include fore-arc basins between the trench and magmatic arc (D, E), back-arc basins between the magmatic arc and continent (D), and foreland basins, depressed by the overloading of the lithosphere as continental crust piles up through thrust-faulting and folding (E).

they trap sediments, which may become sedimentary rocks over time. Basin fill may accumulate to a thickness of a dozen kilometres or more. The figure to the left shows some of the kinds of sedimentary basins that are related to plate tectonics.

Some sedimentary basins develop in association with convergent plate margins. Sometimes a fore-arc basin develops between a magmatic arc and a subduction zone. Basins developing between an island arc and a continent are called back-arc basins. If the rocks in a mountain belt are pushed out over the adjacent lithosphere, the weight of the rocks pushes down and creates a giant moat-like trough, or foreland basin. This basin soon fills with sediment eroded from the nearby mountains. The best example of a foreland basin in Canada is the one underlying large parts of the Western Interior Plains to the east of the Rocky Mountains (Chapter 13).

Passive margins are almost always underlain by deep sedimentary basins. An example is the Scotian Basin off Nova Scotia, where seismic surveys and oil-exploration wells over the past half century have revealed an accumulation of sediment and sedimentary rocks up to 15 kilometres or more thick, ranging in age from about 200 million years to the present. These sediments accumulated on a stretched and thinned lithosphere that cooled and subsided as it became increasingly distant from the Mid-Atlantic Ridge.

Sedimentary basins also occur well within continents, perhaps due to the subtle rippling effect on the continental lithosphere reflecting distant marginal plate collisions, or to activity in the underlying mantle. An example is the ancient Michigan Basin, in which rocks now forming Niagara Falls and the Niagara Escarpment were deposited over 400 million years ago.

Of Supercontinents and the Future

Plate-tectonic processes tend to be cyclical. Over time, supercontinents form when landmasses on separate plates converge, collide, and become accreted to one another. After millions of years, supercontinents break up into new continents, a process probably driven by buildup of convection-related heat beneath the insulating blanket of the supercontinent. The youngest and most famous supercontinent is Pangea, which existed between about 310 and 180 million years ago, from the time of the first reptiles to the time of early mammals. Surrounding Pangea was the superocean Panthalassa, which at its maximum extent covered more than half of the Earth's surface and was the ancestor of our modern Pacific Ocean.

We now know that Pangea is only the latest of several supercontinents. The next oldest, Rodinia, existed from about 1,000 to 750 million years ago. The pre-Rodinian continents collided to make mountain ranges, including those of the Grenville Orogen. Geologists are now recognizing even older supercontinents, rem-

Sediments deposited in the Drumheller area of south-central Alberta in the Western Interior Basin (Chapter 9), a foreland basin east of the thrust-and-fold belt of the Rocky Mountains. JOYCE S. K. CHEW.

These fluvial sedimentary rocks (redbeds) at Burntcoat Head, on the south shore of the Minas Basin, Nova Scotia, were deposited in a rift basin over 200 million years ago (Chapter 9). ANDREW MACRAE.

nants of which can be found in rocks of the Canadian Shield. The process by which supercontinents repeatedly form and break up, each time in different configurations, has been named the supercontinent cycle. Likewise, the cyclical opening and closing of ocean basins is called the Wilson Cycle, after Canadian John Tuzo Wilson.

Today we live at a time between supercontinents. If the plates continue to move on their present trajectories, the North American Plate will eventually grind its way over the Juan de Fuca Plate, which lies off Vancouver Island and Washington. Southwestern and Baja California may move north to become embedded in southern Alaska. Australia will continue to crunch into Indonesia and move northwards. However, we do not know how long the plates will move in their current trajectories. Much may depend on the fate of the oceanic lithosphere beneath the Atlantic Ocean, some of which is close to 180 million years old—among the oldest modern ocean floors. Sometime in the future, the oldest (and thus coldest and heaviest) Atlantic lithosphere may begin to subduct, so that the present passive Atlantic margin would be replaced by a convergent margin. This scenario would dramatically change the global pattern of plate movements and lead to volcanism and earthquakes in Atlantic Canada. But that's millions of years into the future. Whichever way the continents turn, there is a strong likelihood that eventually a successor to Pangea will form. Geologists even have a name for this future supercontinent—Amasia.

3 · IT'S ABOUT TIME

CHAPTER SUPPORTED BY A DONATION FROM ROGER SMITH, SUNCOR ENERGY (RETIRED)

The Principle of Superposition states that younger sediments are always deposited on top of older sediments. Thus, in this scene from the Big Muddy Badlands of southern Saskatchewan, layers (or beds) lower down on the slope are relatively older than the layers higher up. ROB FENSOME.

Discovering Deep Time

Early philosophers despaired of trying to solve the riddle of the age of the Earth. But in 1650, armed with increasing knowledge of the Hebrew calendar and astronomy, James Ussher, the Archbishop of Armagh, felt confident enough to declare in print that the world was born and time began at 6 P.M. on Saturday, October 22, 4004 BC (or BCE). That the Earth was about 6,000 years old was no surprise, for it agreed with the Talmudic prophecy that "The world is to exist 6,000 years". Ussher's date allowed some breathing space, since it meant that the world wouldn't end until about the year 2000 CE. With over three hundred years ahead of them, his contemporaries could not possibly feel threatened. Today, some might mock such concise predictions as Ussher's, but he deserves our respect for examining available records methodically in order to establish a viable chronology for the Earth. Because geology did not gain respectability until well after Ussher's time, he was not obliged to take into account the history told by the rocks. However, this meant he really had no concept of the magnitude of time.

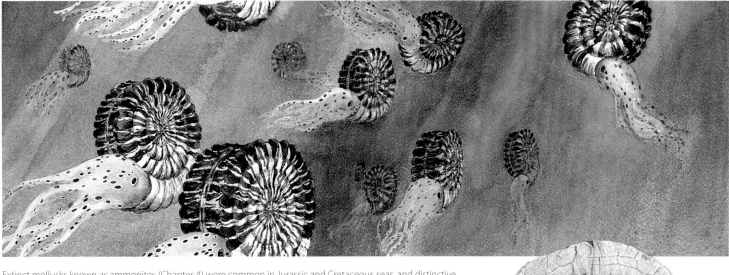

Extinct mollusks known as ammonites (Chapter 4) were common in Jurassic and Cretaceous seas, and distinctive species evolved rapidly, making their fossil shells excellent tools for relative dating. Ammonite fossils were used extensively by the nineteenth-century pioneers of the geological time scale. PAINTING BY JUDI PENNANEN, COURTESY OF THE NEW BRUNSWICK MUSEUM.

How have scientists since Ussher's day unravelled the age of our planet and the tempo of its history—a history that we now know goes back billions (thousands of millions) of years? Perhaps the first step toward an understanding of geological time was the realization that younger sediments or volcanic rocks are always deposited on top of older rocks. This insight, called the Principle of Superposition, was first defined in 1669 by the Dane Nicolaus Steenson—more commonly known by his Latin name, Steno. He also realized that if the fossils of marine animals are found in sedimentary rocks, those rocks must have been formed in the sea.

The Principle of Faunal Succession is a second key tenet. One of the earliest proponents of this was the Englishman William Smith, who surveyed routes for canals across southern Britain in the late 1700s and early 1800s. Smith noted that lower, older strata contained different kinds of fossils than higher, younger strata, and that the changes were consistent over long distances. This was true even if the rock type changed, say from sandstone to shale. Consequently, Smith could keep track of where he was in a vertical succession of rocks by observing the fossils. He used his new discovery to correlate strata across most of Great Britain, creating the first modern geological map, published in 1815. This was *The Map that Changed the World*, according to the title of Simon Winchester's bestselling book about Smith's work.

The late Jurassic to early Cretaceous ammonite *Pseudocraspedites*, found north of Eureka, Ellesmere Island. Such large fossils, clearly visible to the naked eye, are rarely recovered in exploration wells, so geologists rely largely on microfossils for dating subsurface sections. ANDREW MACRAE.

Geologists in the early nineteenth century began to recognize that these orderly and consistent changes in fossil assemblages were generally the same from one country to another, even from continent to continent. Thus, the principles of Superposition and Faunal Succession allowed the global rock record to be subdivided into intervals of time, the basic units of which geologists now call periods. The succession of intervals is known as the geological time scale and the equivalent succession of rocks is known as the geological column. The intervals of the geological time scale don't tell us the actual time

The relative ages of adjacent intrusive rocks, such as these two dykes in Kejimkujik National Park of Canada Seaside, Nova Scotia, can be determined by cross-cutting relationships. That the large dyke trending from bottom left to upper right cuts through and offsets the small dyke running from top left to bottom right shows that the former is the younger of the two dykes. ROB FENSOME.

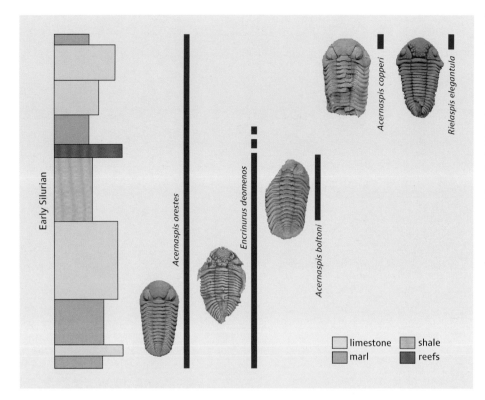

The ranges of several species of trilobite (Chapter 4) in early Silurian rocks as recorded in a sequence on Anticosti Island, Quebec. The coloured column at left shows the rock succession. To the right are illustrations of key trilobite species and the parts of the succession in which they are found—that is, their ranges. The ranges of the different species allow us to date the rocks in a relative sense: for example, a bed containing *Acernaspis orestes*, *Acernaspis boltoni*, and *Encrinurus deomenos* is older than a bed containing *Acernaspis orestes*, *Acernaspis copperi*, and *Rielaspis elegantula*. The use of fossils to date rocks in a relative sense is called biostratigraphy. It is the historical basis for the development of the geological time scale and still plays a vital role. It works because organisms have evolved over time. ADAPTED FROM MATERIALS PROVIDED BY BRIAN CHATTERTON AND ROLF LUDVIGSEN.

Neoproterozoic eras. The Phanerozoic is also divided into three eras: the Paleozoic (ancient life), the Mesozoic (middle life), and the Cenozoic (modern life). Each of the Phanerozoic eras is further subdivided into several orders of units. In this book we will use only the one level of subdivision, periods, for the Paleozoic and Mesozoic; and two levels of subdivision, periods and epochs, for the Cenozoic. These

in years, but they provide us with a relative order and they let us identify, or correlate, rocks of the same age around the world.

The largest units now recognized in the geological time scale are eons, of which there are four: the Hadean (after Hades—the Underworld in Greek mythology), the Archean (pertaining to the beginning), the Proterozoic (time of early life), and the Phanerozoic (time of visible life). Collectively, the Hadean, Archean, and Proterozoic can be referred to as the Precambrian. Fossils are absent from the Hadean and barely known from the Archean, sparse in the Proterozoic, and common in the Phanerozoic. The Proterozoic is divided into the Paleo-, Meso-, and

are shown in figure below. (Technically, the terms eon, era, and period refer to intervals of time, and the equivalent packages of rock are referred to as eonothem, erathem, and system. In this book, for simplicity, we use only the time designations. Similarly, we refer to early, middle, and late parts of an interval informally and loosely, and avoid the terms lower and upper, applied by geologists to system subdivisions.)

Most of the names of Phanerozoic periods reflect some attribute of rocks in Europe, where they were first studied. Several period names reflect regions that were considered to be representative of rocks of that age. Examples are the Devonian, named after the county of Devon in England, and the Jurassic, named after the Jura Mountains of France and Switzerland. Cambrian is from Cambria, the classical name for Wales. Other period names are derived from tribal names of ancient people who lived in areas underlain by rocks of particular ages, for example Silurian for the Silures, who lived in southern Wales. Yet other period names denote a dominant rock type, such as the Carboniferous, named for the coal deposits laid down in Europe during that period.

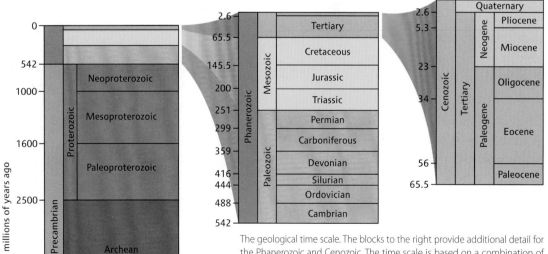

The geological time scale. The blocks to the right provide additional detail for the Phanerozoic and Cenozoic. The time scale is based on a combination of fossil ranges, radiometric dating, and other physical data.

Most North Americans divide the Carboniferous into Mississippian and Pennsylvanian and tend not to use the term Carboniferous. However, as international agreement now accepts the use of Mississippian and Pennsylvanian as subperiods, we will use the internationally accepted period name Carboniferous. The Triassic is an oddity, the name reflecting the threefold division of rocks of that period in northwestern Europe.

The Dating Game

How can we calibrate the geological column with numerical, or absolute, ages? Throughout the eighteenth and nineteenth centuries, scholars attempted to do so, with the ultimate aim of determining the age of the Earth. These scholars included Edmond Halley, Georges-Louis Leclerc (Comte de Buffon), Charles Darwin, and William Thomson (Lord Kelvin). Estimates of our planet's antiquity varied from tens of thousands to a few hundred million years, but none were based on reliable evidence.

The method that revolutionized our knowledge of the age of rocks and our planet is based on the radioactivity of certain elements contained in minerals. In order to understand how radioactivity is involved in the dating of rocks, a basic knowledge of atomic structure is needed. In large part, atoms are made up of protons (which have a positive charge), electrons (which bear a negative charge), and neutrons (which have no charge). Protons and neutrons huddle together, forming the atom's nucleus, and electrons zip around that nucleus. Atoms have a neutral charge because they have an equal number of protons and electrons. Thus a carbon atom, which has six protons, also has six electrons. The number of protons present in an atom of an element is designated

Conodonts, tooth-like mouth parts of early fish-like chordates (Chapter 4), are important microfossils for dating Ordovician marine rocks. The centre specimen is about 1 millimetre across. GODFREY NOWLAN.

as the element's atomic number—so carbon's atomic number is 6. The atomic number is directly related to the unique chemical properties of each element. Neutrons add to the mass of an atom but do not affect the element's unique properties. The total number of protons and neutrons in each atom is the element's atomic mass, and this value can vary for a particular element depending on the number of neutrons. Carbon, for example, can have an atomic mass ranging from 12 to 14 because the number of neutrons ranges from 6 to 8, though no changes occur in the properties of the element. Such types of a single element that have different in atomic masses are termed isotopes. Carbon thus has three isotopes—carbon-12 (with 6 neutrons), carbon-13 (7 neutrons), and carbon-14 (8 neutrons).

Marie and Pierre Curie and Henri Becquerel shared the 1903 Nobel Prize in Physics for showing that certain atoms and/or isotopes can change into other types of atoms and/or isotopes through loss of neutrons and/or protons (termed nuclear decay). This change is radioactivity, a process that produces heat. The first people to realize that radioactive elements could be used to measure geological time were Ernest Rutherford and his assistant, Frederick Soddy. They realized that radioactive decay proceeds at a predictable rate, depending on the isotope. Some isotopes, such as nitrogen-16, last but a few seconds, whereas some uranium isotopes take billions of years to decay to a stable product. Uranium-238, the most abundant isotope of uranium, decays through 17 intermediate steps to an ultimate stable daughter isotope, lead-206. The use of radioactive decay rate to date rocks (radiometric dating) was pioneered in the early twentieth century by Arthur Holmes.

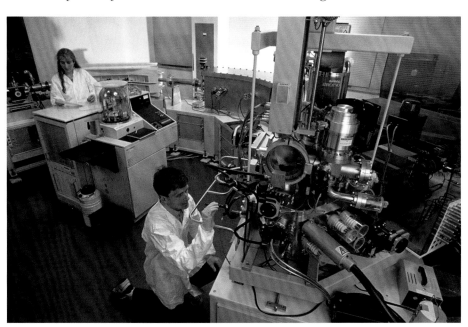

Technicians at the Geological Survey of Canada offices in Ottawa use an instrument called a SHRIMP II ion microprobe to determine the age of tiny mineral crystals. REPRODUCED WITH THE PERMISSION OF NATURAL RESOURCES CANADA 2013, COURTESY OF THE GEOLOGICAL SURVEY OF CANADA.

In measuring the decay rate of isotopes and elements, scientists use the term half-life, which is the time it takes for one half of the atoms of an unstable isotope/element to change through one or more steps to atoms of a stable isotope/element. Uranium-238 decays to lead-206 with a half-life of 4,500 million years. This contrasts with the decay of uranium-235 to lead-207 over a half-life of 700 million years. So after 1,400 million years, the ratio of uranium-235 to lead-207 would be 25 percent to 75 percent.

What are the best rocks/minerals and elements/isotopes to use for reliable dating? Igneous rocks are preferable, because they form directly from a magma. Any date derived from a crystal in an igneous rock will reflect the date of the origin of that rock, unlike the situation for most sedimentary rocks. A favourite mineral for dating is zircon. During its growth, a zircon crystal traps uranium and other radioactive atoms within its atomic structure. The lead that results from the decay of some of the uranium also becomes trapped within the crystal. Thus, zircon crystals are ideal for determining the ages of many igneous rocks. If zircon crystals in older rocks are heated and squeezed to make metamorphic rocks, an outer rim of new zircon may form around the original zircon crystal. By analyzing both the older core and the younger rim of the crystal, both the original igneous event and the subsequent metamorphic event can be dated. Zircon sand grains in sedimentary rocks, referred to as detrital zircons, do not reflect the date of deposition, but are useful in pointing to possible source areas for the sedimentary zircon particles.

Radiometric dating, then, is the analysis of the rate of decay of certain elements and isotopes to determine the age of rocks and minerals. Early analyses of Precambrian rocks by Arthur Holmes yielded ages as old as 2,550 million years. As it is highly unlikely that the Earth's oldest rocks would have survived, such numbers give only a minimum age for our planet. American geochemist Clair Patterson realized that a better line of evidence for dating the Earth's age would come from meteorites, which he reasoned would reflect the age of the Solar System and thus give a date close to that of the planet's origin. In 1953, Patterson measured uranium-238 to lead-206 ratios in meteorites and determined the Solar System to be 4,550 (plus or minus 70) million years old, an age confirmed by many subsequent studies (and is coincidentally about the same as the half life of uranium-238 as it decays to lead-206).

The use of uranium and lead isotopes has been invaluable for dating older rocks, but other elements and isotopes are used for younger timeframes. Radioactive isotopes of potassium and argon routinely provide age control for rocks ranging from many hundreds of millions to as young as 50,000 years old. For even younger ages, we need the famous carbon-14 "clock", used for dating the past few tens of thousands of years. Carbon-14

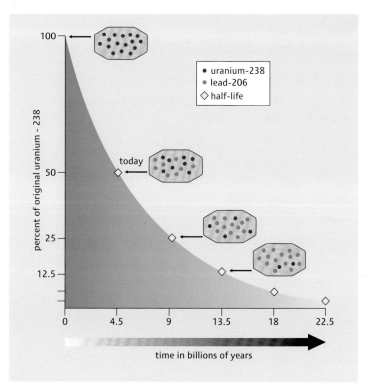

Zircon crystals such as these, set in the eye of a needle for scale, are ideal for the radiometric dating of igneous rocks. Zircons can also provide ages for many metamorphic rocks because these minerals resist changes in composition resulting from high temperatures and pressures. REPRODUCED WITH THE PERMISSION OF NATURAL RESOURCES CANADA 2013, COURTESY OF JULIE PERESSINI.

The rate of change of uranium-238 (red dots) to lead-206 (green dots) over 22,500 million years. One half-life, in this case 4,500 million years, indicates when half of the uranium-238 atoms present at the start of the half-life interval have changed to lead-206 atoms. ADAPTED FROM VARIOUS SOURCES.

(with 6 protons and 8 neutrons) changes to nitrogen-14 (with 7 protons and 7 neutrons) with a half-life of 5,730 years. One advantage of radiocarbon dating is that carbon isotopes occur in organic matter, so dates can be derived from bones, shells, wood, and other animal and plant tissues.

Because of the variance that results from some natural and human-caused effects, "radiocarbon years" (the number of years determined from the decay rate of carbon-14) are not exactly equivalent to the number of calendar years. To make resolution more precise, we can match radiocarbon dates with tree-ring counts, which can be taken back to about 26,000 "tree-ring" years, equivalent to about 21,300 radiocarbon years. Although the tree-ring-year count is closer to calendar years, scientists dealing with the last few tens of thousands of years still tend to use radiocarbon years rather than a converted value because of the present uncertainty in calibration. In this book, we will use the best estimate of calendar years at the time of writing.

Marking Time

Since the 1950s, new techniques for determining ages have developed. One is fission-track dating, which uses crystals of the mineral apatite. A relatively rare mineral in granitic rock, apatite may contain microscopic scars, called fission tracks, produced by the radioactive decay of minor amounts of uranium that disrupt the crystal structure. The scars heal and disappear if the apatite crystal is heated much above 100°C. From the number of scars and the measured concentration of uranium in the apatite, experts can tell how long ago the crystal last cooled below 100°C. This information is useful in determining when rocks, for example in mountain belts, became uplifted to near-surface levels and temperatures—and this helps us understand the history of the mountains.

Another technique is magnetochronology, which involves the magnetic reversals that helped to demonstrate sea-floor spreading. As we discovered in Chapter 2, the Earth's magnetic poles periodically flip-flop and such changes can be measured in rocks both on the ocean floor and on land. The sequence and timing of these flips over the past several hundred million years has been established through radiometric dating. Because of the irregular tempo of these global magnetic reversals, the pattern of an isolated part of the sequence found in a rock can be compared with a standard sequence to determine its age—a bit like the way a bar code identifies a product at a supermarket checkout.

Two recent developments are the thermo-clock and cosmogenic isotope dating methods. The thermo-clock method, which measures the amount of time since a rock cooled below a certain temperature, is based on the decay of radioactive isotopes such as uranium-238 and thorium-292. These isotopes emit small particles from their nuclei, one of of which is the alpha particle. The alpha particle consists of two protons and two electrons—in other words it is a helium atom. When the temperature cools to a critical point within a mineral, the escape of alpha particles stops and the mineral is said to be closed. However, alpha particles will continue to be produced from the isotope, but are trapped within the mineral. The amount of alpha particles (or helium) relative to the parent isotope in the mineral thus reflects the amount of time since the mineral was closed.

Carbon-bearing plant material from archaeological sites can be dated using the carbon-14 method. These cones are from an 11,500-year-old site in Waterloo, Ontario, and were recovered 4 metres below ground level. ALAN MORGAN.

Tree-ring counts have helped calibrate radiocarbon dates back to about 26,000 years ago. This technique, known as dendrochronology, is based on each ring of a tree representing one year's growth. COLIN LAROQUE.

Fission tracks in a grain of apatite. The tracks mark the path of subatomic particles produced by radioactive decay within a crystal. SANDY GRIST AND MARCOS ZENTILLI.

Collecting samples for cosmogenic isotope dating from a glacially deposited boulder near Peggys Cove, Nova Scotia. JOHN GOSSE.

The cosmogenic isotope dating technique helps us measure the timing and rates of processes on the Earth's surface that are triggered by cosmic rays. Cosmic rays are mainly protons, but include alpha particles, from space and the Sun. After penetrating the Earth's magnetic field and atmosphere, these particles collide at the surface with atoms of various elements—and pack a considerable punch. The energy of a single cosmic ray particle can be equivalent (though on a tiny scale) to dropping a heavy sledgehammer on your foot—and more than enough to break the bonds between protons and neutrons in an atom's nucleus. The dislodging of protons and/or neutrons from atomic nuclei leads to the production of rare isotopes and elements, the accumulation of which can be measured. So, just as increasingly dark shades of a tan may indicate the length of time that someone has been exposed to the sun, the abundance of these rare cosmogenic isotopes increases with time of exposure of a mineral at the Earth's surface. An example of the use of this technique comes from Newfoundland, an island covered by thick ice during the last glaciation (Chapter 11). According to measurements of cosmogenic beryllium isotopes in large granite boulders throughout the island, almost all of the ice there melted between 14,000 and 12,000 years ago.

Tying it all together

Historically, geologists have debated the definition of periods and their boundaries, part of the problem being a lack of internationally agreed-upon definitions. To set this right, the international geological community has established committees to select global reference, or stratotype, sections for significant boundaries, for example between periods. Such sections must be accessible, well-exposed, well-studied, and accurately dated. Several stratotype sections have already been chosen. Specific levels (termed golden spikes) within the sections define major boundaries. Most golden spikes are marked by the first appearance of one or more fossil species, supported by other data.

Two established golden spikes reside in Canada. One, at Fortune Head on the Burin Peninsula of southeastern Newfoundland, marks the base of the Cambrian Period (and by extension the bases of the Paleozoic Era and the Phanerozoic Eon), dated at 542 million years ago. This level is defined by the earliest occurrence of the worm burrow *Treptichnus*. The other Canadian golden spike, which marks the base of the Ordovician Period, is at Green Point in western Newfoundland. This golden spike, dated at 488 million years, is based on the first appearance of a conodont called *Iapetognathus fluctivagus*, and supported by the occurrences of other fossils. Two Newfoundland localities are thus critical benchmarks for the geological time scale.

The cyclicity in the layers of these non-marine Triassic sedimentary rocks at Five Islands, Nova Scotia, may have been controlled by astronomical cycles. Such rhythmic bedding can help geologists get a sense of the time that it took for sedimentary sequences to accumulate. ROB FENSOME.

Fortune Head, Newfoundland, the site of the "golden spike" for the base of the Cambrian, and therefore the Precambrian-Cambrian boundary. The boundary, marked by the tip of the geologist's hand occurs within a continuous sedimentary sequence, so enabling the analysis of an uninterrupted series of fossils. SUSAN JOHNSON.

Green Point, western Newfoundland, site of the "golden spike" marking the base of the Ordovician, and hence the Cambrian-Ordovician boundary. JAMES STEEVES, COPYRIGHT PARKS CANADA.

BOX 3 • MAPPING, PROBING, AND SENSING

Beneath our Feet

The oldest-known document displaying geological information is a papyrus map from ancient Egypt showing the location of a quarry containing a desirable ornamental stone and a gold mine. The map, dating from about 1150 BCE, uses different colours to show the different rocks; and the gravels of a wadi (a dry river bed) are portrayed with a variety of coloured dots. Modern geological maps, which come in many different varieties, likewise show where different kinds of materials and features occur. Some show the distribution of different soil types, but these are not generally considered to be conventional geological maps. In Canada, much of the land is covered by the deposits of former ice sheets, as well as recent and modern sediments of rivers and lakes. Maps can be made of such surficial deposits, or of related surface landforms, most of which reflect the past impacts of ice sheets. However, conventional geological maps show bedrock, the hard rock beneath geologically recent and soft surficial materials.

Traditionally, most information for geological maps comes from field mapping. Although helicopters and airplanes have replaced packhorses and canoes, "boots-on-the-ground" field mapping still must be done, and it still takes time and hard work. Geological maps of bedrock provide information on the distribution of different kinds of rock in an area. Such maps generally distinguish rocks that have different origins—for example whether rocks are sedimentary, metamorphic or igneous and, if igneous, whether extrusive or intrusive. Where possible, ages are shown so that the user can visualize how the rocks are related in time. Bedrock maps commonly display the inclination, or dip, of tilted or folded strata. Faults are also shown. Some idea of the timing of deformation can therefore be deduced from the map by observing which rocks are affected by folding and faulting and which are not. Additional information such as vertical cross-sections may also be presented on a geological map. A cross-section depicts the inferred structure and distribution of the rock units below the surface—rather like slicing through an onion.

The next step, after making maps and using them to predict what lies beneath, is to dig holes or drill wells. Rocks and fossils recovered from wells provide information and help refine ideas about the geology of a region. A video camera can be lowered down a well to provide real

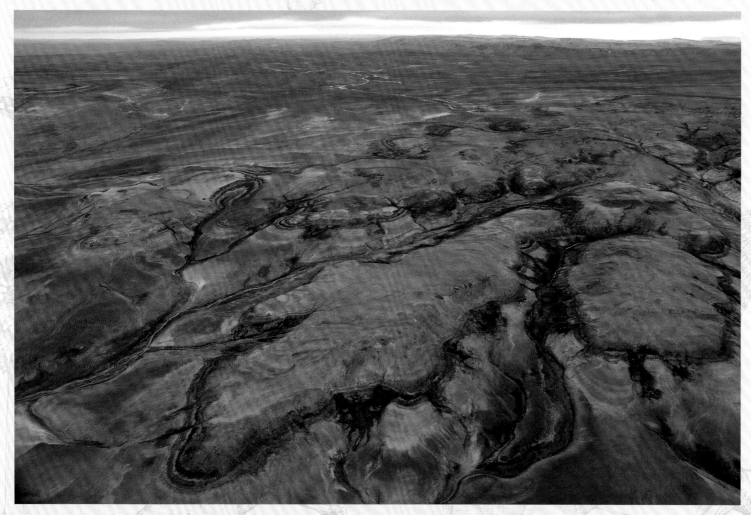

These Cretaceous rocks on Ellef Ringnes Island, Nunavut, are so colourful and distinctive that it is easy to imagine how a geological map can be created from them. CAROL EVENCHICK.

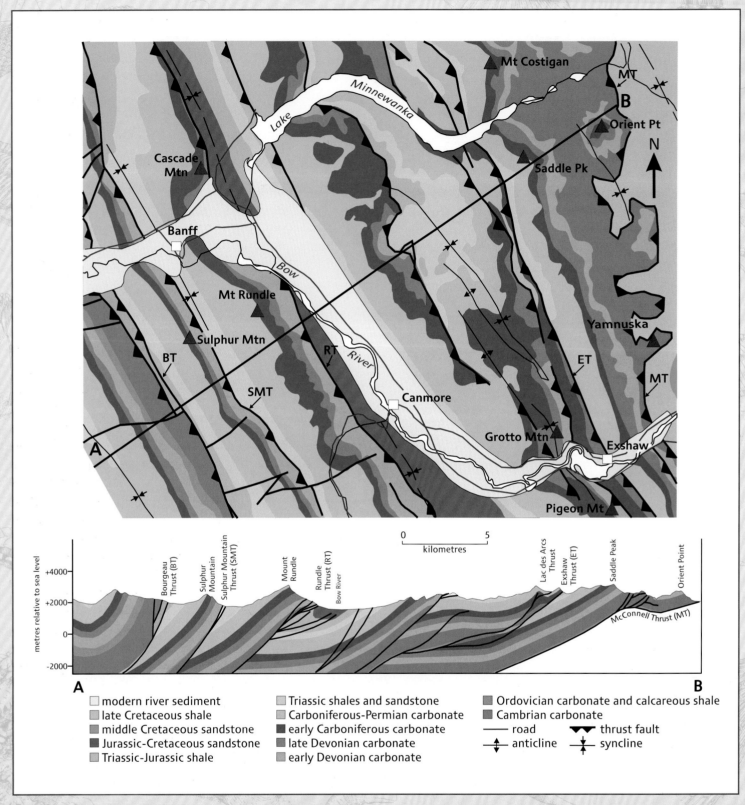

Geological map of part of the Rocky Mountains, a thrust-and-fold belt in the Cordillera. Apart from the modern sediment deposited by the Bow River, the map shows the distribution of bedrock. It also shows symbols that represent faults, anticlines, and synclines. From such information, geologists can predict the structure beneath the surface as shown in the cross-section A–B below the map (the line of section is indicated on the map). To prove these predictions and provide additional information, core holes or remote sensing analyses are needed. ADAPTED FROM A MAP COMPILED BY MARK COOPER, USED WITH PERMISSION OF THE CANADIAN SOCIETY OF PETROLEUM GEOLOGISTS.

Boulder hopping, one of the skills developed by field geologists, in this case over the Current River in northern Ontario, near where Steepledge Lake empties into Ray Lake en-route to Lake Superior. GRAHAM WILSON.

The equipment used for shooting deep seismic lines for the LITHOPROBE Trans-Canada Transect included big trucks affectionately known as "dancing elephants". These trucks generate seismic energy that penetrates deep into the crust. PHIL HAMMER, LITHOPROBE.

images of subsurface rocks. Other instruments can be lowered to record electrical, acoustic, radioactive, magnetic, and other properties of the rocks and fluids adjacent to the well. Such records, known as well logs, provide valuable information on the type, porosity, permeability, and fluid content of rocks at depth, including whether oil or gas is present.

Good Vibrations

The drilling of holes is expensive, and depth of penetration is limited. So we need other ways to "look" into the Earth. Enter seismology, the science of studying the structure of the Earth using acoustic, or seismic, waves. Natural earthquakes produce energy waves that can be used in seismology. These waves travel through the Earth and bounce back, or reflect, off surfaces at depth where the rocks change. The vibrations can also be refracted back to the surface—that is, be bent or deflected as they pass through different subsurface materials with a resulting change of velocity and direction, much as rays of light are deflected, or refracted, when they pass from air to water.

Although energy waves produced by earthquakes can be used in understanding the general structure of the Earth, they are not suitable for targeted research or resource exploration closer to the surface. This is because we don't know when an earthquake will occur, and also because the energy from earthquakes is not at a high enough frequency to provide detailed images. So geophysicists create their own waves where and when they want them. When seismic methods were first introduced in the 1930s, dynamite explosions (shots) were used—and sometimes still are in approved locations. More commonly nowadays we rely on safer and more environmentally acceptable methods, such as large vibrating trucks on land, and ship-towed air guns that release pulses of highly compressed air at sea. On land, reflected or refracted echoes of the small ground vibrations thus set off are detected by sensitive instruments called geophones. At sea, returning vibrations are detected by hydrophones, specially adapted pressure sensors encased in a thin, flexible tube called a streamer, which may be up to 10 kilometres long. Geophones and hydrophones record changes in the waves that arrive at the surface after travelling through different rocks at depth—analogous to medical ultrasound techniques. Seismic data from a single line of geophones will provide a two-dimensional view, like a single thin slice into the Earth. But a three-dimensional image of the subsurface (a 3D seismic survey) can be produced using multiple lines of geophones or hydrophones. The data received from such surveys must be analyzed using sophisticated software, but the end results can reveal stunning details about our planet's subsurface.

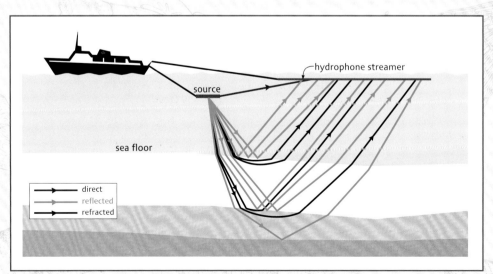

Seismic analyses involve the generation of seismic waves, in this case setting off vibrations at sea, which are reflected or refracted back to the surface. The returning waves are recorded by a hydrophone streamer (or geophones on land). Digital processing of the seismic records produces seismic sections as shown in the figure on the next page.

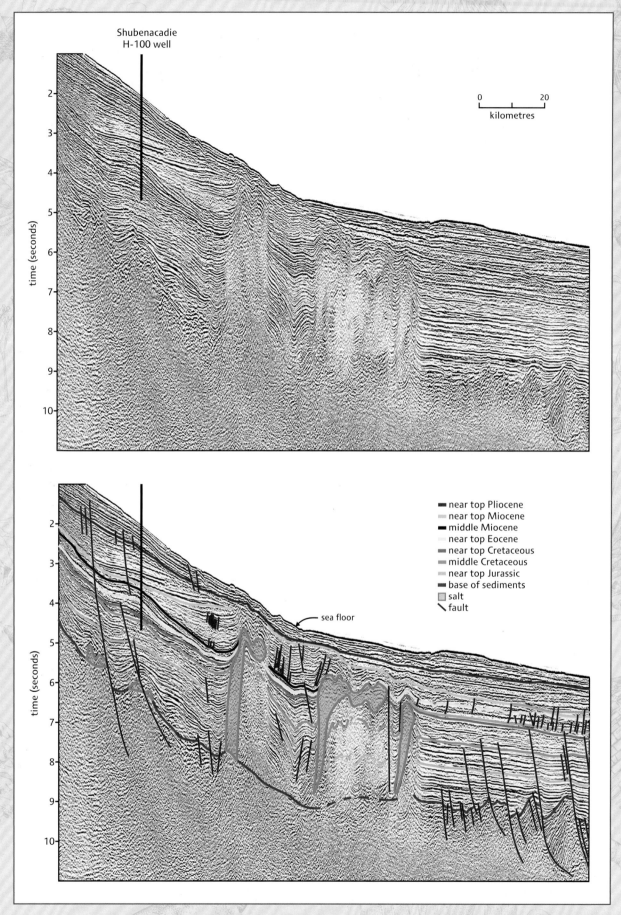

A seismic section from offshore Nova Scotia. The top figure shows the digitally processed but uninterpreted seismic section. The bottom figure shows the same section, with the geological features such as salt structures (some forming overhanging "canopies") and faults interpreted. The interpretation is reliant largely on rock and fossil data retrieved from nearby wells, such as the Shubenacadie H-100 well shown here. The vertical scale represents the time it takes for seismic waves to travel down to a particular surface, then back up again to the recording instrument. COURTESY OF JOHN SHIMELD AND JOHN WADE, GEOLOGICAL SURVEY OF CANADA.

Three-dimensional seismic surveys and sophisticated computer processing can reveal ancient surfaces such as this one, about one kilometre beneath sea level near Sable Island, off Nova Scotia. It shows the top of the late Cretaceous chalk, revealed in digitally enhanced colours: shallower areas are shown in red and yellow, deeper areas in green and blue. The unevenness of the surface is in part due to faulting (shown by the cliff-like features) and gentle folding. The area represented is 12 kilometres long by 7.5 kilometres wide. COURTESY OF ANDREW MACRAE.

Seismic technology was the principal remote sensing method used in the LITHOPROBE Project, a Canadian endeavour to learn more about the lithosphere beneath Canada and its offshore margins. LITHOPROBE, which ran from 1984 to 2005, provided information about rocks and their structure up to depths of 100 kilometres in ten selected areas. One important product of this project was a trans-Canada cross-section of the lithosphere that extends down to 250 kilometres and for 6,000 kilometres from the Juan de Fuca Ridge west of Vancouver Island to the continental shelf off Atlantic Canada. The trans-Canada cross-section was the first of its kind in the world. Many other techniques besides seismic were used to make LITHOPROBE a comprehensive project. LITHOPROBE provided scientifically valuable information about the lithosphere beneath Canada (for example, see Chapter 5), and also contributed to resource exploration and natural hazard research.

More Remote

In addition to seismology, geophysical studies include gravity, magnetic, and electromagnetic methods. Gravity is the universal force of attraction between bodies as a result of their mass. We are most familiar with this force because it keeps our feet on the ground as our bodies are attracted by the Earth's gravitational pull. Rarely do we stop to think that our bodies are also exerting an attractive force on the Earth. The gravitational effect of our bodies is extremely small compared to the mass of the Earth, but this is not true for lofty mountains and deep-sea trenches. Differences in gravitational attraction can also be produced within the Earth by the distribution of rock types with different densities. Such differences can be recorded by a gravimeter, an instrument used to measure variations, or anomalies, in the Earth's gravitational field at or near the planet's surface. These gravity anomalies can be shown on maps using colours or contours. Gravity anomalies may reflect the presence of sedimentary basins, spreading ridges, or igneous intrusions.

Magnetic maps are also useful for interpreting subsurface geology. Measurements of the Earth's magnetic field are made with an instrument known as a magnetometer, operating from a moving platform such as a ship, aircraft, or satellite, or can be hand-held during ground field surveys. Different rocks contain different amounts of magnetic minerals. Although the differences are very small, they are detectable as slight anomalies from the Earth's main magnetic field (Chapter 2). A map showing magnetic anomaly data can help define the geological structure of an area. Detailed magnetic surveys are particularly useful for delineating ore bodies, which commonly contain more magnetic minerals than surrounding rocks.

The measurement of reflected radio waves—a type of electromagnetic wave—was used effectively for the first time in 1940, during the Second World War, to identify the range, speed, and direction of moving objects such as warships and aircraft. The then-new defensive tool was termed Radio Detection and Ranging, which was shortened to the acronym RADAR. Nowadays, a variant of radar known as ground-penetrating radar (GPR) uses radio waves to reveal the character and structure of the shallow subsurface. Radio waves are transmitted into the ground, and some of the waves are reflected and refracted from shallow features back to a receiver on the surface. High-frequency waves provide a detailed image of the first several metres beneath our feet. Lower frequency waves can be used to probe deeper, up to about 20 metres, but as the depth increases the resolution decreases. GPR is a rapid and relatively inexpensive tool for mapping features such as the water table, contaminant plumes, or buried objects, or for determining ice thickness.

LIDAR (Light Detection and Ranging) is a similar technique to radar. It measures the distance to an object based on the time it

Ground-penetrating radar survey, Kluane Lake, Yukon. JOHN CLAGUE.

takes a light pulse to travel to and return from that object. LIDAR uses electromagnetic energy with much shorter wavelengths than radar and is able to detect more subtle features. Repeated LIDAR surveys can be used to measure changes in the surface as small as centimetres, which makes it useful, for example, in measuring uplift in tectonically active areas or rates of cliff erosion.

A technique used in marine and freshwater settings is side-scan sonar. It uses sound energy, derived from an instrument towed behind a ship or boat, to create an image of a lake or sea floor. Irregularities in a lake bottom and variations in the types of bottom sediments produce changes in the amount of energy that returns to the instrument, making it possible to distinguish boulders and rocky areas from muddy ones. Like radar, sonar was originally developed for military purposes, and it was not until 1960 that the first commercial side-scan sonar, a converted echo-sounder, was introduced. Side-scan sonar has since been used to make nautical charts, to study dwindling fish stocks, and in marine archaeological studies (such as the locating of King Henry VIII's flagship, the *Mary Rose*, off southern England in 1967). Data from side-scan sonar surveys are typically shown as greyscale images showing reflections from a lake or sea floor. The images are commonly used in conjunction with maps of bathymetry or surficial sediments.

A related but more sophisticated technique for imaging lake bottoms and sea floors is multibeam bathymetric sounding. Usually referred to simply as multibeam, this technique was made possible with improvements in computing, precision navigation systems using GPS, and sonar equipment. It can be used to map large swaths of seabed in exquisite detail. Multibeam images resemble photographs of the sea floor, but with the water drained and colours artificially enhanced.

The techniques described in this box have all helped in building a better understanding of the Earth's surface and interior. The maps and images are increasingly in digital online form and can be continually updated, even during the data-collection process. Such products are key to resource exploration and development. Equally important is the insight they provide into environmental issues.

Gravity and geomagnetic surveys can be made from the air. COURTESY OF KATIE MCWILLIAMS, ADAM SHALES, AND FUGRO AIRBORNE SURVEYS, A DIVISION OF FUGRO CANADA CORP.

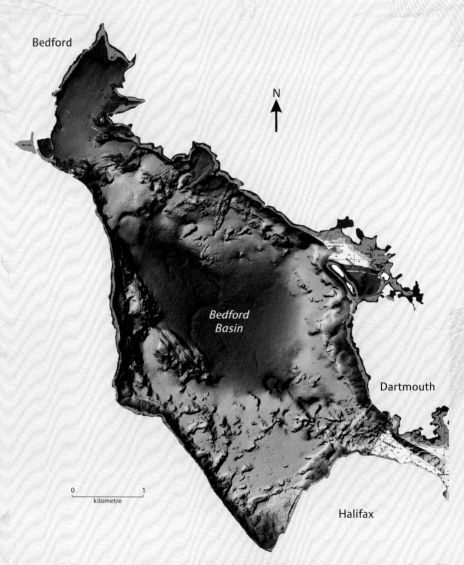

Multibeam image of the Bedford Basin at the head of Halifax Harbour, Nova Scotia. Water depths range from shallow (red and orange) to deep (blue). Maximum depth is 61 metres. Such images show many features of the seabed, including former now-submerged shorelines, bedrock outcrops, deep-water methane gas escape trenches, and anchor marks. COURTESY OF GORDON FADER AND THE CANADIAN HYDROGRAPHIC SERVICE.

4 • FOSSILS AND THE BUSH OF LIFE

CHAPTER SUPPORTED BY A DONATION FROM IAN YOUNG

Among fossil aficionados, Canada is famous for having many sites that have yielded striking specimens of global significance, including four UNESCO World Heritage Sites designated for their paleontological importance. One such site is Miguasha National Park in Quebec, famous for its Devonian fossil fish, such as *Eusthenopteron* shown here. JOHANNE KERR, COURTESY OF MIGUASHA NATIONAL PARK.

Tyrrell's Encounter

On the morning of August 12, 1884, twenty-six-year-old Joseph Tyrrell and his assistant were canoeing down the Red Deer River in the badlands of Alberta. They stopped once in a while to take a closer look at the rocks. At one of these stops, Tyrrell scaled a cliff and, to his great surprise, came face-to-face with a beautifully preserved but fearsome-looking fossil skull. The skull was that of a dinosaur later to be known as *Albertosaurus*, a close relative of the notorious *Tyrannosaurus rex*. This chance find was a harbinger of what would soon become a treasure trove of ancient bones. For Tyrrell it was a truly memorable event, one that he would vividly recall on his ninety-fifth birthday.

Joseph Tyrrell had been assigned by the Geological Survey of Canada to assist to the great geologist George Dawson (Chapter 9). Setting out in the summer of 1883, Tyrrell's main objective was to find coal to power the steam trains of the planned Canadian Pacific Railway. He found coal aplenty, but this success was overshadowed by the discovery of several dinosaur "graveyards". Tyrrell headed back to the badlands in 1884 to look for more dinosaurs, the trip culminating in his *Albertosaurus* encounter. Today, Joseph Tyrrell and his pioneering achievements are commemorated in the name of the Royal Tyrrell Museum of Palaeontology, opened in 1985 in Drumheller, Alberta.

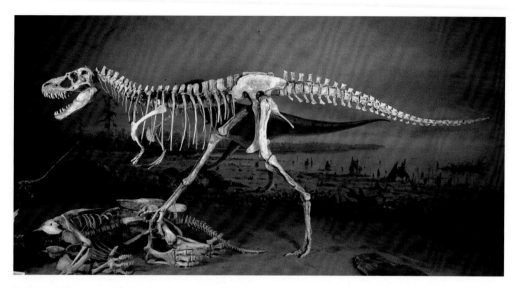

Skeleton of *Albertosaurus libratus*, found in Dinosaur Provincial Park, Alberta, and now at the Royal Tyrrell Museum of Palaeontology in Drumheller. ROYAL TYRRELL MUSEUM.

Over the years, other fossil hunters have made dramatic discoveries of dinosaur bones in the Red Deer River badlands, including many skeletons that today grace museums worldwide. Prominent among these hunters was Lawrence Lambe, who also worked for the Geological Survey of Canada. Lambe started exploring the badlands in 1897 and discovered many new genera and species over just a few years. He is remembered in the name of the dinosaur *Lambeosaurus*. Perhaps the most famous collectors were members of the Sternberg family, Charles H. and his sons George, Charles M., and Levi. Their finds, made in the early twentieth century, formed the basis of what was to become the collection of the National Museum of Canada (now the Canadian Museum of Nature) in Ottawa, home to some of Canada's finest fossils.

Evidence of Ancient Life

Dinosaurs have long caught our imagination, although few people stop to think about the amazing range of life preserved in the fossil record. Indeed, what are fossils? The simplest definition is that they are the remains or evidence of ancient life. The study of such ancient remains is known as paleontology. That an organism will be preserved as a fossil is extremely unlikely, because organic matter readily decays and scavengers are constantly on the prowl for a meal. Only rapid burial will stop the destruction. Most fossil remains are the tough parts of animals and plants, such as shells of invertebrates (animals without backbones), bones of vertebrates (animals with backbones), and wood, leaves, and pollen from plants. Land-dwellers (terrestrial organisms) such as dinosaurs are less likely to be preserved than water-dwellers (aquatic organisms, marine and non-marine), since most sediments are deposited under water. Most fossils are therefore of aquatic organisms, and the overwhelming majority of these lived in the sea. Preservation of terrestrial organisms is relatively rare, which makes rich deposits of dinosaur bones all the more special.

Joseph Tyrrell in his later years. REPRODUCED WITH THE PERMISSION OF NATURAL RESOURCES CANADA 2013, COURTESY OF THE GEOLOGICAL SURVEY OF CANADA.

Fossilization can take many forms. When hard parts are buried and not changed in composition, which is rare, we have complete or unaltered preservation. More commonly, the fossil material, usually a shell or bone, is partly or completely mineralized. Permineralization is when

A pond snail from late Quaternary sediments excavated from a kettle hole (see Chapter 11) near Waterloo, Ontario. The snail shell is made of calcite. Other small clams and snails are also visible. ALAN MORGAN.

The shells of these Silurian nautiloids from Dyer's Bay, Bruce County, Ontario, have been replaced by silica, but the interior cavities are still partly empty. ALAN MORGAN.

In this Silurian rock from Arisaig, Nova Scotia, the shell material of a brachiopod (left) and several gastropods have been dissolved leaving gaps that are molds of the original shell. ANDREW MACRAE.

Fern foliage preserved as a carbonized film in late Carboniferous rocks of the Sydney area of Nova Scotia. HEINZ WIELE, COURTESY OF THE ATLANTIC GEOSCIENCE SOCIETY; SPECIMEN FROM THE NOVA SCOTIA MUSEUM.

A petrified late Carboniferous tree stump from Inverness, Nova Scotia. The tissue of the plant has been replaced by silica. MARTIN GIBLING.

Tetrapod footprints (*Hylopus*) from middle Carboniferous rocks near Parrsboro, Nova Scotia. This view is of the underside of the bed, so the footprints are raised rather than depressed. The many small raised bumps represent small pits in the original sediment, probably the impact marks of rain drops that fell about 320 million years ago. DAVID BROWN; SPECIMEN COURTESY OF ELDON GEORGE.

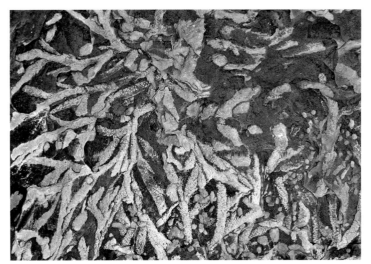
The branching feeding burrows in this organic-rich Ordovician mudstone from eastern Canada (location unknown) are known as *Chondrites*. The burrows typically branch at about 45 degrees and may show evidence of the organism (possibly a worm) probing the sediment and producing the small pits along the burrow walls. DARREL LONG.

new minerals, usually deposited by groundwater, fill empty pore spaces and the original shell, bone, or wood is unaltered. Replacement is when the bone or shell material is mineralized. Petrification is when there is both permineralization and replacement. Plant material is often carbonized, which means that parts are changed to carbon and preserved as black, coaly films or impressions.

Some fossils occur as molds or casts. After a fossil, usually a shell, has been preserved and its host sediment turned to solid rock, groundwater may dissolve away the preserved hard parts leaving behind only a cavity where once existed the original shell. This cavity is a mold. Sometimes the mold is secondarily filled with sediment or minerals transported by groundwater, forming a cast. Soft parts of organisms are rarely preserved as fossils, but spectacular exceptions do occur. One is the 505-million-year-old assemblage of strange and wonderful animals retrieved from the Burgess Shale in Yoho National Park of Canada in southeastern British Columbia (Box 6).

Not all fossils represent parts of animals and plants (body fossils). Some are trace fossils, produced by an organism's behaviour. Trace fossils include burrows, tracks, trails, and footprints left in sediments that may later become sedimentary rocks. By analyzing traces such as trackways, we may determine how big the animal was; whether it was a reptile, an amphibian, or a mammal; or if it was walking, running, or perhaps limping. Trackways may even provide insights into an animal's social life. For example a group of reptile trackways all heading in the same direction and on the same bedding surface have been found in 285-million-year-old rocks at Brule on the Northumberland Strait shore of Nova Scotia. These tracks record the oldest-known evidence anywhere of animals moving in herds, behaviour that perhaps evolved to give individuals more protection against predators. Trace fossils yield many other clues to the ways in which animals lived in the past. They confirm, for example, that dinosaurs generally walked upright; their trackways convincingly show that dinosaurs were unusual among reptiles in moving with their legs under their bodies.

Most trace fossils were made by invertebrates; these fossils include dwelling burrows, borings in shells, and feeding and resting traces, as well as trackways. Anyone who has been to a sandy beach at low tide is aware of the multitude of traces left by animals on and under the surface, such as the depressions that reveal the location of clam burrows below. Plants, too, can leave their traces. Paleontologists give trace fossils Latin names, just as they do for body fossils.

Fossil droppings or dung (geologists use the more sanitized term coprolite) shed some light on the feeding habits of animals. From fossil fragments preserved in its dung, we might determine if a creature was a herbivore or a carnivore. Sometimes, paleontologists can discover specifically what was eaten. An example is a huge tyrannosaur coprolite from the Frenchman Valley in southwestern Saskatchewan, which consists of shattered bone fragments that appear to have come from juvenile herbivorous dinosaurs. Similarly, the diet of New Brunswick's Hillsborough mastodon has been established from the plant remains preserved in its dung. Some fossils from the Burgess Shale actually contain gut contents—the semi-digested remains of last meals.

Getting to the Nucleus

The fossil record provides compelling evidence for evolution, and Canada has many superb fossil localities that add significantly to our understanding. These include four UNESCO World Heritage Sites: the Burgess Shale in southeastern British Columbia, Dinosaur Provincial Park in Alberta, Miguasha in Quebec, and Joggins in Nova Scotia. In the next few sections, we will explore the vari-

Fossil dung or intestinal remains found in the late Cretaceous sediments of the Frenchman Valley in southern Saskatchewan. PROVIDED BY DAVID EBERTH.

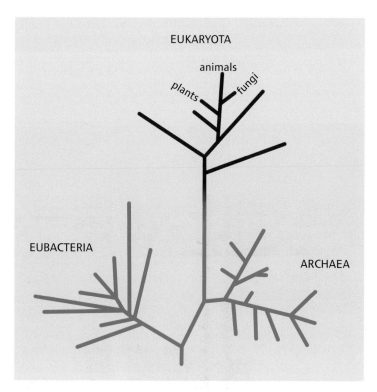

Relationships between the three main branches of life on Earth: prokaryotic Eubacteria and Archaea (both shown by green lines), and the Eukaryota (shown by red lines). Line lengths indicate the degree of difference among gene sequences and do not directly represent time. However, striking parallels can be made between this pattern derived from molecular studies of modern organisms and the timing of first appearances of fossils of the same groups of organisms—a powerful confirmation of evolution. ADAPTED FROM KNOLL (2003).

ety of life and its broad evolution so that Canada's fossil treasury, described in the second part of this book, can be better appreciated.

For both fossils and living organisms, the basic unit of classification is the species. Genera are groups of species. Hence, we humans belong to the genus *Homo* and the species *Homo sapiens*. Traditionally, genera have been organized into families, above which are orders, then classes, and ultimately at highest ranks phyla and kingdoms. Until Charles Darwin's time, organisms were classified largely on the basis of their shape and general anatomy. But since then, relationships and affinities have played a critical role in classification. In recent decades the development of molecular studies (primarily DNA and RNA) has reshaped the classification of life. For example, before molecular studies took off, the division between plants and animals seemed life's most fundamental split. From molecular studies, we now know that the divide between organisms whose cells lack a nucleus and those whose cells possess a nucleus—prokaryotes and eukaryotes, respectively—is a far more fundamental split.

A major surprise in the late 1970s was the recognition of two distinct groups of prokaryotes (or microbes): true bacteria and archaea. Studies showed that archaea were so fundamentally different from bacteria that a whole new level of classification was merited—the domain. Three domains are now recognized: Eubacteria, Archaea, and Eukaryota, the last-named including all organisms whose cells have a nucleus.

As molecular studies have advanced, the tree of life has come to look more like a bush. The diagram on the previous page shows, surprisingly, that complex eukaryote groups such as plants, animals, and fungi are not that far apart on the bush. And it appears that animals are most closely related to fungi. So humans are cousins to mushrooms and not so distant from oak trees. Eukaryotes in general are far removed from all prokaryotes, but show a closer genetic relationship to the archaea than to the bacteria.

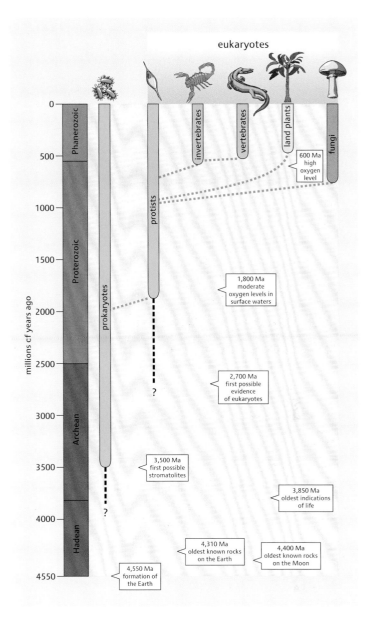

The sequence of major events in the history of life on Earth and the timing of first appearances and ranges of the main groups of organisms. Probable interrelationships are shown by the dashed lines. ADAPTED FROM ATLANTIC GEOSCIENCE SOCIETY (2001).

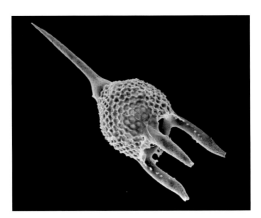

A fossil radiolarian from offshore eastern Canada. Radiolaria are marine protists that have shells made of silica. They live among the plankton, and their skeletal remains cover huge areas of deep ocean floor. REPRODUCED WITH THE PERMISSION OF NATURAL RESOURCES CANADA 2013, COURTESY OF THE GEOLOGICAL SURVEY OF CANADA.

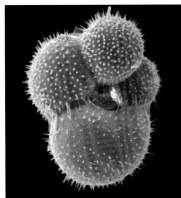

Foraminifera are protists with shells commonly made of calcite. This fossil is from offshore eastern Canada. REPRODUCED WITH THE PERMISSION OF NATURAL RESOURCES CANADA 2013, COURTESY OF THE GEOLOGICAL SURVEY OF CANADA.

Several disc-like coccoliths can be distinguished in this close-up of a piece of chalk from beneath the continental shelf off Nova Scotia (Chapter 9). The prominent coccolith at centre-left is about 6 microns (0.006 millimetres) across. ANDREW MACRAE AND THE SAINT MARY'S UNIVERSITY MICROANALYSIS UNIT.

Besides plants, fungi, and animals, the eukaryote domain includes a diverse group called the protists, which are simple, mostly single-celled organisms such as amoebas. Some protists are plant-like and, with the aid of chloroplasts, manufacture their own food through photosynthesis. Such protists are informally referred to as algae, although they represent diverse evolutionary lineages. The miniscule calcareous scales, or coccoliths, of one group of algae, known as the coccolithophores, form the variety of limestone known as chalk. Other protists, such as foraminifera (amoeba-like single-celled organisms with tiny calcareous shells) and radiolaria (single-celled organisms with tiny shells made from silica), are animal-like and actively ingest food rather than manufacture it. And yet others, such as dinoflagellates (single-celled organisms that use two whip-like flagella for propulsion), can be animal-like and/or plant-like in their feeding habits. Collectively, the fossil remains of microscopic organisms, as well as microscopic parts of larger organisms such as pollen grains, are called microfossils.

	animals with radial symmetry			sponges, cnidarians (corals, jellyfish, sea anemones)
invertebrates	animals with bilateral symmetry (bilaterians)	protostomes	arthropods	insects, myriapods (including arthropleurids), spiders, trilobites, crustaceans (lobsters, shrimps)
			annelids	common worms
			mollusks	bivalves (clams, mussels, scallops)
				cephalopods (octopuses, squids, ammonoids, nautiloids)
				gastropods (snails, slugs)
			brachiopods	
			bryozoa	
chordates		deuterostomes	echinoderms	echinoids (sea urchins, sand dollars), asteroids (starfish), crinoids (sea lilies)
			graptolites	
			hemichordates	
			sea squirts and salps	
			lancelets	
			conodonts	
			vertebrates	fish
				amphibians
				tetrapods — reptiles and their descendants — diapsids: anapsids (possibly including turtles)
				lizards
				dinosaurs, pterosaurs, crocodiles
				birds
				synapsids: mammal-like reptiles
				mammals: monotremes
				multituberculates
				marsupials
				placentals

A quick guide to the names used for the major groups of animals (metazoans).

Spineless Critters

Animals can be defined as eukaryotic multicellular organisms (metazoans) that must actively seek their food. Metazoans have been traditionally divided into invertebrates, which lack a backbone, and vertebrates, those that possess such a structure. The simplest invertebrates are those with radial symmetry, including sponges and cnidarians, the latter group including corals and jellyfish. Sponges lack a nervous system or other tissues, and secrete a stiff framework that keeps their bodies from toppling over. This framework can be made from a tough organic material (which some people use in the bath) or from a network of spiky structures called spicules made of silica or calcium carbonate (not so good in the bath). Sponge skeletons made from spicules are commonly preserved as fossils. Sponges of different sorts have at times been important contributors to reefs. For example, a group of calcareous sponges known as archaeocyathans built the earliest Cambrian reefs. Another group of calcareous sponges, the stromatoporoids were also reef builders from the Ordovician until their demise in the Cretaceous.

Cnidarians have bodies that are more organized than those of sponges. They have a simple nervous system and muscle cells but, in contrast to more complex organisms, cnidarians have only two cell layers. Cnidarians can exist as medusas or polyps. A jellyfish is a medusa; its tentacles hang down and enmesh food, and it swims by muscular contractions of its bell-shaped body. Corals and sea anemones are polyps, which anchor to the sea floor with their mouth uppermost; the mouth is again surrounded

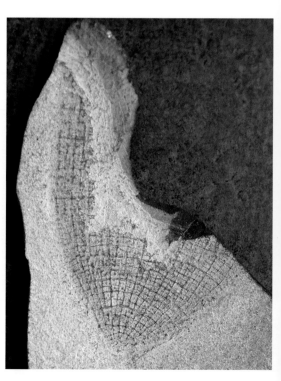

A fossil sponge from Silurian rocks of Cornwallis Island, Nunavut.
MIKE MELCHIN.

Thin section of the solitary horn coral *Streptelasma* from late Ordovician rocks at Fossil Creek, Southampton Island, Nunavut. GODFREY NOWLAN.

A specimen of *Isotelus*, an Ordovician trilobite, from Bowmanville, Ontario. Trilobites were among the first shelled bilaterians. DAVID EBERTH.

Eurypterids were arthropods and top marine predators during parts of the Paleozoic. This specimen of *Eurypterus remipes* came from late Silurian rocks near Ridgemount, Ontario. DAVID RUDKIN.

by tentacles that are used to catch food. Some cnidarians alternate lifestyles between polyp and medusa stages. Since cnidarians are radially symmetrical about a central mouth, they have no head, no left or right sides, and no front or back.

Corals can live as solitary individuals or in colonies. Colonial corals commonly build reefs, the most famous being the Great Barrier Reef. Such reefs are formed from the calcium carbonate skeletons of the corals, most of which are happiest in warmer waters. For many people, the words coral and reef are almost synonymous, but not all corals occur in reefs and, as we have noted, not all reefs in the past were composed of corals.

Animals with identical left and right sides are said to have bilateral symmetry, and so are collectively called bilaterians. We humans are bilaterians and, like other members of the group but unlike sponges and cnidarians, basically have three cell layers. Bilaterians also have a head (or at least a head end), move under their own muscle power, and most have appendages and

The Cambrian sea floor was dominated by invertebrates, as shown in this scene based on fossils found in the Saint John area of New Brunswick. Several species of trilobites can be seen. To the right are white jellyfish-like creatures, whose inclusion in the painting was based on structures now no longer believed to have an organic origin. Several red eocrinoids stand like sentinels to the left, with a group of translucent ostracods (small arthropods) floating by. Behind the eocrinoids are brown sponges, and some small brown brachiopods are about to be steamrollered by the big trilobite. PAINTING BY JUDI PENNANEN, COURTESY OF THE NEW BRUNSWICK MUSEUM.

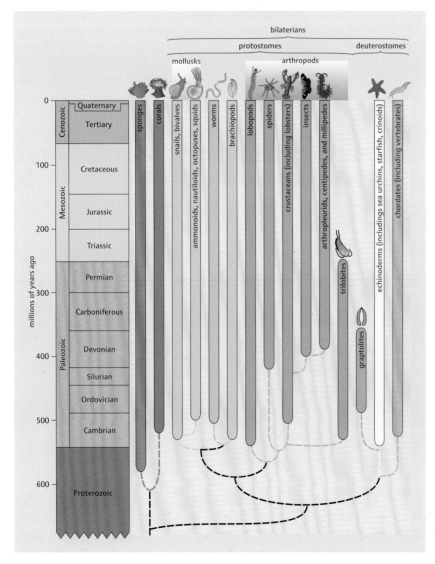

Evolutionary relationships and the timing of first appearances and ranges of the main groups of invertebrates. Probable interrelationships are shown by the dashed lines. ADAPTED FROM ATLANTIC GEOSCIENCE SOCIETY (2001).

at least some complex organs such as a heart, eyes, and brain. The earliest evidence of bilaterians is from trace fossils in the latest Neoproterozoic.

Bilaterians are subdivided into two groups, protostomes and deuterostomes, based on differences in the development of the embryo. Protostomes are all invertebrates and include arthropods, mollusks, and annelids (common worms). Deuterostomes comprise some invertebrates, most notably the echinoderms, and all chordates (including vertebrates).

Arthropods are a major group of protostomes, and encompass all invertebrate animals that have jointed limbs, including insects, spiders, and lobsters. The group also includes trilobites, an icon of the Cambrian Explosion (Chapter 7), which appeared about 520 million years ago and became extinct near the end of the Paleozoic. Because of their calcareous external skeletons, trilobites are commonly found as fossils and exhibit tremendous diversity and rapid evolutionary change throughout much of their range. This makes them useful as time markers. Insects also have a fascinating history, and include some monsters with wings spanning half a metre or more appearing in the Carboniferous, their size perhaps the result of the oxygen-rich atmosphere of the time. A major diversification of insects occurred in tandem with a similar trend for flowers in the past 100 million years or so. Another group of arthropods that developed to huge sizes

The preserved shells of some Cretaceous ammonites from Alberta are used to make so-called ammolite jewelry. ALAN MORGAN.

Asteroids (starfish) are a group of echinoderms that are not common as fossils. This specimen is from Quaternary sediments near Saint John, New Brunswick. HEINZ WIELE, COURTESY OF THE ATLANTIC GEOSCIENCE SOCIETY; SPECIMEN COURTESY OF THE NEW BRUNSWICK MUSEUM.

Specimen of *Mucrospirifer arkonensis*, a Devonian brachiopod from Arkona, Ontario. ALAN MORGAN.

during the Devonian and Carboniferous were the arthropleurids, extinct relatives of centipedes and millipedes.

Mollusks include gastropods (snails and slugs), bivalves (clams, scallops, and mussels), and cephalopods (octopuses, squid, and their kin). Gastropods are found both in the sea and on land. The earliest fossil gastropods are from marine rocks about 530 million years old. Bivalves have two hinged calcareous shells that are generally of the same shape and size. Most bivalves are marine, though a few live in fresh water. The bivalve fossil record extends back about 510 million years. Cephalopods are all marine and most are carnivores. They appeared about 500 million years ago, well after the peak of the Cambrian Explosion. Perhaps the best-known fossil cephalopods are the ammonoids (including the Jurassic-Cretaceous ammonites), some of which had shells up to 2.6 metres in diameter. Although they resemble modern nautiloids, with their chambered and generally coiled shells, ammonoids are only distant cousins of that still-living group and died out with the dinosaurs about 65 million years ago. Nautiloids go back to the Cambrian, and early forms tended to be straight-shelled rather than coiled.

Another group of protostomes, important as fossils but not common today, are the brachiopods (lamp shells), which like bivalves have two hinged shells. This similarity is due to evolutionary convergence, and the two groups are not related. Like trilobites, brachiopods first appeared about 530 million years ago. The group was abundant and diverse in the Paleozoic, but barely survived extinction at the end of the Permian about 250 million years ago. Since then, most of the ecological niches brachiopods once occupied have been taken over by bivalves, showing how mass extinctions can impact the course of evolution. Bryozoa are another group commonly found as fossils. They are typically colonial protostomes whose individuals are tiny, polyp-like filter feeders; their colonies form skeletal structures, sometimes calcareous, that can be found encrusting the external shells or skeletons of other marine organisms.

The most prominent invertebrate deuterostomes are the echinoderms, which include echinoids (sea urchins and sand dollars), asteroids (starfish), and crinoids (sea lilies). All of these animals live in the sea and have a five-fold radial symmetry, developed secondarily from a bilateral form. Most members of this group, such as echinoids and starfish, are mobile, with mouths facing downward. But some, such as crinoids, are anchored to the sea floor by a stalk and have mouths facing upward. The fossil record shows that crinoids appeared about 475 million years ago, in the Ordovician, though similar-looking anchored forms, called eocrinoids, lived in the Cambrian.

Now extinct but distantly related to echinoderms are graptolites, which were prominent planktonic colonial organisms in Ordovician and Silurian seas. Many graptolites look at first glimpse like hieroglyphic writing or scratches in the rock. Indeed, early observers believed them to be inor-

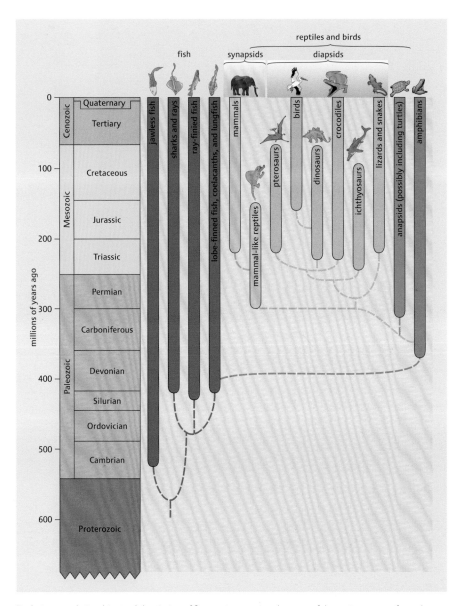

Evolutionary relationships and the timing of first appearances and ranges of the main groups of vertebrates. Probable interrelationships are shown by the dashed lines. ADAPTED FROM ATLANTIC GEOSCIENCE SOCIETY (2001).

ganic; later they were thought to be cnidarians, and are now considered to be deuterostomes, not far removed from our own lineage.

Stiffening Up

During their embryonic development, all vertebrates have a stiffened, rod-like structure called a notochord, which then develops into a true backbone. In a few animals such as lampreys, the notochord never develops into a backbone. Animals with a notochord during some stage of their life cycle, including humans and all other vertebrates, are called chordates. According to recent molecular studies, echinoderms are the closest living invertebrate relatives of the chordates, supporting the similarity in embryonic development of the two groups. The most primitive chordates are the sea squirts and salps, which have a notochord in the larval stage but lose it as adults. Lancelets are a second lineage of chor-

Graptolites from early Silurian rocks at Arisaig, Nova Scotia. HEINZ WIELE, COURTESY OF THE ATLANTIC GEOSCIENCE SOCIETY.

dates; the living *Branchiostoma* has a notochord but no brain, backbone, or limbs. Remains of a similar form, *Pikaia*, have been found in the 505-million-year-old Burgess Shale. The most common fossils of primitive chordates are conodonts. These minute toothlike structures are composed of phosphate and generally between 0.1 and 1 millimetres long. Few fossils of the soft body of the "conodont animal" are known, yet the hard conodonts, probably the rapidly evolving mouth parts, are common in Paleozoic rocks.

Fossils thought by some to be the remains of fish have been found in 525 to 520-million-year-old rocks from China. The earliest fish, like modern lampreys and hagfish, had a skeleton made from cartilage rather than bone and lacked jaws, teeth, and fins. Consequently, their diet and their mobility were limited. During the Ordovician, which lasted from 488 to 444 million years ago, many of these jawless fish evolved armour, especially on their heads, presumably as protection against predators. About 430 million years ago, a group of fish developed jaws, perhaps from what had been the front pair of gill arches in their ancestors. The appearance of fish with jaws—and the teeth that evolved with them—represented a giant leap forward in the evolution of vertebrates. Jaws and teeth made some fish deadly hunters, leading to a revolutionary change in diets with far fewer filter feeders and a new generation of carnivores. An important spinoff was the development of the stomach, a tough, sac-like structure formed from part of the gut.

Another significant evolutionary step in the late Silurian and early Devonian was the appearance of lobe-finned fish, a group that includes modern coelacanths and lungfish. Lobe-finned fish have fin bones arranged in a similar way to the limb bones of tetrapods (animals like us with four limbs), in contrast to the more fan-like fin structure of most fish. Lobe-finned fish also developed functional lungs. *Tiktaalik*, whose fossil remains were recovered from 370-million-year-old rocks on Ellesmere Island, was a lobe-finned fish that had many tetrapod features (Chapter 8). Superb specimens of fossil fish can be seen at many sites in Canada, including the UNESCO World Heritage Site at Miguasha in Quebec.

Lobe-finned fish evolved into amphibians, the earliest tetrapods, which entered the fossil record about 370 million years

In this scene based on fossil finds from late Carboniferous rocks at Joggins, Nova Scotia, an early reptile gazes out of a tree stump as a forest fire approaches. STEPHEN GREB.

ago. Amphibians were the first vertebrates to literally step onto land. Some of the earliest amphibians had limbs that could have been used for manoeuvrability on the sea floor but would not have been strong enough to support the animal's weight on land. The amphibians' lone reign on land was short-lived, for the first reptiles appeared about 313 million years ago. *Hylonomus lyelli*, the earliest-known reptile, was found at Joggins, Nova Scotia. The key innovation of reptiles is the amniotic egg—an egg enclosed in a hard shell, like a kind of liquid-filled "space capsule" that prevents its contents from fatally drying out if laid away from water, and thus freeing the animal to live in dryer habitats. Amniotic eggs in reptiles led eventually not only to birds' eggs, but to the placenta of mammals.

Reptiles and their descendents can be divided into three groups based on the number of extra holes they have on each side of their skulls (besides those necessary for eyes, ears, and other critical functions). The anapsids, possibly represented today by turtles, have no extra holes. The synapsids, including the mammal-like reptiles, have one extra hole. And the diapsids, including crocodiles, lizards, and snakes have two extra holes. Molecular studies have largely confirmed this basic division. Extinct diapsids include: dinosaurs, the largest land animals ever to have lived; those ancient flyers, pterosaurs; and sea dwellers such as ichthyosaurs and mosasaurs. Perhaps the most successful diapsids are the birds, which appeared about 150 million years ago and survived the extinction of their dinosaur kin.

Mammals evolved from the synapsid mammal-like reptiles, animals that had their heyday in the Permian before becoming extinct in the early Mesozoic. The mammal-like reptiles included the sail-backed *Dimetrodon*. As with the reptiles, molecular studies have largely confirmed the long-standing main subdivisions of mammals: monotremes, marsupials, and placentals, all of which are still living, and the extinct multituberculates. Fossil evidence takes the group back to about 220 million years ago, a few million years after the first dinosaurs. We will explore mammal evolution further in Box 8. Plants also have an important fossil record, with several amazing sites in Canada (Box 7).

The most famous impact (pun intended) on the fossil record of dinosaurs is the one that wiped them out at the end of the Cretaceous. Less well known is that the end-Triassic mass-extinction event largely cleared the way for the great Jurassic diversification of dinosaurs, much in the way that their own demise at the end of the Cretaceous led to the Cenozoic dominance of mammals. This scene of a group of prosauropod dinosaurs is based on fossils found in earliest Jurassic sediments near Parrsboro, Nova Scotia (Chapter 9). PAINTING BY JUDI PENNANEN, COURTESY OF THE ATLANTIC GEOSCIENCE SOCIETY.

If the Mesozoic was the age of reptiles, mammals reigned over the vertebrate world of the Cenozoic. This diorama shows a family of mastodons (*Zygolophodon*) in a thicket in the Rockglen area of Saskatchewan, 14 million years ago. IMAGE OF DIORAMA COURTESY OF THE ROYAL SASKATCHEWAN MUSEUM.

The Pattern of Life

Prior to about 2,000 million years ago, life was exclusively microscopic and prokaryotic. Undisputed fossils of eukaryotes appeared in the Mesoproterozoic, perhaps reflecting increased oxygen levels. In the Neoproterozoic, a series of major glacial episodes from about 750 to 580 million years ago was immediately followed by the appearance of the earliest true animals. Within the ensuing 60 million years, culminating in the Cambrian Explosion, the body plans of nearly all living animals appeared.

The interval from the beginning of the Cambrian to the present, the Phanerozoic, is the age of clearly visible life. But even during the Phanerozoic, life has been on a roller-coaster ride, with slow and steady diversification punctuated by abrupt mass-extinction events, of which five are of particular significance. Four of these great dyings occurred at or near the ends of periods: the Ordovician, Permian, Triassic, and Cretaceous. The other took place during the late Devonian. That most of these events were at or near the end of geological periods shows that the founders of the geological time scale recognized dramatic changes in the fossil record, even if they didn't understand their evolutionary implications. The causes of mass-extinction events are still subject to much debate. The end-Cretaceous event, famous for the demise of the dinosaurs and other animals, was the focus of much publicity in the 1980s. Convincing evidence now shows that this event was caused by the impact of an asteroid in the area of the present-day Yucatán Peninsula, as we will see in Chapter 9.

Insights into the cause of the end-Cretaceous event led to increased speculation that perhaps some of the other mass-extinction events could have been caused by extraterrestrial impacts. By extension, did the timing of mass-extinction events correspond to the Earth cyclically passing through an asteroid or comet belt of some sort? In recent years, the general consensus is that each of the five events was the result of unique combinations of circumstances.

Mass-extinction events are not all bad news. True, they toll the end for numerous species, but they also open up new environments and new opportunities for the survivors. Mammals, including ourselves, have risen to success in the Cenozoic, mainly because dinosaurs (except for birds) did not survive the end-Cretaceous extinction. Today, we are in the midst of another mass extinction, this one seemingly driven by humans. Witness the passing of the dodo, the moa, the carrier pigeon, the woolly mammoth, and innumerable other species of whose demise we are ignorant. To quote an old New Zealand folk song: "No moa, no moa/ ... /can't get 'em/ they've et 'em/ they're gone and there ain't no moa." Perhaps the lesson to be learnt from extinctions is that nothing is forever.

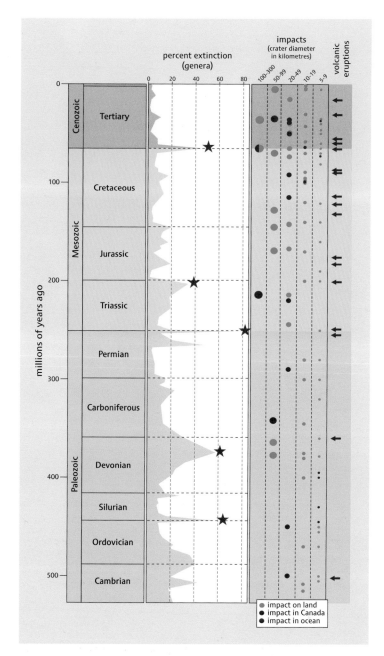

Chart comparing the timing of extinctions with that of extraterrestrial impact structures and major volcanic episodes over the past 542 million years. The column labelled "percent extinction (genera)" also shows (with red stars) the traditionally recognized five major Phanerozoic mass-extinction events. However, as can be seen, other peaks might have comparable claims. The column labelled "impacts" shows the frequency and size of major impact events in Earth history, and whether the bodies plunged into Canada, other countries, or the ocean. As no oceanic crust is older than about 180 million years, all records of impacts in the ocean prior to that time have been lost. Significant episodes of volcanic eruptions are shown in the column to the far right. ADAPTED FROM KELLER (2005), USED WITH PERMISSION OF THE AUTHOR AND THE GEOLOGICAL SOCIETY OF AUSTRALIA.

BOX 4 • SPHERES OF INFLUENCE

The Outer Spheres

The Earth is a dynamically evolving planet in which matter is continuously circulated within and between the terrestrial spheres—a set of spherical layers that enclose each other like a set of Russian matryoshka dolls. In Box 2 we explored the Earth's mostly solid inner spheres—the core, mantle, and crust; collectively these constitute the geosphere. Three other spheres can be identified overlapping the outermost geosphere, but mostly above it. These are the hydrosphere, biosphere, and atmosphere, incorporating the realms of water, life, and air, respectively. Of these, only the atmosphere forms a truly continuous layer. The hydrosphere mainly fills depressions in the Earth's oceanic crust, although it also includes lakes and rivers on continental crust, groundwater within the outer crust, and water vapour in the atmosphere. Earth's biosphere is the realm of the living and occupies the outermost crust, most of the hydrosphere, and the inner part of the atmosphere.

The atmosphere is not as simple as it first appears. It consists of five layers, from the troposphere at the bottom, through the stratosphere, mesosphere, and thermosphere, to the exosphere at the top. We tend to think that life is well protected by the atmosphere, but the troposphere, the only layer of the atmosphere that is oxygenated and warm and in which we and other animals can survive, is a mere 16 kilometres thick at the Equator and considerably less at higher latitudes. Above the troposphere is the stratosphere, home to the ozone layer, which protects us from dangerous ultraviolet rays. Life in the stratosphere would be impossible for humans because it is frigid and has very little oxygen. Temperatures are also numbingly chilling in the mesosphere. Higher still, the thermosphere extends from about 80 kilometres to between 500 and 1,000 kilometres above the Earth's surface, where it merges with the exosphere, the last layer before outer space. The thermosphere is so named because it can heat up to 1,500°C during daylight hours. But you would not feel warm because the atoms that make it up are so dispersed. The exosphere contains concentrations of the Earth's lightest gases, mainly hydrogen, with some helium, carbon dioxide, and oxygen.

Circulation within the atmosphere is driven by the difference in solar heating at equatorial and polar latitudes. At the Equator hot air rises, creating a zone of low pressure. As the air rises it cools, and its contained water vapour condenses and returns to the surface of the Earth as rain or snow—a good example of how the hydrosphere and atmosphere interact. Eventually, at higher latitudes, the air that rose at the Equator returns to the surface, closing a loop in circulation. At the poles, cold surface air flows toward the Equator, replaced from above by warmer air from lower latitudes. This natural circulation of

The atmosphere is our planet's outermost sphere. ROB FENSOME.

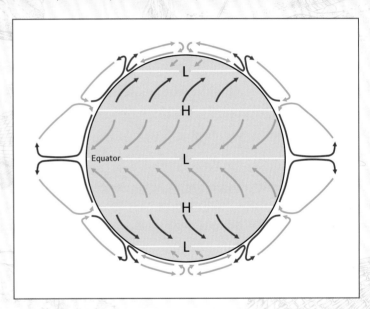

Earth's wind patterns showing the prevalence of easterlies (blue arrows denoting cooler air) in the tropics and at the poles, and westerlies (red arrows denoting warmer air) in temperate latitudes. The loops depict air circulation in the troposphere, with hot air rising in the lower and middle latitudes. L and H refer to latitudes of generally low and high pressure, respectively. ADAPTED FROM VARIOUS SOURCES.

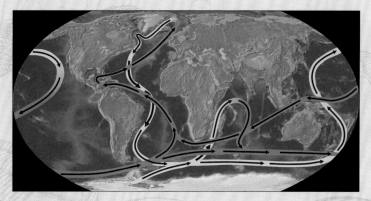

Ocean circulation: warm currents are in red, cold currents are in blue. ADAPTED FROM VARIOUS SOURCES.

the atmosphere, with symmetrical patterns of thermal convection cells, is modified by the distribution of continents and of the oceans and their currents. Circulation and mixing of the atmosphere are relatively rapid, as the dispersal rates of radioactive contaminants and volcanic emissions clearly show.

We take water for granted, but it is a rare commodity in the Universe. For water to exist in a liquid state, the temperature must not vary for long periods outside the range of 0 to 100°C, which remarkably it has managed to do on Earth for over 4,000 million years. That our planet has so much water reflects not only its fortuitous location within the Solar System, but also the complex natural machinery and feedbacks that make this possible. Some water is carried into the mantle at subduction zones and vented back to the surface by volcanoes, and some extraterrestrial water comes from comets. However, most of it has been part of the hydrological cycle, a huge recycling system, since the Earth was young. Water is constantly on the move, circulating between hydrosphere, atmosphere, and lithosphere by processes mostly familiar to us. In the rock cycle, water plays a critical role in processes such as weathering, erosion, transportation, and deposition. And water is vitally important in tectonic processes.

Ocean basins, major reservoirs of the hydrosphere, are typically about 4 kilometres deep and occupy just over 60 percent of the Earth's surface. Water circulates within these gigantic mixing bowls in response to prevailing winds, to variations in density caused by differences in the solar heating at different latitudes, and to differences in salinity. Warm, buoyant, salty surface water flows from lower to higher latitudes, transferring heat, then sinking and eventually migrating at depth back toward the Equator. The effect of such mixing is well illustrated by the North Atlantic's Gulf Stream, which warms western Europe but not Atlantic Canada. Bordeaux in France is much warmer on average than Saint John in New Brunswick, even though the two are almost at the same latitude. Meanwhile water that is cold and less dense because of the influx of polar ice and meltwater flows south near the surface. An example is the Labrador Current, which keeps Atlantic Canada relatively cool in both summer and winter. Ocean currents are strongly influenced by the Earth's rotation and, in the long term, by plate tectonics and the distribution of continents, which control the shapes and locations of ocean basins. But, most importantly to us, liquid water is vital to life on Earth.

Eduard Suess, the great alpine geologist, coined the term biosphere in 1875 and defined it as "the place on Earth's surface where life dwells". Thus, the biosphere extends to the depths of the oceans, to the upper reaches of the troposphere, and down hundreds of metres into the crust, the last being the domain of countless microbes.

Oxygen Revolutions

The composition of Earth's atmosphere has changed over time. In the earliest days, the atmosphere consisted primarily of carbon dioxide, nitrogen, water vapour, methane, and sulphur, a lethal brew for oxygen-consuming organisms. Over time, the mix changed radically: the modern atmosphere consists of about 78 percent nitrogen, 21 percent oxygen, and 1 percent water vapour, with trace amounts of argon, carbon dioxide, and methane.

Variations in oxygen and carbon-dioxide levels in the atmosphere during the Phanerozoic. Both carbon-dioxide and oxygen levels have been much higher in the past—and oxygen sometimes also much lower. Note the strong inverse relationship between the two curves. ADAPTED FROM BERNER (2004).

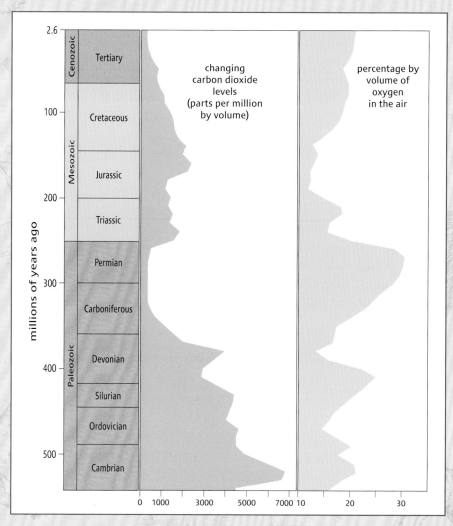

Animal life would not be possible without oxygen in the atmosphere, but the buildup to today's levels was initially slow. How oxygen levels increased and what triggered them is still uncertain. Our planet's earliest atmosphere had little or no oxygen as witnessed by minerals and rocks that could only have formed under anoxic conditions. A big change seems to have occurred 2,500 to 2,400 million years ago, with the appearance of the first redbeds—sediments in which grains have a coating of the red iron oxide mineral hematite, which can only form in the presence of oxygen. Although called the Great Oxidation Event, it involved a rise in oxygen levels from trace amounts to only a few percent, still much less that the modern 21 percent. The rise was probably due to an increase in abundance of the prokaryotes known as cyanobacteria (Chapter 6). From the Great Oxidation Event to about 1,000 million years ago, oxygen levels remained low, but from then to about 540 million years ago, they rose to reach amounts close to present levels. This major change in the composition of the atmosphere may have been a key factor in the Cambrian Explosion.

Throughout the Phanerozoic, atmospheric oxygen levels have varied from as low as about 12 percent to as high as about 33 percent. Evidence shows that oxygen levels reached a peak in the Permian, about 280 million years ago. Increasing oxygen levels in the late Carboniferous, 315 to 300 million years ago, may explain the appearance of enormous dragonfly-like insects and 2-metre-long millipede-like arthropleurids. Plunging oxygen levels during the Permian and Triassic possibly helped shape the physiology of the dinosaurs, whose descendants, the birds, use oxygen very efficiently. The rise of dinosaurs seems to have delayed the rise of the mammals, which suggests that the impact of oxygen levels on evolution has been enormous.

Banded iron formations of Archean and Paleoproterozoic age are evidence that Earth's atmosphere at one time had little free oxygen (Chapter 5). This example, from near Temagami, Ontario, consists of layers of jasper and magnetite. ALAN MORGAN.

Redbeds, such as the Permian strata seen here at Elephant Rock, near Tignish, Prince Edward Island, form only when there is free oxygen in the atmosphere. Since this photo was taken, the "elephant" has lost its trunk. KEITH VAUGHAN.

Never-ending Story

The movement of organic and inorganic carbon between the atmosphere, hydrosphere, biosphere, and geosphere constitutes the carbon cycle. Carbon occurs in the atmosphere as carbon dioxide, and in the biosphere in all plants and animals, including shells and skeletons.

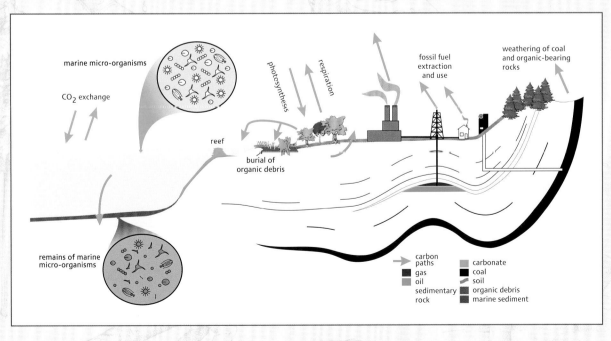

The carbon cycle. In addition to the factors shown in this figure, plate tectonics has a major role: carbon (bound in minerals) is carried by subduction down to the mantle and is returned, for example, via volcanoes.

Coal and limestone store large amounts of carbon. The Elkview Mine, near Sparwood, British Columbia, is a large open-pit coal mine that has produced about 5.5 million tonnes of coal of latest Jurassic and earliest Cretaceous age. ALAN MORGAN

In the geosphere, carbon is a significant component of many rocks, especially carbonates, as well as of soils and fossil fuels. Carbon also occurs in the hydrosphere.

How does the carbon cycle work? And, as the amount of carbon on the planet is remarkably stable, how is equilibrium maintained? All carbon reservoirs are connected by pathways involving chemical, physical, or biological processes. Gains and losses in the different reservoirs can be expressed in terms of a budget. When carbon is gained, the reservoir involved is serving as a sink; when carbon is lost, the reservoir is serving as a source. The atmosphere is a sink when it receives carbon dioxide from the burning of vegetation and fossil fuels. But it is a source when plants and animals grow and extract carbon from the atmosphere.

Over the past 250 years, the amount of carbon dioxide in the atmosphere has increased, reflecting the dramatic rise in the consumption of fossil fuels (Box 9). Although this increase may be responsible for recent global warming, atmospheric carbon dioxide levels have fluctuated widely through time. Levels were highest during the Earth's early history, but also rose markedly during episodes of massive volcanic eruptions.

Much of the carbon dioxide in the atmosphere is a product of respiration in organisms, including animals and plants. In respiration, sugars and oxygen interact to generate energy plus the waste products carbon dioxide and water. Fungi and some bacteria break down organic material of dead plants and animals—they produce carbon dioxide in the presence of oxygen, and methane in its absence. Other sources of atmospheric carbon dioxide, and hence carbon, are upwelling ocean waters, which release carbon dioxide as they warm.

Carbon dioxide is toxic at high concentrations, so life would not survive if the gas continued to build up in the atmosphere. Fortunately, the carbon cycle maintains a balance by removing the gas through organic and inorganic processes. In the biosphere, plants and many bacteria consume carbon dioxide during photosynthesis to produce primarily sugars and oxygen. Some organisms, especially certain archaea, obtain their energy from chemosynthesis—the conversion of carbon dioxide or methane and other nutrients into organic matter, using the oxidation of hydrogen, hydrogen sulphide, or methane as a source of energy. In photosynthesis, the source of energy is the Sun. All organisms use carbon for constructing cells (which is in part why fossil-fuel deposits—peat, coal, oil, and gas—contain vast quantities of carbon). In such cases, the biosphere is serving as a sink.

The hydrosphere contains huge volumes of carbon. Marine organisms such as corals, clams, and coccolithophores use this to produce calcite skeletons and shells, which are deposited on the sea floor when the organisms die. Thus, large quantities of carbon are trapped in the geosphere when carbonate rocks form. Chemical and biochemical weathering of silicate minerals that form much of the Earth's crust also consumes carbon dioxide. Silicate minerals weather to clays, which contain more carbon than the original mineral. Mountain building promotes this process by presenting more rocks to be weathered and eroded, thus producing more clay minerals. The geosphere returns carbon to the atmosphere through the weathering and erosion of carbonate rocks or through the release of large amounts of carbon dioxide during volcanic eruptions.

Exchanges of carbon between sources and sinks are not always a sedate and gradual process. For example methane, a simple hydrocarbon compound (consisting of hydrogen and carbon), may have been periodically released into the atmosphere in huge belches emanating from the oceans. The same gas is also produced in large amounts by flatulent ruminants (including cows), termites, and other organisms.

By far the largest amount of the Earth's carbon—over 99.9 percent—is stored in carbonate rocks and fossil fuels. Of the remainder, over 92 percent resides in ocean waters, about 3.8 percent is in soils, about 1.9 percent is in the atmosphere, and the rest is split among smaller reservoirs such as living organisms and non-carbonate sediments. The relatively small amount of carbon in the atmospheric total may give us a false sense of security about climate warming. But we need to be careful, because there are so many variables and so many unknown factors.

BOX 5 • THE LAY OF THE LAND

Perhaps the first clues that lead toward an understanding of Canada's long geological history come from the shape of the land. The map below shows the topography of Canada and adjacent lands as well as the bathymetry of the neighbouring oceans. The map clearly shows that Canada's lowlands lie around Hudson Bay, as well as in the southern Arctic Islands, the St. Lawrence Lowlands, and the Maritimes. The highest region is the western mountains, or Cordillera, of British Columbia, Yukon, and parts of Alberta and the Northwest Territories.

Relief—the ups and downs of the land surface—is also important. On the map, relief is shown by the degree of uniformity of the colour pattern. The mountains of the Cordillera show a very uneven colouring because of the density of peaks and valleys that made the mountains so formidable to cross for the early explorers. In contrast the area around Hudson Bay shows a uniform shading, testifying to a relatively flat landscape. But higher land does not necessarily

The rugged St. Elias Mountains in southwestern Yukon, northern British Columbia, and southern Alaska contain North America's highest peaks as well as spectacular icefields and valley glaciers. Here, Kaskawulsh Glacier scours the Paleozoic rocks of the valley walls. In the distance are mounts Logan and St. Elias. JOHN CLAGUE.

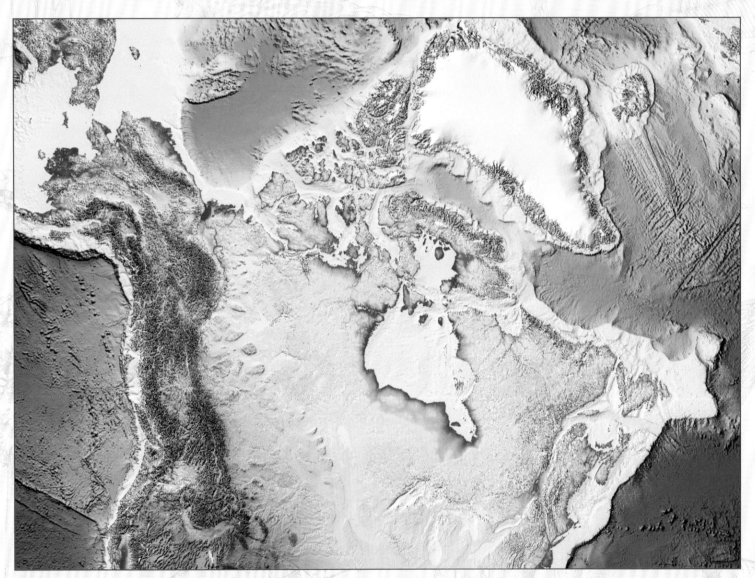

Physiographic (topographic and bathymetric) map of Canada and surrounding lands and seas. To indicate increasing height, land is shaded from dark green through light green and light brown to dark brown; white represents ice. Note the distribution of highlands and lowlands. Shades of blue indicate water bodies, from light blue for shallow water to dark blue for deeper water. Note the generally sharp transition from continental shelves to ocean floor. FROM AMANTE AND EAKINS (2009).

Lake Louise, Banff National Park of Canada, Alberta, is a popular tourist destination in the southern Rocky Mountains. KEITH VAUGHAN.

A view of a flat Prairie landscape near Struan, Saskatchewan. ROB FENSOME.

The Appalachian Mountains reach the sea on the Gaspé Peninsula of Quebec, as seen here at Cap Gaspé, Forillon National Park of Canada. Lower than the Cordilleran mountains, the Appalachians are an older mountain system that has been worn down by weathering and erosion over a much longer time. E. LE BEL, COPYRIGHT PARKS CANADA.

Parts of the eastern Arctic are mountainous and ice covered. Visible in this aerial view are several glaciers descending from a small ice cap on Axel Heiberg Island, Nunavut. ALAN MORGAN.

have high relief, and lower land need not be flat. For example, the region to the east of the Cordillera, known as the Western Interior Plains, is as high as or even higher than regions along the eastern edge of the continent, but the former are relatively flat whereas the latter are hilly. The hills of southeastern Atlantic Canada are part of the Appalachians, those in northeastern Labrador are the Torngat Mountains, and those on the eastern edge of the Arctic Islands are known as the Innuitian Mountains.

Offshore, the platform created by the continental shelves is in striking contrast to the deeper waters of the ocean floors beyond the shelves. These ocean floors generally seem smoother than the land, but are certainly not featureless—note for example the Mid-Atlantic Ridge extending south of Iceland.

One critical feature of Canada's geography that is not readily seen on topographic maps is the Canadian Shield, a distinctive region underlain by some of the oldest rocks in the world. The Shield extends across the northern heart of the continental mainland from the Northwest Territories to Labrador and south into the eastern Prairie Provinces and underlies much of the western Great Lakes, northern Ontario, and northern Quebec. The Shield shows more clearly on a map of agricultural uses or mining activity as its soils are generally poor for cultivation yet in places it abounds with mineral resources.

Another distinctive aspect of Canada is the coastline, which defines the shape of the country and many of its parts. Among present-day features that make an impression (or an indentation) are the Gulf of St. Lawrence and Hudson Bay. Contributing to the length of Canada's record-breaking coastline are its offshore islands, including Newfoundland in the east and Vancouver Island and Haida Gwaii in the west. Readers may be less familiar with the numerous Arctic Islands and their names, but they play such a significant role in our story that we are including a labelled map of them here.

We also need to look more closely at Canada's western mountains. Many people think of the Cordillera as the Rockies, but the Rocky Mountains are only the easternmost ranges. Our map (next page) names other Cordilleran ranges and topographic features, and their distribution is shown clearly in the satellite image to the left of the map.

Why are the western mountains so much higher than the eastern hills? Why is the interior relatively flat? Why is much of the Arctic chopped up into islands? These are some of the questions we seek to answer in the next part of the book.

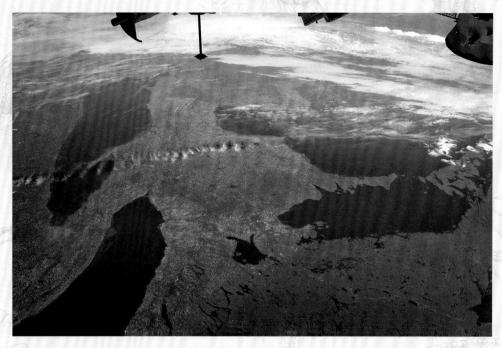

Satellite view with north to the right, Lake Ontario at bottom left (bordered by Toronto to its top right), Lake Erie at top left, and Lake Huron to the right. Canada is to the right of lakes Ontario and Erie, and the United States is most of the area to the left of these lakes. Note the sediment plume entering Lake Ontario from the Niagara River (linking the eastern end of Lake Erie with western Lake Ontario), and the Bruce Peninsula, a peninsula in Lake Huron that owes its existence to the Niagara Escarpment and its Silurian cap rock (Chapter 1). Note also the distinctly different quality of the landscape and tone of green to the bottom right; this area is the infertile and sparsely populated Canadian Shield. COURTESY OF NASA.

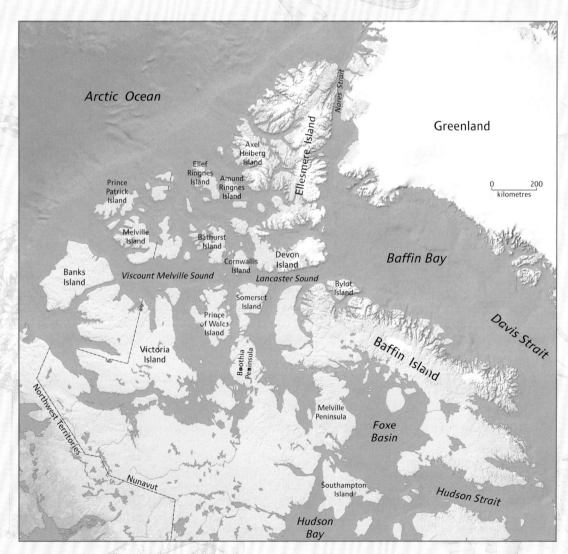

A map of the Arctic Islands and surrounding lands and seas. Together, Viscount Melville Sound and Lancaster Sound are part of the famed Northwest Passage, and the islands to the north of these two seaways are collectively referred to as the Queen Elizabeth Islands. The lightest-coloured areas are ice caps. ADAPTED FROM NATURAL EARTH (NATURALEARTHDATA.COM).

BOX 5 · THE LAY OF THE LAND

A is a physiographic map of the Canadian Cordillera and adjacent areas, with shading as previously explained. *B* names the main physiographic regions of the Canadian Cordillera. The more westerly mountain ranges (pink), including the Coast Mountains, are separated from the more easterly ranges (blue), including the Rocky Mountains, by an interior belt (green) dominated by plateaus.
A FROM AMANTE AND EAKINS (2009); *B* ADAPTED FROM MATHEWES (1986).

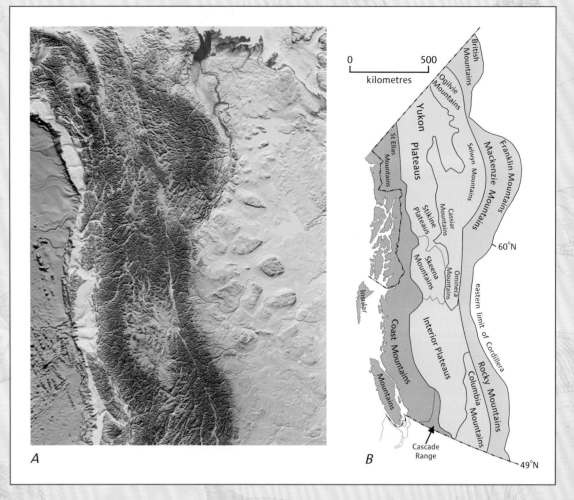

The mountains of the Cordillera reach far inland, especially in the north, as here in the Mackenzie Mountains, Northwest Territories. In the right foreground is Mount Close. LARRY LANE.

PART 2 · THE EVOLUTION OF CANADA

The steep-sided isolated peaks in this scene, at the head of Maitland Creek, Skeena Mountains, British Columbia, are Pliocene volcanic necks of the Stikine Volcanic Belt (Chapter 10). They were intruded into Jurassic sedimentary rocks. CAROL EVENCHICK

5 • FROM STARDUST TO CONTINENTS
before 2,500 million years ago

CHAPTER SUPPORTED BY A DONATION FROM KEN GILLIS, HUDBAY MINERALS INC.

Convoluted Archean gneiss at Ford's Harbour, near Nain, Labrador. This rock, part of what geologists call the North Atlantic Craton, formed under very high pressures and temperatures in the lower continental crust. BRUCE RYAN.

In the Beginning

The prehistory of Canada, and the rest of the Universe, began with the Big Bang. This seminal event is estimated to have taken place 14,000 to 13,500 million years ago. The early days of the Universe saw the formation of lighter chemical elements (those with low atomic numbers such as hydrogen, helium, and lithium). Then, nuclear fusion reactions within stars created heavier elements up to iron, including all those that, in addition to hydrogen, are essential to life. Finally, elements heavier than iron were produced in supernova events, gigantic stellar explosions that were caused by the burning out of the first stars. In the process, the various chemical elements were scattered far and wide into the surrounding Universe as cosmic dust.

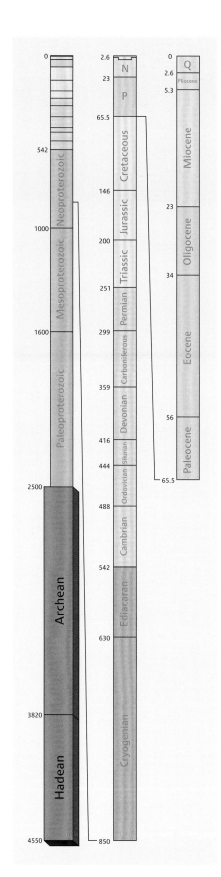

Geologic time scale, showing the interval covered in this chapter. Numbers indicate millions of years ago. P = Paleogene (Paleocene to Oligocene), N = Neogene (Miocene and Pliocene), and Q = Quaternary.

A group of five galaxies known as Stephan's Quintet, viewed through the Hubble Space Telescope. Individual members of the Quintet are prominent, but many of the other "stars" in the background are in fact more distant galaxies. Galaxies are a result of the Big Bang, which occurred 14,000 to 13,500 million years ago. FROM THE HUBBLE-SITE, COURTESY OF NASA AND STSCI.

Our Milky Way Galaxy likely attained its present spiral form within a few billion years of the Big Bang. The Solar System was born some 4,567 million years ago when gas and clouds of cosmic dust became locally dense enough to be pulled together by gravity. Shock waves from nearby supernovae may have provided an impetus, while at the same time seeding new material into the elemental soup that would become the Sun and its satellites. Gravitational attraction rapidly increased, leading to a more condensed mass and segregation of materials. The result was a flat, spinning, and increasingly dense disk of matter. At the centre, in the area of highest temperature and pressure, was the Sun. Dust grains and heat-resistant elements concentrated close to the new star, where they collided and coalesced to produce progressively larger bodies and, eventually, small planets called planetesimals. Present-day asteroids and meteors are leftovers from those early times. Radiometric dating of meteorites, the remains of meteors that land on the Earth, mostly gives ages around 4,550 million years, indicating that the early rocky objects of the Solar System, including the Earth, coalesced quickly and simultaneously at that time.

Some 25 million years after its formation, Earth's gravity was still attracting planetesimals and other space debris, and the collisions generated heat at the planet's

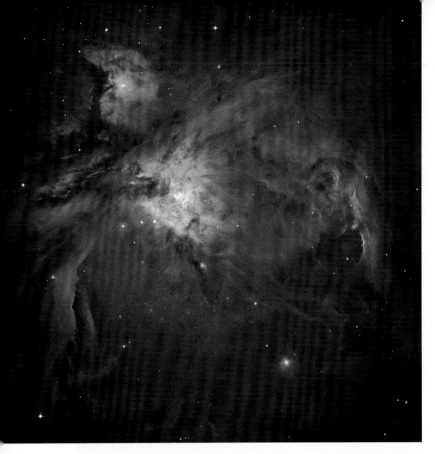

The Orion Nebula, a nebula within our Milky Way Galaxy, as seen through the Hubble Space Telescope. Nebulae are accumulations of gas and dust that glow due to radiation from nearby stars, and are the nurseries for new solar systems. FROM THE HUBBLESITE, COURTESY OF NASA AND STSCI.

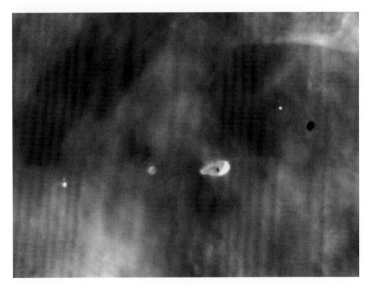

The bright and dark spots respectively shining out from or silhouetted against the Orion Nebula are planetary discs. Such discs are nascent solar systems, clouds of gas and dust surrounding a youthful star. FROM THE HUBBLESITE, COURTESY OF NASA AND STSCI.

The largest intact meteorite ever found in Canada fell near Madoc, Ontario, perhaps in the early nineteenth century, and was discovered by William Logan in 1854. It is a single mass of iron weighing over 150 kilograms, twice the weight of an average adult person. ROB FENSOME, SPECIMEN COURTESY OF NATURAL RESOURCES CANADA.

surface. Around that time, radioactive decay of unstable, short-lived elements generated so much internal heat that our planet became a fiery molten ball, its surface a magma ocean. This episode is known as the Iron Catastrophe. Heavier elements such as iron and nickel sank inward, and lighter elements such as silicon were displaced toward the surface. The Iron Catastrophe left the Earth with a molten iron-nickel core about twice the size of the Moon. Today only the outer part of the core is molten, but this serves as a dynamo that generates our planet's powerful magnetic field—a critical factor for us because it helps deflect harmful extraterrestrial radiation back into space. The Iron Catastrophe lasted for about 50 million years. But for hundreds of millions of years afterwards, even though its surface began to cool and solidify, the Earth would still have been an inhospitable place for us, with continuing planetesimal bombardment and volcanoes spewing noxious gases.

Hell on Earth

The Earth-Moon system, consisting of a small planet with a relatively large satellite, is unique in the Solar System. How did this planetary duet arise? Moon rocks collected during the Apollo missions showed that, although ancient, lunar rocks are not as old as most meteorites. The general composition of lunar rocks is similar to that of the Earth's mantle, suggesting a close relationship between Earth and Moon. This and other evidence led to the idea that the Earth and a Mars-sized planetary body (about 10 percent of the mass of the Earth) collided and partly merged about 4,520 million years ago. The impact sent enough vapour and debris into space to form a ring around the Earth. Because of the huge amount of energy delivered in the collision, the Earth acquired the tilt to its axis that gives us our seasons.

During the Hadean Eon, the ring around our planet re-aggregated to form the Moon. The Moon was at first only about 84,000 kilometres from the Earth, startlingly close compared to today's distance of some 384,000 kilometres, and would have seemed enormous. Each Earth day was then only about 6 hours long. Tidal pull would have been huge, causing our planet's surface to rise and fall by possibly more than 50 metres during each tidal cycle, and

During their early years, the Moon and Earth were bombarded by planetesimals, which left impact craters that can still be seen on the Moon. On Earth, most such features have disappeared due to the movement and consumption of tectonic plates. JOYCE S. K. CHEW.

A thin-section of basalt from the Moon, shown in cross-polarized light. The black mineral is ilmenite, the multi-coloured mineral is pyroxene, and the grey-green striped mineral is plagioclase feldspar. The field of view is 4 millimetres across. BARRIE CLARKE.

slowing the Earth's rotation. Today, the Moon continues to recede from the Earth at a rate of 3 to 4 centimetres per year.

In contrast to the Earth's active surface, the Moon's outer skin has been largely frozen since its early days. This skin is pockmarked by impact structures (craters), with younger ones superimposed on older ones in places. Some of the Moon's more ancient impact structures are over 2,000 kilometres across. The youngest of these huge impacts date from about 3,820 million years ago. Such a phase of giant impacts seems to have represented a final weeding out of the menagerie of orbiting bod-

Specimen of Acasta Gneiss from the Northwest Territories, the world's oldest-known rock at 4,030 million years. The Hadean date is of the original felsic igneous rock, which was metamorphosed to gneiss during the Archean. GODFREY NOWLAN.

General extent of Hadean and Archean rocks at the surface (after glacial deposits are removed) of onshore and offshore Canada. The lighter shaded areas denote either uncertainty or areas where rocks of the particular age have been confirmed but are intimately associated with rocks of other ages and the scale of the map doesn't let us show them separately. ADAPTED FROM WHEELER ET AL. (1996), COURTESY PHIL O'REGAN AND THE GEOLOGICAL SURVEY OF CANADA.

ies, after which just a few stable, regularly spaced planets and their satellites remained to make up the Solar System familiar to us today. The Earth is about 100 times more massive than the Moon—a bigger target with a more powerful gravitational attraction. Therefore, it must have experienced a hellish pummelling prior to 3,820 million years ago, though the direct evidence is now lost. These violent times explain why American geologist Preston Cloud coined the term Hadean to refer to the Earth's earliest days to about 3,820 million years ago. Although not many Hadean rocks remain today at the Earth's surface, we can surmise that the eon was dominated not only by frequent and massive impacts, but also by voluminous, mostly basaltic volcanism as the Earth's magma ocean solidified.

Part of the planet's sparsely preserved Hadean rock record is found in Canada, all of it from the Canadian Shield. Strong contenders for the oldest-known terrestrial rocks are Canadian. The Acasta Gneiss, found northwest of Yellowknife in the Northwest Territories, was originally a felsic igneous rock that radiometric dating tells us formed about 4,030 million years ago. In recent years, a date of almost unbelievable antiquity—4,400 to 4,300 million years—was obtained from a suite of rocks called the Nuvvuagittuq Greenstone exposed on the shore of Hudson Bay in northern Quebec. This date is controversial, however, with the most recent evidence suggesting that the Greenstone may "only" be 3,780 million years old. So the Acasta Gneiss remains the oldest-known, definitively dated terrestrial rock. Evidence of startling antiquity does come, though, from detrital zircons from Australia. The zircon crystals, which occur in a 1,600-million-year-old sandstone from Australia and were derived from the weathering and erosion of now-lost older rocks, have been radiometrically dated to about 4,400 million years. Research on these zircon crystals has yielded the remarkable discovery that Earth had abundant water 4,400 million years ago.

The end of the interval of large impacts 3,820 million years ago heralded the end of the Hadean and the beginning of the Archean Eon.

4,300 to 2,500 million years old. Since 2,500 million years ago, the cratons have remained relatively stable parts of Earth's surface. In Canada, the best-preserved cratons are the Slave, Superior, and North Atlantic (part of which is in eastern Labrador, with other parts now located in Greenland and Scotland). Other cratons, such as the Rae and Hearne, were far more affected by later deformation, metamorphism, and magmatism associated with the welding process that brought the cratons together—best seen in the Paleoproterozoic orogenic belts between cratons, which we will explore in the next chapter.

Hadean and Archean rocks are predominantly metamorphic and igneous. Unmetamorphosed sedimentary strata, which dominate many younger successions, are relatively rare. The rocks of the cratons tend to be well stirred and mixed, making them a fascinating prospect for "hard-rock" geologists, but perhaps daunting for non-specialists to understand. Igneous rocks of the cratons include many types also common in younger times, such as granite and basalt. But they also include types largely restricted to the Archean, such as the ultramafic rock known as komatiite, which is low in silica, potassium, and aluminum content but high in magnesium. The conditions neces-

Archean Cratons

As well as being home to the world's oldest-known rocks, the Canadian Shield is deeply entrenched in Canada's history. In the days of the voyageurs, the myriad lakes, rivers, and streams provided an ideal habitat for the diverse flora and fauna that supported the fur trade, the New World's principal economy from the earliest European colonization to the early 1800s. The Shield's stark beauty has inspired writers and poets, as well as iconic Canadian painters such as Tom Thomson, Frederick Varley, and Lawren Harris. Since the decline of the fur trade, the Shield has been the source of much of Canada's mineral wealth, yielding gold, silver, nickel, and many other metals. Mining on the Shield still thrives, and some of the major ore bodies reside in Archean rocks.

Much of the Canadian Shield and similar features on other continents are formed from so-called Archean cratons (or simply cratons), fragments of Hadean and Archean lithospheric rocks that are

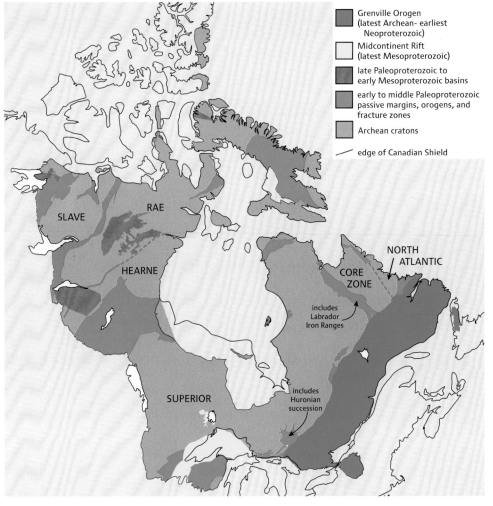

The geology of the Canadian Shield.

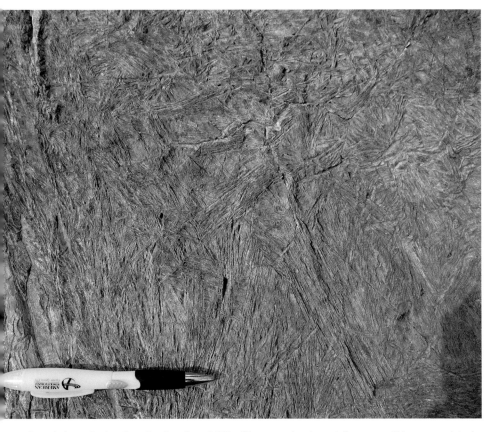

Part of a komatiite lava flow dated to about 2,700 million years, showing spinifex texture. This exposure is in the Abitibi Greenstone Belt of the Superior Craton, near Rouyn-Noranda, Quebec. DAVID CORRIGAN.

Komatiite, dated to around 2,700 million years old, showing columnar jointing. These rocks are from the Rae Craton on northern Baffin Island, Nunavut. MIKE YOUNG.

Archean conglomerate from the James Bay region of northern Quebec. The conglomerate was originally a fluvial deposit that formed in a greenstone belt within the Superior Craton. It is about 2,700 million years old. JEAN GOUTIER.

Folded Archean felsic gneisses (3,300 to 3,100 million years old) and mafic dykes (metamorphosed to amphibolite) of the North Atlantic Craton in the Hopedale area of Labrador. Mafic dykes, originally intruded typically as tabular sheets, bend and as here may break into smaller segments if the host rock is later deformed under high temperature and pressure. BRUCE RYAN.

sary for the formation of komatiite magmas were present only at the very high temperatures that occurred in the mantle before about 2,600 million years ago. Komatiites formed as volcanic flows, dykes, and sills and are mostly of Archean age. They commonly have distinctive large radiating crystals of olivine or pyroxene, which form the so-called spinifex texture because the crystals resemble *Spinifex*, a type of Australian grass with blade-like leaves. The blade-like crystals formed when the superheated magma rapidly cooled as it approached the surface.

Archean sedimentary rocks include the usual range of clastics, but limestones are rare and evaporites and redbeds are absent. Very characteristic of this part of our planet's history are banded iron formations. These distinctive striped sedimentary rocks first appeared around the Hadean-Archean boundary, reached a peak in the late Archean, and mostly faded out by the late Paleoproterozoic. We will delve more deeply into the origin of banded iron formations later in this chapter.

Among the diversity of Archean metamorphic rocks, gneiss and greenstones are the most prominent. Gneiss can be derived from both igneous and sedimentary rocks, and analysis of the rock's chemistry and mineral content will generally reveal a gneiss's derivation. When metamorphic rocks begin to partially melt, a migmatite results. Greenstones are metamorphosed mafic igneous rocks, especially basalts, and are green primarily because of the minerals chlorite and actinolite; further heating and metamorphism will convert these minerals into the amphibole mineral hornblende, producing a black rock called amphibolite. Armed with a general understanding of Archean rocks, let's focus on two cratons, the Slave and Superior, and then discuss what their structure may tell us about evolving tectonic processes.

This view of the North Atlantic Craton on Shuldham Island in Saglek Fiord captures the major elements of the geology of the Nain area of Labrador. All rocks are gneisses; the white-weathering exposures to the right and at centre are derived from granitic rocks, the brown-weathering exposures originated from sedimentary rocks, and the black-weathering exposures to the left are metamorphosed mafic volcanic rocks. BRUCE RYAN.

that came into position after 2,000 million years ago.

The Superior Craton is the largest continuous piece of Archean crust exposed today at our planet's surface, roughly the size of India. It underlies huge areas of Ontario and Quebec, as well as parts of Manitoba, Minnesota, and a small part of western Labrador. Most rocks within the Superior Craton are 2,800 to 2,600 million years old. Because of its mineral wealth and relatively easy access, the geology of the Superior Craton has been well studied. Its rocks fall into three associations: granitic rocks, including felsic gneisses; greenstones; and metasedimentary rocks.

The greenstone belts are varied in composition, although their oldest rocks, metamorphosed basalts (greenstones), are prominent.

Slave and Superior

The Slave Craton underlies the northwestern part of the Canadian Shield, with Yellowknife on its southern margin. The oldest rocks on the Slave Craton are mostly gneisses, including the Acasta Gneiss, which range from 4,030 to 2,850 million years old. Although the gneisses (originally intrusive igneous rocks) formed deep in the crust, by about 2,800 million years ago they were exposed at the surface. We know this because they are directly overlain by volcanic rocks resting on an erosion surface. These now-metamorphosed volcanic rocks are up to 6 kilometres thick and date from 2,730 to 2,700 million years ago. As the volcanic episode came to a close, melting deep in the crust produced a new series of granitic plutons 2,700 to 2,600 million years old. The youngest surface rocks of the Slave Craton are metasedimentary, originally clastic sediments shed from nearby land and deposited in a long-lost sea between about 2,687 and 2,660 million years ago. Thus, four main types of rock make up the Slave Craton: gneissic, metavolcanic, granitic, and metasedimentary. Today, the Slave Craton is bounded by younger orogenic belts and faults, beyond which are other cratons

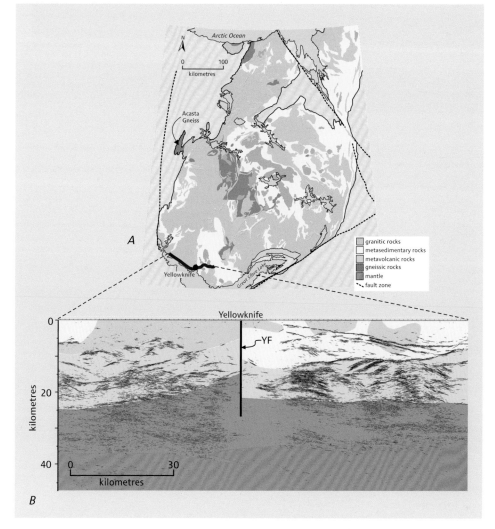

A is a simplified geological map of the Slave Craton (coloured areas). B is a section across part of the southernmost Slave Craton showing basic geology interpreted from a LITHOPROBE seismic section (blue line in A). YF indicates the Yellowknife Fault, a strike-slip fault in which the block to the right has moved toward the reader and the block to the left has moved away. ADAPTED FROM WILSON AND CLOWES (2009), COURTESY OF LITHOPROBE.

Archean pillow lava (top) deposited on cross-bedded sandstone (below), Slave Craton, Yellowknife, Northwest Territories. RAY PRICE.

The youngest surface rocks of the Slave Craton are marine clastic metasedimentary rocks about 2,700 million years old. These rocks bear the characteristic cyclic bedding of turbidites, and are among the oldest-known examples of such deposits. LUKE OOTES.

These metamorphosed basalts are overlain by thick sequences of interlayered pillow basalts and felsic volcanic rocks, which in turn are overlain by metasedimentary rocks. Most features within the greenstone belts suggest a marine origin, such as rift basins, back-arc basins, and magmatic arcs—and are thus reminiscent of more recent plate-tectonic features. Perhaps the most famous, and one of the largest greenstone belts anywhere, is the Abitibi. The Abitibi Greenstone Belt straddles the border between northern Ontario and northwestern Quebec and includes mining centres such as Timmins, Kirkland Lake, Rouyn-Noranda, and Val d'Or. The areas of metasedimentary rocks in the Superior Craton originated mainly as clastic deposits eroded from adjacent lands. Among them are quartz-rich sandstones, now metamorphosed to quartzites, which have sedimentary structures such as cross-bedding similar to those of modern beach sands.

Granite dykes intruding amphibolite within the Superior Craton near Lake Minto, Quebec. JOHN PERCIVAL.

The greenstones and metasedimentary rocks are mostly arranged in roughly east-west orientated linear belts. The areas occupied by the granitic rocks are not so clearly aligned, but some, especially in northern Quebec are vast. Geologists have unravelled a broad history for the Superior Craton. About 2,750 million years ago, the parts of the Craton underlain by granitic rocks (let's informally call them protocontinents for simplicity) were separated by oceans, the width of which we can only guess. Around 2,720 million years ago, two protocontinents in the north (in terms of modern directions) collided, trapping a mixture of oceanic material between them. Around the same

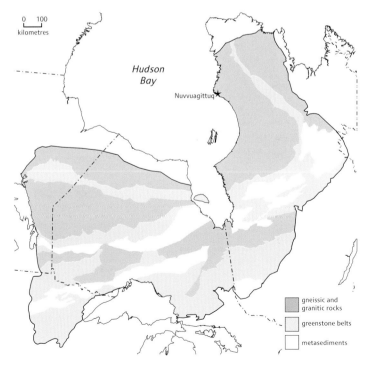

Simplified geological map of the Superior Craton. ADAPTED FROM WILSON AND CLOWES (2009), COURTESY OF LITHOPROBE.

time, other small protocontinents in the south amalgamated and were welded onto the northern block, sweeping up more tracts of oceanic lithosphere in the process. This amalgamation process, during which the oceanic material became compressed into the greenstone belts, continued until about 2,680 million years ago, by which time all the pieces of the Superior puzzle had come together.

Early Tectonics

The geology of the Slave and Superior cratons provides insights into Archean tectonic processes and perhaps the early development of plate tectonics. Very early on in our planet's history, the thin basaltic crust that began to form over the magma ocean would have been vulnerable to stresses. The huge tidal pull of the then-much-closer Moon, convection currents in the magma below, and bombardment of extraterrestrial bodies would have contributed to the breaking up of this new crust into slabs. These slabs slid past, over, and under each other, in the last case sinking back into the hot magmatic interior. A present-day analogy is the surface of a lava lake in an active volcano such as Kilauea on Hawaii, where we can see lava crusts sinking slowly and new lava welling up through cracks, cooling, and then spreading.

By 4,400 million years ago the Earth had cooled sufficiently for a watery ocean to form. Sinking crustal slabs dragged water into the hot but slowly solidifying magma ocean beneath. The water promoted chemical reactions that led to magma with more silica and less magnesium and iron. Because these silica-rich magmas were of relatively low density, they tended to rise up and intrude the overlying basaltic crust, forming large granitic complexes. Locally, granitic magmas would have extruded at the surface to produce felsic lava flows, domes, and pyroclastics that would have contributed to a rising topography.

The relatively light felsic igneous complexes, too buoyant to sink back into the mantle, would have formed ponderous rafts floating among the slowly moving slabs of basaltic crust, and the 4,030-million-year-old Acasta Gneiss is probably a relic of one of these rafts. The felsic rafts must have started to collide and amalgamate into larger slabs—what we called protocontinents above—at least as early as 4,000 million years ago. Additional granitic magmas, rising from below to crystallize near the surface, provided extra buoyancy.

We have learned much about Canada's deep lithospheric structure because of the LITHOPROBE Project (Box 3). LITHOPROBE's seismic line across the southwestern part of the Slave Craton shows that subsurface boundaries between different components of the Slave Craton are almost horizontal. Such horizontal layering suggests that thin slices of lithosphere were perhaps sliding beneath and above one another, rather than diving deep into the mantle as they tend to do at true subduction zones.

In contrast to evidence from the Slave Craton, rocks from the Superior Craton seem to reflect plate-tectonic elements and processes in miniature. LITHOPROBE seismic reflection data across part of the Superior Craton have revealed dipping structures that appear to be "fossilized" subduction zones, indicating that plate-tectonic processes, or something like them, were active in some places 2,700 million years ago. Just how similar these processes were to modern ones is conjectural.

Large areas of the world's Archean cratons date to about 2,700 million years ago, suggesting increased formation and amalgamation of continental crust around that time, perhaps driven by rapid plate-tectonic-like processes and the generation of large bodies of magma above subduction zones.

Chasing Life's Origins

As early tectonic processes were evolving, another type of evolution was stirring—that of life. A vital prerequisite for life is water in its liquid state. Based on evidence from sedimentary rocks and from the chemistry of certain minerals in igneous rocks, we know that water has been consistently present on the planet's surface for well over 4,000 million years. Water's presence reflects an amazing long-term stability in climate and has provided an ideal environment, without which life almost certainly would not have been possible.

Many questions remain about how life began. When and how did the transition occur from inorganic reactions, perhaps already involving complex carbon-based molecules, to the first self-replicating, metabolically active biomolecules that we would consider alive? Did life originate once or several times? Did it occur on the Earth only or also on other planets such as Mars? Were early carbon-based complex molecules in meteorites and cosmic dust responsible for seeding life onto any favourable planetary surface where it could survive and evolve? Was the origin of life a predictable consequence of the complex interactions between reactive mineral surfaces, volcanic gases, and a water-rich atmosphere-hydrosphere? Questions such as these drive much of the exploration of other planets, especially Mars. If we find evidence of life on that planet, would its biochemical framework be identical to that of Earth, suggesting a common origin of life? Or would it be radically different?

We cannot yet answer most of these questions. But some lines of inquiry are proving fruitful. Perhaps the best way to start investigating early life is to look at molecular studies of modern organisms (Chapter 4). The basal lineages of organisms are all prokaryotic, chemosynthetic, and anaerobic (able to live under anoxic conditions). They are also extremophiles—organisms that survive in extreme environments. These observations suggest that the last common ancestor of all living organisms thrived in settings such as hydrothermal vents on the ocean floors (Chapter 12). Such vents, with lots of hydrogen, sulphur, and hydrogen sulphide for chemosynthesis, would have provided ideal environments for early cells.

All the evidence from modern organisms supports the idea that the earliest organisms were prokaryotes. But prokaryotes are tiny, generally less than a few microns across (a micron being one thousandth of a millimetre), and thus hard to detect in the fossil record. The best candidates for the earliest Archean fossils are spherical microstructures found in 3,500-million-year-old rocks from South Africa. These tiny structures, 2 to 4 microns in diameter, have the same shape and size as some modern prokaryotes, and some even seem to be caught in the act of dividing, like modern cells. Experts concur that these structures are not completely convincing as fossils. But 3,200-million-year-old compressed spheroidal microscopic structures, also from South Africa and also thought to be prokaryotes, are more convincing.

Another line of evidence for detecting life in rocks involves determining the ratio of the isotopes carbon-12 and carbon-13. Because it has an extra neutron, carbon-13 weighs more than carbon-12. Consequently, when organisms take in carbon dioxide during photosynthesis, they capture carbon-12 preferentially over the heavier carbon-13. Thus, the ratio of carbon-12 to carbon-13 in organic matter will be different from the ratio found in carbonate minerals precipitated from the same water body. Unaltered sedimentary rocks older than 3,500 million years are extremely rare. Nevertheless, carbon isotopes in metamorphosed shales from southwestern Greenland suggest that life was already present 3,800 million years ago. And from analyses of 3,500-million-year-old rocks, it appears that by that time not only was life present, but the carbon cycle was fuelled by photosynthesis. Whether this was the familiar photosynthesis observed today in plants and algae is uncertain. More likely it was a type of photosynthesis seen today in several groups of bacteria that does not generate oxygen as a by-product. Much progress has been made in recent decades, but the evidence for early Archean life is still largely controversial and confusing.

Perhaps new clues will come from the Canadian Shield given the unsurpassed exposure of Archean rocks there.

Microbes at Work

The earliest-known direct evidence of life in Canada is from stromatolites in Superior Craton rocks about 2,800 million years old at Steep Rock Lake in western Ontario. Stromatolites are thinly laminated structures that range in age from the Archean to the present, and can be in the form of knobs, domes, and single and branching columns. Most stromatolites are only a few centimetres wide, but examples tens of metres high are known. Although

Modern stromatolites at Shark Bay, Western Australia. ANDREW KNOLL.

These large ovoid stromatolites in the western part of the Superior Craton, at Steep Rock Lake, Ontario, are 2,800 million years old. Note the hammer at bottom right-centre for scale. JOHN PERCIVAL.

they lack the elaborate skeletal structures and shells that are so conspicuous in more conventional Phanerozoic reefs, stromatolites are of organic origin. We know this because similar structures, though rare, are forming today, for example in Western Australia's Shark Bay. Stromatolites owe their existence to the ability of some microbes to produce slimy films and mats that trap fine sediment particles. Bacterial processes within the mats cause calcium carbonate to precipitate, either as calcite or aragonite, and this acts as a cement to bind the trapped particles and build up layers to form the laminated stromatolite structure. Modern stromatolites are restricted to places where the water is too salty for grazing or burrowing invertebrates, which would eat the microbes and disrupt the layers.

Sedimentary rocks as old as 3,500 million years contain stromatolite-like features. These structures largely reflect the physical precipitation of carbonate minerals from waters saturated with calcium carbonate. However, microscopic textures of at least some of these structures show that they may have been produced by microbes—which would not be surprising given the carbon isotope evidence for life from rocks of that age. Younger Archean rocks contain stromatolites of unambiguous biological origin, including the specimens at Steep Rock Lake and others near Joutel in the Abitibi Greenstone Belt in Quebec.

Cyanobacteria are one of the groups of prokaryote microbes associated with the formation of stromatolites. Their appearance was one of the great milestones in biological evolution, since oxygen is one of their gaseous waste products and from them evolved the plant organelles known as chloroplasts, the amazing mini-factories that plants use during photosynthesis. Thus, the initial gains in oxygen in the Earth's atmosphere came from cyanobacteria. When these remarkable prokaryotes first evolved, however, is unclear. Certainly, they were around before 2,400 million years ago when the atmosphere began to accumulate oxygen. Unfortunately, microfossils from rocks more than 3,000 million years old have simple shapes that tell us little about their affinities.

The oxygen in the atmosphere is what makes life possible for animals. But the buildup to today's levels was initially very slow and may have increased in steps rather than gradually. How

Archean banded iron formation, about 2,700 million years old, from the Sherman Mine, Temagami, Ontario. This example consists of red jasper interbedded with grey magnetite. BRIAN PRATT.

many steps and what triggered them is uncertain, although there are clues. One is the presence of distinctive rocks known as banded iron formations (also referred to as banded ironstones and commonly abbreviated as BIFs), which consist of alternating layers of chert (sometimes its red variety, jasper) and iron-bearing minerals, including the iron oxides hematite and magnetite and the iron carbonate siderite. Banded iron formations were deposited in ancient seas, starting at least 3,800 million years ago and reaching their peak 2,700 to 2,500 million years ago. Their presence shows that lots of iron existed in the oceans of the time, which is possible only when the waters are largely free of oxygen. The iron minerals could have precipitated when iron in solution came in contact with oxygen-bearing waters near the ocean's surface. However, another reason for such extensive iron precipitation is possible. Early photosynthetic organisms could have used iron to provide electrons for photosynthesis, generating oxidized iron much the way that cyanobacteria produce oxygen. In these early oceans, oxidized iron would have fuelled respiration, much as oxygen does in our own cells. The existence of banded iron formations may therefore suggest a world in which the carbon cycle was closely linked to the cycling of iron through the oceans. Such a world is only possible when oxygen is limited or absent from the Earth's surface. As oxygen levels increased, this world disappeared, as we will see in the next chapter.

6 · LAYING THE FOUNDATIONS
Canada 2,500 to 750 million years ago

CHAPTER SUPPORTED BY A DONATION FROM THE MINING ASSOCIATION OF NOVA SCOTIA

In this view of the front ranges of the northern Mackenzie Mountains, Northwest Territories, Neoproterozoic deltaic deposits of the Amundsen-Mackenzie Mountains Basin are represented by resistant quartz-rich sandstone overlain by brown shaly layers. HENDRIK FALCK.

Stanley Tyler Goes Fishing

It was a fishing trip near Schreiber, Ontario, on a lazy August Sunday in 1953 that was to change the direction of Stanley Tyler's career and open a new window on Precambrian life. Tyler, a geology professor at the University of Wisconsin, was drawn to the northern shore of Lake Superior to study the 1,900-million-year-old Gunflint banded iron formation. Equivalent rocks in northern Minnesota's Mesabi Range were (and still are) an important source of iron ore. On that fateful day, Tyler rented a boat and headed out. While fishing, he spotted an anomalous-looking Gunflint outcrop and decided to take a closer look. His attention had been drawn to a half-metre-thick chert bed, which—unlike the iron-rich reddish cherts elsewhere in the Gunflint—was jet black and waxy-looking. Tyler couldn't resist bagging some samples before getting back to his fishing.

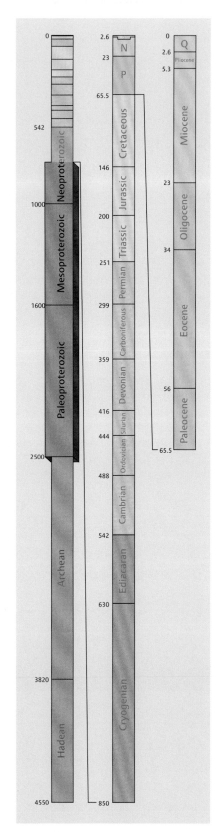

Geologic time scale, showing the interval covered in this chapter. Numbers indicate millions of years ago. P = Paleogene (Paleocene to Oligocene), N = Neogene (Miocene and Pliocene), and Q = Quaternary.

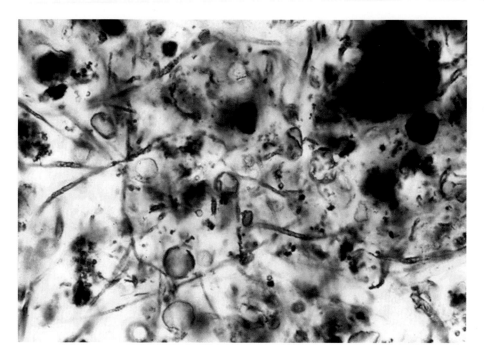

Stanley Tyler set off the search for Precambrian life when he observed microfossils such as these in thin sections of Gunflint chert from northern Ontario. ANDREW KNOLL.

After returning to his lab, Tyler made thin sections of rocks collected during the summer, including the black chert. Studying these sections under the microscope, he soon discovered why the rock was black: it contained fuzzy clumps of dark-brown to black coaly organic matter distributed among the tiny quartz crystals. At higher magnification, the organic matter resolved into myriad thread-like filaments and tiny hollow balls, all less than 10 microns across. Although he had examined many rocks under the microscope in his career, Tyler had never before seen anything like this. The filaments looked like fossils—but what were they doing in rocks 1,900 million years old? They didn't seem to be modern contaminants, and there was nothing like them in the paleontology books of the day; the South African Archean microfossils we noted in the previous chapter would not be discovered for several decades.

At a Geological Society of America meeting in Boston that year, Tyler showed photographs of the supposed Gunflint fossils to Harvard paleobotanist Elso Barghoorn. Barghoorn too was convinced that the Gunflint specimens represented true Precambrian fossils. So the two agreed to collaborate, and published their findings in April 1954. They interpreted the fossils as representing a mixture of protists and fungi, all eukaryotes. Although these particular interpretations did not withstand later scrutiny—all Gunflint fossils probably represent prokaryotes—the results of Stanley Tyler's fishing trip were of fundamental importance. The Gunflint specimens were far older than any fossils then known. And they opened up the possibility that vestiges of early life could be found in Precambrian rocks, a theme we will return to later in this chapter.

Setting the Scene

Stanley Tyler's fossils were from rocks of the Proterozoic Eon, which followed the Archean Eon, with the boundary set at 2,500 million years ago. The Proterozoic is divided into the Paleoproterozoic (2,500 to 1,600 million years ago), Mesoproterozoic (1,600 to 1,000 million years ago), and Neoproterozoic (1,000 to 542 million

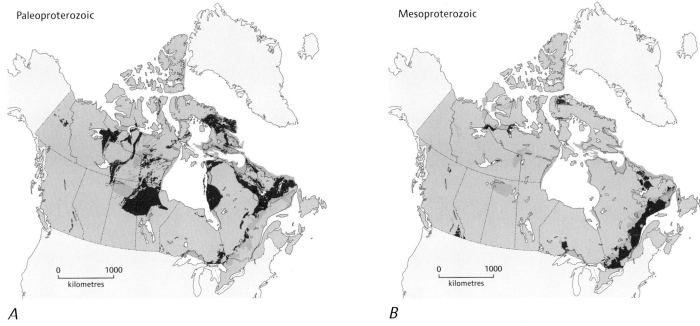

General extent of Paleoproterozoic (A) and Mesoproterozoic (B) rocks at the surface (beneath glacial deposits), onshore and offshore. The lighter shaded areas denote either uncertainty or areas where rocks of the particular age have been confirmed but are intimately associated with rocks of other ages and the scale of the map doesn't allow us to show them separately. ADAPTED FROM WHEELER ET AL. (1996).

years ago). In this chapter, we voyage from the beginning of the Paleoproterozoic to 750 million years ago, around the middle of the Neoproterozoic. This interval, from 2,500 to 750 million years ago, witnessed the consolidation of plate-tectonic processes, the evolution of life from simple prokaryotes such as those in the Gunflint chert to large unicellular eukaryotes, and the stepwise buildup of oxygen in the atmosphere to levels that start to approach those of today.

The early Paleoproterozoic was dominated by the breakup of larger late Archean continental entities whose original size, shape, and location are poorly understood. Archean cratons such as Slave and Superior are identifiable today because they were the results of this breakup; they were generally much smaller than continents of the past billion years but generally much larger than the earlier protocontinents. Some of the earliest-known passive margins developed on the flanks of these cratons. Later in the Paleoproterozoic, the cratons began to come together again, a process that culminated in the formation of a supercontinent known as Nuna by about 1,800 million years ago. From latest Paleoproterozoic time through the Mesoproterozoic, convergent tectonic activity became focussed in the Grenville Orogen along what is now the southeastern margin of the Canadian Shield. On that margin, oceanic lithosphere converged with continental lithosphere until a titanic continent-to-continent collision culminated in the Grenvillian Orogeny (the final event within the Grenville Orogen) some 1,090 to 1,000 million years ago. This episode was part of the assembly of another supercontinent, Rodinia, which dominated the globe from about 1,000 million to 750 million years ago, when it too began to break up.

References to points of the compass in this chapter refer to modern directions, as we do not yet have a clear understanding of Proterozoic geographic orientations.

Ancient Margins

The breakup of continental crust 2,450 to 2,000 million years ago was accompanied by the development of several giant sets, or swarms, of mafic dykes, which radiated out possibly from mantle plumes. An example is the 2,450-million-year-old Matachewan dyke swarm in the southern part of the Superior Craton. Radiating dykes of similar age and composition to the Matachewan swarm occur in the Karelian Craton, now part of Scandinavia and northwestern Russia, and in the Hearne Craton in north-central Canada. Matching such dykes across different modern continents has led some geologists to speculate that the Superior, Karelian, and Hearne cratons had previously been part of a single landmass before breaking apart.

Thick sequences of detritus that eroded from the cratons accumulated in rifts and on passive margins that began to develop, also about 2,450 million years ago. These sequences contain rocks much like those of the passive margin bordering eastern North America today. At the base are rift-related, predominantly coarse-grained clastic sedimentary rocks, in places interlayered with mafic volcanic rocks and intruded by mafic dykes. Above these—providing evidence of a broadening rift, subsidence, the invading sea, and the start of sea-floor spreading—are finer-grained and better-sorted sediments such as now-metamorphosed sandstones and mudstones. The upper parts of such sequences commonly include thick carbonate and

Archean gneisses of the North Atlantic Craton are intruded by Paleoproterozoic mafic dykes, which are the dark bands in this 300-metre-high cliff north of Saglek Fiord, Labrador. BRUCE RYAN.

Early Paleoproterozoic white sandstones in Killarney Provincial Park, Ontario. These rocks are part of the Huronian passive-margin succession. SASKIA ERDMANN.

chert layers with stromatolites, deposited in shallow seas on cratonic (continental) shelves.

One of the best-known passive-margin sequences of the Paleoproterozoic is the Huronian succession in southwestern Quebec and central Ontario, developed along the southern margin of the Superior Craton. The 12-kilometre-thick sedimentary pile is well exposed in road cuts along the Trans-Canada Highway between Sudbury and Sault Ste. Marie, and was deposited between 2,500 and 2,200 million years ago. During early rifting, ash and lava from nearby volcanic eruptions, plus sediments brought in by streams and rivers, began to fill rift basins. These rocks are overlain by a succession of mainly sandstones and conglomerates, and include uranium-rich layers that originated as placer deposits. Placers are accumulations of dense minerals—in this case uraninite (uranium oxide) and pyrite (iron sulphide)—that were carried by fast-flowing streams and selectively deposited where currents slowed down. Uraninite readily dissolves in water in the presence of oxygen, so its preservation in the oldest part of the Huronian succession testifies to the still oxygen-poor nature of the atmosphere and oceans around 2,500 million years ago. From 1955 through the 1970s, Huronian deposits at Elliot Lake in Ontario were mined as a major source of uranium (Chapter 13).

Evidence for a marked increase in oxygen occurs higher in the Huronian succession, within a thick sequence of fluvial to marine quartz-rich sandstones well above the uraninite-bearing strata. These sandstones form striking white exposures in Ontario's Killarney Provincial Park. But within this sequence of white rocks are red sandstones—red because the sand grains are coated with hematite. Such a coating, familiar as the orange tinge of typical desert sand dunes, forms only in an atmosphere with free oxygen. The red sandstones occur in sedimentary sequences of a similar age worldwide, evidence for the Great Oxidation Event between 2,500 and 2,400 million years ago. That was when the Earth's atmosphere changed from an anoxic to a slightly oxic condition. Previously, oxygen made up a fraction of 1 percent of the atmosphere; afterwards it may at times have been a few percent—still low compared to today's 21 percent, but enough to oxidize iron-rich minerals. The later

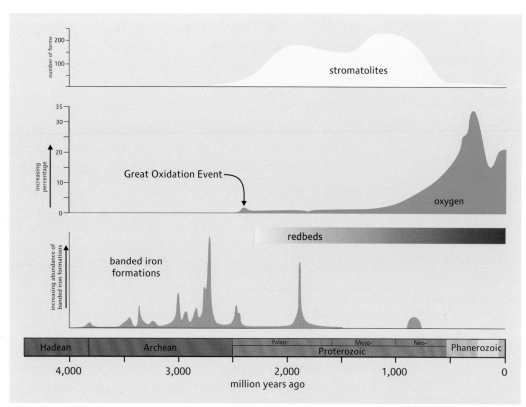

Distribution through time of banded iron formations, redbeds, and stromatolites, and the changing atmospheric concentrations of oxygen. ADAPTED FROM VARIOUS SOURCES INCLUDING, ILYIN (2009), LYONS AND REINHARD (2009), RASMUSSEN ET AL. (2012), AND WALTERS AND HEYS (1985).

This Paleoproterozoic conglomerate near Elliot Lake, Ontario, is part of the Huronian passive-margin succession and is probably a debris-flow deposit. ROB RAINBIRD.

Tillite from the Paleoproterozoic Huronian succession, near Elliot Lake, Ontario. The alignment of the rock fragments is evidence that they were deposited beneath a flowing glacier. DARREL LONG.

Huronian redbeds and their equivalents elsewhere heralded a new oxygen-enriched era on the Earth, eventually setting the stage for the evolution of more complex life forms.

Within the Huronian succession are layers of rock consisting of angular boulders and smaller fragments set in a matrix of hardened mud. Because they resemble consolidated versions of the jumbled mixtures of rocks and mud familiar to most Canadians as glacial till, such rocks are called tillites. These Huronian deposits are indeed interpreted as among the earliest evidence for extensive glaciations, supported by the presence in the tillites of some pebbles and boulders whose surfaces are polished and striated, presumably because of glacial processes (Chapter 11). The Huronian tillites and associated deposits evidently represent three separate glaciation events that occurred between 2,450 and 2,200 million years ago.

Remains of other early to middle Paleoproterozoic sedimentary basins are preserved in Canada. One such basin, the Labrador Trough in Labrador and northeastern Quebec contains rocks that accumulated 2,170 to 1,870 million years ago on the eastern side of the Superior Craton. The rocks formed in the Labrador Trough occur in a belt that extends for about 1,200 kilometres from the Manicouagan area in Quebec, through western Labrador, northward to Ungava Bay. The succession includes thick banded iron formations, which have been intimately linked to the economic development of northern Quebec and Labrador. Although known since the mid-nineteenth century, the first scientific investigation of these iron deposits was by Albert Low toward the end of that century. He criss-crossed this vast area by boat and canoe, by dog-train, and on foot, taking particular note of the deposits in the area where Schefferville, Quebec, is now located. These iron deposits sustained Schefferville from 1954 to 1982, and are still the raison d'être for Fermont, where production began in 1975. In Labrador, both Labrador City and Wabush have owed their economic well-being to iron ore since the 1960s.

In this scene at Lac Guillaume-Delisle, northern Quebec, the flat-topped hills in the middle distance are underlain by undeformed, 2,000-year-old, rift and passive-margin strata on a flank of the Superior Craton. The lower slopes are underlain by Archean gneiss. JEAN-YVES LABBÉ.

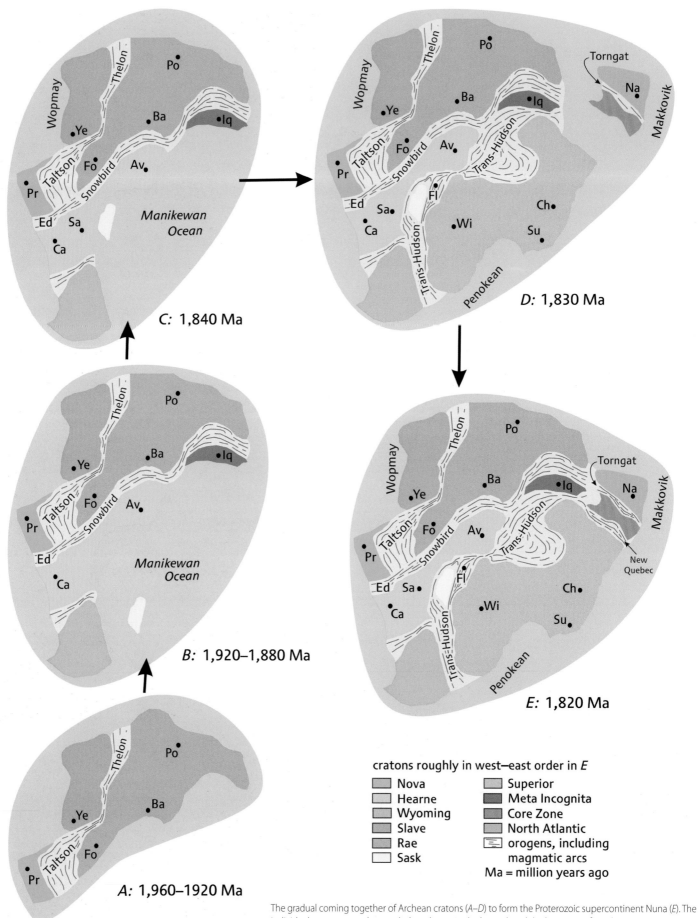

The gradual coming together of Archean cratons (A–D) to form the Proterozoic supercontinent Nuna (E). The individual cratons are colour coded, as shown in the legend, and the locations of modern communities are shown for orientation: Av = Aviat, Nunavut; Ba = Baker Lake, Nunavut; Ca = Calgary, Alberta; Ch = Chibougamau, Quebec; Ed = Edmonton, Alberta; Fl = Flin Flon, Manitoba; Fo = Fond du Lac, Saskatchewan; Iq = Iqaluit, Nunavut; Na = Nain, Labrador; Po = Pond Inlet, Nunavut; Pr = Prince Rupert, British Columbia; Sa = Saskatoon, Saskatchewan; Su = Sudbury, Ontario; Wi = Winnipeg, Manitoba; Ye = Yellowknife, Northwest Territories.

Meeting at Nuna

As cratons drifted apart during the early Paleoproterozoic, new oceanic lithosphere developed beyond the passive margins. Spreading ridges, magmatic arcs, and oceanic plateaus all had a place in these ancient oceans, just as they do in modern oceans. Beginning about 1,980 million years ago, the cratons began to converge once more as new subduction zones and magmatic arcs developed. Magmatic arcs that formed within the ocean basins were eventually accreted to craton margins when the basins closed. By about 1,800 million years ago, the cratons and magmatic arcs had become welded together to form the supercontinent of Nuna. LITHOPROBE's seismic sections (Box 3) have provided stunning images of oceanic lithosphere "frozen" as it was being subducted within Nuna about 1,800 million years ago. In rare cases during tectonic collisions, pieces of ocean floor were thrust onto craton margins as ophiolite suites, preserving some of the oldest-known fragments of oceanic lithosphere. An example is the Purtuniq Ophiolite in northern Quebec. This slab of oceanic lithosphere was thrust in several successive layers onto the Superior Craton, together with fragments of magmatic arcs that had evolved nearby.

Sandstone with giant cross-beds on Melville Peninsula, Nunavut. These rocks originated as eolian or fluvial sediments, which were deposited about 1,850 million years ago on Archean rocks of the Rae Craton during formation of the supercontinent Nuna. DAVID CORRIGAN.

The coming together of cratons to form Nuna is best seen in the series of maps on the previous page. Collisions between cratons resulted in orogenic belts such as the Taltson-Thelon Orogen between the Slave and Rae cratons, and possibly the Snowbird Orogen between Rae and Hearne. In the far west, the Wopmay Orogen is well developed, but we know only the eastern partner in the collision, the Slave Craton. The identity of the western partner is unknown because it is hidden under younger sedimentary rocks.

Most residents of Saskatchewan and Manitoba would be surprised to learn that, around 1,900 million years ago, their provinces were roughly separated by a huge expanse of water, perhaps as wide as the modern Pacific, called the Manikewan Ocean. This ocean was to leave a substantial mark on Canada. About 1,900 million years ago, this great water body lay between the combined Rae-Slave-Hearne-Meta Incognita "supercraton" to the northwest and the Superior Craton to the southeast. Closure of the Manikewan Ocean between 1,900 and 1,800 million years ago would produce North America's best-preserved Proterozoic orogen, the Trans-Hudson Orogen, which stretches from the southwestern United States (where it underlies younger rocks) to Nunavut, and can even be traced all the way to Greenland and Scandinavia.

The Manikewan Ocean and its borderlands had much in common with the present-day southwestern Pacific Ocean, with its numerous island arcs separated by back-arc basins from adjacent cratons. Arcs in the Manikewan Ocean included the La Ronge-Lynn Lake and Flin Flon arcs of Saskatchewan and Manitoba, and the Tasiuyak and Aillik arcs in Labrador. Former sedimentary basins associated with arcs around the Manikewan Ocean are represented by the Thompson Belt in Manitoba and the Cape Smith Belt in northern Quebec. As the ocean closed, the arcs tended to pile up on one another and adjacent crustal material.

These mountains in Auyuittuq National Park of Canada on southeastern Baffin Island are formed from granitic rocks of the Paleoproterozoic Cumberland Batholith. The Batholith was intruded along a subduction zone during the Trans-Hudson Orogeny 1,900 to 1,800 million years ago. J. POITEVIN, COPYRIGHT PARKS CANADA.

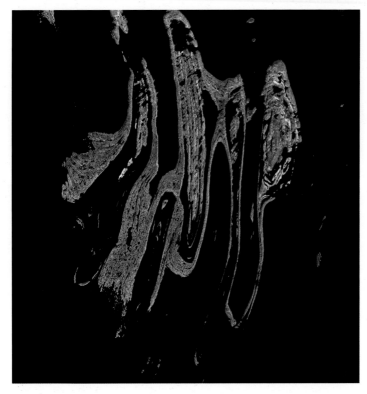

Satellite view of the Belcher Islands in Hudson Bay. The shapes of the islands reflect the sweeping folds in the rocks, a result of compression during the Trans-Hudson Orogeny 1,900 to 1,800 million years ago. COURTESY OF PAUL BUDKEWITSCH, CANADIAN CENTRE FOR REMOTE SENSING.

Intensely deformed conglomerate and amphibolite at Amisk Lake, northern Saskatchewan. The conglomerate originated as a river deposit 1,840 million years ago above rocks of the Flin Flon magmatic arc in the heart of the Nuna supercontinent. The amphibolite was originally a mafic sill that intruded the conglomerate. Some of the clasts are stretched, whereas others are only slightly distorted. LEN GAL.

One puzzle to geologists for many years was why the Trans-Hudson Orogen is so wide in northern Saskatchewan and Manitoba. A LITHOPROBE seismic investigation provided the answer. It revealed that a small piece of continental crust, the Sask Craton, was sandwiched between larger cratons as they collided. The Sask Craton was possibly a breakaway fragment of one of the larger cratons that, about 1,840 million years ago, found itself trapped within the closing Manikewan Ocean. A chain reaction ensued. The Sask Craton ploughed into the magmatic arcs and sedimentary basins in the narrowing ocean, causing them to fold and buckle. As convergence continued, arcs and basins were further deformed in the tectonic vise between the Sask and Hearne cratons, producing imbricated (repetitively overlapping) thrust sheets of rocks now clearly seen in seismic sections. This stacking lifted magmatic-arc rocks onto the margin of the Sask Craton, where they have been preserved for 1,900 million years. Today, rocks of the Flin Flon Arc remain so well preserved that we can clearly recognize features such as pillow lavas and volcanic bombs. In many of the arc settings, mineral-rich hydrothermal systems developed among the volcanic rocks below the sea floor, and lead, zinc, copper, and gold ores precipitated from black smokers similar to those of today (Chapter 12). These deposits are the basis for some of the world's richest metal-producing mines, such as those at Flin Flon and Snow Lake.

Even today, only a small part of the Sask Craton is exposed, though geophysical data suggest that it broadens substantially at depth, reaching about the size of Newfoundland and Labrador some 45 kilometres below the surface. The presence of the Sask Craton prevented the Superior and Hearne cratons from colliding, preserving so freshly the rocks and structures of the margins of the Manikewan Ocean beneath today's boreal forest.

All the cratons now exposed on the Canadian Shield between Slave and Superior were welded together by 1,830 million years ago. The final act leading to Nuna's construction took place to

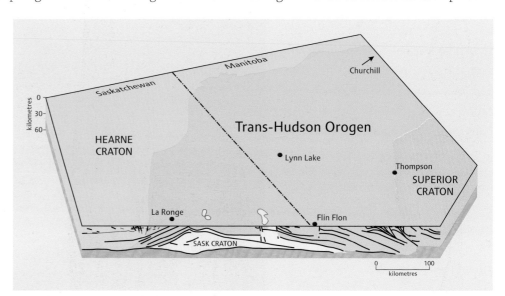

Block diagram showing how the Sask Craton is caught between the western Superior and Hearne margins and accounts for the greater width of the Trans-Hudson Orogen in parts of northern Saskatchewan and Manitoba. ADAPTED FROM WILSON AND CLOWES (2009), COURTESY OF LITHOPROBE.

the east of the Superior Craton. There, two small pieces of continental lithosphere, the North Atlantic and Core Zone cratons, had amalgamated by collision 1,860 million years ago creating the Torngat Orogen, now exposed in Labrador and northeastern Quebec. About 1,820 million years ago, this amalgam collided with the Superior margin to produce the New Quebec Orogen, which today underlies parts of western Labrador and northeastern Quebec. This last collision shuffled, folded, and compressed into a linear belt the rocks that had accumulated in the Labrador Trough.

The coming together of Archean cratons to produce Nuna was a fundamental step towards the eventual creation of the core of modern North America—so fundamental that the resulting amalgamation has been referred to as the "United Plates of America".

New Horizons

Once Nuna was formed, plate convergence shifted from the interior to the southeastern side of the new supercontinent, forming a tectonic belt, the Grenville Orogen, that extends from northwestern Mexico to southeastern Canada. It underlies much of southern Ontario, Quebec (including the Laurentian Highlands and a large area north of the St. Lawrence River), and Labrador. Its continuation even farther northeast is revealed by rocks in Greenland and Scandinavia. Most of the evidence for the history of the Grenville Orogen is buried or has been removed by erosion, leaving only the hard-to-interpret, intensely deformed, and

Ice-sculpted, 1,650-million-year-old granite and gneiss on the south side of Philip Edward Island, Georgian Bay. These rocks formed in magmatic arcs of the Grenville Orogen on the southeastern side of Nuna. CHRISTOPHER HARRISON.

metamorphosed roots of what must have been a vast mountain chain similar to the present-day Cordillera. Nevertheless, geologists are beginning to unravel its complicated history.

Several episodes of deformation have been identified in the Grenville Orogen. The episodes between 1,850 to 1,200 million years ago reflect convergence between Nuna to the northwest and accreting island arcs to the southeast. Some of the best evidence for arcs and associated sedimentary basins can be seen in the Laurentian Highlands, north and east of Shawinigan, and along the north shore of the St. Lawrence River northeast of Sept-Îles. Similar exposures are found in Ontario, from Georgian Bay to the Rideau River. Other remnants are preserved as

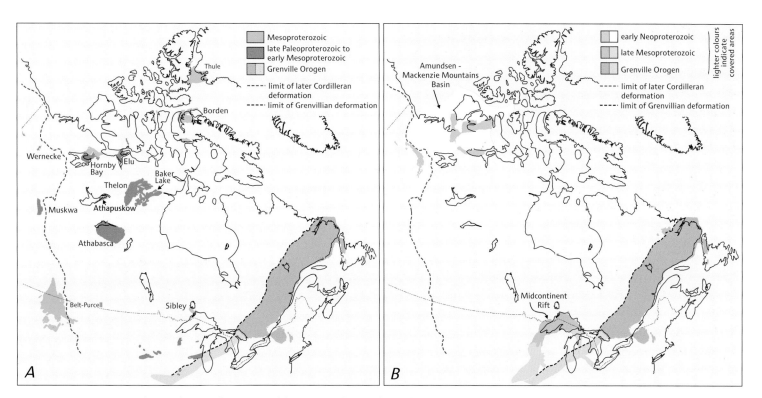

Map showing the Grenville Orogen and the continental-interior basins that existed in what was to become part of North America during the intervals from 1,800 to 1,300 million years ago (A) and 1,300 to 750 million years ago (B). ADAPTED FROM VARIOUS SOURCES.

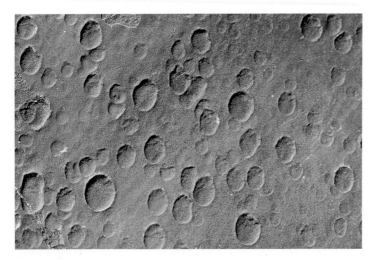

Paleoproterozoic raindrop prints in redbeds of the Baker Lake Basin, Nunavut. ROB RAINBIRD.

View north along the eastern flank of Adams Sound, south of the community of Arctic Bay on Baffin Island, Nunavut. Here, Mesoproterozoic lava flows (the steep face in the middle part of the cliff) unconformably overlie Paleoproterozoic gneisses (the gentler slopes near the base of the cliff). The red rocks near the top of the cliff are Mesoproterozoic marine sandstones of the Borden Basin. DARREL LONG.

Mesoproterozoic red argillite viewed from the Carthew-Alderson Trail, Waterton Lakes National Park of Canada, Alberta. W. LYNCH, COPYRIGHT PARKS CANADA.

Red Rock Canyon in Waterton Lakes National Park of Canada is aptly named for the red argillite (1,500 million years old), which was formed from iron-rich sediments deposited on ancient tidal mudflats. JOHN WILLIAM WEBB.

a remarkable series of plutons and batholiths, once vast magma reservoirs and today the source of many kitchen-counter tops. Rich mineral deposits are associated with some of the arcs—for example from rocks of a 1,450-million-year-old back-arc basin at Montauban-les-Mines, a small community in the Laurentian Highlands northeast of Trois-Rivières.

Extensive remains of a back-arc basin on the southeastern edge of the continent, the Elzevir Basin, extend from Kingston in Ontario, through the Gatineau Hills north of Ottawa, to Mont-Laurier in Quebec; they date from 1,300 to 1,200 million years ago. The Elzevir Basin contains volcanic rocks as well as quartzites and marbles, once sandstones and carbonates. The marbles contain unusual deposits of apatite, a calcium phosphate mineral. The apatite-rich layers, mined as a source of phosphate in the early twentieth century, may have been precipitated with the carbonates in shallow seas under hot and dry climatic conditions. Metamorphism of these unusual sediments, combined with their intermingling with felsic magma and mineral-rich hydrothermal fluids, led to the growth of rare gem-quality minerals such as apatite, sodalite, and tourmaline, particularly in the Bancroft area of Ontario.

While much of the action involving converging plates during the Mesoproterozoic had switched to Nuna's southeastern margin, the mountain belts in the continent's interior, such as the Trans-Hudson, were wearing down through weathering and erosion. Sediments from these belts accumulated in basins such as the Thelon and Athabasca in the continental interior. These two basins now underlie parts of northeastern Alberta, northern Saskatchewan, the Northwest Territories, and Nunavut. Today we can walk on the surface of these two basins and still see ripple marks left by wave action in shallow waters all those eons ago. By about 1,700 million years ago, the interior mountains had worn down to plains. Evidence from detrital zircons supports the idea that the sediment accumulating in the Athabasca and Thelon basins by that time came from erosion of the Andes-like mountains building on Nuna's southeastern margin. The detritus was probably carried northwestward via a huge river system even bigger than the modern Amazon (and not constrained by vegetation). The Athabasca Basin (not related to the much younger Athabasca oil sands) contains some of the largest uranium deposits in the world.

As these basins filled up, some sediments were carried even farther, probably by the same river system, to basins on the far-northwestern side of the Canadian Shield. These depressions are preserved today as the Muskwa and Wernecke basins of northern British Columbia and Yukon, and developed between 1,650 and 1,600 million years ago. Deposits within these basins include deep-water turbidites, suggesting that they accumulated on a passive margin.

Somewhat younger are rocks of the Belt-Purcell Basin, preserved today in the southern Rocky Mountains. Interpreted as remnants of a rift basin (later thrust eastward, as we will discover in Chapter 9), the Belt-Purcell rocks are up to 20 kilometres thick. The older parts of the sequence are mainly deep-water turbidites, interlayered with numerous mafic sills, some of which intruded while the sediments were still soft. The turbidite sequence now hosts the major lead and zinc ores of the Sullivan Mine at Kimberley in southeastern British Columbia. Within the upper part of the Belt-Purcell sequence are shallow-water clastics and carbonates, as well as basalts dating to about 1,450 million years ago. These ancient rocks make the modern Rocky Mountain scenery south of the Crowsnest Pass dramatically different from that formed by the younger rocks north of the Pass.

While magmatic arcs and back-arc basins dominated the southeastern margin of Nuna, its northeastern side, in what was to become the Nain region of Labrador, was undergoing different tectonic processes. There, the lithosphere was being stretched and torn apart once again, allowing magmas to penetrate upwards from the mantle into the old continental suture of the Torngat Orogen. Between 1,360 and 1,290 million years ago, numerous plutons of granite and anorthosite (an unusually feldspar-rich rock formed at very high temperatures) crystallized from these magmas. Magnificent exposures of anorthosite occur on the coastline near Nain. Some feldspars in the anorthosite are of the beautiful labradorite variety, which shimmer blue and green when rotated (Chapter 14). One anorthositic

The Coppermine Basalts, here exposed near Kugluktuk, Nunavut, were extruded 1,240 million years ago above a hot spot centred on Victoria Island. They are associated with the Mackenzie Dyke Swarm (Chapter 1). The Coppermine Basalts, named for the native copper they contain, consist of about 150 individual lava flows, each 10 to 25 metres thick. HANS WIELENS.

Mount Lister is an 850-metre-high bald massif on the north side of Tikkoatokak Bay, west of Nain, Labrador. It is underlain by massive pale grey anorthosite of the Nain region, intruded between 1,360 and 1,290 million years ago. BRUCE RYAN.

intrusion, at Voisey Bay near Nain, became famous in mining circles in 1994 when it was found to contain a large and incredibly rich copper nickel-cobalt deposit.

Toward Grenvillian Peaks

Around 1,210 million years ago, the ocean to the southeast finally closed as another continent, incorporating part of present-day South America, began to collide with Nuna. This titanic collision culminated in the Grenvillian Orogeny, which lasted from 1,090 to 1,000 million years ago. The collision, part of a major amalgamation of continents that formed the supercontinent Rodinia, caused the continental lithosphere to thicken through thrust faulting and folding, so that in places the crust became more than 70 kilometres thick. This great thickness resulted in deep crustal roots and a surface welt that was carved by erosion into mountains of possibly Himalayan proportions. Vestiges of Grenvillian crustal roots—which locally contain a rare green and red rock called eclogite, formed only under great pressure—can be seen today in the Manicouagan and Mont-Tremblant regions of Quebec and in Ontario's Algonquin Provincial Park. But the best exposures are along the windswept eastern shore and islands of Georgian Bay, near Parry Sound in Ontario. Such rocks provide vital clues about tectonic processes in the deep crust—indeed, it's fascinating to think that these Grenville rocks reflect what we might see beneath the modern Himalayas if a giant bulldozer could remove the top 35 kilometres of crust.

About 1,100 million years ago, local crustal stretching and mafic volcanism in the southern part of Nuna produced a feature called the Midcontinent (or Keweenawan) Rift. Now mostly buried but long known from gravity and magnetic surveys, this rift can be traced 2,000 kilometres from Kansas to Michigan. Associated basalts can be seen near Thunder Bay, and an offshoot of the rift extends into the Nipigon area of northern Ontario. Ship-borne seismic analyses undertaken in the 1980s on Lake Superior as part of the joint Canadian-American Great Lakes International Multidisciplinary Program on Crustal Evolution (GLIMPCE) project provided a spectacular cross-section through the rift, demonstrating that in places it almost completely ruptured the continental crust and that it contains a succession of latest Mesoproterozoic sedimentary and volcanic rocks more than 30 kilometres thick.

Where did all the material eroded from the Grenvillian Mountains go? Clues come from early Neoproterozoic rocks in the Amundsen-Mackenzie Mountains Basin of northern mainland Yukon, Northwest Territories, and adjacent parts of Victoria and Banks islands. In the late 1970s, geologists working in this northern basin surmised that most of the sandstones came from the deposits of rivers that had a consistent northwesterly flow. These rocks are about 1,000 to 700 million years old, and could have come from erosion of the Grenvillian Mountains, 3,000

These rocks on the western side of Franklin Island in Georgian Bay, Ontario, originated between 1,680 and 1,400 million years ago and were metamorphosed and deformed to folded gneiss during the Grenvillian Orogeny about 1,000 million years ago. CHRISTOPHER HARRISON.

Layering in this rock from Parry Sound, Ontario, is not sedimentary but was produced by intense heating and compression of many igneous intrusions deep within the crust during the Grenvillian Orogeny. DAVID CORRIGAN.

Neoproterozoic sandstones and shales of the Amundsen-Mackenzie Mountains Basin in the Backbone Ranges, Mackenzie Mountains, Northwest Territories. HENDRICK FALCK.

Redbeds at the base of this road cut on the Trans-Canada Highway, just east of Nipigon, Ontario, were deposited in the small Sibley Basin 1,550 to 1,400 million years ago. Above the redbeds is a mafic sill, intruded about 1,100 million years ago during formation of the Midcontinent Rift. GRAHAM WILSON.

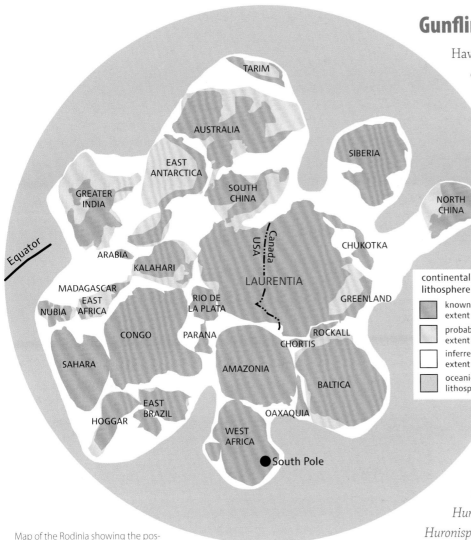

Map of the Rodinia showing the possible distribution within that supercontinent of later continental elements. ADAPTED FROM LI ET AL. (2008).

Gunflint and Beyond

Having begun this chapter with Stanley Tyler's discoveries in the Gunflint chert, let's end it with an exploration of Proterozoic life to 750 million years ago. At 1,900 million years old, the Gunflint specimens are still among Canada's oldest-known fossils. The Gunflint chert was deposited in the seas off the Superior Craton along with banded iron formations and argillite. Some of the cherts are black and form stromatolites with abundant microfossils. We're not sure whether these microfossils represent the stromatolite builders or organisms that fell onto the stromatolite surfaces. The most common microfossil is *Gunflintia*, specimens of which consist of thin, iron-coated tubes 1 to 2 microns across. *Gunflintia* broadly resembles modern cyanobacteria, but is also like iron-loving bacteria found today wherever oxygen comes into contact with iron-rich waters. Other microfossils in the cherts include the spherical *Huronispora*, a few thousandths of a millimetre across. *Huronispora* most resembles modern iron-loving bacteria.

Of about the same age as the Gunflint assemblage, but discovered more recently, are spherical microfossils called *Eoentophysalis* from the Belcher Islands in Hudson Bay. These fossils resemble modern mat-forming cyanobacteria, suggesting that the Belcher Island fossils are the remains of cyanobacteria that lived in ancient tidal-flat environments.

Gunflint and similar fossils are generally considered to represent prokaryotes. Molecular studies of living organisms suggest that the last common ancestor of all modern eukaryotes lived about the time that the Gunflint sediments formed, and fossils thought to be of eukaryotic origin occur in a banded iron formation in Michigan of roughly the same age as the Gunflint deposits. Still earlier evidence for eukaryotes has been claimed in rocks up to 2,700 million years old containing remnants of molecules known to be produced only by eukaryotes. However, debate persists as to whether these molecules truly accumulated 2,700 million years ago, or whether they represent younger eukaryote material introduced into the rock by percolating groundwater.

Some of Canada's oldest-known eukaryotes come from a sequence of 1,200-million-year-old cherts and shales from Baffin Island and nearby Bylot Island. The assemblage contains

kilometres to the southeast—an idea supported in the 1990s by studies of detrital zircons. Results showed that more than 50 percent of the zircon grains in the Basin's fluvial deposits are of the same age as rocks generated during the Grenvillian Orogeny. What's more, the zircons are well rounded, supporting the idea of a long journey across a continent. Thus, a great volume of detritus from the Grenvillian Mountains was apparently dispersed over a huge area.

Marine sedimentary rocks above the thick fluvial sequence show that later in its history the Amundsen Mackenzie Mountains Basin was invaded by a shallow sea. As the sea expanded, sand and then mud were deposited, above which is a sequence of up to 500 metres of shallow-water carbonates, mainly dolostone with extensive stromatolites. At the top of the succession, the carbonates alternate with thick deposits of evaporites (gypsum and anhydrite), indicating repeated episodes of drying out and replenishment of the Basin's waters. The evaporite deposits are strikingly light-coloured, and the gypsum includes a soft recrystallized variety known as alabaster, which is used by local Inuit sculptors.

a variety of hollow, organic-walled, cell-like fossils. The smaller of these fossils probably represent cyanobacteria that lived in coastal habitats. Larger ones, up to 1 millimetre across, are more complex and are probably the remains of eukaryotic cells. Even more convincing as eukaryotes are fossils found in 1,200-million-year-old rocks from Somerset Island, also in Nunavut. The fossils, named *Bangiomorpha*, consist of multicellular filaments. Based on similarities with living forms, these fossils are clearly red algae—the oldest eukaryotic fossils whose affinity we know. But even more exciting is the fact that different preserved life stages of *Bangiomorpha* testify that this red alga reproduced sexually—meaning that Canada can claim the world's oldest direct evidence of sex!

Further light is shed by 800-million-year-old deposits from the Amundsen-Mackenzie Mountains Basin in the Northwest Territories. A collection of fossils known as the Little Dal assemblage includes some of the first fossils, aside from stromatolites, readily visible to the naked eye. Because of their size, they have been interpreted as eukaryotes. The Little Dal fossils, associated with stromatolitic reefs up to 300 metres high, consist of spheroidal forms (*Chuaria*) and sausage-shaped forms (*Tawuia*) composed of organic matter and compressed flat on the surfaces of shale beds. The sausage-shaped *Tawuia* can be straight or curved, 1 to 2 millimetres wide, and up to 25 millimetres long. Specimens of *Chuaria* are up to 4.5 millimetres in diameter. These fossils may represent algae, but we have no convincing evidence for their affinity.

Another exciting find from the Amundsen-Mackenzie Mountains Basin is the world's earliest-known example of biomineralization (shell or skeleton formation) in eukaryotes. The find involves scale-like microfossils, probably associated with protists, from rocks about 810 to 715 million years old on Yukon's Mount Slipper. These microfossils, which include *Characodictyon*, are unusual in that they are formed of calcium phosphate.

So, although evidence is sparse, enough exists for us to conclude that eukaryotes were well established by the Neoproterozoic. And the indication of sexual reproduction in *Bangiomorpha* hints at the coming explosion of life.

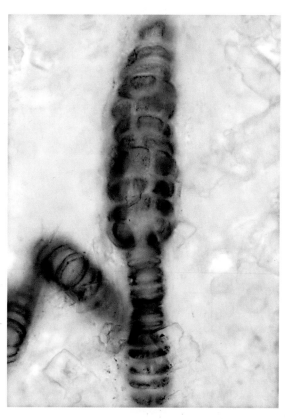

Bangiomorpha pubescens, a red alga from Mesoproterozoic (1,200-million-year-old) rocks on Somerset Island, Nunavut. The specimen is just under 0.2 millimetres long. NICK BUTTERFIELD.

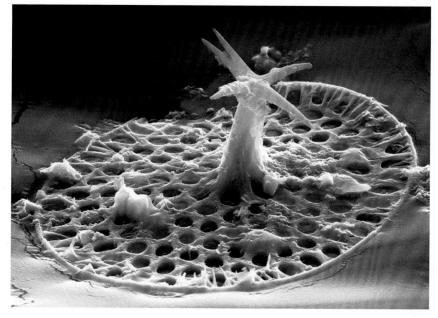

The microfossil *Characodictyon*, composed of calcium phosphate, from Yukon's Mount Slipper in the Amundsen-Mackenzie Mountains Basin. PHOEBE COHEN.

7 • SOUTHERN SOJOURN
Canada 750 to 444 million years ago

CHAPTER SUPPORTED BY A DONATION FROM JOHN BOYD, RPS BOYD PETROSEARCH

Aerial view near Keels, Newfoundland, of Cambrian fluvial clastic rocks deposited on Avalonia, a microcontinent associated with the ancient Iapetus Ocean. SEAN O'BRIEN.

No Mistake at Mistaken Point

The first inkling that Newfoundland's Avalon Peninsula had something special to contribute to our knowledge of the history of life came in the late nineteenth century. In a report published in 1881, the first Director of the Geological Survey of Newfoundland, Alexander Murray, remarked on a discovery in rocks previously thought to be barren of fossils. Murray noted, "I have long had some obscure forms in my possession, collected [by the Rev. Mr. Harvey] in the neighbourhood of St. John's, which were suspected to be organisms of a low type, but which I could not venture to pronounce to be such without palaeontological reference."

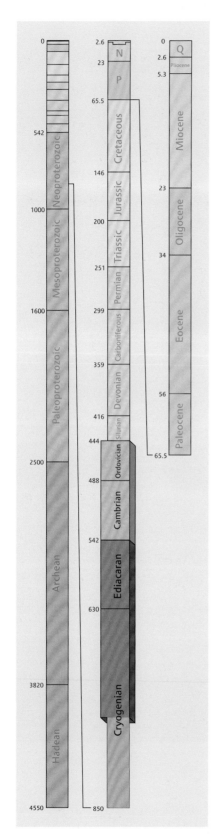

Geologic time scale, showing the interval covered in this chapter. Numbers indicate millions of years ago. P = Paleogene (Paleocene to Oligocene), N = Neogene (Miocene and Pliocene), and Q = Quaternary

Elkanah Billings, Canada's first professional paleontologist. REPRODUCED WITH THE PERMISSION OF NATURAL RESOURCES CANADA 2013, COURTESY OF THE GEOLOGICAL SURVEY OF CANADA.

A specimen of *Aspidella* from the Avalon Peninsula, Newfoundland. DOUG BOYCE.

That "palaeontological reference" was Elkanah Billings, a lawyer whose passion was fossils. In 1852, Billings gave up his law practice to become editor of the *Ottawa Citizen* and in 1856 launched a monthly periodical called *Canadian Naturalist and Geologist*. These endeavours caught the attention of William Logan, Director of the recently founded Geological Survey of Canada, who consequently appointed Billings as Survey paleontologist. It was natural, then, that Murray would send Reverend Harvey's finds to Billings, Canada's first professional paleontologist.

Billings examined the dime-sized discs and, in an 1872 publication, named them *Aspidella terranovica* (Newfoundland's little shield). He recognized that the rocks containing these fossils were older than trilobite-bearing Cambrian strata, even though Precambrian fossils were then practically unknown. In fact, many Victorian geologists wouldn't accept *Aspidella* as a fossil, preferring to consider it an inorganic artifact.

Aspidella remained a curiosity until 1967, when Shiva Balak Misra, a graduate student at Memorial University of Newfoundland, discovered the impressions of soft-

Ediacaran strata at Mistaken Point, Newfoundland. ALAN MORGAN.

FOUR BILLION YEARS AND COUNTING: CANADA'S GEOLOGICAL HERITAGE

Charniodiscus, a rangeomorph, from Mistaken Point, Newfoundland. BRIAN PRATT.

bodied animals, including *Charniodiscus*, at Mistaken Point, about 100 kilometres from St. John's. These fossils turned out to be even older than the *Aspidella* from St. John's. The Mistaken Point specimens, which we will discuss in more detail later in this chapter, were the first sign of a diverse fauna of complex Precambrian fossils in the New World. These finds were all the more dramatic because the fossils are large (centimetre to metre scale), and are abundant over broad surfaces that represent ancient sea floors.

Misra's published report on the Mistaken Point fossils, co-authored with his thesis supervisor Michael Anderson, provided convincing evidence that metazoan life abounded in late Neoproterozoic oceans. These fossils and assemblages of similar age around the world reflect an important transition for our planet. At that time, the Earth changed from a world harbouring predominantly microscopic life to one in which large multicellular organisms thrived. In the following pages, we will discover how life and the Earth evolved during this remarkable transition.

Setting the Scene

This chapter spans the interval from 750 to 444 million years ago. It includes part of the Cryogenian Period (from 850 to 630 million years ago), the Ediacaran Period (630 to 542 million years ago), the Cambrian Period (542 to 488 million years ago), and the Ordovician Period (488 to 444 million years ago). The Cryogenian and Ediacaran periods end the Neoproterozoic Era and the Cambrian and Ordovician begin the Paleozoic Era.

Until about 780 to 750 million years ago, the supercontinent Rodinia evidently remained largely intact. But then cracks began to appear. The continents that were to result from the breakup that followed are unfamiliar to us—any hints of our modern continents were still more than 500 million years in the future. Rodinia's offspring included Laurentia (most of present-day North America plus Greenland), Amazonia (which underlies

A

B

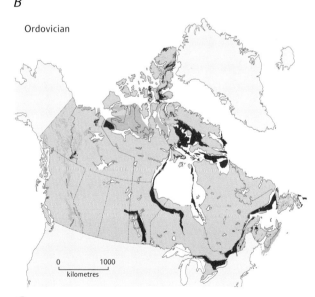

C General extent of Neoproterozoic (*A*), Cambrian (*B*), and Ordovician (*C*) rocks at the surface (beneath glacial deposits), onshore and offshore. The lighter shaded areas show either uncertainty or areas where rocks of the particular age have been confirmed but are intimately associated with rocks of other ages and the scale of the map does not allow us to show them separately. ADAPTED FROM WHEELER ET AL. (1996),.

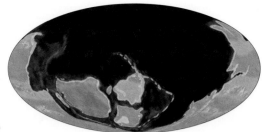

Global paleogeography 550 million years ago, during the Ediacaran. Colours as for previous figure.

Paleogeography of what was to become North America and adjacent regions in the late Ediacaran, 550 million years ago. Land is shown in brown, with shading showing topography. The lighter blue areas represent possible coastal or nearshore areas, darker blue represents deeper ocean waters, and black indicates trenches. Aspects of modern geography (including provincial and state boundaries) are shown for orientation.

much of present-day Brazil), Baltica (most of present-day Scandinavia, plus western Russia), Siberia, and South China. These continents-to-be would have been unrecognizable as individual entities within Rodinia, at least until it began to break up. How this breakup progressed and its impact on the planet's paleogeography is not clear. However, most interpretations show that what was to become Laurentia was located at high to mid-southern latitudes near the middle of Rodinia and was rotated 90° such that what is today eastern North America then lay to the south—or, as geologists say, to the paleosouth. Abutting Laurentia on its paleosouthern and paleoeastern sides were the future continents of Amazonia and Baltica, as well as a smaller plate that would become the Chukotka region of far-eastern Russia. About 600 to 570 million years ago, Amazonia, Baltica, and Chukotka drifted away, with new oceanic lithosphere forming the Iapetus Ocean to the paleosouth of Laurentia and the Ural Ocean to the paleoeast.

What continents lay on the paleonorthern (present western) side of Laurentia is unclear, though South China, Australia, and Siberia are all possibilities. Rifting started along Laurentia's future paleonorthern margin as early as 780 million years ago, but it is uncertain just when drifting began. Some geologists consider that an ocean, Protopanthalassa, began to develop between Laurentia and its paleonorthern neighbour by 750 million years ago and that it had become a wide ocean by the start of the Cambrian. This scenario is reflected in maps accompanying this chapter. Other geologists contend that drifting did not commence and Protopanthalassa did not form until the Cambrian.

All experts agree that by 520 million years ago Laurentia had broken free from its neighbours and had become an isolated continent at low southern to tropical paleolatitudes: during the Cambrian, its paleosouthern margin lay at about 20°S. Laurentia would remain at these latitudes throughout the early Paleozoic, and would retain its recumbent orientation until the Ordovician, when it began to rotate counterclockwise. Laurentia would also remain in isolation until the advent of major continental collisions on its paleosouthern (modern eastern) margin in the early Silurian. However, earlier convergence with smaller arcs and terranes had begun by the early Ordovician, initiating the Appalachian Orogen. A broad similarity of rocks and structures along the length of the Orogen in eastern Canada led to the designation in the 1970s of so-called zones, best defined in Newfoundland. These zones, from northwest to southeast, are the Humber (the paleosouthern continental margin of Laurentia), Dunnage (remnants of oceanic crust, volcanic arcs, and back-arc basins now collectively located southeast of the Humber Zone), Gander, Avalon, and Meguma. The last three zones equate roughly with three microcontinents (Ganderia, Avalonia, and Meguma) that will enter our story later in this chapter. Now that geologists know more about the evolution and paleogeography of the Appalachian Orogen, this zonal terminology is not as helpful as it once was, and for this reason we have chosen not to use it in this book.

Snowball or Slushball?

We begin our explorations of Canada's late Neoproterozoic to early Paleozoic history in the paleonorth—today's west. As noted, geologists debate the broad evolution of this margin-to-be of Laurentia. But the evidence shows that a stack of Neoproterozoic sedimentary and volcanic rocks accumulated from California to Alaska in a broad, elongate marine trough, which we'll call the Windermere Seaway, along what is now the eastern Cordillera. This succession, up to 8 kilometres thick, includes deep-water clastics and carbonates. Windermere rocks can be seen in roadcuts and natural exposures along the Yellowhead Highway between Jasper, Alberta, and Tête Jaune Cache, British Columbia. They are also exposed in road cuts near Lake Louise, Alberta. The Windermere rocks were deposited either in an extremely long-lived rift basin or a rift basin evolving into a passive margin, depending upon how the broader debate about the separation of paleonorthern Laurentia is resolved.

The Windermere succession includes tillites, interpreted as glacial deposits. Associated with the tillites are layers of fine-grained deposits with isolated cobbles and boulders, probably dropstones. Dropstones are chunks of rocky debris melted out of icebergs or ice shelves and dropped onto the sea floor. Modern examples have been found in sediments beneath Antarctica's Ross Ice Shelf. Some of the dropstones are flat-sided and striated just like boulders in modern tills that have been shaped and scratched by other rocks carried by the ice. Examples of beds with dropstones can be seen along Highway 3 west of Creston, British Columbia.

In the Windermere succession, at least two intervals, 740 to 670 and 635 to 600 million years old, contain tillites and dropstones. Some geologists have equated these intervals with global ice ages, the earlier Sturtian and the later Marinoan, from a comparison with similar-aged deposits elsewhere. Surprisingly, paleomagnetic data

Deep-water Neoproterozoic strata of the Windermere Seaway exposed at Cushing Creek, British Columbia. These strata, dominated by thick sandstone beds, are interpreted as turbidites.
BILL ARNOTT.

Outcrops in the headwater region of Castle Creek, British Columbia, expose a deep-sea channel that formed on the flank of the Windermere Seaway during the Neoproterozoic. The channel deposits are lighter-coloured and steeply tilted.
BILL ARNOTT.

Neoproterozoic strata with dropstones in a roadside exposure along Highway 3 near Creston, British Columbia.
HU GABRIELSE.

indicate that some of these glacial deposits formed within 10 degrees of latitude of the paleoequator. This and other evidence have led to the idea that the Sturtian and Marinoan glaciations were so severe that the planet was completely covered by ice—the Snowball Earth hypothesis.

If the Earth did indeed fall into periodic and overwhelming deep freezes in the Neoproterozoic, what triggered melting? The most likely explanation involves volcanoes, which would have continued to emit carbon dioxide. Extensive ice cover would have prevented this gas from being absorbed by oceans and used up during weathering processes, so it would have stayed in the atmosphere. Because carbon dioxide is a greenhouse gas, this would have led to warming and, at a certain threshold, rapid melting of the ice, followed by acid rain and intense chemical weathering of exposed land. The presence of so-called cap-carbonate deposits, which tend to lie directly above the tillites, supports this scenario. Once oceans were ice-free, excess atmospheric carbon dioxide would have combined with excess calcium in seawater to precipitate extensive carbonate deposits.

The idea of a Snowball Earth is intriguing, but some geologists are skeptical. We still have much to learn about the peculiar sequences of glacial sediments that were deposited at tropical latitudes during the late Neoproterozoic, and about the events they represent. The evidence seems to point to a scenario with tracts of open ocean in which organisms would be able to survive. Because survive they surely did!

Animals Appear

Although there is some evidence for multicellular organisms before 580 million years ago, life up to that time was dominated by prokaryotes and single-celled eukaryotes. Metazoans—multicellular animals with tissues—before this date are unknown. However, during the later Ediacaran, centimetre- to metre-scale fossils representing metazoans appeared. This assemblage is called the Ediacara biota and its remains are called Ediacara-type fossils. The Ediacara biota is known from more than 30 localities around the world, including Mistaken Point. All Ediacara-type organisms were soft-bodied; hard parts in the form of skeletons or shells did not appear until near the end of the Ediacaran. Most Ediacara-type fossils are preserved as impressions on bedding surfaces of sandstone or volcanic ash and consist of the remains of disc-shaped, segmented, and frond-like animals whose affinities have been hotly debated. Some of them may represent early versions of modern animals. Others appear to represent evolutionary dead ends.

The Mistaken Point fossils were preserved as impressions because repeated volcanic eruptions blanketed the living communities on the sea floor with ash. Today, weathering of these

Mountain peak formed of Cryogenian dolostone, Shale Lake area, Mackenzie Mountains, Northwest Territories. NOEL JAMES.

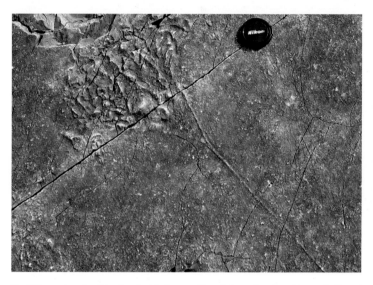

The Ediacara-type fossil *Ivesheadia lobata*, sometimes informally referred to as a "tethered pizza disc", from Daley's Cove, Newfoundland. ALAN MORGAN.

ash beds has exposed hundreds of bedding surfaces littered with specimens. In muddy sequences that lack ash beds, assemblages are less diverse and are dominated by *Aspidella*, the disc-like fossils that Elkanah Billings first described. About 75 percent of the Mistaken Point specimens are exquisitely preserved, centimetre-scale, branching organisms called rangeomorphs, whose smallest "twigs" are less than a tenth of a millimetre long. They include the 2-metre-long frond *Charnia wardi*, the oldest-known large eukaryote. The relationship of these fossils to living animals is still unknown. Rangeomorphs are less common in later Ediacaran rocks and disappeared before the end of the period.

In western Canada, the earliest Ediacara-type metazoan fossils occur in strata more than 1,000 metres above rocks that probably record the Marinoan Glaciation. The best-preserved Ediacaran fossils in this region occur in the Mackenzie Mountains of the Northwest Territories, in strata formed in deep Windermere Seaway waters. The most common and widespread fossil in

Ediacaran sea-floor scene based on fossils found at Mistaken Point, Newfoundland. Most of the organisms are rangeomorphs, an extinct early group of soft-bodied animals that lived attached to the sea bottom in this deep-water setting. The elegant, metre-long fronds waving gently in the currents are specimens of *Frondophyllas*. Smaller, leaf-like forms are specimens of *Charniodiscus* and *Charnia*. The spindle-shaped rangeomorph *Fractofusus* and small dendritic forms, *Primocandelabrum*, somewhat resembling bonsai trees formed the basal tier in the Mistaken Point ecosystem. This diorama can be viewed at the Oklahoma Museum of Natural History. DIORAMA BY CHASE STUDIO, INC., USA; ALL RIGHTS RESERVED.

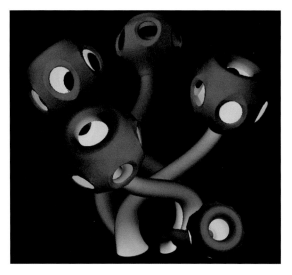

Reconstruction of *Namacalathus*, one of the first animals to produce a skeleton formed of calcium carbonate, but of the mineral aragonite rather than the more usual skeletal mineral calcite. The stalk, which attached the animal to the sea floor, is about 2.5 centimetres long. The cup can be up to 2.5 centimetres wide but is usually less than 1 centimetre. *Namacalathus* occurs in the late Ediacaran rocks of British Columbia. The colours shown are not intended to be a true reflection of the colour of *Namacalathus*. FROM GROTZINGER ET AL. (2000); REPRODUCED WITH PERMISSION OF WES WATTERS, JOHN GROTZINGER, ANDREW KNOLL, AND THE JOURNAL *PALEOBIOLOGY*.

Ediacaran rocks of western Canada is *Aspidella*. Some specimens of *Aspidella* have stems attached to them, supporting the current view that this fossil represents the base of frond-like organisms.

The Cambrian Explosion

The appearance of the Ediacara biota 580 million years ago was the opening salvo of one of the most remarkable episodes in the history of life—an incredible diversification that continued into the Cambrian. As we've noted, before this episode began, few organisms would have been visible to the naked eye. By about 520 million years ago, the fossil record shows that the ancestors of most major types of animal alive today were living in the Cambrian seas. This geologically short interval, when animals evolved at dizzying rates (geologically speaking), culminated during the earliest Cambrian in what has been called the Cambrian Explosion.

Events in the diversification of animal life during the latest Neoproterozoic and early Cambrian unfolded globally in a specific order, and evidence for most of the stages are found somewhere in Canada. Conclusive evidence of mobile animals suddenly appeared worldwide about 555 million years ago, in the form of trace fossils, which represent the development of bilaterians—animals with a head end that allowed them to move forward. Most of the sessile organisms (those attached to the sea floor) of the Ediacaran biota became extinct at or near the end of the Ediacaran Period. We don't know whether bilaterians caused that demise, but immobile mouthless organisms would probably have been easy prey for newly evolved wormy wanderers grazing on the sea floor.

Then, about 550 million years ago, came the earliest-known animal skeletons in the form of *Cloudina* and *Namacalathus*. Numerous examples of these fossils are found in latest Ediacaran carbonate rocks in southeastern British Columbia. Specimens of *Cloudina* are 1 to 5 centimetres long and consist of

Trace fossil known as *Psammichnites gigas* from earliest Cambrian rocks near St. Martins, New Brunswick. The animal that made this trace must have been a soft-bodied creature similar to a marine slug. HEINZ WIELE, COURTESY OF THE ATLANTIC GEOSCIENCE SOCIETY; SPECIMEN COURTESY OF THE NEW BRUNSWICK MUSEUM.

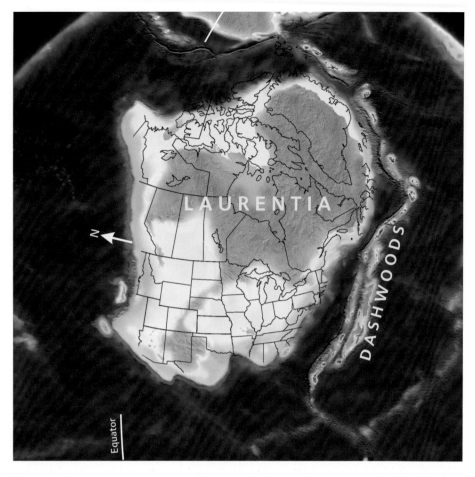

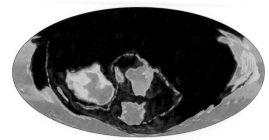

Global paleogeography 500 million years ago, during the Cambrian. Colours as for previous figure.

Paleogeography of what was to become North America and adjacent regions in the Cambrian, 500 million years ago. Land is shown in brown, with shading showing topography. The lighter blue areas represent possible coastal or nearshore areas, darker blue represents deeper ocean waters, and black indicates trenches. Aspects of modern geography are shown for orientation.

ing the Mistaken Point fossils in Newfoundland record such a transition. This increase alone may have caused metazoan evolution to accelerate and animals generally to have become larger. New and revolutionary morphological features that appeared immediately preceding and during the Cambrian Explosion included heads, eyes, mouths, teeth, segments, and the first appendages such as legs and antennae. A combination of any of these evolutionary steps would have had a major impact on predatory and grazing animals alike, and could have set off a chain reaction. One legacy of such an "arms race" might have been protective and readily preserved shells—the traditional icons of the Cambrian Explosion.

nested, gently curved cups; those of *Namacalathus* look like tiny lanterns that were attached to stalks. *Cloudina* and *Namacalathus* were possibly related to worms, sponges, or cnidarians.

The increase in abundance and variety of trace fossils in the very latest Ediacaran and earliest Cambrian rocks reflects ever more complex development and behaviour. The earliest occurrence of the worm burrow *Treptichnus* at Fortune Head, on Newfoundland's Burin Peninsula, has been designated as the marker for the base of the Cambrian (Chapter 3), and is about 542 million years old. Small shelly fossils (the "small shelly fauna") appear just above the base of the Cambrian. Together with trace fossils, these are the only remnants of life preserved in rocks representing several million years of earliest Cambrian time. After that came the earliest brachiopods and archaeocyathans, followed a few million years later by the first trilobites, echinoderms, and mollusks.

Why did metazoan evolution speed up so much in the late Ediacaran and early Cambrian? One possibility is the increase in oxygen levels from about 15 to more than 50 percent of their present values during the late Neoproterozoic. Geochemical analyses of sedimentary rocks immediately below those contain-

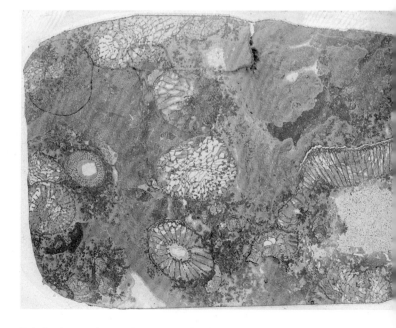

Early Cambrian archaeocyathans encrusted with dark-grey calcite precipitated through the activity of microbes and set in a reddish mudstone, Pointe Amour, southern Labrador. BRIAN PRATT.

Cambrian Life

The middle Cambrian trilobite *Paradoxides davidis*, from Manuels River, Newfoundland. RICCARDO LEVI-SETTI.

Part of the filtration feeding structure of a branchiopod crustacean (not to be confused with brachiopods), preserved in middle Cambrian rock retrieved from a subsurface core in Saskatchewan. The specimen is about 0.15 millimetres long. NICK BUTTERFIELD.

Cambrian organisms included representatives of groups living today, but many differed in fundamental ways from their modern cousins and some had interesting transitional features. For example, Cambrian mollusks were small, and groups that were later to become bivalves, gastropods, and cephalopods were poorly differentiated until late in the period. Similarly, echinoderms were represented by a remarkable variety of exotic forms somewhat similar to later crinoids. The range of Cambrian invertebrates is demonstrated by the amazing faunas of the Burgess Shale (Box 6) and similar deposits elsewhere, which provide a sampling of the soft-bodied creatures that lived between 525 and 505 million years ago. In post-Explosion Cambrian marine deposits in which only hard parts are preserved, trilobites are the most conspicuous fossils. Trilobites, which evolved rapidly and hence are important Cambrian timekeepers, mostly lived on the sea floor and left characteristic scratching, digging, and walking traces. Brachiopods were also common, but primitive and small; one early Cambrian brachiopod, *Lingulella*, is very similar to the "living fossil" *Lingula*. The reef-building archaeocyathans were common in the early Cambrian, but became extinct about 515 million years ago. Conodonts, ostracods (small, shelled arthropods still common in today's seas), and organic-walled acritarchs (microscopic, single-celled, commonly spiny fossils of unknown affinity) round out the Cambrian menagerie.

Trackways in marine rocks include some large and very odd forms. The most unexpected Cambrian trace fossils were found in a building-stone quarry north of Kingston, Ontario. Rocks in the quarry represent large windblown (eolian) dunes about 500 million years old. Terrestrial invertebrates supposedly did not appear for tens of millions of years, so when trackways were found in these rocks in 1983 it was quite a surprise. The tracks, referred to as *Protichnites*, appear to have been made by arthropods that left behind the impressions of eight pairs of legs, as well as a central groove representing the animal's tail or midbody. Later research has confirmed that the tracks were indeed made in the open air on a windblown surface. The nature of the trackways reveals that the animals walked like swimming creatures. We don't know why the arthropods that made these tracks took a probably brief and temporary excursion out of the sea. Regardless, the specimens from the Kingston-area quarry remain by far the earliest-known definitive evidence anywhere of animal activity on land.

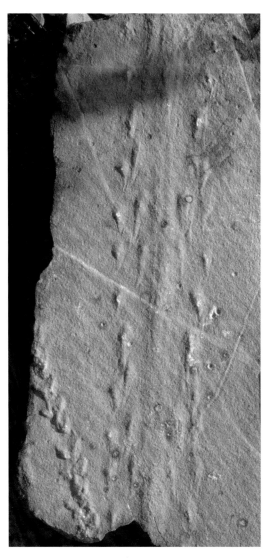

The Cambrian trackway *Protichnites* from near Kingston, Ontario. ROBERT MACNAUGHTON.

Flanking Iapetus

We now turn to Laurentia's paleosouthern margin, remnants of which are found in present-day eastern Canada. As noted earlier, about 600 million years ago rifting started to separate Laurentia from Amazonia and Baltica, leading to the formation of the Iapetus Ocean, from which the oldest preserved oceanic lithosphere is about 570 million years old.

As Laurentia broke away from Amazonia, crustal stretching produced a series of rifts on Laurentia. Several of these underlay modern valleys such as those of the St. Lawrence and Ottawa rivers, and so have had a fundamental influence on eastern Canada's modern geography. Many of the rifts flanking Iapetus were roughly parallel to the continental margin, and at least one fragment of Laurentia's paleosouthern edge broke away and drifted into the ocean. This fragment was the Dashwoods Microcontinent, a sliver that separated about 550 million years ago and was in some ways similar to present-day New Zealand, which has drifted away from Australia-Antarctica over the past 90 million years. The Dashwoods Microcontinent was separated from Laurentia by an arm of Iapetus called the Taconic Seaway.

For the first time in the planet's history, thanks to the Cambrian Explosion, shelled organisms contributed to reef structures. Early Cambrian reefs were metre-high submarine mounds dominated by archaeocyathans. Excellent examples of early Cambrian limestones with such reefs can be seen in southern Labrador and western Newfoundland. After an interlude when clastic deposition dominated, carbonate deposition

Resistant quartz-rich Cambrian-Ordovician sandstones form distinctive hills and islands, such as the one in the distance, near Notre-Dame-du-Portage along the South Shore of the St. Lawrence River in Quebec. The sandstone was deposited in deep marine channels cut into the continental slope of Laurentia's passive margin. In the foreground is a glacial boulder that is probably of the same rock. ROB FENSOME.

returned in the middle Cambrian to form a carbonate bank that extended from the northwestern tip of Newfoundland to the southeastern United States. This bank reminds geologists so much of today's Great Barrier Reef off Australia that they call it the Great American Bank. As is normal for reef complexes, the Bank was characterized by a landward platform-like area and a steeper offshore slope.

While the Dashwoods Microcontinent and the Great American Bank were developing on the paleosouthern margin of Laurentia in low latitudes, three fragments of lithosphere destined to contribute to Atlantic Canada were evolving near the other shore of Iapetus. These three exotic terranes are Ganderia, Avalonia, and Meguma. Each is partly underlain by continental lithosphere and hence, like Dashwoods, can be considered a microcontinent. In contrast to the Laurentian margin, carbonate deposits, which are indicative of warmer latitudes, are rare on the three terranes. This suggests that they were located at higher paleosouthern latitudes, especially during the Cambrian and Ordovician.

Today, remnants of Ganderia underlie parts of New England, Atlantic Canada, and western Europe. The European vestiges are now separated from their North American counterparts by the North Atlantic Ocean, which opened relatively recently, in the Mesozoic. Ganderia likely rifted from Amazonia about 500 million years ago, after Amazonia

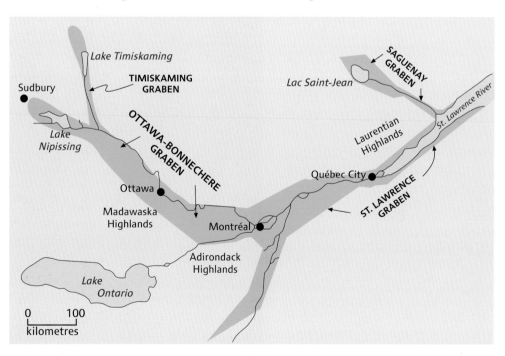

As a prelude to the opening of the Iapetus Ocean, crustal stretching and the resulting normal faulting produced a series of rifts (or grabens) near what was to become the paleosouthern margin of Laurentia. Several of these late Neoproterozoic and earliest Paleozoic rifts underlay modern valleys, such as those of the St. Lawrence and Ottawa rivers. The modern southeastern margin of the St. Lawrence Graben was subsequently overridden by the thrust-and-fold belt of the Taconic Orogeny, as we will discover later in the chapter. ADAPTED FROM VARIOUS SOURCES.

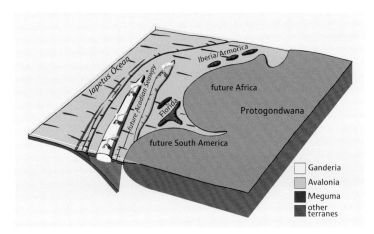

Initial stages of drifting of the three terranes—Ganderia, Avalonia, and Meguma—across the Iapetus Ocean about 490 million years ago.

Vertically tilted, early to middle Cambrian turbidites at L'Islet-sur-Mer on the South Shore of the St. Lawrence River, Quebec. The turbidites accumulated on the passive margin off Laurentia. ROB FENSOME.

had amalgamated with several other former pieces of Rodinia to form a continent we call Protogondwana. Remnants of later Cambrian magmatic arcs and back-arc basins formed on Ganderia's leading edge (the margin that was to collide with Laurentia) can be seen along the Gander River in east-central Newfoundland.

Avalonia is named after the Avalon Peninsula in Newfoundland, and so encompasses the Mistaken Point succession. Today, Avalonian rocks form a belt generally southeast of Ganderian rocks, extending from the Boston area of Massachusetts, under the Gulf of Maine, through the Maritime Provinces and eastern Newfoundland, to northwestern Europe. Avalonian rocks are prominent in southeastern Cape Breton Island, as around the lighthouse at Louisbourg, and in New Brunswick where they underlie Fundy National Park of Canada. Avalonia is dominated by Neoproterozoic rocks representing magmatic arcs and oceanic plateaus similar to those in the present-day Tonga-Kermadec-New Zealand region in the southwestern Pacific. Paleomagnetic evidence suggests that Avalonia formed near the South Pole and may have been originally attached somewhere along the

Ediacaran diorite intruded by late Cambrian granite at Middle Head, Cape Breton Highlands National Park of Canada, Nova Scotia. These rocks were part of Ganderia. P. ST. JACQUES, COPYRIGHT PARKS CANADA.

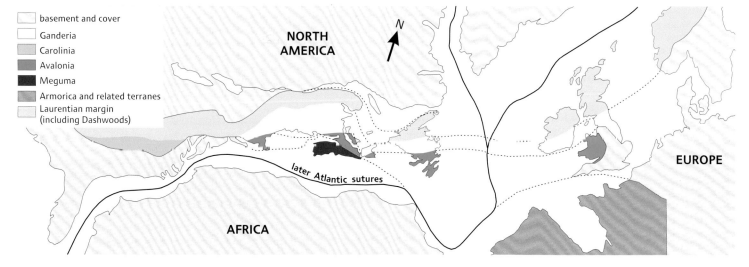

Geological collage of terranes as distributed today in North Atlantic borderlands, but with later oceanic lithosphere removed to better represent their relationships in the late Paleozoic. This distribution has been complicated by latest- and post-Paleozoic tectonic activity, such as movement along strike-slip faults, so the match in not perfect. And we do not have a full understanding of the extension of the terranes beneath the cover of young rocks. ADAPTED FROM VAN STAAL AND BARR (2012).

CHAPTER 7 · SOUTHERN SOJOURN

Neoproterozoic marine sedimentary rocks of Ganderia, exposed at Swallowtail Head, Grand Manan Island, New Brunswick. FRANCIS KELLY, COURTESY OF FISHERIES AND OCEANS CANADA.

Early Ordovician metasedimentary rocks of the Meguma Terrane at Blue Rocks, near Lunenburg, Nova Scotia. The original bedding is reflected in the lighter and darker bands. These rocks were tilted and metamorphosed during the Neoacadian Orogeny (Chapter 8). ROB FENSOME.

Amazonian or African margin of Protogondwana, from which it started to rift away about 540 million years ago.

Throughout the Cambrian and Ordovician periods, Ganderia and Avalonia migrated slowly northward toward Laurentia, separated from each other by an arm of Iapetus known as the Acadian Seaway. Behind Avalonia was a new and growing ocean, the Rheic Ocean. Within that ocean arose the third terrane whose remnants are now found in Atlantic Canada—Meguma. Located to the southeast of Avalonia, Meguma outcrops only in southern mainland Nova Scotia, but underlies much of the continental shelf from off Cape Cod in Massachusetts to the southernmost Grand Banks of Newfoundland. Cambrian and Ordovician Meguma rocks are dominated by a clastic succession likely deposited in an evolving rift between Protogondwana and Avalonia. Later, possibly about 480 million years ago, Meguma detached fully from Protogondwana and became a separate microcontinent within the Rheic Ocean.

A Third Margin

As the region that was to become Atlantic Canada was evolving during the late Neoproterozoic and Cambrian, so was the Arctic. Marked changes are reflected in rocks of northwestern Canada about 780 to 750 million years ago. In the western part of the Amundsen-Mackenzie Mountains Basin, coarse-grained sedimentary rocks and mafic intrusions reflect a new rifting event. Farther east in the Basin, over the sedimentary succession is draped a blanket of 720-million-year-old basalts, originally perhaps more than 2 kilometres thick. These basalts appear to have resulted from a large igneous episode that injected massive mafic sills into the Basin's sedimentary succession, as well as the mafic dykes of the Franklin Dyke Swarm, which extends 2,000 kilometres from the Canadian mainland near the Alaska border to Baffin Island. All of these features reflect an episode of crustal stretching, an early regional harbinger of the breakup of Rodinia. Sea-floor spreading began about 600 million years ago as Laurentia's paleoeastern (present northern) margin started to drift from its continental neighbours. The continents became separated by the new Ural Ocean, which probably had a broad connection with Iapetus to the paleosouth. It may also have merged with the Windermere Seaway and/or Protopanthalassa in the paleonorth.

Laurentia's paleoeastern edge was thus initially a passive margin, which lasted from about 600 million years ago to about 480 million years ago. This margin had a broad, shallow, conti-

The dark band seen in this aerial view of Brock River, Tuktut Nogait National Park of Canada, Northwest Territories, is a mafic sill associated with the 720-million-year-old Franklin Dyke Swarm. The sill cuts through later Neoproterozoic sediments of the Amundsen-Mackenzie Mountains Basin. HANS WIELENS.

in the Grenville Orogen, which extended into present-day Scandinavia. Researchers believe that this northern tip represents part of a microcontinent called Pearya. During the Cambrian and early Ordovician, Pearya was probably one of a number of microcontinents lying to the paleoeast of Laurentia in the Ural Ocean, which was reaching its maximum size. Other continental fragments within the Ural Ocean likely included parts of western Svalbard, Arctic Alaska, and Chukotka. Some time between about 480 and 445 million years ago, a subduction zone and accompanying magmatic arc developed within the Ural Ocean paleoeast of Laurentia, which would eventually lead to the accretion of Pearya during the later Paleozoic.

nental shelf in which sediments accumulated, thickening oceanward. Carbonates accumulated over wide areas of the shelf and are now preserved as flat-lying strata over much of the southern Arctic Islands. Conditions may have been quite similar to those of the modern Persian Gulf. At times, the sea was so shallow that circulation was restricted. This, in combination with the tropical sun, led to the evaporation of water faster than it could be replenished from the open ocean, leaving extensive deposits of evaporites. Beyond the shelf edge, deeper water environments prevailed, represented today by shale and chert deposits from northern Ellesmere Island to northwestern Melville Island.

What was going on out in the middle of the Ural Ocean during the early Paleozoic? The geology of the northernmost tip of Ellesmere Island gives us a clue. Rocks there are quite different from those elsewhere in the Arctic Islands. Some are about 1,000 million years old and show evidence that seems to place them

Rocks northeast of Ayles Fiord at the northern tip of Ellesmere Island, Nunavut, record the geological history of Pearya, a continental fragment that drifted within the Ural Ocean during Cambrian and Ordovician times. This outcrop shows an anticline, with latest Neoproterozoic or Cambrian schist and quartzite overlying Mesoproterozoic gneiss. HANS TRETTIN.

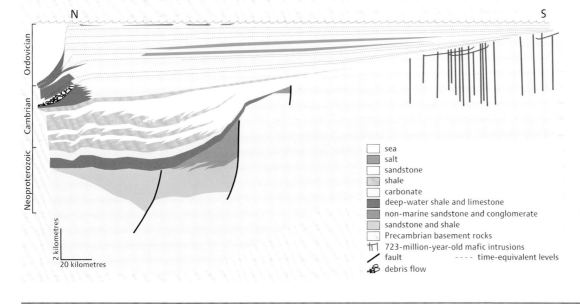

Section from north to south through the Neoproterozoic to Ordovician rocks of Ellesmere Island, reconstructed to show how they would have appeared at the end of the Ordovician. Notice that older rocks were deposited in less extensive, fault-bounded rift basins and younger rocks are distributed more broadly over a passive margin, with strata thickening oceanward. This pattern is similar to that along the early Paleozoic Cordilleran margin and the Atlantic margin off eastern Canada today.

CHAPTER 7 • SOUTHERN SOJOURN

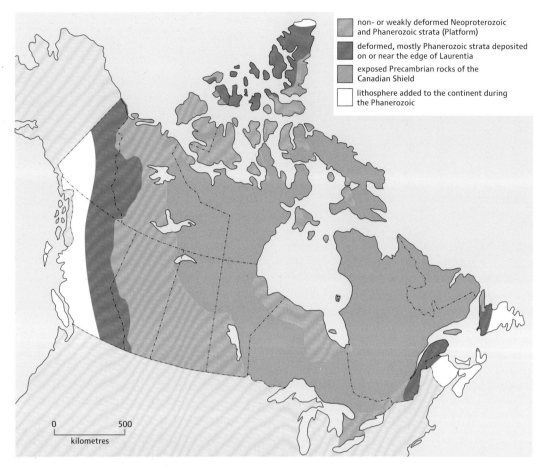

non- or weakly deformed Neoproterozoic and Phanerozoic strata (Platform)

deformed, mostly Phanerozoic strata deposited on or near the edge of Laurentia

exposed Precambrian rocks of the Canadian Shield

lithosphere added to the continent during the Phanerozoic

The modern geological structure of Canada in part reflects early Paleozoic geography, with a large part of the former continent of Laurentia central to the story. The white areas to the far west, far east, and at the tip of Ellesmere Island in the far north are the only parts of Canada not underlain by the former continent. These areas are largely composed of terranes that have accreted during the past 550 million years, crumpling the edges of the former Laurentian continent in the process, as shown by the dark green areas. Light green represents areas of relatively undeformed strata formed during the past 550 million years. Pink represents the Canadian Shield, the exposed core of Laurentia, which was once also covered by a blanket of Phanerozoic sediment. ADAPTED FROM VARIOUS SOURCES.

The Core of Laurentia

We have so far explored three margins of Laurentia. The fourth, paleowestern, margin lies beneath parts of the southern United States and northern Mexico. At the beginning of the Cambrian, 542 million years ago, the deeply eroded roots of Archean cratons such as the Slave and the Superior, and Proterozoic orogens such as the Trans-Hudson and the Grenville, were exposed at Laurentia's surface. These ancient cratons and orogens collectively formed the continent's "bedrock" or basement. (To geologists, the term basement usually denotes crystalline Precambrian rocks, commonly below a Phanerozoic sedimentary succession; the term is also used by petroleum geologists to mean deep rocks not considered to contain oil or gas.) Today, the former extent of Laurentia stretches from within the Appalachians to within the Cordillera, and from Arizona to the northern Arctic Islands. Part of the Laurentian basement, now exposed because the younger sedimentary-rock cover has been stripped by erosion, forms the Canadian Shield. Erosion continues today, exposing more of the basement and thus enlarging the Canadian Shield.

At different times since the Cambrian, Laurentia's margins have been buffeted by the colossal tectonic forces that gave rise to the marginal orogens (Appalachian, Cordilleran, and Innuitian, as we will discover). However, the continental interior has been relatively stable since the Grenvillian Orogeny, as reflected by the mostly flat-lying or gently folded Phanerozoic sedimentary rocks between the Shield and the younger marginal orogens. This stable area is called the North American Continental Platform (or simply, the Platform). Most Platform deposits underlie the plains and lowlands that are home to the majority of Canada's people, cities, and farms. The largest expanse, known as the Western Platform, lies beneath the Western Interior Plains, an area southwest of the Shield. Platform regions underlie the Hudson Bay area and the Arctic Plains. South of the Shield are other, smaller Platform areas—the heavily populated Great Lakes Lowlands and the St. Lawrence Lowlands. The oldest deposits covering the basement in these eastern regions are latest Ediacaran to Cambrian quartz-rich sandstones. These rocks, which reflect erosion of the continental core, are mainly sandstones of shallow marine and intertidal origin

Ordovician strata overlying crystalline rocks of the Grenville Orogen, Trenton, Ontario. PIERRE JUTRAS.

Cambrian carbonates form the precipitous cliffs of Castle Mountain in Alberta. The snow-clad, more thinly bedded sequence on the lower slopes is mainly composed of Cambrian quartz-rich sandstone. RON GARNETT / AIRSCAPES.CA.

This aerial view, 40 kilometres southeast of Mount Field, British Columbia, shows black and rusty-weathering Cambrian calcareous shale abutting Cambrian carbonate of the Cathedral Escarpment. Fault movements more than 500 million years ago created the escarpment and, in places, caused the edge of the paleonorthern continental shelf of Laurentia to collapse in huge underwater landslides. BRIAN PRATT.

with evidence for occasional exposure above water, as shown by the odd subaerial trackways of the Kingston area.

Driving over the Edge

Laurentia's paleonorthern margin can perhaps be appreciated best by taking a scenic drive along the Trans-Canada Highway from Calgary, Alberta, to Revelstoke, British Columbia. On this journey we travel westward successively through the Foothills, the Front Ranges, and the Main Ranges of the Rocky Mountains, across the Southern Rocky Mountain Trench and into the Columbia Mountains. The early Paleozoic rocks deposited on this passive margin are now deformed by slicing thrust faults and sweeping folds that formed 155 to 60 million years ago during tectonic upheavals that we will highlight in Chapter 9. This relatively recent episode of deformation moved some Paleozoic rocks 200 kilometres or more to the northeast from the site of their original deposition. (In this section, for simplicity, we will use modern directions.)

The boundary between continental and oceanic lithosphere in the early Cambrian was somewhere west of the present-day Columbia, Omineca, and Cassiar mountains, which lie immediately west of the Rocky Mountains (Box 5). This margin was at paleolatitudes of about 10 to 15°N. As in parts of Ontario and Quebec, the oldest rocks above the basement are quartz-rich latest Ediacaran to earliest Cambrian sandstones made from sediment eroded from the Laurentian interior. Above this sandstone blanket are Cambrian to Silurian sedimentary rocks—mainly carbonates and shales—that record east-to-west changes from shallow- to deep-sea environments. These changes can be seen in the rocks as we drive westward.

On the Western Platform in the vicinity of Calgary and in the Foothills to the west of the city, the entire Paleozoic succession is just over 1 kilometre thick and buried beneath about 3 kilometres of mainly Mesozoic clastics. This relatively thin Paleozoic succession underlies the western Plains, Foothills, and Front Ranges, extending westward as far as Banff. Rocks of this succession form the towering cliffs of the Rocky Mountain Front, about 75 kilometres west of Calgary. There, hard, cliff-forming Cambrian carbonate has been thrust over soft, readily eroded Cretaceous shale.

A far thicker, more complete Paleozoic succession occurs in the Main Ranges west of Banff. For example, Cambrian strata are 5,000 metres thick near Lake Louise, compared to 600 metres around Banff. Although the early Paleozoic succession is much thicker in the Main Ranges, it is still predominantly composed of shallow-water carbonates, as seen in the fortress-like cliffs of Castle Mountain, about halfway between Banff and Lake Louise. Clearly, deposition kept pace with subsidence of the continental margin as it became increasingly distant from Protopanthalassa's spreading ridge. An analogous situation has prevailed off the margin of Nova Scotia for the past 150 million years or more. There the stable inshore LaHave Platform, essentially a submerged continuation of Nova Scotian lithosphere, has a relatively thin sedimentary cover, whereas farther offshore in the Scotian Basin, much thicker deposits have accumulated (chapters 9–10). The succession in the Rocky Mountains is also reminiscent of the early Paleozoic, thin to thick, onshore to offshore passive margin sequence that we explored in the Arctic earlier in the chapter.

Earliest Cambrian quartz-rich sandstone beds in a road cut on the Trans-Canada Highway west of Golden, British Columbia. These beds overlie the Purcell Thrust, which carried Neoproterozoic and mostly non-calcareous, deep-water early Paleozoic strata eastward over early Paleozoic calcareous shale and carbonate. JIM MONGER.

Folded Cambrian calcareous shale in the Solitude Range of the Columbia Mountains, about 100 kilometres northwest of Golden, British Columbia. LEN GAL.

The inspiring scenery around Lake Louise and Moraine Lake is carved largely from early Cambrian quartz-rich sandstone and overlying middle Cambrian carbonate. A little farther west, near Field in British Columbia, the Cambrian rocks change abruptly from shallow-water carbonate to deeper-water calcareous shale. The boundary between the two environments was marked by a break in slope in the Cambrian sea floor known as the Kicking Horse Rim, which can be traced along much of the length of the Main Ranges. Locally the Rim is represented by the Cathedral Escarpment, which formed part of a 70-metre-high submerged carbonate cliff. Preserved in the shale below this ancient cliff is the Burgess Shale fauna (Box 6). The shale-dominated succession continues westward to the town of Golden, in the Southern Rocky Mountain Trench, and for a short way up the western wall of the Trench beyond Golden. The Southern Rocky Mountain Trench is a great valley that marks the western boundary of the Rocky Mountains. To the west of it are the Columbia Mountains.

Farther west the story is complicated yet again by jumbling caused by later faulting. Instead of heading into deposits of the great ocean Protopanthalassa, as would be expected from the onshore to offshore trajectory that we have been following, on the Southern Rocky Mountain Trench's western wall we encounter a major fault called the Purcell Thrust. This fault carries Neoproterozoic Windermere strata and Cambrian shallow-water quartz-rich sandstone over early Paleozoic rocks dominated by deep-water shale. The Cambrian sandstones above the Purcell Thrust form the rugged peaks just east of the Rogers Pass summit in Glacier National Park of Canada. Above the sandstones are small discontinuous limestone bodies, mostly formed as archaeocyathan reefs; these reefs are relics of the outermost margins of the early Cambrian continental shelf. During the later Cambrian and the Ordovician, the shelf and its reefs were buried by black, deep-water organic-rich mud, now metamorphosed to argillite. This rock can be seen in outcrops on the Trans-Canada Highway's long descent west of Rogers Pass. As we head farther west along the Highway we finally leave former Laurentian lithosphere.

Collision Course Act One

In contrast to the relatively quiet times on Laurentia's paleonorthern passive margin, things were heating up "down paleosouth" (today's east). Subduction zones with magmatic arcs developed within Iapetus as early as the Cambrian, especially in the Taconic Seaway and around the Dashwoods Microcontinent. As intervening ocean floor began to be consumed in these subduction zones, several of the arcs, as well as the Microcontinent, began to converge and collide with Laurentia's paleosouthern margin in the early Ordovician, about 480 million years ago. These events are recorded in the rocks as the Taconic Orogeny, the first in a series of mountain-building events that collectively formed the Appalachian Orogen. Taconic tectonic uplift exposed the Great American Bank above water, destroying its marine communities.

During the Taconic Orogeny, former passive-margin strata were pushed toward and over the continental margin in a thrust-and-fold belt whose deformation reached as far inland as present-day Québec City. The limit of that deformation is recognized as Logan's Line (Box 10). The onset of Taconic collision between the Dashwoods Microcontinent and Laurentia would have also caused seismic activity, which produced debris flows and turbidity currents that transported material from

Early Ordovician breccia at Cow Head, Newfoundland, originally a debris flow at the edge of a carbonate bank on the Laurentian margin. The debris flow may have been triggered by earthquakes associated with the Taconic Orogeny. MICHAEL BURZYNSKI.

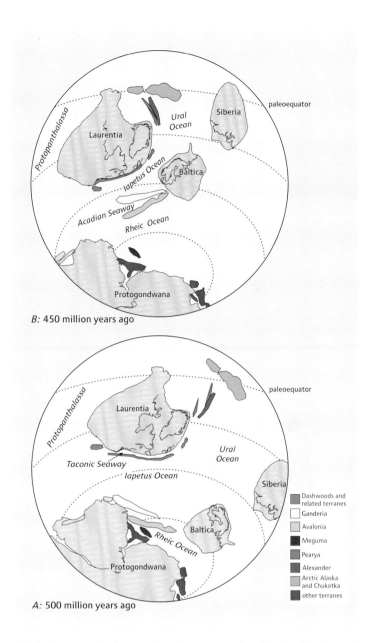

Global paleogeography, from a paleosouthern perspective, of the late Cambrian, 500 million years ago (A), and of the late Ordovician, 450 million years ago (B). The northern part of Laurentia straddles the paleoequator in both figures. In A, the Dashwoods Microcontinent is immediately off southern Laurentia, but a narrow remnant of the Taconic Seaway remains. Ganderia and Avalonia lie immediately to the north of Protogondwana in high southern latitudes. The two terranes are separated from Laurentia by the broad Iapetus Ocean and from Protogondwana by the developing Rheic Ocean. Meguma and adjacent terranes are still close to or attached to Protogondwana. Protopanthalassa, the Ural Ocean, and Iapetus surround Laurentia. In B, the Taconic Seaway has closed and the Dashwoods Microcontinent has accreted to the southern margin of Laurentia. Fast approaching Laurentia from the south are Ganderia and Avalonia, reflecting the rapid closing of Iapetus and the widening Rheic Ocean. Meguma, which now forms part of Nova Scotia, and adjacent terranes, which now form parts of western Europe, are still close to or attached to Protogondwana. Also shown in A and B are the possible positions of Pearya and the Alexander Terrane. ADAPTED FROM VARIOUS SOURCES.

shallow to deep water. Impressive limestone breccias made up of enormous angular fragments, some hundreds of cubic metres in volume, accumulated at the toe of the continental slope, including the debris-flow deposit preserved at Cow Head in western Newfoundland. The Taconic Orogeny was also responsible for development of the ophiolite suites of Thetford Mines in Quebec and the Bay of Islands area in western Newfoundland, part of which is exposed in Gros Morne National Park of Canada.

Parts of the Laurentian continental edge were pushed down beneath the load of the advancing thrust-and-fold belt and of sediments eroded from the emerging mountains. The depressed areas formed foreland basins that were filled with clastics eroded from both the old continent and the rising mountains. During the middle to late Ordovician, these depressed areas included the Anticosti Basin in what is now southeastern Quebec and the Appalachian Basin (not to be confused with, and not within, the Appalachian Orogen). The Appalachian Basin lies beneath lakes Ontario and Erie and extends through New York State to Kentucky. Subsidence in these basins was rapid, especially closer to the mountains, and seas commonly flooded in. Initially, basin sediments were shallow marine carbonates, with some local metazoan reefs, examples of which can be seen in the Mingan Archipelago on the North Shore of the St. Lawrence, within strata deposited in the Anticosti Basin. But as the mountains shed more sediment and the basins sub-

Tilted 550-million-year-old pillow lavas and overlying strata at Green Gardens, near Gros Morne National Park of Canada, western Newfoundland. The tilting occurred during the Taconic Orogeny. MICHAEL BURZYNSKI.

This sequence of early Cambrian turbidites, at chutes-de-la-Chaudière, Charny, Quebec, was deposited on the continental slope off Laurentia. Thin mudstone layers separate sandstone beds. The beds were thrust toward the continent and tilted during the Taconic Orogeny. Note the person at bottom left for scale. ROB FENSOME.

sided, deeper-water sediments accumulated. From the middle Ordovician, the main clastic sediment source also gradually switched from the continental interior to the paleonorth to the new Taconic Mountains to the paleosouth. Regardless of the source of sediment, life thrived in the sea, and late Ordovician sedimentary rocks of present-day southwestern Quebec and southern Ontario abound with shelly fossils.

The Iapetus Ocean continued to close during the Ordovician, with subduction zones and magmatic arcs associated with both of its margins and the intervening microcontinents. About 450 million years ago, a magmatic arc ahead of Ganderia converged with what was formerly the Dashwoods Microcontinent but by then was part of Laurentia's paleosouthern margin. Meanwhile, another part of Ganderia was docking with Baltica. These two collisions closed much of the Iapetus Ocean, leaving remnants such as the Acadian Seaway between Avalonia and Ganderia.

Between Ganderia and its leading magmatic arc, now accreted to Laurentia, was a large back-arc basin, the Tetagouche-Exploits Basin, which survived as an echo of now-closed Iapetus for another 20 to 30 million years. Today, remnants of this basin underlie large parts of northern New Brunswick and central Newfoundland and are home to important metallic ore deposits such as those in the Bathurst area of New Brunswick (Chapter 12).

Ordovician limestone eroded into pillars at La Grandelle, Mingan Archipelago, Quebec. This limestone was deposited in the Anticosti Basin. P. ST. JACQUES, COPYRIGHT PARKS CANADA.

Early to middle Ordovician metavolcanic rocks of the Tetagouche-Exploits Basin, on the banks of the Nepisiguit River, south of Bathurst, New Brunswick. ROB FENSOME.

More Momentous Steps for Life

The earliest Ordovician heralded a new and perhaps even greater diversification of marine metazoan life than had occurred in the Cambrian. Brachiopods proliferated. Bottom-dwelling mollusks such as snails and clams developed new approaches to living on and within the sea floor. The nautiloids rapidly rose to prominence, though most had straight shells in contrast to the coiled shells of modern nautiloids. Some middle Ordovician nautiloids were up to 4 metres long, making them formidable sea-going predators. But despite the new competition, trilobites continued to thrive and diversify, as did echinoderms, especially the crinoids. As the Ordovician progressed, other groups with little or no Cambrian record became firmly established, particularly on shallow carbonate banks where the first skeletal reefs since the demise of the archaeocyathans developed. Stromatoporoids built calcareous mounds and sheets, accompanied by true sponges and early corals. New or vastly expanded lineages in the Ordovician included two other colonial groups. One was the graptolites, whose planktonic way of life meant that they could migrate widely. Their widespread distribution and rapid evolution makes these fossils ideal timekeepers. The other colonial group was the bryozoa with their extraordinarily varied calcareous skeletons.

This diorama, on display in The Manitoba Museum, depicts the diverse late Ordovician marine life preserved at sites north of Winnipeg. The assemblage includes corals, tall columnar stromatoporoid sponges, crinoids, trilobites, brachiopods, gastropods, and nautiloids. GRAHAM YOUNG, COURTESY OF THE MANITOBA MUSEUM.

So much for the marine realm, but what was happening on land? Perpetually damp areas had probably been colonized as early as the Neoproterozoic by crusts of cyanobacteria and fungi, but landscapes were rocky and barren before the emergence of land plants. Without plants as anchors, loose rock debris had few nutrients, dried out readily because of a lack of shade, and was rapidly eroded and transported by wind and water. Thus, prior to the Ordovician, well-developed soils with significant organic content did not exist. Microscopic plant debris, including spores, began to appear in middle Ordovician marine deposits about 470 million years ago (Box 7).

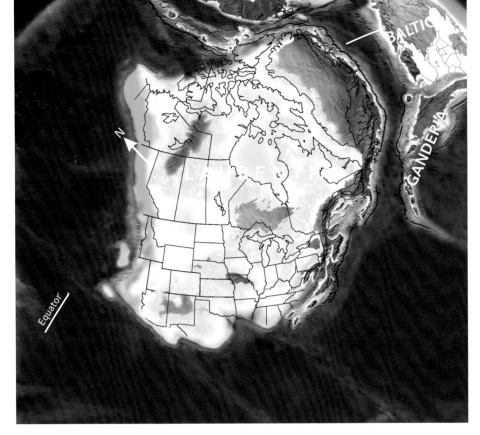

Paleogeography of what was to become North America and adjacent regions in the middle Ordovician, 450 million years ago. Land is shown in green and brown, with shading showing topography. The lighter blue areas represent possible coastal or nearshore areas, darker blue represents deeper ocean waters, and black indicates trenches. Aspects of modern geography are shown for orientation.

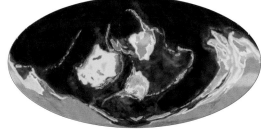

Global paleogeography 450 million years ago, during the late Ordovician. Colours as for previous figure.

On a Young Continent

Driven by global sea-level rises and falls, shallow seas advanced (transgressed) and retreated (regressed) during the Paleozoic, covering and uncovering vast areas of the continent's interior. Intervals during which seas advanced are represented by extensive marine deposits. Intervals during which the seas retreated to the continent's margins are recorded in a few places by non-marine deposits. However, they are more usually represented by gaps in the sedimentary succession, commonly as unconformities, which reflect times when erosion and perhaps gentle tectonic deformation predominated. Such extensive changes in sea level may reflect the growth and melting of huge ice caps. Or they may result from changes in sea-floor spreading rates and spreading-ridge volumes (Chapter 2).

Although most early Paleozoic Platform strata are more or less flat-lying, their distribution and thickness were greatly influenced by positive and negative tectonic features—arches and basins that also tended to be topographic (or bathymetric) features. Arches and basins periodically rose and sank, with some lasting longer than others. These gentle warpings of the Laurentian interior may have been caused by weakly transmitted forces from distant plate margins, by vertical movements in the underlying mantle, or perhaps by a combination of the two factors. In places, the trends of ancient orogenic belts in the underlying basement clearly control features in the cover rocks. One example is the Boothia Uplift, an arch more than 500 kilometres long that brings Precambrian basement to the surface, flanked on both sides by Paleozoic strata. This feature underlies the northward-pointing "finger" of the Boothia Peninsula in Nunavut and reflects the alignment of the Paleoproterozoic Thelon-Taltson Orogen.

The first major Paleozoic invasion of the sea (transgression) over Laurentia lasted from about 510 to 472 million years ago—from the middle Cambrian into the early Ordovician. During that time, the sea progressively covered the continent from the paleonorth and the paleosouth. Deposition on the Western Platform during this invasion was focussed in two basins, the Lloydminster and Hay River embayments, which were separated by the Peace River Arch. As we saw during our drive along the Trans-Canada Highway, the thin Platform succession underlying the Prairies passes into a much thicker sequence between Banff and Field. In the opposite direction, the sea reached progressively farther toward the heart of the continent. Marine clastics eroded from Laurentia's interior were the earliest deposits shed into the transgressing sea. Carbonate deposition followed, spreading to southwestern Saskatchewan in the late Cambrian, about 490 million years ago. Rocks formed during this transgression are now deeply buried beneath the Prairies.

Specimens of the graptolite *Rhabdinopora flabelliforme* on an early Ordovician bedding surface in Meguma Terrane rocks near Wolfville, Nova Scotia. ROB FENSOME, SPECIMEN COURTESY OF CHRIS WHITE.

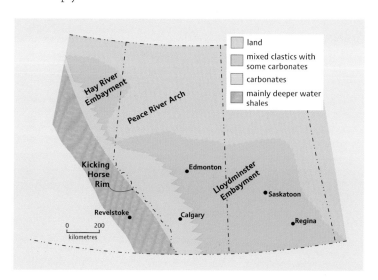

Present distribution of Cambrian and Ordovician strata in the southern Canadian Western Interior and major features that influenced their deposition. Note that the rocks reflect a general trend from shallower to deeper water in a modern southwesterly direction. ADAPTED FROM HEIN AND NOWLAN (1998).

A fall in sea level from roughly 472 to 460 million years ago left a gap in the sedimentary record of the continental interior. This regression was followed by a second transgressive phase during the late Ordovician, from about 460 to 444 million years ago. At its maximum, about 450 million years ago, almost all of Laurentia was inundated by the sea. Initially, the sea spread paleonorthward from the foreland basins behind the rising Taconic Mountains, and across parts of the continent's interior, where subsidence was just beginning in four big basins: Michigan, Hudson Bay, Moose River, and Williston. These basins were similar to modern-day Hudson Bay in that they were broad and shallow and were hundreds of kilometres across; they subsided gradually for tens of millions of years. Thicker accumulations of sediment are preserved in the basins' centres than on the flanks, reflecting both depositional patterns and later erosion. Seas in these interior basins were connected from time to time with the Appalachian-associated basins to the paleosouth.

Major upward flexures (arches) and downward flexures (basins) in the lithosphere of the continental interior that persisted through much of the Paleozoic. ADAPTED FROM VARIOUS SOURCES.

On the Western Platform, the earliest sediments of the second transgressive phase were clastics, deposited around 460 million years ago. For the remainder of the Ordovician, deposition of carbonates dominated Laurentia from Greenland to New Mexico, extending across much of the region now occupied by the Canadian Shield. Isolated outcrops such as those around Manicouagan and Lac Saint-Jean in Quebec and Lake Timiskaming on the Ontario-Quebec border are remnants of those deposits.

Evidence exists of isolated islands that poked above the surface of the continent-wide late Ordovician sea. Sedimentary rocks at sites in central and northern Manitoba indicate lagoon and tidal-flat environments along ancient shorelines around 445 million years ago. These deposits contain diverse fossils, including eurypterids (sea scorpions), jellyfish, and some of the earliest known examples of horseshoe crabs. These animals lacked hard mineralized shells or skeletons, so their preservation is extraordinary. One site has yielded the remains of the gigantic *Isotelus rex*, one specimen reaching a length of 72 centimetres, the largest complete trilobite known.

The most famous rock from this second Ordovician transgressive phase is the Tyndall Stone, formed in the Williston Basin about 450 million years ago. This beautiful carbonate, quarried at Garson in Manitoba, was used in the construction of the Parliament Buildings in Ottawa, and for many other projects (Chapter 14). Tyndall Stone contains numerous fossils—including receptaculitids (fossils that may represent an extinct type of calcareous algae), stromatoporoids, and other

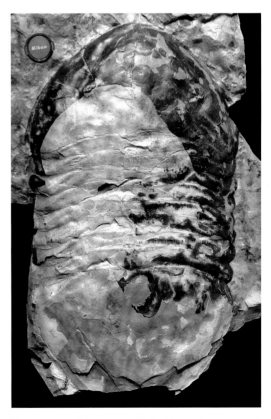

This specimen of *Isotelus* from Ordovician rocks at Churchill, Manitoba, is the largest known trilobite. Note the lens cap at top left for scale. ALAN MORGAN.

CHAPTER 7 · SOUTHERN SOJOURN

Aerial view of the shoreline of Hudson Bay, near Churchill, Manitoba. The boulders are embedded in Ordovician carbonate and are remnants of the ancient shoreline of an island within the late Ordovician continental sea covering Laurentia. Paleoproterozoic metasandstone deposited in the Trans-Hudson Orogen was the source of the boulders. GRAHAM YOUNG.

Close-up of the rocks shown in the previous photo. GRAHAM YOUNG.

Nautiloid in a polished slab of Ordovician Tyndall Stone from Garson, Manitoba. GRAHAM YOUNG.

sponges, corals, brachiopods, cephalopods, and gastropods—buried among myriad fossil burrows that produce the stone's distinctive mottled texture.

Stinky and Icy Times

Some of the warmest climates of the Phanerozoic occurred during the Cambrian and persisted until near the end of the Ordovician, when the rock record shows dramatic environmental changes. Many aspects of these changes remain puzzling, although ocean waters may hold a clue. Modern oceans are well oxygenated, even at depths, because of active mixing of seawater by the global conveyor belt of ocean currents (Box 4), which in turn are kept going by the strong temperature gradient between the Equator and the poles. At certain times in the past, when the Earth's climate was more equable from equator to poles, the oceanic conveyor belt shut down and oceans became stratified. Surface waters were well oxygenated, but deeper waters were unrefreshed and anoxic, enriched in gases such as hydrogen sulphide, as in today's Black Sea. On the floors of these seas, putrid organic-rich muds accumulated that later became black shales.

The latest Ordovician was one such time of anoxic bottom waters, and black shales were deposited in the foreland basins to the paleosouth and in interior basins. Anoxic bottom waters hampered the decay of organic material that sank from surface waters as organisms that lived there died. The resulting organic-rich sediment, with burial and time, would become the source of the petroleum discovered during the early years of oil exploration in the American Midwest and adjacent parts of Canada, for example at Oil Springs in Ontario (Chapter 13).

Between 445 and 444 million years ago, temperatures dropped dramatically, leading to a series of glaciations in high latitudes on Protogondwana. The ice reached its maximum extent near the end of the Ordovician. Evidence for these ice ages includes tillites in northern Africa and elsewhere. Growing ice caps caused a dramatic drop in global sea level, which led to widespread exposure of the continents above sea level. The increased temperature gradient between polar and tropical regions caused ocean circulation to be temporarily renewed and bottom waters to become briefly re-oxygenated. These massive and complex environmental changes preceded one of the largest mass-extinction events in the Phanerozoic. Extinction appears to have happened in two major steps, one at the onset of peak glaciation and the other near the end of glaciation, just before the Ordovician-Silurian transition about 444 million years ago.

The first step of the end-Ordovician mass-extinction event, about 445 million years ago, may have resulted from a loss of diverse shallow-water habitats, perhaps exacerbated by migration of many marine organisms from high latitudes seeking warmer water. During this first step, about 100 families of marine organisms died out, including a third of brachiopod families; numerous graptolite, trilobite, and conodont species also disappeared, as temporarily did reefs.

Low sea levels, cooler temperatures, and well-circulated deep-ocean waters persisted for about a million years, during which time some marine organisms appear to have adapted to the new conditions. But the recovery was partial and short-lived. Global warming, sea-level rise, and the return of anoxic deep-ocean waters as glaciers began to wane brought about a second wave of extinctions. It took marine life several million years during the early Silurian to recover after this double whammy. But that's a story for the next chapter.

These late Ordovician carbonates with impressive stromatolites can be seen at low water levels near the Quebec shore of the Ottawa River, Gatineau, Quebec. Much of Laurentia was covered by a continental sea during the late Ordovician. WOUTER BLEEKER.

Black shales within a sequence of clastics and carbonates marking the Ordovician-Silurian boundary on Ellesmere Island, Nunavut. The boundary is probably at the base of the thickest band of black shale. MIKE MELCHIN.

BOX 6 • WALCOTT'S LEGACY: THE BURGESS SHALE

From Cambrian Mud

Canada is home to one of the world's most important geological treasures, the Burgess Shale. The fossil-rich strata of the original discovery site are perched about 2,500 metres above sea level, near the town of Field, British Columbia, in Yoho National Park of Canada. Set amidst the stunning scenery of the Rocky Mountains, the site is renowned for the exceptional preservation of middle Cambrian (505-million-year-old) marine soft-bodied animals, including arthropods, sponges, and worms. The Burgess Shale locality became a UNESCO World Heritage Site in 1980. Today, it continues to inspire and fascinate scientists and the general public alike, providing valuable insights into the morphology, ecology, and evolution of some of the planet's first animals.

The Walcott Quarry of the Burgess Shale, Yoho National Park of Canada, British Columbia. ALAN MORGAN.

The Burgess Shale story began near the end of the nineteenth century, when a major influx of European settlers in western Canada promoted a burgeoning new tourism industry. Numerous fossils were spotted on Mount Stephen in the 1880s, possibly by workers building the Canadian Pacific Railway through the Rocky Mountains. In 1886, Richard McConnell of the Geological Survey of Canada confirmed the presence of abundant Cambrian fossils on Mount Stephen, at a locality known today as the Trilobite Beds. In the late nineteenth century, Canadian paleontologists first described many of the fossils from the Burgess Shale that would become iconic.

As an authority on Cambrian trilobites, American paleontologist Charles Walcott was well aware of the tantalizing studies of his Canadian colleagues when he first visited the Trilobite Beds in 1907. In 1909, Walcott explored for similar sites on the west side of a ridge between Wapta Mountain and Mount Field, about 6 kilometres northwest of the Mount Stephen locality. Along the way, he and his family, who had accompanied him, discovered numerous fossils of beautifully preserved soft-bodied animals on fallen loose slabs. Walcott returned in 1910 and confirmed the source horizons for these extraordinary fossils. He subsequently named the rock unit the Burgess Shale after nearby Burgess Pass and Mount Burgess. On expeditions to the Burgess Shale between 1910 and 1924, Walcott and his family collected more than 65,000 specimens, mostly from a 2-metre-thick section he called the Phyllopod Bed in an excavation that has subsequently been called the Walcott Quarry. At the time of his initial Burgess Shale explorations and in the years that followed, Walcott was Secretary of the Smithsonian Institution in Washington, to which he shipped the fossils. Walcott went on to describe many species from the site.

In 1930, after Walcott's death, more modest collections were made by Percy Raymond of Harvard University from beds above the Walcott Quarry. His excavation is now known as the Raymond Quarry. The Geological Survey of Canada re-opened both quarries in 1966 to 1967, and a major re-examination of the Burgess Shale biota led by Harry Whittington and his team from Cambridge University started soon afterward. From 1975 to 2000, Desmond Collins's Royal Ontario Museum crews discovered abundant soft-bodied animal fossils in more than a dozen localities throughout the southern Rocky Mountains. The Royal Ontario Museum's Burgess Shale holdings now dwarf all previous collections combined, with an estimated 150,000 specimens, including many new species.

Until recently, all known Burgess Shale sites were located next to the Cathedral Escarpment (Chapter 7). Recent Royal Ontario Museum field studies, however, show that similar fossils are present in areas away from the Cathedral Escarpment, in completely different types of paleoenvironments. Additional fossils might surface in more unexpected places, perhaps once again challenging our views about this celebrated biota. Today, the Walcott Quarry is still the most important of all Burgess Shale-type localities for its abundance, diversity, and quality of fossils, although other sites have been found not only farther afield in the Rocky Mountains but around the world, including several in China, most notably at Chengjiang.

An Ancient Menagerie

The Burgess Shale has so far yielded more than 200 species of animals, most of which lack hard shelly parts, along with a few algal and microbial forms. The fossils are usually preserved in exquisite anatomical detail, and occasionally even have gut contents! This degree of preservation provides great insights into the morphology, ecology, and affinities of the organisms. Some Burgess Shale fossils look familiar to us and belong to major groups (phyla) of animals, such as

Herpetogaster, a recently described animal from the Burgess Shale. JEAN-BERNARD CARON, COPYRIGHT ROYAL ONTARIO MUSEUM.

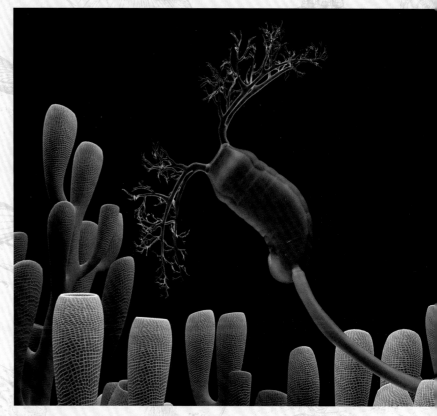

Reconstruction of *Herpetogaster*, swimming among a colony of sponges. COPYRIGHT ROYAL ONTARIO MUSEUM.

sponges, still alive today. Others are more puzzling, with combinations of features not seen in any modern animals. Among the strangest forms is the five-eyed *Opabinia*, which had an extended proboscis, a segmented body about 5 centimetres long, delicate gills, but no legs.

How the true morphologies of some Burgess Shale animals came to light makes fascinating reading. In some cases, various elements previously regarded as separate types of animal, have been revealed as parts of the same creature. For example, almost a century passed between the naming of *Anomalocaris* and an understanding of its complete body shape. What was first described as *Anomalocaris canadensis* ("Canada's odd shrimp") was thought to be a prawn-like crustacean preserved without a head. But further discoveries and research have shown that the original find represents one of the animal's two feeding claws. A segmented ring-shaped structure, resembling a slice of pineapple and once interpreted under a separate name as a jellyfish-like animal, turns out to represent mouthparts of *Anomalocaris* and related forms. Thus, *Anomalocaris* is now known to have been a large arthropod, probably up to 1 metre long, evidently an efficient swimmer, and probably the biggest predator or scavenger in Cambrian seas. In a similar example, thanks to discoveries of complete articulated specimens, the many small spines and ridged scales found scattered throughout the Burgess Shale are now known to have constituted the overlapping body armour of the mollusk *Wiwaxia*.

Anomalocaris and *Wiwaxia* provide examples of how individual animals have been re-interpreted over time. But our changing understanding of the fauna as a whole is also fascinating. Walcott believed the animals were simply ancient members of living phyla. This vision remained mostly unchallenged until the late 1960s, when detailed re-examination of Walcott's specimens by Harry Whittington and his students revealed that many species had features that did not easily match those of modern animals, and thus may have been members of extinct phyla.

The weirdness of many Burgess Shale fossils led Harvard paleontologist Stephen Jay Gould in his 1989 bestselling book, *Wonderful Life*, to suggest that more animal phyla were present in the Cambrian than exist today, and that many of those must have gone extinct after the Cambrian. Current views are rapidly changing through the discovery of new fossils, bolstered by detailed studies of existing specimens, as well as new insights based on, for example, molecular studies of modern organisms. We now understand that the individual weirdness of some Burgess Shale animals is much less important than the features that they had in common. This new view has important implications for our understanding of the Cambrian Explosion. For example, we now realize that the diversity of marine animal forms in the Cambrian is probably closer to modern levels than we once thought. This conclusion does not undermine the uniqueness of the Cambrian Explosion—animal body plans must have still evolved very quickly around the beginning of the Cambrian to match today's diversity. But perhaps fewer unique body plans exist than was suggested by Gould.

Despite some ongoing controversies, species previously regarded as members of long-vanished phyla are now interpreted as extinct early branches of still-surviving animal lineages. For example, *Hallucigenia* is now accepted as a primitive lobopod (distantly related to today's velvet worms); *Nectocaris*, *Odontogriphus*, and *Wiwaxia* as primitive mollusks; and *Anomalocaris* and *Opabinia* as primitive arthropods. These fossils are especially important because they show unique suites of morphological features not retained in living relatives, and some of these features in turn suggest links between what now seem to be distinct phyla. Many questions remain, however. We still

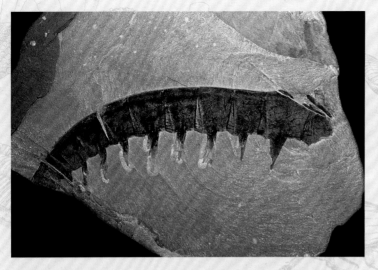
A claw of *Anomalocaris* from the Burgess Shale. JEAN-BERNARD CARON, COPYRIGHT ROYAL ONTARIO MUSEUM.

The mouthpart of an anomalocarid from the Burgess Shale. JEAN-BERNARD CARON, COPYRIGHT ROYAL ONTARIO MUSEUM.

Anomalocaris specimen from the Burgess Shale. JEAN-BERNARD CARON, COPYRIGHT ROYAL ONTARIO MUSEUM.

Wiwaxia specimen from the Burgess Shale. JEAN-BERNARD CARON, COPYRIGHT ROYAL ONTARIO MUSEUM.

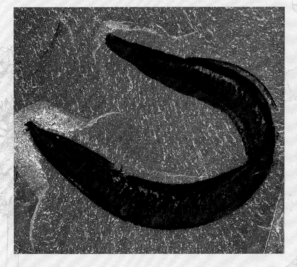
Pikaia specimen from the Burgess Shale. JEAN-BERNARD CARON, COPYRIGHT ROYAL ONTARIO MUSEUM.

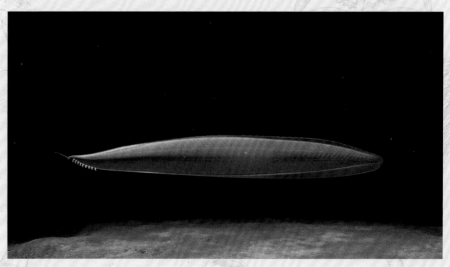
Reconstruction of *Pikaia*. COPYRIGHT ROYAL ONTARIO MUSEUM.

do not know for sure where some of the more bizarre Burgess Shale animals fit; examples are *Amiskwia*, *Chancelloria*, and *Dinomischus*.

Although invertebrates dominate the Burgess Shale fauna, one fossil, *Pikaia*, is thought to be a primitive chordate (Chapter 4). First described by Walcott in 1911 as an annelid worm, its chordate affinity is suggested by its apparent possession of a notochord and myotomes. (Myotomes are repeating blocks of muscle—when you check that a fish is properly cooked, the desired flakiness is caused by separation of these blocks.) Less than 5 centimetres long and laterally flattened, *Pikaia* resembles living lancelets, but differs from modern chordates in having a tiny head with two tentacles and several pairs of appendages.

A Question of Survival

How did the Burgess Shale fossils come to be so exquisitely preserved? Although still something of a puzzle, a few points are widely accepted. Most Burgess Shale organisms were benthic (bottom-dwelling) and probably lived on the slope or at the foot of the Cathedral Escarpment. The Burgess Shale was deposited in a relatively deep-water environment, but the remains of photosynthetic green and red algae show that the sea floor received at least some sunlight or that some of the organisms were transported from shallower water. The mix of both complete and broken remains, as well as their jumbled orientations, suggests that currents, storms, or earthquakes periodically set off mudslides, burying both living and dead creatures more or less in place. Animals were probably buried and killed instantly, because no trace-fossil evidence exists of escape attempts. The fine mud sealed the fresh carcasses, probably in an anoxic environment, preventing decay. Early compaction of the sediment, changing it from mud to shale, would have further limited disturbance of delicate soft tissues. Over the ensuing millions of years, the deposits and fossils came to be buried beneath some 10 kilometres of younger deposits, further consolidating the rock and partially replacing the original composition of the fossils. However the mysteries of their wonderful preservation are ultimately resolved, the Burgess Shale fossils, now exposed at the Earth's surface thanks to a combination of tectonics and erosion, are a magnificent legacy of Cambrian life that still surprises and delights us more than a century after they were first discovered.

Reconstruction of the Burgess Shale community, with the Cathedral Escarpment in the background. Included in this view are groups of tubular sponges, a spiny *Wiwaxia* on the sea bed, and a cruising *Anomalocaris*. COPYRIGHT ROYAL ONTARIO MUSEUM.

BOX 7 • THE GREEN REVOLUTION

Early Days

Earth is commonly referred to as the blue planet, reflecting the dominance of water. But our planet could also be thought of as green because of the abundance of plants over much of the land. Through the process of photosynthesis, plants (in the broad sense including algae and cyanobacteria) are indispensible for sustaining life on Earth. Each year, plants trap almost ten times more energy than we humans use. Ultimately, like other animals and the fungi, we rely on them completely for sustenance.

Photosynthesis occurs in chloroplasts, the organelles in plant cells responsible for generating sugars from water, carbon dioxide, and sunlight. These organelles first evolved as prokaryotic cyanobacterial cells that became chloroplasts in eukaryotic cells through symbiosis. Not all cyanobacteria became swallowed up in this way and they are still with us as independent prokaryotic organisms. Because oxygen is released as a waste product of photosynthesis, we might think of these free-living photosynthesizers as the world's first polluters. But without them and the oxygen-rich atmosphere they produced, animals and fungi would not have evolved at all.

Several lineages of algae developed many-celled forms independently. As well, different lineages of algae developed or symbiotically acquired different pigments to trap sunlight, as part of their evolving photosynthetic "machinery". Thus, red, brown, and green algae evolved along different lines. The red and brown algae are most familiar to us as "seaweeds", although the green algae are also represented in that slippery ensemble. It was the green algae, however, that gave rise to plants as we know them. Green plants predominantly use chlorophyll, inherited directly from ancestral green algae.

Green algae were (and are) mostly aquatic single-celled or multicellular organisms. Evidence shows that freshwater forms of green algae invaded the land on many different occasions, but only one group made a real success of this new habitat. The successful group, known as the embryophytes, includes all familiar land plants, from creeping liverworts to towering redwoods.

Land Ahoy

The real challenge for plants was not the invasion of the land, but how to adapt to life in the open air. In water, plants have many advantages, not the least of which is that food and water can be freely acquired from the surrounding medium, and waste products can be readily expelled into it. Moreover, plants in water are shielded from harmful ultraviolet light and they enjoy neutral buoyancy. The first land plants had to devise clever engineering solutions to these formidable challenges. Key to their success was the evolution of the vascular system, a series of rigid tubes connecting the roots to the green surfaces of the plant, where photosynthesis takes place. These tubes have two functions: to provide mechanical support and to allow the transport of water and food. The transport system is driven by evaporation from the plant's external surface, which in turn sucks up water from the roots. Natural valves called stomata prevent extensive water loss during drought, but allow carbon dioxide, a requirement for photosynthesis, to diffuse into the plant's tissues.

When did embryophytes (or plants, as we will henceforth call them) make landfall? Evidence comes from spores, simple cell-like reproductive structures of early plants and many of their modern descendants. Spores, which are preserved as fossils because they have a tough outer organic wall, first appeared in the Ordovician, around 470 million years ago. A comparison with modern spores suggests that the earliest plants were bryophytes, a group that includes mosses, liverworts, and hornworts. Early bryophytes had only a rudimentary vascular system, and so still needed to remain in or on moist soils.

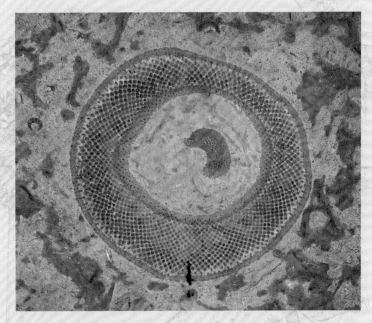

A receptaculitid from the Ordovician Tyndall Stone (Chapter 8). Receptaculitids may represent an extinct type of calcareous alga. ROB FENSOME, SPECIMEN COURTESY OF THE UNIVERSITY OF SASKATCHEWAN.

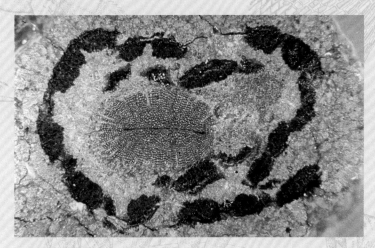

Cross-section of the stem of a Devonian plant akin to *Psilophyton*, an early tracheophyte from Dalhousie Junction, New Brunswick. PATRICIA GENSEL.

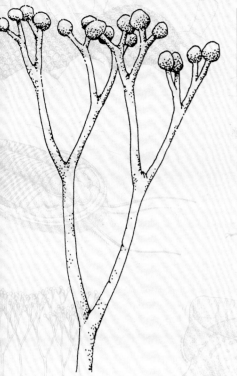

Sketch of *Cooksonia*, an early vascular plant found in Devonian strata at Dalhousie Junction, New Brunswick. CHRISTOPHER HOYT AND THE NEW BRUNSWICK MUSEUM.

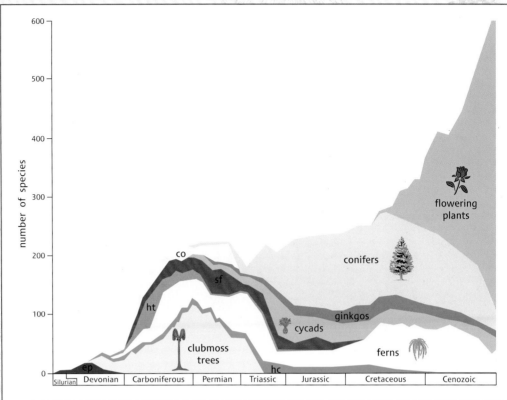

Distribution of the main groups of vascular plants through time. Abbreviations: co = cordaites, ep = early plants, hc = herbaceous clubmosses, ht = horsetails, and sf = seed ferns. ADAPTED FROM NIKLAS ET AL. (1985).

However, in the middle Silurian, about 425 million years ago, plants with a superior vascular system appeared. These were the tracheophytes, the group that encompasses plants most familiar to us, including ferns and oak trees. Early tracheophytes are found in late Silurian and Devonian strata on Bathurst Island in the Arctic and in Devonian strata at Dalhousie Junction in New Brunswick (Chapter 8). One form was *Cooksonia*, whose vascular system allowed it to stand taller—some specimens even reached the giddy height of 10 centimetres.

Three further innovations occurred in the Devonian that paved the way for the diverse plant communities we know today: leaves, wood, and seeds. Leaves are so familiar to us that it seems strange to think that the first plants lacked them. Nevertheless, early forms such as *Cooksonia* simply had green photosynthetic stems with knob-like sporangia (spore-bearing organs) at the tips—but no leaves. The development of leaves may have been a response to declining atmospheric carbon-dioxide levels during the Devonian. We're not sure why carbon dioxide declined then, but with less of it in the air, plants had to increase their surface area to get enough of the gas to sustain the levels necessary for photosynthesis. Leaves were the perfect answer.

The second innovation was wood—or more specifically a zone of actively dividing cells that produces wood on its inner surface and bark on its outer surface. Known as the vascular cambium, this zone of dividing cells allowed plants to increase in girth. With thicker and

Scene from a late Carboniferous forest. The dominant trees are clubmosses, with many specimens of *Lepidodendron* (centre) and, in the distance at right, *Sigillaria*. To the left is a grove of the early gymnosperm *Cordaites*. PAINTING BY JUDI PENNANEN, COURTESY OF THE NEW BRUNSWICK MUSEUM.

The tallest plants were thus the most successful because they could trap the most sunlight and grow the fastest.

Succeeding with Seeds

Climate was arid for parts of the Devonian, and such conditions may have promoted the development of seeds, the third innovation of that period. All the plants discussed so far reproduced by microscopic spores, as do modern ferns, horsetails, and clubmosses, all of which require damp conditions to grow. Spores can only germinate in the presence of water, so early land plants were restricted to moist environments. Some spore-bearing plants had spores all of the same size (homospores). Others developed spores of two types, with the smaller ones, the microspores, being the male gametes and the larger ones, the megaspores, being the female gametes. Microspores, which were about the same size as homospores, evolved into pollen grains. In the late Devonian, some plants started to retain their megaspores in a capsule (the ovule), like many modern plants do. A microspore or pollen grain, dispersed by the wind, fertilizes a megaspore to produce an embryo, the latter and its protective ovule becoming a seed. The seed benefits greatly from being attached to the parent plant, with a ready supply of food and water. Thus, seed plants are better adapted to cope with seasonally dry conditions. This evolutionary step in plants parallels the development from amphibians to reptiles in the animal world: both seeds and shelled eggs free their respective organisms from a life near water.

Although seed plants had an ecological advantage, spore-bearing plants continued to dominate the plant world during the Carboniferous. The tropics were warm and wet during this period—and spore-bearing plants thrive under such conditions. Some spore-bearers grew to be giants. For example, the clubmosses *Lepidodendron* and *Sigillaria*, common as fossils in the Maritimes, could be 40 metres tall. Others such as *Calamites*, a horsetail, may have reached 15 metres. Generations of these giant plants lived, died, and were buried to form the widespread coal seams across the equatorial regions of the day, in a belt extending from what are now the western Appalachians of the United States to central Europe, and including the coalfields of Pennsylvania, the Maritimes, Britain, and the German Ruhr Valley. These deposits represent some of the first equatorial rainforests.

About 300 million years ago, around the Carboniferous-Permian boundary, the Earth's climate became much drier, and the moisture-loving spore-bearers declined dramatically. Waiting in the wings were the seed plants. In particular, two groups of seed plants proliferated: the conifers and the cycads and their allies. When we think of conifers we imagine fir, spruce, and pine trees. But early conifers were more like the monkey-puzzle tree, a native of the Chilean Andes today. One conifer that was a contemporary of the dinosaurs was dawn redwood (see photos on page 194). This tree has an unusual claim to fame. It was first described from fossil remains in the Canadian Arctic in 1941. Then, less than a decade later, living trees were found in China—a striking example of a "living fossil". Even more amazingly, fossil specimens of *Metasequoia* tell us that these trees lived close to the North

Whorl-shaped leaves known as *Asterophyllites* represent the foliage of the fossil horsetail, *Calamites*. Several branches with leaves are preserved in this specimen from Minto, New Brunswick. HEINZ WIELE, COURTESY OF THE ATLANTIC GEOSCIENCE SOCIETY; SPECIMEN COURTESY OF THE NEW BRUNSWICK MUSEUM.

Sketch of the horsetail tree *Calamites*. CHRISTOPHER HOYT AND THE NEW BRUNSWICK MUSEUM.

Archaeopteris is among the earliest known plants to have had wood, the evolution of which gave rise to trees and the first forests. *Archaeopteris* has been found in late Devonian strata, for example, at Miguasha, Quebec, and in the Canadian Arctic. STEPHEN GREB.

Reconstruction of the clubmoss tree *Sigillaria*. ANDREW MACRAE

sturdier woody stems, or trunks, plants could grow taller. By the late Devonian, lofty forest canopies had replaced knee-high thickets. One of the earliest trees was *Archaeopteris*, whose 30-metre-long trunks are found in rocks on the Quebec shores of the Gulf of St. Lawrence. What drove this amazing race for the skies? As the land became more crowded, presumably plants began to cast shadows over one another.

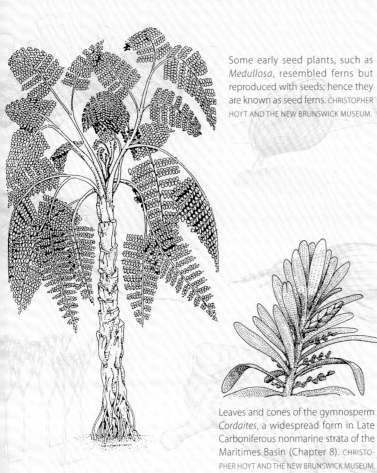

Some early seed plants, such as *Medullosa*, resembled ferns but reproduced with seeds; hence they are known as seed ferns. CHRISTOPHER HOYT AND THE NEW BRUNSWICK MUSEUM.

Leaves and cones of the gymnosperm *Cordaites*, a widespread form in Late Carboniferous nonmarine strata of the Maritimes Basin (Chapter 8). CHRISTOPHER HOYT AND THE NEW BRUNSWICK MUSEUM.

The unparalleled success of the flowering plants reflects, in part, their dependence on insect pollination. This strategy, in contrast to a reliance on wind or water, increases the odds of being fertilized. Nevertheless, grass, one of the most widespread of all flowering plants, is mainly wind-pollinated. Grass is one of the marvels of plant evolution. The first possible grass fossils date from the late Cretaceous, but pollen evidence suggests that grass didn't become abundant and widespread until the Oligocene, 25 million years ago, perhaps adapting to the extreme seasonal conditions of continental interiors. But when grass took off, it was responsible for the great grasslands of Eurasia, North America, and elsewhere, and promoted the evolution of many types of hoofed mammal. Today, in the guise of cereals, grasses are keeping people happy at the breakfast table.

A spruce cone from Eocene fluvial deposits on Axel Heiberg Island, Nunavut. BEN LEPAGE.

Pole 100 million years ago. They were still there (or had returned) 45 million years ago. Cycads, the other prominent early Mesozoic group, had short, stumpy trunks (like giant pineapples) topped with a mop of palm-like leaves.

Flowering plants, or angiosperms, are the most successful of all plants today, with about 250,000 species (compared to only 700 species of modern conifers). Yet the origin of flowering plants remains an unsolved problem. Charles Darwin called it "an abominable mystery". What we do know is that the distinctive pollen of flowers, distinguished by one or more regularly arranged slits and/or holes and a complex wall, first appeared in the early Cretaceous, about 140 million years ago.

Flower calyx of *Florissantia quilchenensis* from Eocene lake deposits at Quilchena, British Columbia. ROLF MATHEWES.

 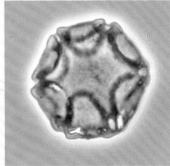

Microscopic spores and pollen are far more common as fossils in many sedimentary rocks than larger plant fossils such as leaves and stems. Spores and pollen have played a significant role in dating rocks, unravelling past environments and revealing plant evolution. These artificially stained examples from Atlantic Canada are a Cretaceous liverwort spore (top left), a Cretaceous fern spore (top right), a Tertiary conifer pollen grain (bottom left), and a Tertiary angiosperm pollen grain (bottom right). ROB FENSOME.

BOX 7 • THE GREEN REVOLUTION

8 • CROSSING THE EQUATOR
Canada 444 to 251 million years ago

CHAPTER SUPPORTED BY A DONATION FROM MICHAEL ROSE, FOUNDER, BERKLEY PETROLEUM, DUVERNAY OIL, AND TOURMALINE OIL

Cape Tryon, Prince Edward Island, where the cliffs consist of Permian red fluvial sandstone. RON GARNETT / AIRSCAPES.CA.

A Nation Founded on Coal

One of British North America's main concerns in the mid-nineteenth century was the availability of a guaranteed energy supply for the emerging industrial powers of Upper and Lower Canada. Bringing large quantities of coal from Britain was expensive and time-consuming. What's more, potential coal resources in the centre of the continent had been lost to the Americans after their War of Independence. A source of easily accessible coal was sorely needed.

And so, when legislative independence was granted to a combined Upper and Lower Canada in the 1840s, an early decision of the new government was to hire Montréal-born geologist William Logan, who had

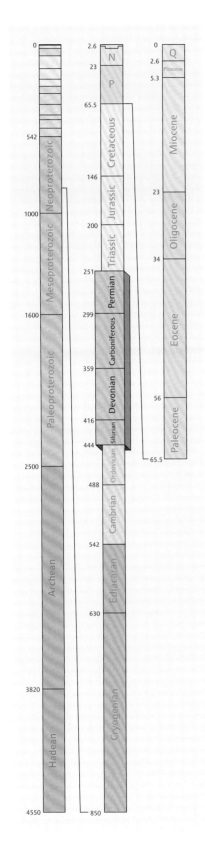

Geologic time scale, showing the interval covered in this chapter. Numbers indicate millions of years ago. P = Paleogene (Paleocene to Oligocene), N = Neogene (Miocene and Pliocene), and Q = Quaternary.

William Logan in his laboratory in Montréal. REPRODUCED WITH THE PERMISSION OF NATURAL RESOURCES CANADA 2013, COURTESY OF THE GEOLOGICAL SURVEY OF CANADA.

developed practical expertise in the Welsh coal fields. Logan's initial mandate, as first director of the Geological Survey of Canada in 1842, was to find a cheap Canadian source of energy other than water and wind power. In the nineteenth century, that meant easily mineable coal. Logan knew, from maps by Charles Lyell and American geologist James Hall showing the extent of North American coal deposits, that the likelihood of such a discovery in central Canada was remote. But reports by Nova Scotian polymath Abraham Gesner suggested that plenty of coal was to be found in the east. Logan travelled to Nova Scotia, then a separate colony, leaving colleague Alexander Murray to survey the Silurian limestone of Upper Canada. After his trip, Logan wrote a detailed description of the Carboniferous coastal section at Joggins, where at least 76 coal seams outcropped in a succession now considered to be about 313 million years old.

Following Logan's expedition to the east, and the confirmation that central Canada had no coal, politicians such as John A. Macdonald realized that if the industrial heartland was to flourish, a future confederation of provinces had to include the energy-rich Maritime colonies. How the Maritimes came to be energy-rich is a key part of the story told in this chapter.

Tilted late Carboniferous rocks are exposed in the cliffs and on the beach at Coal Mine Point, Joggins, Nova Scotia. Note the coal seam extending from bottom right. ROB FENSOME.

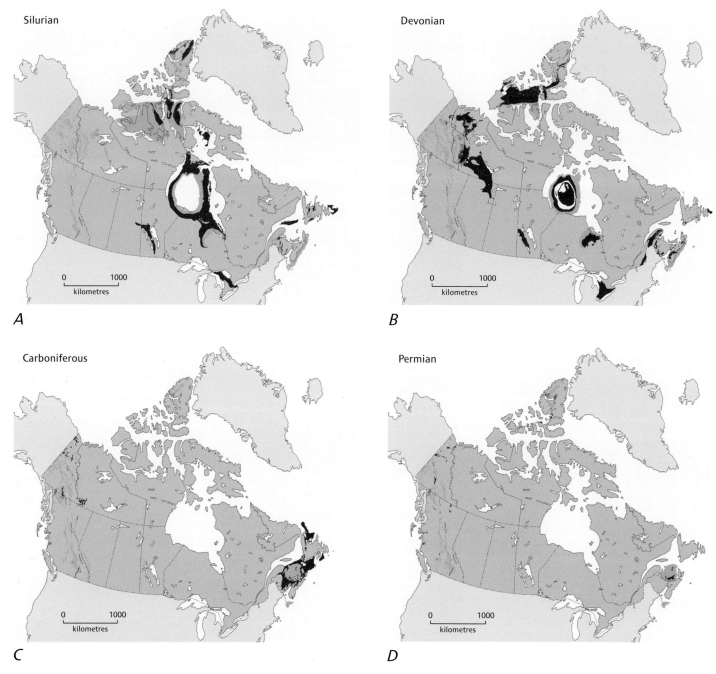

General extent of Silurian (A), Devonian (B), Carboniferous (C), and Permian (D) rocks at the surface (beneath glacial deposits), onshore and offshore. The lighter shaded areas denote either uncertainty or areas where rocks of the particular age have been confirmed but are intimately associated with rocks of other ages and the scale of the map doesn't allow us to show them separately. ADAPTED FROM WHEELER ET AL. (1996).

Setting the Scene

This chapter spans four periods: the Silurian (444 to 416 million years ago), the Devonian (416 to 359 million years ago), the Carboniferous (359 to 299 million years ago), and the Permian (299 to 251 million years ago). During the Silurian, Laurentia fused with Baltica and Ganderia to form the continent of Euramerica, which began to rotate counterclockwise so that, by the middle Devonian, the present eastern margin faced paleosoutheast. Euramerica, like Laurentia before it, was still essentially a southern hemisphere continent. Its paleoeastern and paleonorthern margins became increasingly active tectonically, though both flanks still mostly faced open oceans. (From here on, we will use modern co-ordinates, but readers should keep in mind that what was to become North America was actively rotating during the interval covered.)

During the Carboniferous, Euramerica generally drifted northward, so that by 315 million years ago, the paleoequator extended from what is today the southwestern United States, through Atlantic Canada to central Europe. By about 310 million years ago, Euramerica and Protogondwana had crunched together and fused, an event considered to mark the birth of the supercontinent Pangea. But Siberia did not become a part of the supercontinent until about 270 million years ago with the final closure of the Ural Ocean. Surrounding Pangea was

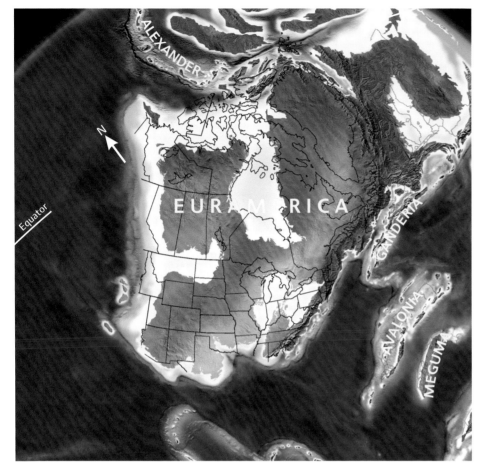

Paleogeography of what was to become North America and adjacent regions in the Silurian, 420 million years ago. Land is shown in brown, with shading showing topography. The lighter blue areas represent possible coastal or nearshore areas, darker blue represents deeper ocean waters, and black indicates trenches. Aspects of modern geography are shown for orientation.

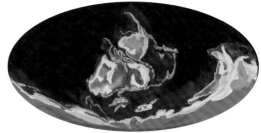

Global paleogeography 420 million years ago, during the Silurian. Colours as for previous figure.

Panthalassa, the vast world ocean that was the successor to Protopanthalassa. (Panthalassa is by definition the vast watery counterpart of Pangea, which is why we have been referring to its pre-Pangean precursor as Protopanthalassa.) A huge embayment of Panthalassa, known as Paleotethys, penetrated Pangea's eastern side. Evidence for Paleotethys is found in the rocks that extend across central Asia to the Mediterranean region.

By the Permian, the North American part of Pangea had edged northward, with the paleoequator passing through Florida and Texas. By the end of the Permian, future North America had largely crossed into the Northern Hemisphere.

Collision Course Act Two

Much of the Iapetus Ocean had closed by the end of the Ordovician, 444 million years ago. But the Tetagouche-Exploits Basin started to close only about 440 million years ago. This latter closure led to the collision of Ganderia with Laurentia, generating an episode of mountain building known as the Salinic Orogeny that ended about 422 million years ago, completing the amalgamation of Laurentia, Baltica, and Ganderia to form the new continent of Euramerica.

While all this was happening, the Acadian Seaway, another late remnant of Iapetus, still lay between Avalonia and Ganderia. Closure of this stretch of oceanic crust continued the tectonic mayhem from about 420 to 390 million years ago. The collision of Avalonia with the Ganderian margin of Euramerica led to the large-scale uplift and mountain building of the Acadian Orogeny. The Acadian Seaway disappeared through subduction beneath the Ganderian margin of Euramerica, and remnants of the accompanying magmatic arc on the margin are found today as a "bloom" of magmatic activity in former Ganderia, including the Rose Blanche Pluton in Newfoundland and parts of the St. George Batholith in southern New Brunswick.

Remains of the seas in and around Avalonia are best seen today in the superb Silurian and earliest Devonian succession at Arisaig in Nova Scotia. There the rocks consist mainly of fine clastic sediments that harbour diverse marine invertebrate fossils, especially bottom-dwelling organisms such as brachiopods, clams, crinoids, and arthropods. Among the arthropods

Colonies of the Silurian coral *Favosites* in strata at Quinn Point, near Jacquet River, New Brunswick. ROB FENSOME.

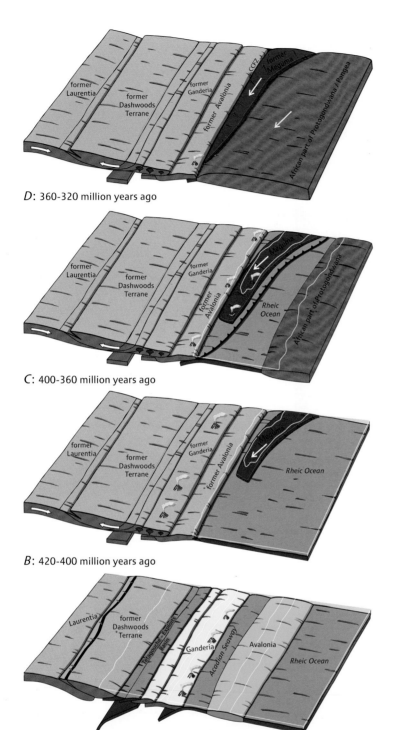

D: 360–320 million years ago

C: 400–360 million years ago

B: 420–400 million years ago

A: 442–423 million years ago

The relationships of the various Appalachian terranes and seaways with Laurentia during part of the early Silurian (*A*); and with Euramerica during the late Silurian to early Devonian (*B*), the middle to late Devonian (*C*), and the early Carboniferous (*D*). The white lines indicate sea level.

Sugarloaf Mountain near Campbellton, New Brunswick, is an eroded volcanic plug, reflecting magmatic activity in Ganderia during the early Devonian. ROB FENSOME.

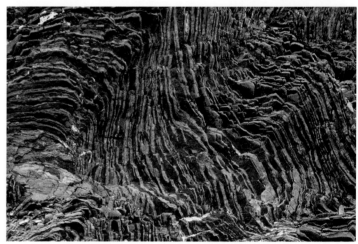

These late Ordovician to early Silurian muddy limestone beds are exposed in the gorge of the Saint John River at Grand Falls, New Brunswick. The original sediments were deposited in a sedimentary basin on former Ganderia, and were deformed between about 440 and 390 million years ago, during the Salinic and Acadian orogenies. ROB FENSOME.

were large, ferocious-looking eurypterids. Corals are rare and not diverse in Silurian rocks at Arisaig, which suggests cooler climatic conditions, muddier water, or perhaps both.

From the late Silurian to early Devonian, the newly risen Appalachians, largely the legacy of the Salinic and Acadian orogenies, were the North American part of a vast mountain system that stretched from the southern United States through eastern Canada (collectively, the Appalachian Orogen) to eastern Greenland, Scotland, and Scandinavia. Those sections from Greenland, through northwestern Britain to Scandinavia are known as the Caledonian Orogen. In the Appalachian Orogen and British parts of the Caledonian Orogen, the collisions that created the mountains were oblique and involved several intermediate terranes in addition to the major "bookend" continents—we thus might think of the entire collisional episode as being relatively "soft". But in the Scandinavian parts, collisions were mostly head-on "hard" impacts directly between colossal continents, without intervening terranes to soften the effects. The crashes between these continental titans built mountains that were probably thousands of metres high, whereas mountains associated with the "softer" crunches were probably less lofty.

Avalonia now formed the outer margin of Euramerica and faced the shrinking Rheic Ocean. Somewhere within that ocean was Meguma. The travels of this terrane during the late Silurian and early Devonian are uncertain, but we know that it was not yet part of the Appalachian Orogen or of Euramerica.

Devonian carbonates at Cap Bon Ami, Forillon National Park of Canada, Gaspé Peninsula, Quebec. E. LE BEL, COPYRIGHT PARKS CANADA.

Ediacaran strata deposited on Avalonia and tilted during the Acadian Orogeny, St. John's, Newfoundland. KEITH VAUGHAN.

In the Shadows of Mountains

Collisions in the Appalachian Orogen during the Silurian and early Devonian led to a complex interfingering of mountains, sedimentary basins, and seaways. Mountains were pushed up and eroded, and basins subsided and filled, all in mesmerizing succession, particularly in the belt extending from Ganderia to the Anticosti and Appalachian basins. Perhaps the best modern analogue is the region today that encompasses Australia, Indonesia, and southeastern Asia. Although Australia is colliding with southeastern Asia, a glance at a map of the region reveals not a high monolithic mountain range but a complex interplay of islands, some with mountainous terrain such as New Guinea and some with active island arcs, all separated by water bodies such as the Banda and Celebes seas. Today's map of that region is a snapshot in time that gives a false impression of stability—if we were to come back in just a few million years, the arrangement of seas, islands, and mountains would look very different.

Silurian and Devonian rocks of the Gaspé Peninsula of eastern Quebec provide us with an ancient example. The early Silurian sea there was deep and cold, probably as a result of the late Ordovician glaciation, and clastics were the main deposits. But a warming trend led to shallow marine carbonate sedimentation through much of the rest of the Silurian. In the late Silurian, a carbonate platform developed, with reefs built by sponges, corals, and microbes. The platform was also home to other sea creatures, such as brachiopods, eurypterids, crinoids, and primitive fish. Toward the end of the Silurian, relative sea level temporarily fell as Ganderia finally docked, and the region became exposed above water; consequently, the reef builders died off.

During the last 3 million years of the Silurian, relative sea level rose again over a wide area of the Appalachian Orogen. A new carbonate platform developed that stretched over 1,000 kilometres

Life in a sheltered Silurian lagoon about 420 million years ago. Shallow, warm embayments were havens for eurypterids (sea scorpions), swimming predators up to 2 metres long and among the largest arthropods known. Their segmented bodies were usually streamlined and tapered to a tail end variably developed into a narrow spike or broad leaf-like structure. Their jointed limbs, which bore spines, pincers, and paddles, were attached beneath the head shield. Also shown are several small shrimp-like crustaceans, swimming across the path of the eurypterids, and a cluster of snails grazing among clumps of green algae on the seabed. PAINTING COPYRIGHT MARIANNE COLLINS, ARTIST.

By the middle Devonian, large areas of the Appalachian Orogen were land and washed by rivers that left redbeds as their legacy, as here west of Miguasha, on Quebec's Gaspé Peninsula. MIKE MELCHIN.

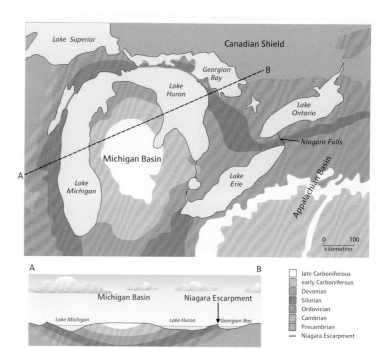

At the top is a geological map of the Michigan Basin and adjacent areas showing the ages of the rocks exposed at the surface (after glacial deposits have been removed) and the course of the Niagara Escarpment. Below is a section across the basin from west of Lake Michigan to the Canadian Shield as shown by the line from A to B in the map. ADAPTED FROM DOYLE AND STEEL (2003).

from western Newfoundland to southern Quebec and New York State, with the best exposures now found in the Port-Daniel area of southern Gaspé. Reefs dominated by stromatoporoids produced structures up to 600 metres from top to bottom. Though this dimension represents a cumulative buildup through time and not a great hill on the sea floor, it does demonstrate the ability of the reef builders to stay abreast of rising sea level for long intervals. But as the sea continued to deepen in the earliest Devonian, eventually the reef builders succumbed. The reefs became buried under an apron of clastic sediment eroded from the rising Acadian Mountains. As uplift continued, marine deposits were replaced by a thick sequence of river-borne detritus shed from the new mountains, including a major delta complex in the Appalachian and Michigan basins known as the Catskill Delta. Thick, terrestrial Catskill sequences of red clastic sedimentary rocks now underlie much of New York State and adjacent areas. The outer, muddy to sandy marine parts of the delta complex underlie some areas of southern Quebec.

Basins and Arches

The ups and downs of sea level in the Appalachian Orogen during the Silurian and early Devonian were largely regional events driven by tectonic activity. Changes in global sea level are best seen away from tectonic influences. As we noted in the previous chapter, global sea level fell dramatically in the latest Ordovician because of all the water locked up in ice caps, but rose equally dramatically in the early Silurian as the ice melted. The warming temperatures and low latitude led to the deposition of carbonates over much of the continental interior of Euramerica.

At times, seas probably connected basins in and near the Appalachian Orogen to the Michigan Basin, though any sediments deposited from these waters in western Quebec and

Looking out through the entrance of a cave cut in Silurian carbonate of the Niagara Escarpment. This cave faces Georgian Bay at Halfway Rock Point, Bruce Peninsula National Park of Canada, Ontario. D. A. WILKES, COPYRIGHT PARKS CANADA.

An aerial view of the Niagara Escarpment in Bruce Peninsula National Park of Canada, Ontario. The Escarpment is formed of erosion-resistant Silurian carbonate deposited in the Michigan Basin. J. BUTTERILL, COPYRIGHT PARKS CANADA.

eastern Ontario have since been eroded away. However, on the flanks of the Michigan Basin in western Ontario, middle Silurian seas left the carbonate that now forms the erosion-resistant caprock of the Niagara Escarpment, and thus of Niagara Falls with its awesome 50-metre plunge. The Niagara Escarpment marks the Michigan Basin's curved edge, which extends from Niagara through Hamilton and Owen Sound, and sweeps up the Bruce Peninsula to Manitoulin Island. Beds that make up the Escarpment dip at a low angle toward central Michigan, leaving the steeper slope facing outward.

Renewed subsidence in the Michigan Basin initially promoted the growth of reefs around its margin. But isolation of the Basin's waters and an arid climate led to periodic drying up of the seas around 420 million years ago, so that enormous thicknesses of gypsum and halite were deposited. As a consequence, commercial deposits now underlie parts of Ontario, with halite (salt) mined at Goderich and Windsor, and gypsum at Caledonia and Hagersville.

In the west, the Tathlina Arch in the western Northwest Territories, the West Alberta Ridge, the Peace River and Sweetgrass arches in Alberta, and the Severn Arch in Manitoba were all

Specimens of the Silurian brachiopod *Virgiana* surrounding a coral, in a carbonate rock from east of Churchill, Manitoba, within the ancient Hudson Bay Basin. GRAHAM YOUNG.

uplifted during the late Silurian to early Devonian. Such areas were commonly topographic highs, from which huge volumes of Cambrian to Silurian sediment were eroded. Despite the presence of these sources of detritus, Silurian rocks preserved in the eastern Rocky Mountains and Western Interior Plains are predominantly carbonates.

Hidden beneath western Saskatchewan and eastern Alberta is a major late Silurian to early Devonian landscape feature, the roughly east-west trending Meadow Lake Escarpment. Now buried beneath several hundred metres of younger sediment, this feature developed in much the same way as the modern Niagara Escarpment. The regions on either side of the Meadow Lake Escarpment were exposed above sea level during most of the early Devonian. But about 400 million years ago, a sea advanced from the northwest as far south as the Meadow Lake Escarpment, as shown by marine deposits of that age in northern Alberta, northwestern Saskatchewan, and westernmost Northwest Territories.

Caledonian Echoes

In the previous chapter, we left Laurentia's paleoeastern margin (today's Arctic margin) on the edge of a vast Ural Ocean littered with microcontinents and other potential terranes. One of these, Pearya, was converging with the continental margin. The Ural Ocean was bordered far to the southeast by the rising Caledonian Mountains. Clastic sediment was shed into the Ural Ocean from these mountains, forming submarine fans that, by 436 million years ago, began to encroach onto both Pearya and the continental margin of Laurentia. By 426 million years ago, deep-water clastic sediments were blanketing what is now central and eastern Ellesmere Island, and had reached Bathurst and Melville islands by 420 and 408 million years ago, respectively. But by about 405 million years ago, warm, shallow seas once more deposited limestone over the present-day Canadian Arctic, including reefs built by stromatoporoids, corals, and microbes.

Like the interior echoes of the Appalachians farther south, Caledonian collisions caused buckling in the Arctic region, as well as some locally significant thrust faulting. During the late Silurian and early Devonian, areas such as the Inglefield and Boothia uplifts rose repeatedly above sea level to form promontories or islands with steep cliffs plunging into deeper water; evidence is provided locally by giant blocks of limestone transported by submarine slides into finer-grained, deeper-water deposits. Due to their position on the flanks of the Boothia Uplift, Silurian sediments on eastern Bathurst Island have never been deeply buried or deformed. As a result, they contain a remarkable and complete sequence of graptolites and radiolaria. The exquisitely preserved fossils represent organisms that lived in the surface waters of the Ural Ocean. Other sediments of the same succession—shallow-water limestones and deeper-water mudstones—provide a unique window into the evolution of fish and, as we will discover, some of the world's earliest land plants.

Stromatolite in Silurian carbonates, Prince Leopold Island, off the northeastern corner of Somerset Island, Nunavut. NORA SPENCER.

Silurian carbonates deposited on the Euramerican margin of the Ural Ocean, now exposed on Prince Leopold Island, off the northeastern corner of Somerset Island, Nunavut. NORA SPENCER.

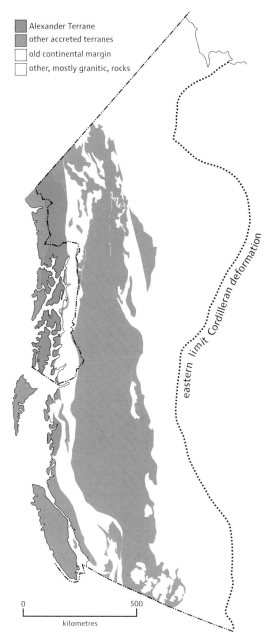

Modern distribution of rocks of the Alexander Terrane. ADAPTED FROM MONGER AND BERG (1987), COURTESY OF THE US GEOLOGICAL SURVEY.

A Northwest Passage

A contemporary of Pearya is the Alexander Terrane, found today in the far-western Cordillera and separated from the old Laurentian/Euramerican margin by a broad collage of late Paleozoic and early Mesozoic terranes. Alexander Terrane rocks are best and most extensively preserved in the Alexander Archipelago of southeastern Alaska, but also occur as scattered outcrops from the St. Elias Mountains on the Alaska-Canada boundary southward to near Bella Coola in the western Coast Mountains of British Columbia. The Alexander Terrane contains a fragment of an unknown early Paleozoic continental margin, and a sequence of Ediacaran to Silurian magmatic-arc rocks. The latter are the only known magmatic-arc rocks from this interval in the entire Canadian Cordillera.

The Alexander Terrane is clearly exotic, but where did it come from? Hypotheses range from as far away as eastern Australia to the Barents Sea region north of Norway and northeasternmost Russia. Alexander Terrane rocks were deformed in two episodes, 500 to 475 and 425 to 415 million years ago. These episodes correlate well with orogenies in the Caledonian Orogen that resulted from the collision of Baltica with Iapetan terranes and Laurentia. Moreover, some red-brown clastic rocks exposed in the Alexander Archipelago are reminiscent of rocks of the classic western European Devonian "Old Red Sandstone"—and are of about the same age. And paleomagnetic and fossil evidence indicate that in the early Devonian the Alexander Terrane lay about 14° north of the paleoequator, in the Ural Ocean between Euramerica and Siberia (see figure on page 115). Armed with such evidence, most geologists now think that the Alexander Terrane originated in the area of the present-day Barents Sea in the Caledonian Orogen. From there it migrated across what is now the Arctic region through the Ural Ocean in the middle Paleozoic—as one of the litter of microcontinents in that ocean, perhaps even connected to Pearya for a while—and had entered Panthalassa by the late Carboniferous.

Mud cracks in Devonian mudstone of the Alexander Terrane on Prince of Wales Island, southeastern Alaska. This rock is similar to parts of the Devonian "Old Red Sandstone" of northwestern Europe. Metamorphosed equivalents of these strata occur in the southwestern Coast Mountains of British Columbia. JIM MONGER.

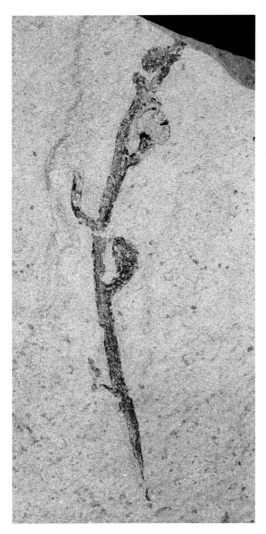

An early land plant from late Silurian rocks on Bathurst Island, Nunavut. JIM BASINGER.

Taking Root

While Pearya and the Alexander Terrane were traversing the Ural Ocean, plants were taking root on land. Plant fossils from eastern Bathurst Island in the Arctic and from sites bordering Chaleur Bay in New Brunswick and Quebec have played an important role in unravelling the story of early land-plant evolution (Box 7). The Bathurst Island fossils are found along the western margin of the Boothia Uplift, in relatively deep-water mudstones and muddy limestones. They date from the late Silurian to early Devonian. Some plants are preserved whole but most are fragments, including root bundles, smooth, scaly, and thorny stems, and clusters of sporangia pregnant with spores. The plants were vascular but small, leafless, fern-like, and herbaceous (lacking woody tissues). By about 410 million years ago, plants locally formed a low ground cover, the tallest growing to about 2 metres. Most common were *Bathurstia*, *Zosterophyllum*, and *Distichophyton*, all of which are possible ancestors of the late Devonian and Carboniferous clubmosses.

An absence of more extensive land-derived organic debris such as coal in the late Silurian marine sediments suggests that the neighbouring land, likely on the Boothia Uplift, was largely barren. The plants probably lived mostly in damp habitats, perhaps along shorelines and the banks of lakes, rivers, estuaries, and tidal channels. From there they could have been easily washed out to sea to come to rest in marine sediment.

The nature of early vascular plants is also well illustrated by fossils from the shores of Chaleur Bay. Early Devonian strata at Dalhousie Junction, New Brunswick, were deposited in the estuary of a river flowing off of former Ganderia into one of the complex of marine basins of the region. Although the plants found in these rocks would look strange to us, they include the forerunners of later forms, both with and without seeds. *Psilophyton* and *Sawdonia* had branches but no leaves. Thus, their chloroplasts may have been in stems and branches, making for a vivid

An early Devonian millipede, *Gaspestria genselorum*, from Point La Nim, near Dalhousie Junction, New Brunswick. This specimen is one of the oldest-known land animals in North America. RANDALL MILLER, COURTESY OF THE NEW BRUNSWICK MUSEUM.

green plant. Some of the branches of these early plants had bulbous endings or lateral growths that were early sporangia.

As plants proliferated and soils became more nutrient-rich with rotting organic material, natural selection favoured those invertebrates that could use the new resources. The earliest-known body fossils (as opposed to trace fossils) of land animals in North America are from early Devonian rocks at Dalhousie Junction. The remains are all of arthropods, and include scorpions as well as millipedes and their giant arthropleurid cousins. Most impressive was an arthropleurid that was perhaps 15 to 20 centimetres long—large enough to be scary even today. And so life had a toehold on land by the early Devonian.

Cordilleran Conception

While plants were taking root, an orogen was in the throes of conception. By the middle Devonian, a passive margin similar to that off eastern Canada today had persisted along the western (originally paleonorthern) margin of Laurentia/Euramerica for at least 150 million years. However, about 390 million years ago, the continent's western edge became an active plate boundary. Imagine the present-day eastern margin of North America, say off Nova Scotia or Newfoundland, developing a subduction zone and all the earthquake and volcanic activity that would ensue. It's hard to picture, but this is exactly what happened in the west in the Devonian. And the margin has remained an active one since then, as witnessed by volcanoes and earthquakes along North America's Pacific rim today. We don't know why the margin changed from passive to active, but it may have been in response to events on the eastern margin of Euramerica. The timing of the transition around 390 million years ago coincides with the end of the Acadian Orogeny—Avalonia could be stuffed no farther beneath the Ganderian Margin of Euramerica. Perhaps the pressure was taken up by a break on the western side of the rigid continent.

In the middle Devonian, the oldest, coldest, and densest oceanic lithosphere of eastern Protopanthalassa flanked western Euramerican continental lithosphere. The latter had been stretched and thinned during the breakup of Rodinia and was laced with old rift-related normal faults. As compression began and the crust in the region had to shorten rather than stretch, these former normal faults became reactivated as reverse faults. Eventually, the lithosphere ruptured and the ocean floor began to subduct. Evidence for compression in the region comes from middle Devonian faulting and folding in the Columbia Mountains of southeastern British Columbia. Evidence for subduction and magmatic arcs comes from middle and late Devonian granitic plutons and volcanic rocks in the Columbia Mountains, seen for example along the Trans-Canada Highway about 25 kilometres east of Revelstoke, as well as from late Devonian volcanic ash beds in the Rocky Mountain Front Ranges at Exshaw in Alberta.

At the end of the Devonian, the magmatic arc on the new convergent margin started to move away from the continent, much as Japan migrated away from Asia in the Cenozoic. Between the arc and the continent, a back-arc basin, called the Slide Mountain Basin, developed. This basin, floored in part by oceanic lithosphere, is analogous to the modern Sea of Japan. Basalt and sedimentary rocks of the Slide Mountain Basin are preserved as the Slide Mountain Terrane and can be seen near Sliding Mountain, along the highway west of the former gold-mining town of Barkerville in eastern British Columbia.

Mountains Come and Mountains Go

By the late Devonian, the Caledonian Mountains of eastern Greenland and Scandinavia had been eroding away for more than 25 million years. The plains in front of these mountains in what is now central Greenland were crossed by rivers carrying their sediment to deltas that covered large parts of present-day Arctic Canada. From about 385 million years ago, Pearya began to collide with

View to the northwest of Devonian strata in the Columbia Mountains of British Columbia. Detailed mapping of these Devonian strata has yielded evidence of compression, marking the change from a passive intraplate margin to an active plate margin on the west side of Euramerica about 390 million years ago. KEVIN ROOT.

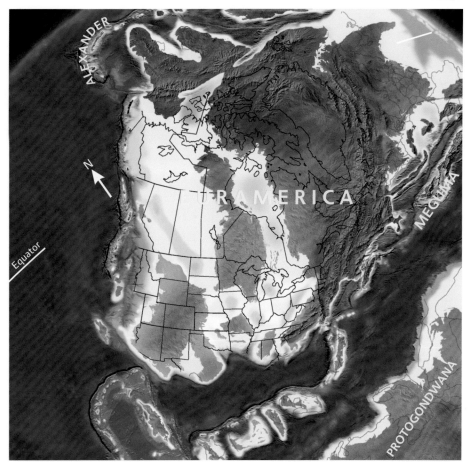

Paleogeography of what was to become North America and adjacent regions in the middle Devonian, 385 million years ago. Land is shown in brown, with shading showing topography. The lighter blue areas represent possible coastal or nearshore areas, darker blue represents deeper ocean waters, and black indicates trenches. Aspects of modern geography are shown for orientation.

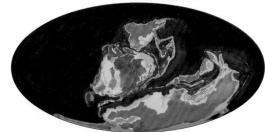

Global paleogeography 385 million years ago, during the middle Devonian. Colours as for previous figure.

Euramerica, creating new highlands during the late Devonian. The episode of deformation caused by the docking of Pearya is known as the Ellesmerian Orogeny, and the mountains that it produced are known as the Franklinian Mountains, whose roots can now be found across northern Ellesmere Island and adjacent areas. The combined influxes of detritus from the nearby Franklinian and distant Caledonian mountains produced an apron of middle and late Devonian fluvial, estuarine, and deltaic sediments up to 10 kilometres thick, now preserved on southern Ellesmere Island, from Devon Island to Banks Island, and in northern Yukon.

Buried in this thick pile of sediment are remains of some of the earliest trees, known as *Archaeopteris*. This majestic tree probably rose from a forest floor carpeted with herbaceous clubmosses, and with thickets of larger horsetails growing along riverbanks and shores. The presence of scales, teeth, and bone fragments in the sediments show that fish thrived in the local waters.

Convergence between Pearya and Euramerica continued until almost the end of the Devonian. Tectonic compression is reflected by thrust faulting and majestic folds, the latter clearly visible on satellite images of Bathurst and Melville islands.

What caused such distinctive folding? The shove came from Pearya to the north, but more important in explaining the sweeping nature of the folds are Ordovician evaporite layers 3 to 5 kilometres underground and mobile Devonian shale much closer to the surface. These easily deformed layers provided lubrication for the thrusting and folding of a thick, stiffer limestone sandwiched in between them. The Ellesmerian Orogeny also squeezed fluids such as oil and metal-bearing brines out of early Paleozoic shales. Precipitation from some of these brines led to the formation of zinc and lead ores, including those mined at the now-closed Polaris Mine on Little Cornwallis Island.

Once the continental lithosphere of Pearya could go no farther beneath␣Eurameriĉa, collision and uplift stalled and erosion wore down the Franklinian Mountains.

Satellite image of Bathurst Island, Nunavut, showing looping folds produced by Devonian deformation during the Ellesmerian Orogeny. FROM GIBSON ET AL. (2010); COURTESY OF CANADA CENTRE FOR REMOTE SENSING.

Late Devonian strata at Miguasha National Park, Quebec. These rocks contain beautifully preserved fossil fish, and other animals and plants. JEAN-PIERRE SYLVESTRE, COURTESY OF MIGUASHA NATIONAL PARK.

A Finny Story

Silurian and early Devonian fish, armed with jaws, teeth, and fins, were much more sophisticated than their Cambrian and Ordovician forebears (Chapter 4). Among the Silurian jawed types were ray-finned fish, a group still with us today—indeed, the vast majority of living fish, from cod and salmon to sea horses and sticklebacks, are ray-finned fish, the most successful vertebrates on the planet. Each individual has pairs of pectoral and pelvic fins, supported by pectoral and pelvic girdles. Lobe-finned fish, unlike ray-finned fish, had functional lungs and fin bones arranged like the limb bones of tetrapods. First appearing during the late Silurian, lobe-finned fish represent a critical step in vertebrate evolution.

Beautifully preserved late Devonian fish fossils can be seen at Miguasha National Park, on the shores of Chaleur Bay in southeastern Quebec. This site was designated a UNESCO World Heritage Site in 1999, in recognition of its outstanding fish fossils. The rocks at Miguasha were laid down under tropical conditions, probably in an embayment connected occasionally to the ocean, with the craggy peaks of the Acadian Mountains towering nearby. Organisms washed into the sediments included land-dwelling scorpions and spiders, as well as *Archaeopteris*

Fern-like foliage of *Archaeopteris halliana*, an early tree from late Devonian strata at Miguasha National Park, Quebec. *Archaeopteris* trees were up to 7 metres tall and formed the first forests. STEVE DESCHÊNES, COURTESY OF MIGUASHA NATIONAL PARK.

A Devonian underwater scene based on fossil fish found at Miguasha National Park in Quebec. Swimming toward the surface, to the right of centre and above the grass-like plants, is *Scaumenacia*, a member of the group of lobe-fins called lungfish. Dominating the scene and in the process of finishing a snack, *Eusthenopteron* is circling in search of additional prey. *Eusthenopteron* is also a lobe-fin but belongs to the group that gave rise to the tetrapods. Hugging the estuary bottom to the right and relatively small is the jawless *Escuminaspis*, and above it the much larger *Plourdosteus*, a voracious carnivore with jaws. FROM CLOUTIER (2001), USED WITH PERMISSION; PAINTING BY FRANÇOIS MIVILLE-DESCHÊNES.

Specimen of the Devonian lobe-finned fish *Eusthenopteron foordi* from Miguasha National Park, Quebec. This fish had pectoral fins similar to tetrapod limbs, lungs and gills, and internal nostrils that allowed it to breathe with its mouth closed. JEAN-PIERRE SYLVESTRE, COURTESY OF MIGUASHA NATIONAL PARK.

foliage. But most remarkable were the fish. One important find was *Scaumenacia*, a member of a group of lobe-fins called lungfish. *Eusthenopteron*, first described and best known from Miguasha, belongs to another group of lobe-fins that gave rise to tetrapods. Exquisitely preserved three-dimensional specimens of *Eusthenopteron* not only show the tetrapod-like fin structure, but also that the skull structure was very similar to that of tetrapods. Some of the *Eusthenopteron* skulls from Miguasha are so well preserved that they even show the traces of blood vessels.

Although *Eusthenopteron* gives us a glimpse of how our closest fish ancestors may have looked, we now know of a more advanced form, which was discovered in 2004 in late Devonian rocks slightly older than those at Miguasha. The lobe-fin *Tiktaalik* was found on southern Ellesmere Island, in the thick package of sediments shed from the rising Franklinian Mountains and deposited 370 million years ago. *Tiktaalik* is a lobe-finned fish, but a very special one. Its pectoral fins contain wrist bones and fingers, indicating that it could prop itself up in the water, even in fast-flowing streams. It also had a sturdy ribcage and a flexible neck. No self-respecting fish has a flexible neck, but tetrapods do.

Braincase of the earliest-known shark, *Doliodus*, from early Devonian rocks at Atholville, New Brunswick. RANDALL MILLER, COURTESY OF THE NEW BRUNSWICK MUSEUM.

Part of the skeleton of *Tiktaalik roseae*, showing its flexible neck. This animal fills the gap between fish and land animals. This landmark fossil was found in late Devonian rocks, about 375 million years old, from Ellesmere Island, Nunavut. COURTESY OF TED DAESCHLER, ACADEMY OF NATURAL SCIENCES OF PHILADELPHIA (FOR IMAGE OF THE SPECIMEN).

Artist's reconstruction of *Tiktaalik roseae* in top and side views. ARTWORK BY FLICK FORD, RECONSTRUCTION COURTESY OF THE ACADEMY OF NATURAL SCIENCES OF PHILADELPHIA.

Collision Course Act Three

By the late Devonian, over 100 million years of plate convergence had led to crustal thickening and uplift in the Appalachian Orogen. The northern parts of the Orogen, embracing Miguasha, now lay well within Euramerica. To the south along the outer Avalonian margin of the continent, the Appalachians faced the Rheic Ocean. This ocean narrowed during the Devonian as Protogondwana drew closer. By about 390 million years ago, Meguma was colliding with the Avalonian margin of Euramerica. Meguma appears to have pushed sideways against Avalonia in an event known as the Neoacadian Orogeny, during which Meguma's mainly clastic Cambrian to Devonian rocks were folded and metamorphosed. An ancient transform plate boundary, called the Cobequid-Chedabucto Fault System and akin to today's San Andreas Fault System in California, represents the zone along which Meguma docked. It is the most fundamental geological structure in Nova Scotia, contributing significantly to the province's distinctive shape.

Crustal melting during the late Devonian produced 380-to 360-million-year-old granitic plutons and batholiths that underlie large areas of mainland Nova Scotia and the adjacent continental margin offshore. Anyone who has visited the lighthouse at Peggys Cove, near Halifax, has set foot on the South Mountain Batholith, the largest granitic intrusion in the Appalachian Orogen. Originally emplaced some 6 to 10 kilometres beneath the surface, the granitic magma first intruded along fractures and then melted and assimilated large volumes of the metasedimentary host rock. This process has been "frozen in motion" along the shoreline of Kejimkujik National Park of Canada Seaside, near Liverpool, Nova Scotia, where the margin of a smaller pluton is well exposed.

The melting that formed these granitic bodies may be explained by a subducting remnant of the Rheic Ocean south of the Meguma Terrane, although evidence for this is likely hidden today under the thick cover of much younger sedimentary rocks on the continental margin off Nova Scotia. The margin between the Rheic Ocean and Meguma may have been a transform one, as the main mass of Protogondwana began to dock with Euramerica and proceeded to slide southwestward to culminate, 340 to 260 million years ago, in the Alleghanian Orogeny recorded in the rocks of the southern Appalachians.

Reefs and Salty Seas

While the edge of the continent west of the modern Rocky Mountains was transitioning from a passive to an active margin, the region that is now the eastern Rocky Mountains and Western Interior Plains was relatively quiet tectonically. It was generally covered by shallow seas as the waters of Protopanthalassa lapped onto the continent. Today, Devonian and Carboniferous

Aerial view of a road following the surface trace of the Cobequid Fault near Port Greville, Nova Scotia. This fault is a component of the Cobequid-Chedabucto Fault System along which the Meguma microcontinent docked with the Avalonian margin of Euramerica during the Devonian and Carboniferous. HOWARD DONOHOE.

Cambrian-Ordovician metasedimentary rocks of the Meguma Terrane at Feltzen South, near Lunenburg, Nova Scotia. The strata were folded and metamorphosed during the Neoacadian Orogeny. ROB FENSOME.

The shoreline at Prospect, Nova Scotia, is underlain by, and littered with boulders of granite from the Devonian South Mountain Batholith. ROB FENSOME.

The west half of Cirrus Mountain (middle distance, viewed from the highway leading up to Sunwapta Pass in Alberta) exposes the classic threefold structure of the Front Ranges of the southern Rocky Mountains: lower cliffs of Devonian carbonate, middle slopes of Carboniferous shale, and upper cliffs of Carboniferous carbonate. CHRIS YORATH.

rocks outcrop around the edges of this region. For example, to the west, excellent exposures of late Devonian and Carboniferous rocks can be seen in many places in the Rocky Mountains, as on the steep slopes of Cascade Mountain and Mount Rundle near Banff and at Roche Miette and Roche à Perdrix near Jasper. In the Crowsnest Pass area of southwestern Alberta, visitors to the Frank Slide site can walk among blocks of early Carboniferous limestone that toppled from Turtle Mountain (Chapter 17). To the north, outcrops of middle to late Devonian strata occur in the southwestern Northwest Territories, for example to the south of Great Slave Lake, where they are flat-lying and rest unconformably on Precambrian rocks. To the east undeformed, flat-lying Devonian strata outcrop near Lake Winnipegosis in Manitoba.

Devonian and Carboniferous rocks continuous with these exposures to the west, north, and east now lie deep beneath the Mesozoic and Cenozoic rocks of the Western Interior Plains. Despite their deep burial, these rocks have been mapped and studied in detail because of numerous boreholes drilled during petroleum and potash exploration and production. Information gleaned from both buried rocks and exposures has revealed the Devonian and Carboniferous history of western interior Euramerica. During

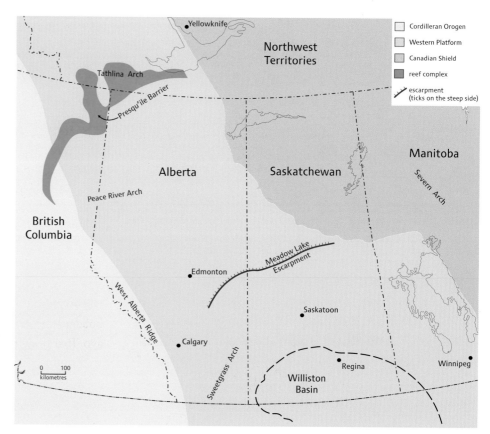

Some important paleogeographic and tectonic features on the Western Platform and adjacent areas during the Silurian to Permian. The Canadian Shield and Cordilleran Orogen are shown for orientation, but these areas represent more recent geological developments and were not present during the late Paleozoic. ADAPTED FROM VARIOUS SOURCES.

the early Devonian, the sea was largely restricted to the region north of the Meadow Lake Escarpment. But about 390 million years ago, the sea started to advance from the northwest across the Escarpment, eventually reaching present-day North Dakota. As the sea was advancing, a carbonate reef complex known as

Mountainside exposing two late Devonian reefs, Flathead Range, southwestern Alberta. BRIAN PRATT.

the Presqu'ile Barrier developed, trending from north-central Alberta to the southwestern Northwest Territories. North of the Presqu'ile Barrier mostly mud (now shale) was deposited, but to the south of the Barrier mainly carbonates accumulated, including many reefs.

As the Presqu'ile Barrier continued to grow, seawater circulation to its south became restricted. In combination with the hot, dry climate, this restriction led to extensive accumulations of evaporites, including Saskatchewan's rich potash deposits. Following deposition of these evaporites, the sea again retreated, only to transgress and regress many more times during the remainder of the Devonian. Several hundred metres of preserved carbonates and marine shales confirm the predominance of marine conditions. Presumably, the sea sometimes covered large areas of what is now the Canadian Shield, but subsequent erosion has removed any evidence of this.

Potash ore from Devonian subsurface deposits in Saskatchewan. The ore, an evaporite rock called sylvinite, consists of a mixture of halite, sylvite (potassium chloride), carnallite (a complex mineral that includes potassium, chlorine, and magnesium), and other minerals. Sylvite is the principal ore mineral and source of potash. The red colour reflects iron oxide staining. DAN KOHLRUSS, SASKATCHEWAN GEOLOGICAL SURVEY.

Part of a diorama at the Royal Tyrrell Museum of Palaeontology, Drumheller, Alberta, depicting a late Devonian reef community. The animals include feathery pink crinoids (bottom left and centre), corals of various types, small brown brachiopods attached to the substrate, a jellyfish (upper left), and a loosely coiled nautiloid. ALAN MORGAN, DIORAMA COURTESY OF THE ROYAL TYRRELL MUSEUM OF PALAEONTOLOGY.

An example of a late Devonian reef is the Swan Hills complex. Now deeply burried beneath younger rocks, this reef complex began to develop about 385 million years ago on the northeastern flank of the West Alberta Ridge. Other examples, built about 380 million years ago, form a complex chain extending north and south of Edmonton; petroleum reservoirs in the Leduc and Redwater oil fields are part of this chain. A stromatoporoid-coral reef that developed at about the same time as the Leduc reefs may be seen on a short hike up the Grassi Lakes Trail above the town of Canmore in Alberta. Reef complexes continued to be important until near the end of the Devonian, but were much fewer and smaller after their main builders, the stromatoporoids, almost died out 375 million years ago. That time marks the late-Devonian mass-extinction event, when almost 60 percent of all marine genera were lost. The time span that the extinction encompassed is much debated, with estimates ranging from a geologically dramatic 500,000 years to a much more sedate 25 million years.

During the latest Devonian and earliest Carboniferous, mainly clastic sediments were deposited by seas on the Canadian part of western Euramerica. Carboniferous seas continued to advance and retreat over the region, probably due in part to the advance and retreat of ice sheets on Protogondwana, which lay over the South Pole at that time. Warm temperatures are reflected by the return of carbonates, sometimes interfingered with clastics and evaporites. However, Carboniferous seas in

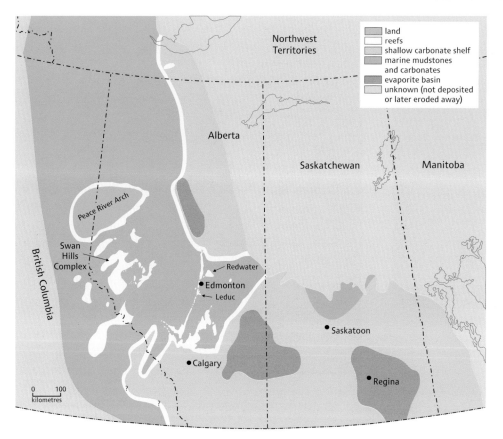

Reefs and other important features on the Western Platform and adjacent areas during the late Devonian. The locations of the reefs now found in the Rocky Mountains Foothills and mountains are shown on this map in their probable original locations before Paleozoic rocks were thrust eastward during the Mesozoic (Chapter 9). ADAPTED BY KEN POTMA FROM WEISENBERGER AND POTMA (2001).

this part of Euramerica generally became shallower and less extensive with each advance.

Devonian and Carboniferous rocks have provided much wealth in the Prairie Provinces and adjacent areas. The generally porous nature of the middle and late Devonian reefs makes them ideal reservoir rocks for petroleum (Chapter 13), and Devonian and early Carboniferous organic-rich shales are important petroleum source rocks. And mining should not be forgotten: in addition to the rich potash deposits of Saskatchewan, reefs in the Presqu'ile Barrier host important zinc and lead ores, including those of the now-closed Pine Point mining district south of Great Slave Lake.

Ranges and Basins

As Protogondwana evidently slipped past the Meguma margin of Euramerica along a series of right-lateral strike-slip faults, the Rheic Ocean closed progressively southwestward in a zipper-like fashion. In what is now southeastern Canada, this complex plate interaction involved compression in some places and stretching in others, leading to the development of a series of fault-bounded uplands and depressions. Collectively, these depressions (we'll call them sub-basins) and intervening uplands are known as the Maritimes Basin. From drill cores and seismic data, we know that late Devonian to Permian rocks of the Maritimes Basin underlie the modern sea floor in most of the central and southern Gulf of St. Lawrence and the southern part of the Grand Banks of Newfoundland, as well as many adjacent onshore areas.

The sub-basins and uplands of the Maritimes Basin had a major impact on regional paleogeography in the late Devonian to Permian and also greatly influence the modern landscape. For example, the Minas and Cumberland sub-basins are still topographic lows and largely underwater, whereas the Gaspé, Miramichi, Caledonia, Cobequid, Antigonish, and Cape Breton highlands echo Carboniferous uplands. So, the modern landscape (and seascape) of southeastern Canada strongly reflects the ancient Carboniferous landscape. In parts of the Maritimes Basin, up to 12 kilometres of sediment accumulated, though with stretches of time unrepresented. The preserved sedimentary rocks thus represent short-lived episodes of subsidence and deposition that were interrupted by intervals of compression, uplift, and erosion.

The oldest strata in the Basin are late Devonian coarse-grained, red clastic deposits, locally interbedded with volcanic rocks. After a long dry period, wetter conditions returned briefly in the earliest Carboniferous. Evidently, rivers transported large volumes of sediment from the highlands across broad inland

Late Devonian conglomerate (red rocks in foreground and to the left) and mafic volcanics (grey rocks), Ballantynes Cove, Nova Scotia. ROB FENSOME.

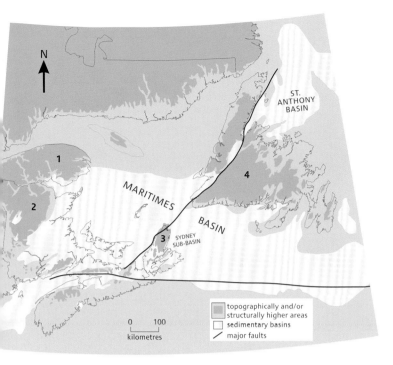

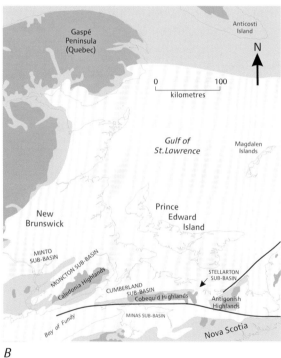

A. The late Devonian to Permian sedimentary basins and sub-basins of Atlantic Canada and surrounding areas. The highlands include: 1, Gaspé Highlands; 2, Miramichi Highlands; 3, Cape Breton Highlands; and 4, Western Newfoundland Highlands. B. Detail of the western Gulf of St. Lawrence and surrounding areas. ADAPTED FROM VARIOUS SOURCES.

plains and into both ephemeral and longer-lasting lakes, to leave accumulations of grey mud and sand. One important site is Blue Beach, near Hantsport in Nova Scotia; fossils found there provide a unique view on terrestrial and freshwater life around 355 million years ago. Although scattered earlier tetrapod tracks are known, Blue Beach is our first real window on how diverse the tetrapod community had become by the earliest Carboniferous.

About 340 million years ago the Maritimes Basin lay about ten degrees south of the paleoequator and had a hot, dry, tropical climate. During this dry spell, the Basin was invaded by the Windsor Sea, which waxed and waned intermittently over the next 15 million years, finally withdrawing about 325 million years ago. At its height, the sea flooded all but the crests of the highlands. The Windsor Sea periodically dried up to precipitate evaporites, interbedded with red mudstone. As evaporation progressed, gypsum, halite, and potash were successively deposited. When the sea returned it laid down marine carbonates, sometimes as local reefs. As with the repeated transgressions and regressions over the Western Platform, these Windsor cycles may reflect retreats and advances of the ice sheets in Protogondwana.

Windsor evaporites are widespread and economically important in Atlantic Canada. They include potash beds near Sussex in New Brunswick, halite deposits at Pugwash in Nova Scotia, and widespread gypsum in the Codroy Valley of Newfoundland and across Nova Scotia. Gypsum forms the unusual cliffs seen from Highway 101 near Windsor, Nova Scotia. Gypsum also underlies much of Bras d'Or Lake—indeed, dissolution of gypsum and halite may be a reason why this deep body of modern seawater reaches into the interior of Cape Breton Island.

Gypsum cliffs along the St. Croix River, near Windsor, Nova Scotia. The gypsum was deposited in the early Carboniferous Windsor Sea. WAYNE GARLAND.

An early Carboniferous underwater lake scene. Three types of early ray-finned fish are shown: silvery purple *Rhadinichthys* swimming in a school toward the left; three larger, brownish-grey *Elonichthys* on the left; and a deep-bodied form akin to *Eurynotus* at right. In the background, coloured green, is the large lobe-finned fish *Latvius*. The colours are guesses, but the morphology is based on fossil finds from rocks at Albert Mines, New Brunswick. PAINTING BY JUDI PENNANEN, COURTESY OF THE NEW BRUNSWICK MUSEUM.

As later deposits piled above the Windsor succession, their growing weight caused the thick, low-density evaporites to flow upward, sometimes along faults, disrupting the overlying younger rocks. In extreme cases, huge columns of halite, called salt diapirs, formed within the sediment pile, some more than 7 kilometres from base to top. A vast field of diapirs underlies the eastern Gulf of St. Lawrence, and the Magdalen Islands exist because they sit atop diapiric salt structures.

A New Arctic Passage

During the early Carboniferous, as the compressive forces of the Ellesmerian Orogeny relaxed and the Franklinian Mountains subsided and eroded, small rift basins developed in northern Ellesmere Island. The change in tectonic regime is also reflected across much of Arctic Canada in a major unconformity separating deformed Devonian and older rocks from less deformed younger rocks. This unconformity is well exposed amid stunning scenery in Auyuittuq National Park of Canada on northern Ellesmere Island.

The small early Carboniferous basins in northern Ellesmere Island contain lacustrine sediments interfingered with fluvial deposits shed from the eroding mountains. Preserved wood in these rocks implies that the climate there 330 million years ago was hot and damp, as would befit its low mid-paleolatitude position. The lacustrine muds, now dark shales, are rich in organic matter, which tells us that the surface waters were oxic and hence a good place for life, but that the bottom waters were anoxic, hampering decay.

By 325 million years ago, Euramerica had migrated northward and what is now Arctic Canada had drifted into hot, increasingly arid climes of paleolatitudes 10 to 40° north. The small basins evolved into an extensive belt of larger rift basins and uplifts that became filled with lacustrine and fluvial deposits. Tidal-flat muds and tongues of shallow-marine limestone and evaporites indicate occasional invasions of the sea. And volcanic rocks on northern Ellesmere Island show that rift-related volcanoes erupted intermittently.

The rift valleys became wider and deeper after about 320 million years ago, with marine incursions becoming more frequent

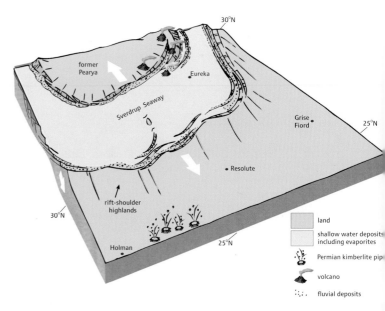

Paleogeography of what was to become the Canadian Arctic in the middle Carboniferous to Permian, 325 to 285 million years ago. The approximate locations of Holman, Resolute, Grise Fiord, and Eureka are included to provide a guide for orientation, as well as paleolatitudes. The focus for the region, now approaching mid-latitudes, was the Sverdrup Seaway, a broad rift basin in which shallow-water deposits, including evaporites, accumulated. The arrows show the direction of crustal tension. Kimberlite pipes, the source of diamond deposits, are discussed in Chapter 12.

and longer lasting. These incursions ultimately led to the formation of a marine passage known as the Sverdrup Seaway. For the next 10 million years, shallow briny seas alternated with intervals when evaporation predominated. A great halite accumulation, possibly over 1 kilometre thick, is preserved in a 50-kilometre-wide belt that runs southwest-northeast for over 800 kilometres from northern Melville Island to western Axel Heiberg Island, and as far as the Barents Sea region north of Norway—a distance then, of course, not as great as now because the Arctic Ocean did not form until the Mesozoic.

After 310 million years ago, the Sverdrup Seaway deepened. Black shales and dark, muddy limestones were now being deposited in open-water areas where, not too long before, halite had accumulated. Carbonate banks formed over structural highs along the flanks of the Seaway. The lateral transition from pale shallower-water limestone to dark deep-water shale is strikingly exposed in towering cliff faces north of Hare Fiord on northern Ellesmere Island. Fossil communities formed carbonate mounds, patch reefs, shelf-edge reefs, and carbonate banks that can be up to 1,500 metres thick.

This cliff near Hare Fiord on Ellesmere Island, Nunavut, reveals changing Carboniferous environments. Redbeds near the base are sequentially overlain by evaporites (lighter bands in the lower part of the cliff), shales (thick brown-grey sequence), and carbonates (light-grey sequence near the top). CHRISTOPHER HARRISON.

Colourful but complex geology is evident in this photograph of Strand Fiord, Axel Heiberg Island. At the upper right is a diapir of white Carboniferous salt, originally deposited in the Sverdrup Seaway. In the foreground are near-vertical, grey, fine-grained clastic Jurassic rocks; and on the slopes to the left are Triassic redbeds. CHRISTOPHER HARRISON.

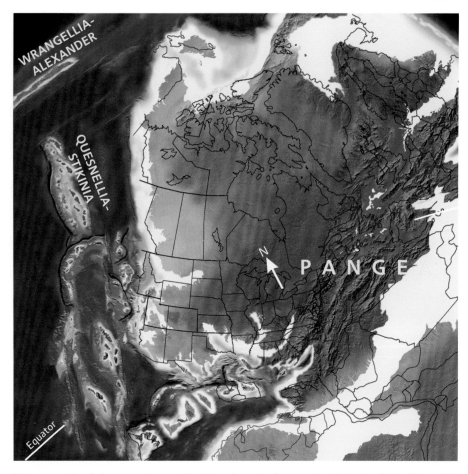

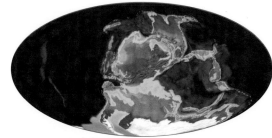

Global paleogeography 315 million years ago, during the late Carboniferous. Colours as for previous figure.

Paleogeography of what was to become North America and adjacent regions in the late Carboniferous, 315 million years ago. Land is shown in brown, with shading showing topography. The lighter blue areas represent possible coastal or nearshore areas, darker blue represents deeper ocean waters, and black indicates trenches. Aspects of modern geography are shown for orientation.

Into the Coal Age

By the late Carboniferous, Euramerica and Protogondwana had collided to form Pangea. As part of Pangea, the Maritimes Basin had drifted northward to straddle the paleoequator by about 318 million years ago. As a result, the Basin's climate changed from hot and dry to hot and humid. Heavy rains must have been the norm, with runoff from the highlands channelled by rivers ultimately emptying into the open sea in what is now western Europe. Throughout the late Carboniferous, depressions such as the Sydney and Cumberland sub-basins in Nova Scotia and the Minto Sub-basin in New Brunswick were subsiding and filling with repeated sequences of peat, flood-plain mud, lacustrine or possibly marginal marine carbonate mud, and fluvial sand. Such repeating sequences are termed cyclothems and occur in Europe as well as Atlantic Canada. Like the cycles recorded in the Windsor Sea deposits and on the Western Platform, they probably reflect the rise and fall of global sea level in response to advancing and retreating ice sheets on former Protogondwana.

Beautifully preserved cyclothem sequences can be seen in the sea cliffs between Boularderie Island and Port Morien in the Sydney Sub-basin of Cape Breton Island, and in the Clifton and Cape Enrage areas of New Brunswick. However, the most famous such cliffs are on the shores of Chignecto Bay in Nova Scotia. The Joggins "fossil cliffs" constitute one of the world's classic late Carboniferous sections. Like Miguasha, Joggins is a UNESCO World Heritage Site,

Late Carboniferous fluvial conglomerate and sandstone form the "flowerpot" sea stacks at Hopewell Cape, New Brunswick. The powerful Bay of Fundy tides undercut the rocks to form the unusual shapes. KEITH VAUGHAN.

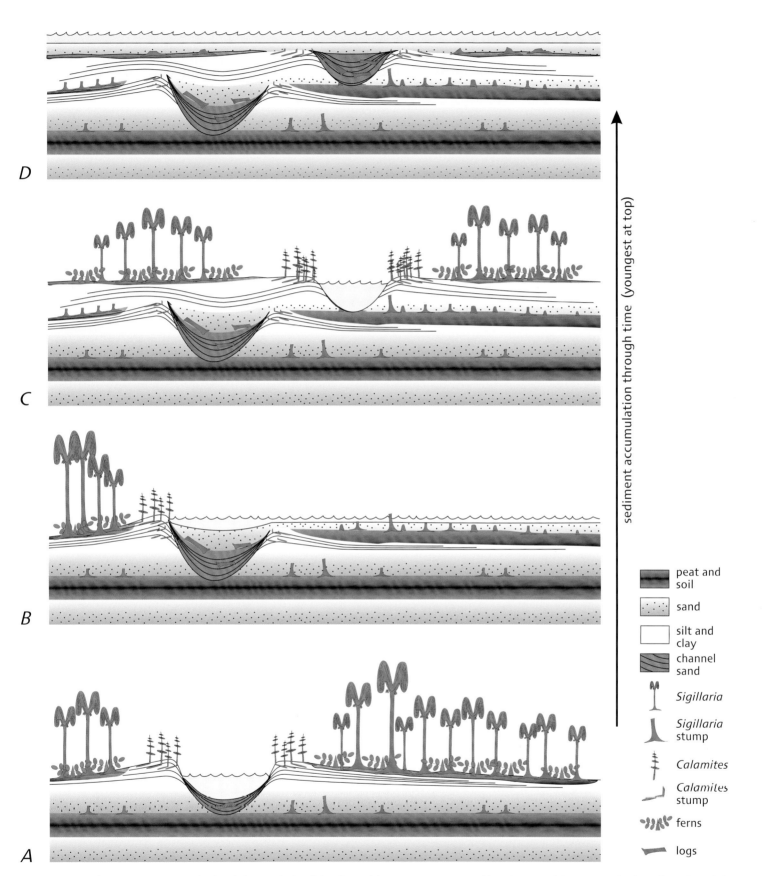

Deposits and environments associated with cyclothems in the late Carboniferous. *A* depicts a river cutting into older sediments and depositing sand and gravel in its channel. The horsetail *Calamites* is growing along the banks, beyond which is a flood plain supporting a forest of clubmoss (*Sigillaria*) trees and ferns. Organic debris from the forest is building up a layer of peat that may eventually form coal. *B* shows the river overflowing its banks during a flood. Sand is deposited by the flood waters, choking the forest, covering the peat, and filling the old channel. Only the stumps of the trees are preserved, protected by the new layer of flood-deposited sand. In *C*, a new river channel is established and a new forest is growing on the flood plain. Periodically this cycle is broken by more general aquatic flooding, as shown in *D*, with possibly brackish waters marginal to a sea whose level rose because of melting ice in distant former Protogondwana; during this aquatic phase, thin, dark limestones full of bivalves (so-called clam coals, not shown) were sometimes deposited. Through time, this process builds up repeating layers of peat, sand, silt, clay, and limestone to produce the cyclothems. ADAPTED FROM ATLANTIC GEOSCIENCE SOCIETY (2001).

Late Carboniferous foliage including horsetail (*Sphenophyllum*; centre), clubmoss branches (right), and a fern (lower left) from Clifton, New Brunswick.
RANDALL MILLER, COURTESY OF THE NEW BRUNSWICK MUSEUM.

receiving this designation in 2008. Charles Lyell was an early visitor. In a letter to his sister dated July 30, 1842, Lyell wrote, "We have just returned from an expedition of three days to the Strait which divides Nova Scotia from New Brunswick, whither I went to see a forest of fossil coal-trees—the most wonderful phenomenon perhaps that I have seen .... This subterranean forest exceeds in extent and quantity of [fossil] timber all that have been discovered in Europe put together."

Forests of the late Carboniferous were different from those of today. No flowering plants existed, and most plants produced spores rather than pollen (Box 7). Preserved lower trunks of once-giant clubmoss trees, such as *Lepidodendron* and *Sigillaria*, are common in the Joggins cliff face, as are their dimpled roots, called *Stigmaria*. Horsetails, recognizable by their jointed stems, were also abundant plants at Joggins. The modern horsetail, *Equisetum*, is a small, widespread non-woody plant that grows in moist places and poor soils. Coal Age horsetails had jointed bamboo-like stems (*Calamites*) and narrow whorled leaves (such as *Asterophyllites*) and could be up to 15 metres tall. These plants favoured the banks of streams and sandy soils, where drainage was better than in the lower-lying wetland habitat of the clubmosses. Other common plants were the true ferns, the now-extinct seed ferns, and an early gymnosperm, *Cordaites*, which had large, strap-shaped leaves. With a reproductive strategy involving pollen and seeds, *Cordaites* was ahead of its time.

Trackways reveal that animal life flourished on the forest floor, and included the myriapod *Arthropleura*, which could be 2 or more metres long. Imagine this wolf-sized multi-legged bug crawling into your sleeping bag! Fortunately, they were plant-eaters. From fossils found elsewhere, we know that the Carboniferous coal forests were also home to the equally intimidating dragonfly-like *Meganeura*, which hovered aloft on 70-centimetre-long wings. The enormous size of these arthropods

Clubmoss tree stump in the cliff at Joggins, Nova Scotia.
ANDREW MACRAE.

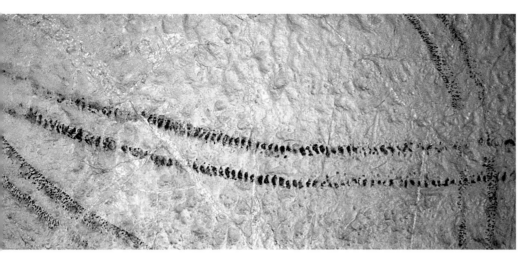

Artificial cast of trackways of the giant myriapod *Arthropleura* from late Carboniferous strata at Joggins, Nova Scotia. This myriapod was a giant, up to 2 metres long. LAING FERGUSON.

Model of *Arthropleura*, on display at the Fundy Geological Museum, Parrsboro, Nova Scotia. ANDREW MACRAE, SPECIMEN COURTESY OF THE FUNDY GEOLOGICAL MUSEUM.

J. William Dawson from Pictou, Nova Scotia. Among many illustrious accomplishments, he was one of the first to interpret fossil plants in an ecological context. REPRODUCED WITH THE PERMISSION OF NATURAL RESOURCES CANADA 2013, COURTESY OF THE GEOLOGICAL SURVEY OF CANADA.

was probably due to the high oxygen content of the atmosphere. This appears to have resulted from unprecedented active photosynthesis (which yields free oxygen) paired with restricted decay (which would consume oxygen) due to the long-term burial of peat. Regardless of its origin, such a richly endowed atmosphere meant that oxygen could diffuse deeply through an arthropod's body, stimulating rapid and continuous growth for long periods, making relative giants possible.

While the arthropods were exploring new-found frontiers, vertebrates were also evolving rapidly. Tetrapods crawled among the trees and rotting logs, leaving behind their bones as fossils. Most were amphibians, such as *Dendrerpeton*, whose well-preserved skulls have been found at Joggins. But the most exciting finds at Joggins were bones of *Hylonomus lyelli* (Lyell's wood mouse), the remains of the earliest-known reptile. The species was named by William Dawson, who made the find together with Lyell in 1852. Dawson, a native of Pictou, Nova Scotia, was to become one of Canada's greatest Victorian scientists, with a career that eventually led him to be Principal of McGill University in Montréal.

Slivers off the Edge

During the late Carboniferous and Permian, the part of Panthalassa that lay off northwestern Pangea probably resembled the present western Pacific Ocean, with its many magmatic island arcs. These arcs had slivered off the Euramerican/Pangean margin when the Slide Mountain Basin opened, and now form parts of Cordilleran terranes known as Quesnellia, Stikinia, and Yukon-Tanana. Another terrane, Wrangellia, is of unknown origin but by the Carboniferous was also in northern Panthalassa. Evidence from their rocks and fossils suggests that the four terranes represent a single arc system that lay off the convergent northwestern margin of the Euramerica/Pangea plate above a subduction zone (or zones) dipping mostly toward the continent.

Earlier in the chapter we left the Alexander Terrane entering Panthalassa by the late Carboniferous after its epic trek through the Ural Ocean. A further clue about its history comes from a group of 309-million-year-old plutons now in southern Alaska. They intrude both the Alexander Terrane and Wrangellia, suggesting that by that time the Alexander Terrane had fused with Wrangellia.

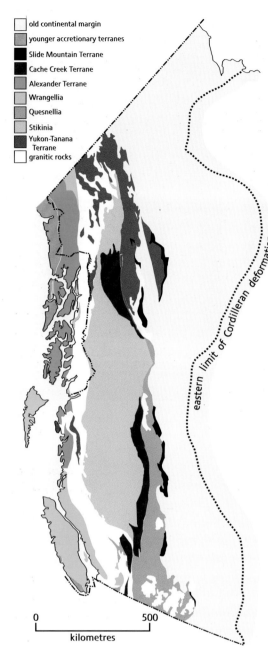

The present-day Cordilleran collage of terranes. ADAPTED FROM MONGER AND BERG (1987), COURTESY OF THE US GEOLOGICAL SURVEY.

One more group of Carboniferous to Permian rocks now occur embedded in the Cordillera. These rocks can be seen in a road cut on the western side of the village of Cache Creek in south-central British Columbia. The exposure is of a mixed-up material called a mélange. The mélange in the road cut has a dark matrix of faulted and broken-up Permian and Triassic radiolarian chert and mudstone. In this matrix are blocks (clasts) of chert and limestone containing late Carboniferous and early Permian fossils. Elsewhere, the clasts include pillow basalt and early Permian carbonate. In natural outcrops (in contrast to road cuts), the softer matrix erodes readily and is covered by soil and vegetation, so that only the blocks stand out. The resulting lumpy surface looks a bit like landslide debris or a swarm of glacially transported boulders. We now know that this messy collection of rocks represents disrupted ocean-floor material that may have originated partly as submarine landslide debris, but was further chewed up in a subduction zone and by later tectonic deformation. The assemblage forms part of the Cache Creek Terrane, which extends along the axis of the Cordillera from south-central British Columbia to south-central Yukon. Bordering the Cache Creek Terrane are the island-arc terranes of Quesnellia to the east and Stikinia to the west, with the Yukon-Tanana Terrane wrapped around its northern end. The intimacy of this association and the age of the rocks suggest a close relationship and history. But the Cache Creek Terrane contains rocks and Permian fossils that are strikingly different from those in the arc terranes.

One distinctive characteristic of the Cache Creek Terrane is the presence of great masses of shallow-water limestone, some many tens of kilometres across and over 1,000 metres thick. In places, the limestone masses have basalt foundations, the chemistry of which suggests an origin within an oceanic plate (like Hawaii today), rather than at a spreading ridge. It seems that the limestones accumulated in shallow water as carbonate caps, or atolls, on volcanic seamounts or oceanic plateaus that had built up to near sea level. The Cache Creek assemblage is reminiscent of parts of today's tropical western Pacific, where many volcanic islands are surrounded by coral atolls. Modern atolls are mostly small, but a few are of impressive size. The Kwajalein Atoll in the Marshall Islands in the western Pacific Ocean is over 100 kilometres long and 40 kilometres wide. Some limestone bodies in

This road cut in Cache Creek, British Columbia, exposes jumbled rocks of the Cache Creek Terrane. JIM MONGER.

A 500-metre-high cliff on the north side of Marble Canyon, British Columbia, exposes massive middle Permian to late Triassic carbonate, part of the Cache Creek Terrane. JIM MONGER.

Dark brown oceanic or back-arc-basin rocks of the Slide Mountain Terrane cap a ridge near Cassiar, British Columbia. The rocks have been thrust over light-coloured middle Devonian carbonate deposited on the Euramerican continental margin. MAURICE COLPRON.

Eclogite containing a sliver of blueschist from the Cache Cheek Terrane, near Fort St. James, British Columbia. These distinctive metamorphic rocks result from deep burial in an accretionary complex. MIKE CHURKIN.

the Cache Creek Terrane were once of similar dimensions, and one Carboniferous to middle Permian body near Atlin, British Columbia, has an age span of about 50 million years.

The birthplace of some of the rocks in the Cache Creek Terrane has been unravelled in part through the study of fossils, especially giant foraminifera called fusulinids. Middle Permian assemblages from the Cache Creek Terrane are characteristic of western Panthalassa and Paleotethys, that long marine embayment whose remains are found across central Asia and as far west as the Mediterranean region. In contrast, fusulinids of the same age found in Stikinia and Quesnellia belong to forms that lived in seas from Kansas to Peru, seas that spilled onto the continent from eastern Panthalassa. In the Permian, Paleotethys and eastern Panthalassa were on opposite sides of Pangea. Just how remains of these faunas came to be within about 100 kilometres of one another today will be revealed in the next chapter.

Today, all of these terranes have a jumbled distribution in the western Cordillera, reflecting post-Paleozoic accretion to the old continental margin and subsequent tectonic shuffling along it.

Shifting Climates

Toward the end of the Carboniferous, Pangea drifted northward, and during the Permian the supercontinent extended almost from pole to pole. The presence of such a huge landmass would have led to extreme seasonality in its interior. Drying may also have been enhanced by global warming and expansion of subtropical arid zones as glaciation on former Protogondwana came to a close. One region severely affected was Atlantic Canada, deep within Pangea, where desert-like conditions with sparse vegetation prevailed. This "great drying" is reflected in the latest Carboniferous and Permian redbeds of Prince Edward Island, the Magdalen Islands, northern Nova Scotia, and parts of New Brunswick. Many of the sediments were deposited by rivers that were swollen with sand and mud during monsoonal flooding, but shrank during the dry season. On the Magdalen Islands, we can still see large cross-beds formed by the migration of windblown Permian sand dunes.

Such a climate was not conducive to the luxuriant, shallow-rooted vegetation that flourished in the coal forests, and vegetation came to be dominated by drought-resistant trees, especially early conifers such as *Walchia*, which resembled the living Norfolk Island pine. Preserved trunks and branches of

At Kildare Capes, Prince Edward Island, red cliffs expose the deposits of Permian rivers that flowed in the region soon after it had crossed into the Northern Hemisphere. ROB FENSOME

Walchia are exposed in Permian rocks along the south shore of the Northumberland Strait at Brule in Nova Scotia. In the same rocks are footprints of small lizard-like animals and larger creatures that may have resembled the sail-backed mammal-like reptile *Dimetrodon*. There are few body fossils in these redbeds, but the skull and upper jaw of a mammal-like reptile named *Bathygnathus* were found in the 1850s by a farmer digging a well near New London, Prince Edward Island. And in 2012, the remains of a sail-backed creature was found along the Northumberland Shore of Nova Scotia.

While the great drying was underway farther south, Arctic Canada—in the vanguard of Pangea's northward migration—was emerging from drier latitudes and climes. Until about 285 million years ago, warm-water carbonates and deep-water mudstones continued to accumulate along the Sverdrup Seaway, as did continental redbeds and conglomerates on its margins. But then climates cooled across Arctic Canada, and within 5 million years warm-water carbonates were replaced by limestones containing mostly brachiopods, bryozoa, and sponges more tolerant of cooler water.

The tectonic setting was also changing, with the array of fault-bounded rifts and uplands that had formed the Sverdrup Seaway merging into a single linear basin, the Sverdrup Basin. About 270 million years ago, rift-related faulting ended, but the Sverdrup Basin continued to subside for the next 200 million years. Later Permian deposits there are fluvial clastics and thin coal seams that suggest wet mid-latitude climates. Isolated stones, probably ice-rafted dropstones, in

Permian chert in the Otto Fiord area, northwestern Ellesmere Island, Nunavut. BENOIT BEAUCHAMP.

marine sediments on Ellesmere Island indicate seasonal ice cover. The only fossils found in these rocks are sponges, whose skeletons were composed of silica rather than calcite.

In the Western Interior Plains, seas that had covered vast areas during the late Devonian and Carboniferous mainly withdrew during the Permian. Indeed, much of what was to become Canada lay above sea level at the end of the Permian and there are few records of continuous sedimentation into the following Triassic Period.

When Life Nearly Died

The end of the Permian witnessed the greatest mass-extinction event in Earth's history. It has been estimated that 70 percent of all land species and a staggering 96 percent of all marine species (with shells or skeletons) met their doom. Among invertebrates, colonial organisms such as corals and bryozoa were hit particularly hard. The two major groups of Paleozoic corals (tabulate and rugose) became extinct. Brachiopods and echinoderms, both diverse and abundant groups in the late Paleozoic, were substantially reduced. Ammonoids also suffered severely. Among arthropods, the very last of the eurypterids and trilobites disappeared. Among vertebrates, mammal-like reptiles, which included most of the large herbivores of the time, were devastated. Canada has a sparse record of these reptiles, but in South Africa their impressive remains fill museums. Some groups of fish were severely affected. Among plants, important changes had occurred earlier in the Permian, so that by the end of the period the majestic clubmoss and horsetail trees of the Carboniferous were already gone, though these groups survive as small herbs to the present day.

The underside of a sandstone block from Lord Selkirk Provincial Park, Prince Edward Island, reveals the trackway of a Permian tetrapod, dating from about 290 million years ago. MATT STIMSON.

According to recent research, much of the devastation happened over a short geological interval. But although many theories have been advanced as to the cause, no consensus has yet been reached. Some scientists have claimed that a big meteorite impact devastated life, but most experts agree that little evidence exists for such an event. Greater significance has been placed on the potency of volcanic activity, prompted by the eruption, 251 million years ago, of the enormous volumes of basalt known as the Siberian Traps, which now cover large areas of Arctic Russia. The volume of pyroclastic debris and sulphurous gases in the atmosphere could have drastically altered the climate, destroyed terrestrial ecosystems, and upset ocean chemistry, causing catastrophic dying. Although some scientists question whether the eruptions were sufficient cause, in association with their environmental impacts they just might have been. Certainly, the coincidence of the eruptions with the timing and speed of the extinction is highly suggestive of a connection. Another hypothesis is that the rapidly declining oxygen levels forced animals and plants into lower paleolatitudes, with consequent overcrowding and competition that proved fatal to many species. Whatever the reason for the mass extinction, it provided the survivors with unique opportunities, as we will discover in the next chapter.

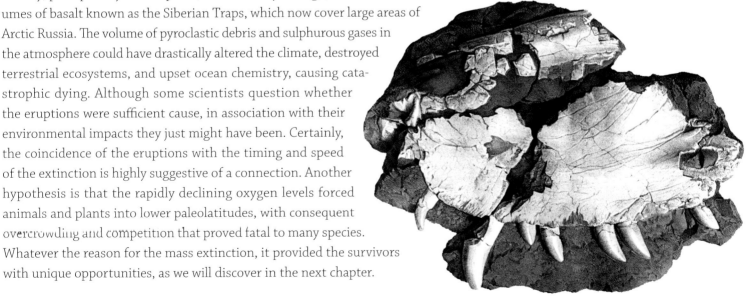

A large skull fragment of *Bathygnathus* found in Permian strata at New London, Prince Edward Island. The specimen was once thought to be Canada's oldest dinosaur, but it is now known to be a mammal-like reptile. FROM LEIDY (1854), WITH PERMISSION FROM THE ACADEMY OF NATURAL SCIENCES, PHILADELPHIA.

9 • PANGEA BREAKS UP AND MOUNTAINS RISE
Canada 251 to 65.5 million years ago

CHAPTER SUPPORTED BY A DONATION FROM DALE LECKIE, NEXEN ENERGY ULC

Although the prominent, resistant rocks of the Rockies are of Paleozoic age, it was Mesozoic tectonic forces that created the distinctive thrust faults and folds that are largely responsible for the modern grandeur of these mountains. This aerial view is of the "Big Bowl" on Mount Inflexible, Kananaskis Country, Alberta. MARILYN GARNETT / AIRSCAPES.CA.

A Western Giant

The geological exploration of Canada has involved some extraordinary personalities, one of whom was George Dawson. Born in Pictou, Nova Scotia, in 1849, Dawson spent part of his childhood learning natural history and geology from his famous father, William Dawson, whom we met in the previous chapter. Although small in stature—as an adult George described himself as no taller than a lad of twelve—and disabled due to a childhood bout of tuberculosis of the spine, he became one of Canada's foremost geologists. His passion for geology and natural history was reinforced by an education in Britain under the direction of Thomas Huxley and Charles Lyell, and matched by a dogged determination to overcome his disability.

In Dawson's first professional field assignment as a member of the International Boundary Commission of 1873–1874, he explored and studied the natural history of a 1,200-kilometre-long strip along the 49th

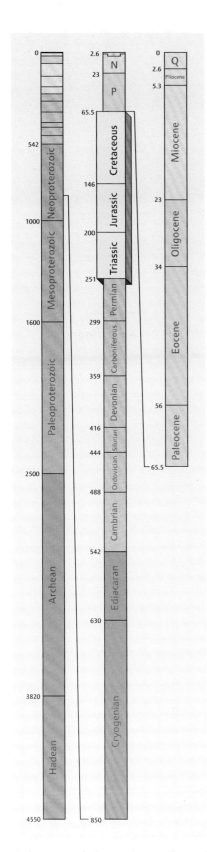

Geologic time scale, showing the interval covered in this chapter. Numbers indicate millions of years ago. P = Paleogene (Paleocene to Oligocene), N = Neogene (Miocene and Pliocene), and Q = Quaternary.

George Mercer Dawson (third from the left) during his International Boundary Commission days. REPRODUCED WITH THE PERMISSION OF NATURAL RESOURCES CANADA 2013, COURTESY OF THE GEOLOGICAL SURVEY OF CANADA.

parallel, from the Red River in Manitoba to Waterton Lakes in southwestern Alberta. Dawson routinely put in sixteen-hour days, covering vast distances on foot, on horseback, or by wagon, and amassed important collections, including some of the country's earliest-found dinosaur remains. His 1875 report was hailed as a classic for both its theoretical and practical achievements. That same year, Dawson joined the Geological Survey of Canada, which allowed him to continue his love affair with western Canada for the rest of his life. He explored and mapped extensive mineral-rich regions of British Columbia, and he documented the sedimentary rocks, including coal, of the Rocky Mountains and Prairies.

Dawson's contributions went beyond geology. He recorded flora and fauna, as well as Aboriginal languages and customs. The dictionary of the Haida language that he compiled and his reports of West Coast customs were so groundbreaking that he is known as the father of Canadian anthropology. He went to then-virtually-unknown northern British Columbia and Yukon in 1887 where, with Richard McConnell, he travelled over 2,000 kilometres in about 4 months. Their reports and maps were among the few reliable records for the region at the time of the Klondike Gold Rush in 1897. Dawson's contributions to Canadian exploration and geology were honoured when, in 1897, Dawson City in Yukon was named after him. His name was also given to a northeastern British Columbia creek in 1879, on which the town of Dawson Creek was later established around 1919.

Dawson was appointed Director of the Geological Survey of Canada in 1895, a position he retained for the rest of his life. In 1900, he became President of the Geological Society of America, following in the footsteps of his father. His presi-

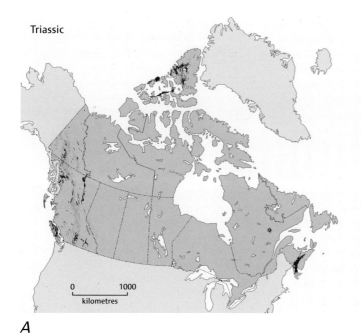

A

B

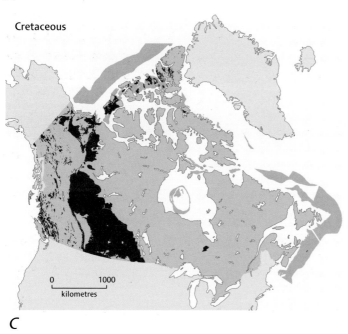

C

General extent of Triassic (*A*), Jurassic (*B*), and Cretaceous (*C*) rocks at the surface (beneath glacial deposits), onshore and offshore. The lighter shaded areas denote either uncertainty or areas where rocks of the particular age have been confirmed but are intimately associated with rocks of other ages and the scale of the map doesn't allow us to show them separately. ADAPTED FROM WHEELER ET AL. (1996).

dential address published in 1901, the year of his death, was a sweeping synthesis of the geology of the Canadian Cordillera that laid a firm foundation for later studies. With his vast knowledge, drive, and passion, George Dawson showed how western Canadian geological resources were tied to the country's future growth and success, a far-sighted vision that continues to have an impact.

Setting the Scene

George Dawson was especially interested in rocks of the Mesozoic Era, the interval covered in this chapter. Nineteenth-century geologists divided the Mesozoic into three periods: the Triassic (251 to 200 million years ago), Jurassic (200 to 145 million years ago), and Cretaceous (145 to 65.5 million years ago).

At the start of the Triassic, Pangea and Panthalassa dominated the planet's surface. Paleotethys was disappearing beneath what is today Kazakhstan in central Asia. As Paleotethys closed, a new ocean, Tethys, born in the late Permian, was expanding to the south. The region that was to become Canada formed northwestern Pangea and was rotated slightly clockwise relative to its modern orientation. It lay roughly between paleolatitudes 10 and 50°N, well south of its present position. The southeastern margin of future North America nudged against northwestern Africa, and its northeastern margin was continuous with Greenland, with Eurasia well connected to the other side of Greenland.

By about 175 million years ago, Tethys had grown to largely split Pangea into two huge continents, Laurasia to the north and Gondwana to the south. As part of this process, a western extension of Tethys was developing into the Atlantic Ocean. By 150 million years ago, sea-floor spreading had created a long, narrow ocean between what were to become North America and Africa, though the two would remain integral parts of Laurasia and Gondwana respectively for millions of years yet. Once Pangea no longer existed, neither by definition did Panthalassa. So from about 175 million years ago, we will start to call the vast ocean west of Laurasia and Gondwana (and eventually the Americas) the Pacific Ocean. What was to become North America and the western North Atlantic did not become a separate plate until well into the Cenozoic, so through the Cretaceous, Canada remained part of Laurasia.

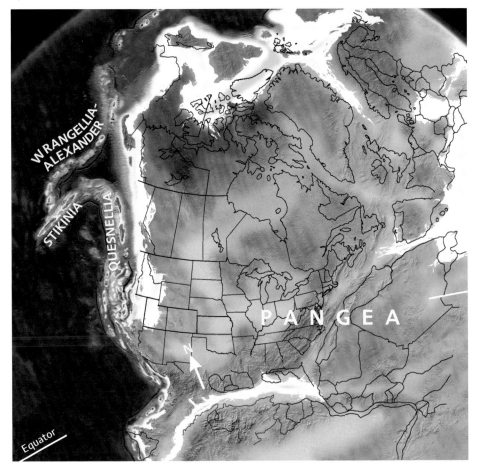

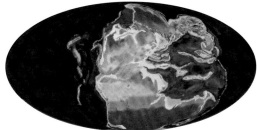

Global paleogeography 245 million years ago, during the Triassic. Colours as for previous figure.

Paleogeography of what was to become North America and adjacent regions in the Triassic, 245 million years ago. Land is shown in brown, with shading showing topography. The lighter blue areas represent possible coastal or nearshore areas, darker blue represents deeper ocean waters, and black indicates trenches. Aspects of modern geography are shown for orientation.

Bouncing Back

Life's recovery from the end-Permian mass-extinction event was slow at first. However, its delayed rebound in the middle and later Triassic was dramatic, and included a variety of new microscopic plankton, key basal members of the food chain. Triassic rocks contain the oldest coccolith and dinoflagellate fossils; foraminifera recovered and radiolaria flourished. These tiny organisms are not familiar to most of us, but their exploding diversity in the early Mesozoic profoundly changed the ecology of the oceans. Also during the recovery, many groups of invertebrates began to more closely resemble modern forms. The scleractinian corals, which first appeared about 240 million years ago and are primary reef builders today, have a hard, skeletal architecture completely different from that of Paleozoic corals. Brachiopods survived into the Mesozoic but did not recover their Paleozoic diversity. In contrast, mollusks did well, with clams taking over many of the environmental niches formerly the domain of brachiopods. Although ammonoids had almost become extinct at the end of the Permian, they diversified and evolved rapidly in Triassic seas.

Among vertebrates, synapsids in the form of mammal-like reptiles survived but never regained their former Permian diversity and dominance. The Triassic was to become the age of diapsids: dinosaurs, crocodiles, pterosaurs (flyers unrelated to birds), and the swimming ichthyosaurs and plesiosaurs are all members of this group of reptiles (Chapter 4). All first appeared in the Triassic, when atmospheric oxygen levels were perhaps half of today's 21 percent. That their dinosaur ancestors evolved during this time may explain why birds process oxygen so efficiently. Some modern birds soar happily over the Himalayas, while nearly all humans need oxygen tanks to climb Mount Everest.

The dominant Triassic and Jurassic plants were ferns, cycads, and some conifers, though not pines, which came later. Tree-sized club-mosses were gone, and flowering plants were not to arise until the Cretaceous.

Triassic clastic rocks in the Kamenka stone quarry in the Bow Valley, just outside Banff National Park of Canada, Alberta. This quarry produces the commercial stone known as Rundlestone (Chapter 14). DIXON EDWARDS.

Some of the best Triassic fossils in Canada come from a succession more than 1,200 metres thick exposed in the Rocky Mountain Foothills of northeastern British Columbia. These rocks, formed on the submerged northwestern continental margin of Pangea but today lying about 500 kilometres inland from the coast, have yielded some superb specimens. For example, digs around Wapiti Lake have produced the excellently preserved remains of fish, including *Whitea* (a coelacanth), *Bobastrania*, and *Albertonia*. *Bobastrania*, a primitive ray-finned fish and perhaps an ancestor of the modern sturgeon, was up to a metre long. *Albertonia* may belong to an ancestral group of teleosts, which are the most successful of all ray-finned fish and include most of our common food fish.

Reptile remains are also common in the Triassic sections of northeastern British Columbia, including coelurosauravids (early flying vertebrates), thalattosaurs, and ichthyosaurs. Thalattosaurs were enigmatic animals that likely lived on land but got their food from the sea, like modern iguanas. They had sharp, slender teeth at the front of their jaws and low, bulbous teeth in the back, a combination that probably allowed them to feed on both fish and shellfish. However, our prized treasures from these ancient seas were the ichthyosaurs. Nicely preserved specimens of these Mesozoic undersea terrors have been found in the Front Ranges and Foothills of the northern Rocky Mountains. Most impressive is the 21-metre-long late Triassic *Shastasaurus*, the largest-known marine reptile, found in the bank of the Sikanni Chief River in northeastern British Columbia.

Bobastrania, an early Triassic fish from Wapiti Lake, northeastern British Columbia. COURTESY OF THE ROYAL TYRRELL MUSEUM OF PALAEONTOLOGY.

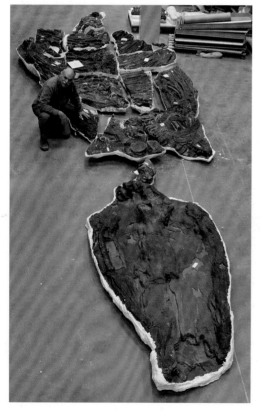

Stretching 21 metres, the late Triassic ichthyosaur *Shonisaurus* is the largest-known marine reptile, past or present. This monster, found on the banks of the Sikanni Chief River in northeastern British Columbia, is now on display at the Royal Tyrrell Museum of Palaeontology in Drumheller, Alberta. COURTESY OF THE ROYAL TYRRELL MUSEUM OF PALAEONTOLOGY.

An undersea Triassic scene from the submerged northwestern margin of Pangea, showing two ichthyosaurs, *Shastasaurus*, based on fossils found in northeastern British Columbia, several hundred kilometres from today's continental margin. Ichthyosaurs are reptiles, and their dolphin-like appearance is a result of convergent evolution. Dolphins (and modern mammals in general) were not to evolve for about another 100 million years. COURTESY OF THE ROYAL TYRRELL MUSEUM OF PALAEONTOLOGY.

The "Flower Pot", at the northwest corner of Graham Island, Haida Gwaii, is composed of Wrangellian Triassic-Jurassic volcanic rocks. CATHERINE HICKSON.

Late Triassic volcanic breccia containing a basalt pillow in a road cut on the Trans-Canada Highway near the west end of Kamloops Lake, British Columbia. Such rocks are typical remnants of the magmatic arcs of Quesnellia. The breccias formed when volcanic material slid down the flank of a submarine volcano. JIM MONGER.

Scraping, Erupting, and Wrapping

The remains of *Shonisaurus* fell to the sea floor from waters covering the old continental margin in what is now northeastern British Columbia. To reach the open ocean of Panthalassa beyond the Triassic Pangean plate margin, *Shonisaurus* would probably have had to swim westward several hundreds of kilometres, past a chain of volcanic islands. We suspect this scenario because we see no evidence of volcanic eruptions, such as ash layers, in the fossiliferous Triassic strata of the Canadian eastern Cordillera. Yet evidence from the rocks of Quesnellia and Stikinia confirms that arc volcanoes were erupting vigourously along the western Pangean plate margin at the time.

In the previous chapter, we encountered the Cache Creek Terrane with its exotic fossils as part of the Permian floor of Panthalassa. Triassic and early Jurassic rocks are also mixed among the older rocks of the Cache Creek Terrane, represented mostly by deep-sea mudstone and chert, the latter made largely from the skeletons of radiolaria. In places, these sediments are interbedded with volcanic detritus. As the floor of Panthalassa was subducted beneath northwestern Pangea during the Triassic and early Jurassic, Cache Creek rocks were scraped off the ocean floor and accreted to the western margin of the overriding Pangean plate.

Mélanges such as those in the Cache Creek Terrane are found along many convergent plate margins. Although some mélanges may be the remains of giant submarine landslides, most formed during subduction when ocean-floor rocks were scraped off, squeegee fashion, from the subducting plate and transferred to the overriding plate. In the process, the accreting rocks become highly deformed, disrupted, and in many places metamorphosed as shown by the local presence of blueschist among Cache Creek Terrane rocks. The name of this distinctive metamorphic rock comes from the blue mineral glaucophane, known from laboratory studies to form at high pressure but relatively low temperature. Blueschists form in the upper parts of subduction zones, where relatively less dense, buoyant ocean-floor rocks (mainly sediments), after having been dragged down to depths of perhaps 40 kilometres or more, recrystallize under high pressures while still relatively cold. Then, because of their buoyancy, the now-metamorphosed ocean-floor rocks are churned rapidly back almost to the surface to be preserved in accretionary complexes such as the Cache Creek Terrane.

A Triassic remnant of a different sort forms parts of Wrangellia, which now underlies areas of south-central Alaska, southwestern Yukon, and coastal British Columbia. Wrangellia's trademark is the vast volume of basaltic volcanic rocks erupted between 230 and 225 million years ago, possibly above a mantle plume. On Vancouver Island the succession, known as the Karmutsen Volcanics, is over 6 kilometres thick and is well exposed in road cuts along Buttle Lake, west of the town of Campbell River. Triassic sediments deposited on top of the volcanics record a gradual subsidence of the region.

About 203 million years ago, a new magmatic arc, the Bonanza Arc, started to form on Wrangellia and remained active for about 30 million years. Its remnants are found today from Vancouver Island to southwestern Alaska. On Vancouver Island, deeper parts of the Bonanza Arc are represented by granitic rocks called the Island Intrusions, which underlie the island's prominent mountain backbone. On Haida Gwaii, the Bonanza Arc is represented by Jurassic pyroclastic rocks and mudstone.

The Bonanza Arc and arcs of the same age on Quesnellia, Stikinia, and Yukon-Tanana were probably parts of an island-arc chain that lay off northwestern Pangea during the early Mesozoic. However, if the arcs did form a single linear system, why does the scraped-off oceanic remnant represented by the Cache Creek Terrane now lie sandwiched between Quesnellia and Stikinia, with the Yukon-Tanana Terrane wrapped around its north end? One idea is that, because of complex plate interactions during the Triassic and early Jurassic, the part of the arc represented now by Stikinia and the western part of the Yukon-Tanana Terrane rotated in a counterclockwise direction. As a result of this swivel, the Cache Creek Terrane became enclosed between the Quesnellian and Stikinian arcs by the early Jurassic. During the final stages of the process, in the early Jurassic, the opposing subduction zones collided, trapping the last remnants of Cache Creek Terrane within a hairpin bend in the arc system. A similar situation exists today in the northwestern Pacific, where the western end of the Aleutian Arc joins the eastern end of the Kamchatka Arc, the two arcs forming an inverted V above the subducting Pacific Plate. If this scenario for the Triassic Cordilleran margin is correct, and if we can add the Bonanza Arc into the mix, then a paleogeographic map for 190 million years ago would show the arcs arranged like a giant mirror-image of the letter N. The westernmost limb of the reversed N was the Bonanza Arc, which faced the open waters of Panthalassa.

Rifting, Belching, and Dying

While arcs were undergoing contortions in the west, developments were also afoot in the east along what was to become the Atlantic margin of North America. Here, though, the story is one of rifting rather than of convergence. The first signs of rifting, about 235 million years ago, included doming and cracking along old lines of weakness within the Paleozoic Appalachian Orogen. A complex of rift basins and adjacent rift-shoulder highlands developed, aligned roughly southwest to northeast. During the later Triassic and earliest Jurassic, clastic sediments were eroded from the highlands and deposited in rivers and lakes in the rift basins.

Paleomagnetic evidence shows that the early Mesozoic rift basins along what is now the North American eastern seaboard extended from around the paleoequator (present-day Florida) to about paleolatitude 15°N (today's Maritimes and adjacent offshore). The largest of these rifts was the Fundy Basin, now mostly covered by the waters of the Bay of Fundy. The oldest visible Triassic rocks in the Fundy Basin form colourful red cliffs, like those at Economy in Nova Scotia. Rare fossil finds in these rocks include the bones of the mammal-like reptile *Hypsognathus* and the crocodile-like amphibian *Metoposaurus*. Footprints of small bipedal dinosaurs, possibly *Coelophysis*, also occur in the late Triassic rocks on the Fundy shore.

These land-dwelling animals and their marine counterparts were living in a world that had largely recovered from the end-Permian crisis. But life was about to undergo another crisis, although not as severe as the devastation some 50 million years earlier. The Triassic, like the Permian, ended with a mass-extinction

The possible disposition of terranes on the Cordilleran margin during the Triassic (*A*) and early Jurassic (*B*), showing how the Cache Creek Terrane might have ended up sandwiched between Quesnellia and Stikinia. The Chugach Terrane is an accretionary wedge that accrued as oceanic crust of Panthalassa subducted beneath Wrangellia. The white lines represent sea level.

At Five Islands Provincial Park, Nova Scotia, the cliffs to the right expose red lake and sand-flat deposits, white fluvial deposits and, at the top, North Mountain Basalt, all of late Triassic age. At left the basalt is now at beach level due to faulting. ANDREW MACRAE.

Late Triassic scene in the Fundy Basin, about 220 million years ago. A family of the mammal-like reptile *Hypsognathus* browse among the ferns, hidden from the predatory crocodile-like amphibian *Metoposaurus*. In the background, a dinosaur-like rauisuchid drinks from a channel in the braided-stream network. The animals in this scene are based on fossil finds around the Bay of Fundy in Nova Scotia. The plants, dominated by ferns, the horsetail *Neocalamites* (to the left of the *Metoposaurus*), and gymnosperm araucarian trees (such as the large tree shading the *Hypsognathus* family), are drawn from the general flora known from the time. PAINTING BY JUDI PENNANEN, COURTESY OF THE ATLANTIC GEOSCIENCE SOCIETY.

Wasson Bluff, near Parrsboro, Nova Scotia. The rocks are: dark grey-brown, latest Triassic North Mountain Basalt at the right; light lacustrine limestone, lower middle; red eolian sandstone, left; and brown paleotalus ("fossil" scree) composed of basalt boulders, upper centre. All the sedimentary rocks are probably of earliest Jurassic age. ROB FENSOME.

Side view of the skull of the tuatara-like reptile *Clevosaurus*, from early Jurassic deposits at Wasson Bluff, near Parrsboro, Nova Scotia. HEINZ WIELE, COURTESY OF THE ATLANTIC GEOSCIENCE SOCIETY; SPECIMEN COURTESY OF THE FUNDY GEOLOGICAL MUSEUM.

event, this one resulting in the loss of nearly a quarter of all marine families and a fifth of all terrestrial families. Almost all ammonoids died out, leaving a single family to give rise to all later Mesozoic ammonoids—the true ammonites. Other mollusk groups declined, including gastropods and bivalves. Reef organisms, particularly corals and sponges, were devastated and brachiopods were again severely depleted. Conodonts, which had been in decline through the latest Triassic, disappeared for good around the time of the Triassic-Jurassic transition. But some groups, such as echinoderms, survived almost intact. Terrestrial plants and vertebrates were also decimated at the Triassic-Jurassic boundary. Among vertebrates, the crocodile-like reptiles that had been prominent during the later Triassic mostly succumbed, leaving the dinosaurs room to diversify and achieve a dominance that would last some 135 million years.

What might have caused the extinctions that provided the dinosaurs with such an opportunity? Suspicion falls on a short-lived but vast outpouring of basaltic lava around 200 million years ago, an episode that has been described as a "mantle belch". Remnants of this episode are found today in parts of eastern North America, northern South America, western Africa, and southwestern Europe. In the Maritimes, this outpouring is represented by the North Mountain Basalt, which forms dramatic cliff scenery in and around the Bay of Fundy—for example at Cape Split and Cape d'Or in Nova Scotia, and on the western side of New Brunswick's Grand Manan Island. The latest evidence suggests that extinctions and volcanism both occurred within the very latest Triassic, with the most intense wave of extinctions coinciding with the earliest eruptions. This timing suggests that the volcanic outpourings were implicated in the crisis among organisms, but the exact relationship is not yet clear.

Immediately above the North Mountain Basalt on the Fundy shore near Parrsboro, Nova Scotia, exposed in a cliff section known as Wasson Bluff, is a mixture of fluvial, lacustrine, and eolian clastic deposits, as well as mud and boulder flows of volcanic rocks. These sedimentary rocks are only slightly younger than the basalt flows, and appear to straddle the Triassic-Jurassic boundary. They have yielded a unique fossil vertebrate assemblage, the discovery of which in 1985 made international news. Among the bones are those of the lizard-like reptile *Clevosaurus*, whose closest living relative is the tuatara of New Zealand. Other finds include bones of the "sabre-toothed" crocodilian *Protosuchus* and the mammal-like reptile *Pachygenelus*. In recent years, a rich bone-bed of prosauropod dinosaurs—the earliest dinosaur remains yet recovered in Canada—has been unearthed at Wasson Bluff in water-lain sediments surrounded by eolian sand-dune deposits. The articulated skeletal remains of numerous individuals may have been buried rapidly in a flash flood between ancient dunes. Prosauropods were the forerunners of the sauropods, the largest land animals that have ever lived and which include the leviathans *Brachiosaurus* and *Argentinosaurus*.

A lava tube and dyke in volcanic rocks that are equivalent to the latest Triassic North Mountain Basalt, Seven Days Work Cliff, Grand Manan Island, New Brunswick. GREGORY MCHONE.

A Shifting Balance

So far in this chapter we've explored rocks and fossils in western and eastern Canada, the two regions in which most of the country's Triassic and earliest Jurassic rocks are to be found. Before changing our focus to later Mesozoic rocks and events, we need to delve a little more deeply into aspects of plate-tectonic theory that are critical to an understanding of fundamental changes in the tectonic history of the Cordillera.

An inspection of the modern globe reveals significant differences between the geography of the eastern Pacific and that of the western Pacific, even though both are dominated by plate convergence. On the western side are numerous chains of volcanic island arcs, such as Japan, that are separated from mainland Asia by back-arc basins such as the Sea of Japan. In contrast, the eastern Pacific is flanked by the great mountain belts of the Andes and the Cordillera that carry continental arcs. We've long known that oceanic lithosphere is subducting around most of the Pacific's margins, giving rise to the volcanoes of the famous Pacific Ring of Fire. So why are there such big differences between the two sides of the ocean?

As oceanic lithosphere cools, becomes denser, and sinks into the mantle at subduction zones, the trench (the surface trace of the subduction zone) retreats back toward the spreading ridge. If the advance of the overriding plate doesn't keep pace with the retreat of the trench, the lithosphere behind the arc becomes stretched and thinned, causing a back-arc basin to form. Today in the western Pacific, trenches are retreating eastward faster than the overriding Eurasian Plate is advancing in the same direction, and so back-arc basins have formed. In contrast, along the eastern Pacific margin, subduction zones are retreating more slowly than the Americas are advancing westward. The result is that in the eastern Pacific the oceanic and American plates are converging *and* colliding. Hot, weak arc lithosphere, caught between converging, cold, strong continental lithosphere and even stronger oceanic lithosphere, is squeezed and thickened and so rises to form the Andes and the Cordillera.

In light of these observations from around the modern Pacific, let's now look at the history of the Cordillera, which means going back to the Devonian when a convergent plate margin formed along the western flank of Euramerica. The Euramerican plate margin and its accompanying arc became separated from the continent in the earliest Carboniferous, probably because the trench in eastern Protopanthalassa was retreating faster than the Euramerican plate was advancing toward it. The arc (now preserved in older parts of Quesnellia and Stikinia) was "towed" oceanward, causing a back-arc basin, the Slide Mountain Basin, to open behind it, just like the modern Sea of Japan. During the early Mesozoic, the balance changed: the rate of retreat of the trench became slower relative to the rate of advance of the overriding Pangean plate. This caused the back-arc basin to close and the arcs to move toward the continent. By the middle Jurassic, the former offshore arc terranes had been accreted to the old margin, and mountains were starting to rise as oceanic and continental plates both converged and collided.

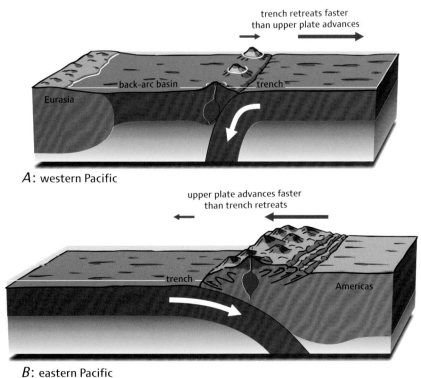

A: western Pacific

B: eastern Pacific

Comparison between aspects of plate convergence in the modern western Pacific Ocean (*A*) and the present-day eastern Pacific Ocean (*B*).

Early Jurassic rocks from the Bighorn Creek area, southwestern Alberta. Thin, light-coloured volcanic ash layers are interbedded with brown-weathering shales. These ash beds record the approach of Quesnellian magmatic arcs to the old continental margin. RUSSELL HALL.

For most of the Triassic, a significant, albeit perhaps narrowing, sea existed between the western margin of the continent and the offshore arc terranes. The absence of Triassic volcanic ash in the Rocky Mountains supports this scenario, even though such rocks abound in the arc terranes farther west. Toward the end of the Triassic, this picture started to change. The axis of the magmatic arc in southern Quesnellia, as represented by granitic rocks south of Kamloops, shifted about 100 kilometres eastward between 210 and 195 million years ago. In the easternmost Cordillera, 185-million-year-old volcanic ash beds are interlayered with black shales bearing early Jurassic ammonites. These ash beds provide the first evidence of arc-related volcanic activity close to the western continental margin since the late Devonian.

Birth of an Ocean

As the balance of plate interactions was shifting in the west, new ocean floor was starting to form in the east. Although the Fundy Basin is the largest and best-exposed Canadian rift basin associated with the initial breakup of Pangea, other basins now 100 to 300 kilometres east of today's coastlines are better placed to reveal the Atlantic Ocean's prehistory. The geology of these offshore basins is known mostly from geophysical exploration and from offshore wells drilled in the search for oil and gas. It was among these basins and their counter-

Mesozoic-Cenozoic sedimentary basins of offshore eastern Canada. Depths are in metres. ADAPTED FROM VARIOUS SOURCES.

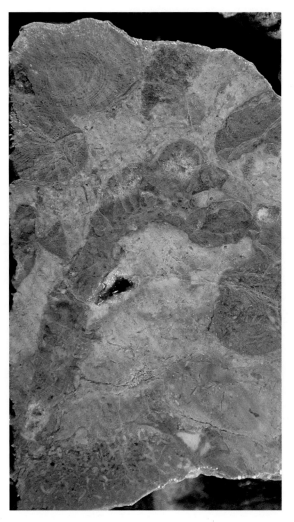

Fossil sponges in a drill core of Jurassic carbonate reef rocks from the Demascota G-32 well in the Scotian Basin off Nova Scotia. LESLIE ELIUK.

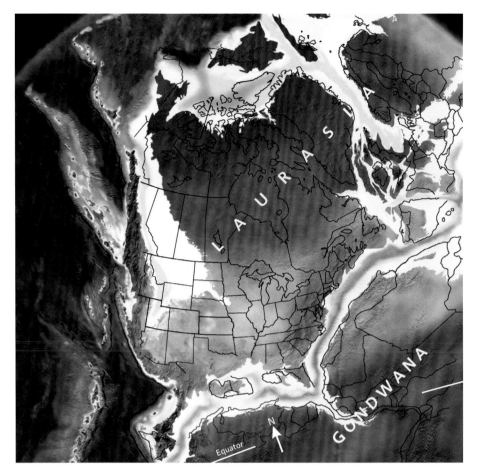

Paleogeography of what was to become North America and adjacent regions in the Jurassic, 170 million years ago. Land is shown in brown, with shading showing topography. The lighter blue areas represent possible coastal or nearshore areas, darker blue represents deeper ocean waters, and black indicates trenches. Aspects of modern geography are shown for orientation.

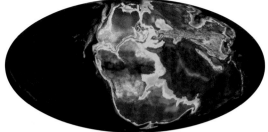

Global paleogeography 170 million years ago, during the middle Jurassic. Colours as for previous figure.

parts off northwest Africa and Iberia (Spain and Portugal) that the North Atlantic was born.

During the latest Triassic and earliest Jurassic, redbeds accumulated in narrow rift basins similar to the Fundy Basin, and evaporites (mainly halite) were deposited from seawater that intermittently flowed in from the western tip of Tethys. The change from continental rifting to oceanic sea-floor spreading (drifting) first occurred about 185 million years ago between the southeastern United States and northwestern Africa, and a few million years later between Nova Scotia and northwestern Africa. The sea evidently flooded into this new ocean basin from Tethys to the northeast.

As the ocean expanded during the middle Jurassic, the small rift basins off what is now southeastern Canada gave way to a large sedimentary basin known as the Scotian Basin. Oil and gas exploration in this basin has revealed a Mesozoic-Cenozoic sedimentary succession up to 20 kilometres thick. Despite this great thickness, sediments were mostly deposited in shallow marine and non-marine environments, and rates of sedimentation must have generally matched subsidence rates to keep the Basin filled. This situation is reminiscent of the thick shallow-water accumulations on the early Paleozoic margin of Laurentia, now in the Main Ranges of the southern Rocky Mountains.

During the middle and late Jurassic, warm subtropical waters led to the carbonate buildup known as the Abenaki Bank along the southeastern margin of the Scotian Basin. When sea level fell substantially, as in the latest Jurassic, the area of the present-day shelf was exposed for 100 kilometres or more east of the modern shoreline. Sediments were brought in by rivers draining parts of Atlantic Canada and deposited in a series of deltas, such as the Sable Delta now preserved beneath Sable Island and the surrounding sea floor. These buried deltaic deposits contain the gas now being extracted around Sable Island.

By the end of the Jurassic, about 145 million years ago, unzipping of the North American part of Laurasia from the African part of Gondwana was well advanced, extending from Florida and West Africa to Nova Scotia and northwest Morocco. Farther north, the Grand Banks of Newfoundland was still firmly attached to Iberia, but sedimentary basins such as the Jeanne d'Arc were forming, in part beneath shallow Jurassic seas.

Continental Bulldozer

Some early proponents of continental drift recognized that the opening of the Atlantic Ocean was happening at the same time as the earliest mountain building on the western side of the Americas, and suggested that the two events were related. Our present understanding of plate-tectonic processes and more precise timing of events supports this idea. It seems that rifting and initial sea-floor spreading in the east moved Pangea/Laurasia westward faster than Panthalassan/Pacific lithosphere could subduct, as predicted by the shifting balance scenario discussed earlier. The consequent squeezing and thickening of the weak lithosphere underlying the arcs off the continent's western margin raised the first Cordilleran mountains.

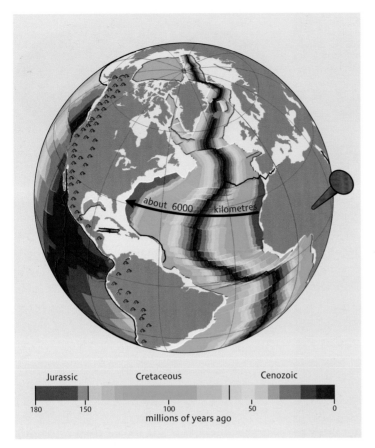

The Atlantic Ocean and bordering continents showing how, over the past 180 million years, this ocean has grown to be over 6,000 kilometres across. The darker grey areas represent modern land, the lighter grey areas represent submerged continental margins underlain by continental crust, and the coloured areas represent oceanic crust of various ages, as indicated in the legend. The black lines represent, in part, plate boundaries. ADAPTED IN PART FROM A GRAPHIC BY WALTER ROEST.

Late Jurassic deltaic sedimentary rocks of the Bowser Basin, southeast of Taft Creek, Skeena Mountains, British Columbia. These rocks were folded during the Cordilleran collisions of the Cretaceous. MARGOT MCMECHAN.

The initial continent-ocean collision created a belt of intensely deformed and high-grade metamorphic rocks that mark the zone where, about 180 million years ago, Quesnellia had been thrust over the continental margin. The timing of this event we know from fossils and from the age of the oldest plutons that intrude both the old margin and Quesnellia. The zone of collision, marked by the boundary between the old margin and Quesnellia, lies south to north along the line of the present Columbia, Omineca, and Cassiar mountains in the eastern Cordillera, just west of the Rocky Mountains. There, the outer part of the old continental margin was heated and softened, folded and faulted. Some rocks of the old continental margin were forced down to depths of over 30 kilometres, where they were metamorphosed and partly melted, and other rocks were elevated well above sea level and sculpted by erosion into new mountains.

Current evidence suggests that by around 175 million years ago, the outer arc terranes—Alexander and Wrangellia—were loosely linked to the inner arc terranes—Quesnellia and Stikinia—and all four were located somewhere along the western Laurasian continental margin. Some of the earliest sedimentary evidence for Cordilleran highlands also dates from around that time, as the Cache Creek Terrane in northern British Columbia was thrust westward over Stikinia, in the process rising above sea level where it was exposed to erosion. The resulting clastic sediments were deposited in the Bowser Basin, which developed on top of Stikinia and today underlies the Skeena Mountains and covers more than 50,000 square kilometres of interior northwestern British Columbia. From the middle Jurassic to the early Cretaceous, a succession of marine and non-marine sediments, rich in material eroded from the Cache Creek Terrane and at least 6 kilometres thick, accumulated in the Bowser Basin.

About 170 million years ago, the locus of magmatic activity shifted from the Bonanza Arc on Vancouver Island eastward into what was eventually to become the southwestern Coast Mountains. What had been a system of loosely assembled offshore island arcs now became the foundation of a continental arc system built within and upon terranes newly accreted to the western edge of the old continent. As part of this process, a multitude of middle Jurassic to early Cretaceous granitic plutons intruded the terranes and the old continental margin. By the middle Cretaceous, all but marginal parts of the Cordilleran Orogen had emerged above sea level.

Thus, by moving westward, Pangea/Laurasia had acted as a continental bulldozer, scraping up the mush of offshore arc terranes. But how far westward did the continent move? In fact, it is not possible to establish paleolongitudes without making some big and much-debated assumptions. However, if we take hotspots as relatively fixed reference points, then what was to become North America probably moved over 3,000 kilometres westward between the middle Jurassic and earliest Cenozoic. If Africa is used as a fixed reference, the total westward drift is

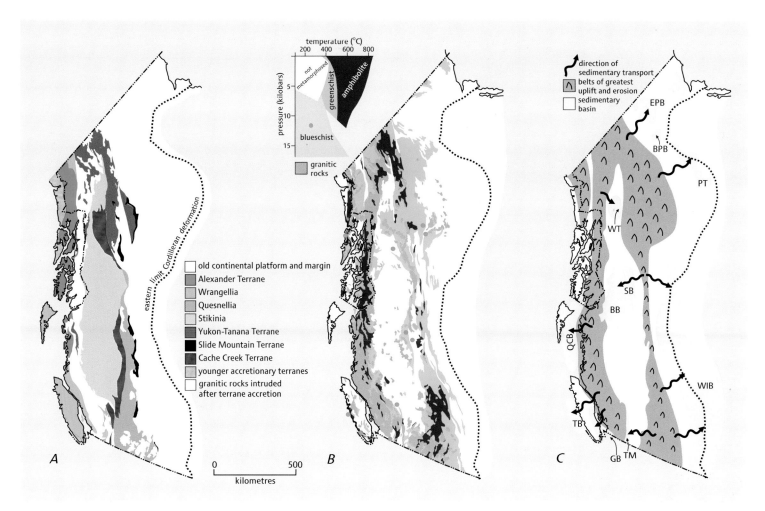

The effects of Mesozoic and early Cenozoic collisions in the Canadian Cordillera and the adjoining part of southeastern Alaska. A. The distribution of terranes in the Canadian Cordillera (see Chapter 8). B. The distribution and nature (see inset graph) of Mesozoic metamorphic rocks and granitic intrusive igneous rocks produced by Mesozoic collisions, showing their grouping mainly in two largely separate belts. The eastern belt underlies the Columbia, Omineca, and Cassiar mountains of the eastern Cordillera, and the western belt underlies the Coast Mountains. Note how the eastern belt corresponds to the boundary between the inner terranes (including Quesnellia, Stikinia, and Yukon-Tanana) with the old continental margin, and the western belt to the boundary between the inner terranes and the outer terranes (Wrangellia and Alexander). Yellow shading on map represents areas of non-metamorphosed to weakly metamorphosed rocks. C. Location of areas of long-term uplift and long-term subsidence: the uplift coincides with areas of metamorphism in B, where once deeply buried rocks are now at the surface; and the subsidence is recorded by sedimentary basins. The sedimentary basins include: Bonnet Plume Basin (BPB), Bowser Basin (BB), Eagle Plains Basin (EPB), Georgia Basin (GB), Peel Trough (PT), Queen Charlotte Basin (QCB), Sustut Basin (SB), Tofino Basin (TB), Tyaughton-Methow Basin (TM), Western Interior Basin (WIB), and Whitehorse Trough (WT). A IS ADAPTED FROM MONGER AND BERG (1987), COURTESY OF THE US GEOLOGICAL SURVEY; B IS FROM READ ET AL. (1991); C IS ADAPTED FROM YORATH (1992).

twice that. Despite the uncertainties, it is clear that such a great trek generated major changes, causing the continent's western margin to mop up terranes and build mountains.

Beneath the Midnight Sun

So far we have focussed on what were to become Canada's western and eastern margins. What of its northern edge? During the Triassic, the Sverdrup Basin was a triangular depression off northern Panthalassa and remained the region's principal feature. A sea covered much of the Basin's western part, where organic-rich marine muds accumulated. To the east, however, deltas built out into this sea throughout the Triassic, leaving an apron of clastic sediment so large that the inflowing rivers must have been draining huge parts of Pangea. The deltas expanded progressively westward, so that by the late Triassic they nearly filled the Sverdrup Basin. The Basin was apparently bordered by a landmass to the northwest, possibly northern Siberia, since some sediments did arrive from that direction, though direct evidence for such a landmass is limited.

The Sverdrup Basin was mainly under the sea through the Jurassic and Cretaceous, generally connected to the Western Interior Seaway (see page 182) and possibly to a polar tongue of the northeastern Pacific Ocean (formerly Panthalassa) that extended onto what was to become the Siberian Continental Shelf. We still have much to learn about the history of the Sverdrup Basin, but it never developed oceanic lithosphere and was not a precursor of the Arctic Ocean.

Even though evidence indicates that the Arctic has been cooler than southern Canada since the Triassic, the diversity of Jurassic fossils found in the region—including ichthyosaurs, plesiosaurs, ammonites, bivalves, and belemnites—suggests that conditions were comparable to warm-temperate modern climes. This is even more surprising considering the dearth of daylight

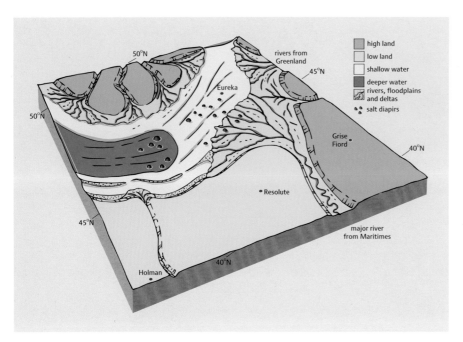

Paleogeography of what was to become the Canadian Arctic in the middle late Triassic to early Jurassic, 215 to 195 million years ago. The approximate locations of Holman, Resolute, Grise Fiord, and Eureka are included to provide a guide for orientation, as well as paleolatitudes. The main feature was the Sverdrup Basin, now dominated by clastic deposition, especially from large deltas.

the origins and history of our most northerly ocean has been revitalized in recent years. These studies have confirmed that the Arctic Ocean did not originate as a simple spreading ridge in the centre, but has a complex structure. It has broad shelves, especially on the Russian side, and is divided down its centre by the Lomonosov Ridge, which extends from off northern Greenland and northeastern Ellesmere Island on one side, to the western Siberian continental shelf on the other. The geological structure of the Lomonosov Ridge still needs to be fully resolved, but it does not appear to be a spreading ridge. More likely it is a sliver of continental lithosphere. The Lomonosov Ridge splits the Arctic Ocean into eastern and western deep-water basins, and the latter is called the Canada Basin. The eastern basin is the easier to explain: magnetic stripes on either side of its central Gakkel Ridge suggest that it was formed by conventional sea-floor spreading, the Gakkel Ridge being essentially a Cenozoic continuation of the Mid-Atlantic Ridge (Chapter 10).

in winter. As Arctic seas became more isolated from the Pacific Ocean during the later Mesozoic, faunas there became more distinct from those farther south. Still, terrestrial Cretaceous deposits in the Arctic contain the remains of hadrosaur and theropod dinosaurs, as well as types of fish and plants that we would expect today in warmer climes; even coals accumulated.

Today, the Arctic is dominated by the Arctic Ocean. Less accessible to research vessels and less attractive to commercial interests than the Atlantic, the Arctic Ocean still remains something of an enigma, but we do know that, like the Atlantic, it dates back to the Mesozoic. Because of the need to define and identify national economic zones under the United Nations Convention for the Law of the Sea (UNCLOS), research into

The Canada Basin has a much different story. Onshore evidence around its borderlands suggests that continental lithosphere began to stretch and rifts developed during the middle and late Jurassic on the North Slope of Alaska, Yukon, and along the western margin of the present-day Arctic Islands. As with the Atlantic, some of the cracks related to the prehistory of the Arctic Ocean did not develop beyond the rift stage. It seems that sea-floor spreading began in the Canada Basin (the southern end of which underlies the Beaufort Sea) during the early Cretaceous, and may have extended by way of a spreading axis that ran northward for 900 kilometres. In this interpretation, the spreading had

Sea cliffs west of Smith Creek, northwest Ellesmere Island, Nunavut, expose early Triassic light-coloured sandstone and siltstone of shallow marine origin, overlain by dark mudstone of deeper marine origin. ASHTON EMBRY.

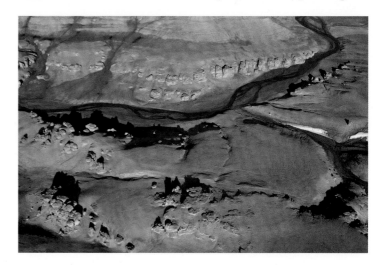

Aerial view of cliffs and hoodoos developed in early Cretaceous deltaic sandstone, Ellef Ringnes Island, Nunavut. CAROL EVENCHICK.

a pivot near the Mackenzie Delta, meaning that this growing ocean developed as a northerly widening wedge.

Magmatic episodes occurred in the Canadian Arctic around 130 to 125 and 95 million years ago and are marked by extensive mafic intrusions and lava flows, including those that form a complex ridge system composed mainly of the Alpha and Mendeleev ridges. This system is a vast undersea mountain range that extends across the northernmost Canada Basin south of and roughly parallel to the Lomonosov Ridge. In the Sverdrup Basin, many mafic intrusions invaded diapirs and related structures formed of Carboniferous halite. Also in the mix were hydrothermal fluids, oil, and gas. The chemical soup produced from the interaction of magma and salty brines precipitated out as metal sulphides, native sulphur, halite, gypsum, and tufa mounds made up of celestite, barite, and calcite. Some features, now preserved as carbonate mounds, even hosted special worm and clam faunas typical of cold deep-sea seeps. The instability and continuous shifting of salt produced large local variations in the thicknesses of most late Triassic to Paleocene rocks in Arctic Canada. The salt structures are exposed today as long ridges, or circular to kidney-shaped ranges of hills (see photos on pages 151 and 268).

Modern bathymetry of the Arctic Ocean and topography of its borderlands. MAP FROM AMANTE AND EAKINS (2009).

Changing Direction

Having explored the evolution of three of Canada's ocean-facing margins, we can now add a twist—literally. The paleomagnetic record shows that Pangea/Laurasia migrated about 20 degrees of latitude northward during the Mesozoic until around 110 million years ago. Then the continent went into reverse and migrated southward by about 5 degrees of latitude. However, recall that the continent was also moving westward. If we combine latitudinal and longitudinal components, we can infer that the continent was actually moving in a northwesterly

Tilted brown Cretaceous lava flows and pyroclastic rocks overlie lighter-coloured Cretaceous sediments, Kanguk Peninsula, Axel Heiberg Island, Nunavut. ANDREW MACRAE.

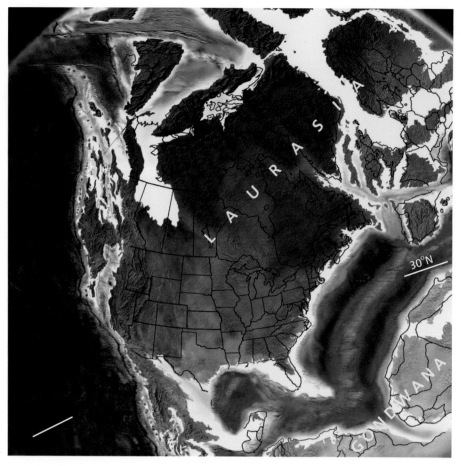

Paleogeography of what was to become North America and adjacent regions in the Cretaceous, 130 million years ago. Land is shown in brown, with shading showing topography. The lighter blue areas represent possible coastal or nearshore areas, darker blue represents deeper ocean waters, and black indicates trenches. Aspects of modern geography are shown for orientation.

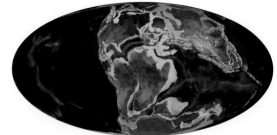

Global paleogeography 130 million years ago, during the early Cretaceous. Colours as for previous figure.

direction until about 110 million years ago, and then changed to a southwesterly trajectory. The timing of the change in direction roughly coincides with that of magmatic activity in the Arctic and very loosely with the opening of the Arctic Ocean—and, as we will see, with a change in orientation of spreading in the Atlantic and the opening of the Labrador Sea along a northwest-southeast axis. Perhaps all these factors were related as plates jostled one another.

The most obvious impact of the continent's movements was in the Cordillera. Convergence and collision was not head-on but at an angle. We can perhaps refine our bulldozer analogy and compare the movement of the continent to that of a snowplough, moving relentlessly forward but with the blades slanted so that the snow is pushed to the side of the road. Cordilleran rocks were not just squeezed and piled up vertically, but also smeared laterally along the margin—first southward relative to the continental interior as the latter moved northwestward, then northward relative to the continental interior as the latter moved southwestward. Studies of strike-slip faults in the Cordillera support this interpretation. The evidence indicates that the southwestern part of the present Coast Mountains, Wrangellia, and the Alexander Terrane moved southward by as much as 800 kilometres between middle

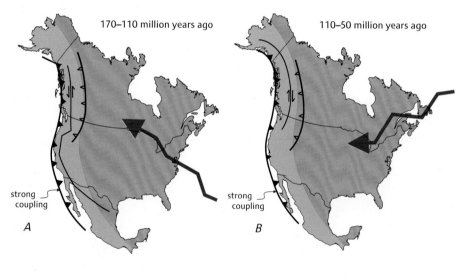

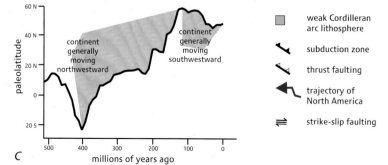

Major structural features of the Cordillera and their relationship to the overall motion of tectonic plates. The fault symbols represent the general situation, not specific faults. The strong compressional force produced by the convergence of the North American Plate on the one side and the plates underlying the Pacific Ocean on the other causes the plates at the boundary to become stuck together (or strongly coupled). As a consequence, the strain was largely taken up by strike-slip faulting within the weak western edge of the North American Plate. From 170 to 110 million years ago, this strike-slip faulting was left lateral, reflecting northward movement of the continent relative to the subduction zone (A); from 110 million years ago to the present, faulting has been right-lateral, reflecting southward movement of the continent (B). C shows paleolatitudes determined from paleomagnetic studies of the continental interior, calculated using the 20- to 30-million year running averages for a point in northwestern Montana (48°N and 115°W).

A middle Cretaceous granitic pluton forms the Stawamus Chief (better known as "The Chief"), a rock climbers' mecca, near Squamish, British Columbia. PAUL ADAM.

Cretaceous sandstone of the Georgia Basin, Tumbo Island, Gulf Islands, British Columbia. C. CHEADLE, COPYRIGHT PARKS CANADA.

Jurassic and middle Cretaceous time. Following the continent's about-face, evidence from younger rocks shows that from the middle Cretaceous much of the Canadian Cordillera moved northward along the continental margin. The terranes came to lie at their present latitudes relative to the continental interior by movements on big right-lateral strike-slip faults that slice along the Canadian Cordillera, with the western sides moving north. Some of these faults, such as the Tintina Fault in Yukon, were active in the late Cretaceous and Paleocene. Others such as the Fraser Fault were active later in the Cenozoic, and one, the Denali Fault in southwestern Yukon, is active today. Similar relative movement is reflected in the San Andreas Fault in modern times as it moves southwestern California, attached to the Pacific Plate, northward against the North American Plate (Chapter 2).

Further evidence for this displacement comes from paleo-magnetic and fossil data, both of which show that during the early Jurassic, all of the Cordilleran terranes, plus a small sliver of the continental margin now in north-central British Columbia, were located between a few hundred and almost 2,000 kilometres south of where they are today relative to the continental interior.

Final Crunches

By 105 million years ago, Wrangellia, the Alexander Terrane, and parts of the new continental arc began to crunch against the inner terranes, themselves now firmly attached to the western margin of Laurasia. In the process, some rocks were buried to depths of at least 30 kilometres, and others were raised up to

The Tombstone Range, seen from Talus Lake, Yukon, has been eroded from the middle Cretaceous Tombstone Pluton. WALTER LANZ.

Richard McConnell was the first person to recognize the thrust-faulted and folded structure of the Rocky Mountains. REPRODUCED WITH THE PERMISSION OF NATURAL RESOURCES CANADA 2013, COURTESY OF THE GEOLOGICAL SURVEY OF CANADA.

make the ancestral Coast Mountains, mostly between 95 and 85 million years ago. This revitalized collision was probably also responsible for the thrusting and folding in the Rockies that we explore in the next section. Between 90 and 45 million years ago, the continental arc migrated eastward, injecting magma that became the granitic rocks that underlie most of the present-day Coast Mountains. The Coast Mountains crunch also caused the strata of the Bowser and neighbouring Sustut basins to be folded and thrust eastward.

Detritus eroded off the newly raised Coast Mountains was shed westward onto former Wrangellia, where it is preserved in a series of depressions including the Queen Charlotte and Georgia basins. The Georgia Basin, which mostly underlies the lowlands of eastern Vancouver Island and the Gulf Islands, contains a succession of up to 2,500 metres of late Cretaceous, non-marine, and shallow- to deep-marine clastics. In the Nanaimo and Comox areas, parts of this succession are non-marine and deltaic, with extensive bituminous coal deposits that for many years sustained the economy of Vancouver Island.

As these basins filled, clastic sediments were also shed beyond Wrangellia onto the floor of the Pacific Ocean. These deposits are now preserved as the Pacific Rim Terrane, a narrow strip of fault-bounded Jurassic and Cretaceous sedimentary rocks on southern and western Vancouver Island, for example along the shore in Pacific Rim National Park Reserve of Canada. Rocks of the Pacific Rim Terrane include mélange containing blocks of chert, carbonate, and basalt. Like other parts of the western Cordillera, the Pacific Rim Terrane may have formed somewhere to the south along the continental margin and migrated northward (right-laterally) relative to Wrangellia as the continent moved south.

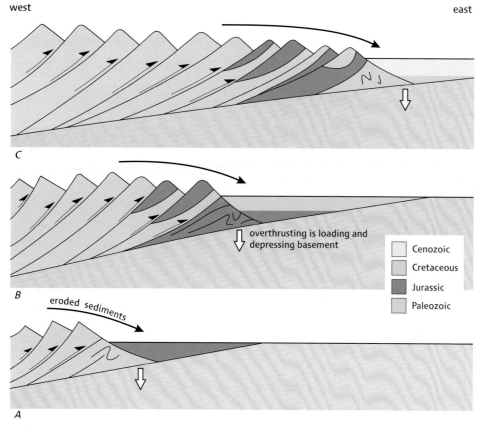

The progressive evolution from Jurassic to Paleocene (A to C) of the foreland basin in front of the advancing Rocky Mountains. ADAPTED FROM YORATH AND GADD (1995), REPRODUCED WITH PERMISSION OF THE AUTHORS, DUNDURN PRESS, AND NATURAL RESOURCES CANADA

Rocky Foundations

Designated collectively as a UNESCO World Heritage Site, Canada's Rocky Mountain parks contain breathtaking scenery that attracts visitors from all over the world. But the Rocky Mountains are also a paradise for geologists who, working since the late nineteenth century, have learned enough to make these mountains probably the best-known example of a thrust-and-fold belt. Visitors to the Rocky Mountains can quickly gain an appreciation of the sedimentary strata that form the towering peaks. Less obvious are the numerous thrust faults that carry older rocks northeastward over younger ones. The distinctive structure of the Rocky Mountains was first recognized by Richard McConnell of the Geological Survey of Canada, who in 1887 identified gently folded, mainly southwest-dipping thrust faults near what is now the Trans-Canada Highway west of Calgary. Along these faults, slabs of Cambrian to Cretaceous strata have been thrust upward and northeastward to form numerous overlapping thrust-fault slices. The result is that today, after uplift and erosion, the many northwest-southeast trending ranges, formed of erosion-resistant Paleozoic carbonates, are separated by valleys eroded in younger, softer, and less resistant Mesozoic shales and sandstones.

During the twentieth century, deep drilling and seismic investigations carried out in the search for oil and gas, together with geological mapping of the Foothills and the Front Ranges, revealed that the thrust faults flatten with depth and merge into a master thrust fault called the basal detachment zone. This master thrust is located close to the base of the oldest strata that were deposited upon the underlying crystalline Precambrian basement: Cambrian strata in the Foothills and Front Ranges, and Proterozoic strata farther west.

Thrust faulting and folding in the southern Rocky Mountains has horizontally shortened and vertically thickened the Proterozoic to Cretaceous rocks in the region. The amount of shortening can be estimated by theoretically unfolding the folds and pulling back the thrust faults so that rocks on one side of a fault have a matching counterpart of the same age on the other side. Such an exercise reveals that shortening along the Bourgeau Thrust Fault, which intersects the Trans-Canada Highway near Banff, is 30 kilometres. The rocks now directly above the Bourgeau Fault originated 30 kilometres southwest of the rocks that now are immediately beneath them. However, the Bourgeau Fault is but one of many thrust faults. If displacements on all of the thrust faults are taken into account, the

An aerial view along the Trans-Canada Highway near Exshaw, Alberta. To the left is Door Jamb Mountain and in the right centre Yamnuska (Mount Laurie). The surface trace of the McConnell Thrust is located at the break in the slope on both mountains. This fault places Cambrian carbonate over Cretaceous shale and sandstone (the vegetated areas). RON GARNETT / AIRSCAPES.CA.

cumulative northeastward displacement across the southern Rocky Mountains is at least 200 kilometres.

The Rocky Mountains consist of sedimentary strata deformed by thrust faulting and folding as they were scraped along the top of the underlying Laurasian Plate. From the ages of the rocks affected, we know that most of the deformation occurred between 110 and 60 million years ago, as the continental bulldozer continued to plough southwestward.

Weighed Down

As part of the process that created the structure of the Rocky Mountains, rocks were driven up the old continental margin and onto the western edge of the continental platform. The weight of this great pile of rock caused the underlying continental lithosphere to flex and bend downward, creating a vast moat-like foreland basin that extended for several hundred kilometres east of the mountains. The earliest rocky record of the development of this foreland basin comes from latest Jurassic to earliest Cretaceous marine and non-marine strata, including coals, of the Kootenay Basin in southeastern British Columbia and southwestern Alberta. Then, during the late Cretaceous and earliest Cenozoic, as thrusting and folding

Typical topography of the Rocky Mountain Front Ranges, seen in this view to the northwest from above the Trans-Canada Highway in the Sundance Canyon area, near Banff, Alberta. The landscape reflects the repetition of erosion-resistant Paleozoic carbonate strata thrust over more-easily-eroded Mesozoic clastic rocks in the valleys. RAY PRICE.

Tilted late Devonian carbonate strata of The Ancient Wall, a mountain in Jasper National Park of Canada, are highlighted by low sunlight. These rocks were moved eastward tens to hundreds of kilometres by thrust faulting during middle Cretaceous to Paleocene times. WALTER LANZ.

Geologists examine tilted Jurassic shale and sandstone on the Trans-Canada Highway east of Banff, Alberta. Cascade Mountain in the background consists of folded Devonian and Carboniferous carbonate strata, which have been thrust over the Jurassic strata. JIM MONGER.

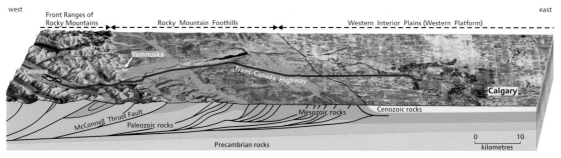

The Rocky Mountain thrust-and-fold belt in southern Alberta, showing a generalized section through the Rocky Mountains and Western Interior Plains around and west of Calgary. The section shows the deformed edge of the former continental margin now underlying the Rocky Mountains, transitioning eastward to relatively undeformed rocks beneath the Plains. Paleozoic rocks (blue) thicken significantly westward; eastward they continue across the Plains beneath a cover of Mesozoic (green) and Cenozoic (yellow) strata. The mountains coincide generally with the area where the harder Paleozoic rocks reach the surface, a region bounded to the east by the McConnell Thrust Fault (orange line). The Foothills, between the Fault and Calgary, are underlain by deformed, but relatively easily eroded, Mesozoic strata. ADAPTED FROM POULTON ET AL. (2002).

in the emerging Rocky Mountains expanded northeastward, the foreland basin migrated in the same direction.

Today, deposits of the Mesozoic foreland basin are thickest adjacent to the Rocky Mountains thrust-and-fold belt, where the succession is up to 6 kilometres thick. The deposits thin progressively eastward and peter out in Manitoba. The more easterly deposits are not strictly part of the foreland basin but were left by seas that intermittently flooded the continental interior. However, for simplicity, we will refer to the Jurassic to early Cenozoic foreland basin and the depositional area farther east on the Western Platform together as the Western Interior Basin. (Readers may also come across the term Western Canada Sedimentary Basin, which geologists apply to the entire Phanerozoic sedimentary succession beneath the eastern Cordillera and Western Interior Plains.)

During the early Cretaceous, more than 120 million years ago, seas began to invade the Western Interior Basin from the Arctic and the Gulf of Mexico, ultimately meeting to form the north-south-trending Western Interior Seaway that lay just east of the Rocky Mountains. For 50 million years, the Seaway alternately expanded and contracted because of changing sea

South of Calgary, Alberta, plains and rolling foothills are underlain by less resistant, but still deformed, Mesozoic clastic sediments. On the horizon are distant peaks of the Front Ranges underlain by more resistant Paleozoic carbonates. Also evident are signs of two of Alberta's major industries, petroleum and farming. ROB FENSOME.

level, and because of sedimentary and tectonic factors. As the shorelines marched restlessly back and forth, beach sands and marine muds interfingered with fluvial sediments. The various advances of marine waters in the Seaway are each given their own names. An example is the Bearpaw Sea, which lasted from about 75 until about 70 million years ago.

The lighter-coloured sandstone in this roadcut at Pink Mountain, British Columbia, represents the channel deposits of a river that flowed in the Western Interior Basin sometime during the middle Cretaceous. DARREL LONG.

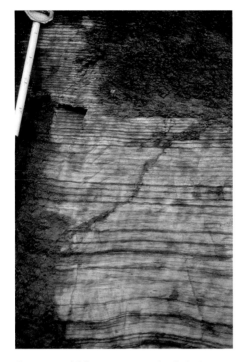

Cretaceous tidal deposits associated with the Bearpaw Sea, near Drumheller, Alberta. The rhythmic layering of the sediment reflects daily tidal cycles. ANDREW MACRAE.

Tilted late Jurassic strata along the Trans-Canada Highway, just east of Banff, Alberta. These rocks were deposited in the Western Interior Basin. They represent a transition from marine to non-marine environments and are part of the first major influx of clastic sediments eroded from the rising Cordillera to the west. JIM MONGER.

Late Cretaceous strata in Dinosaur Provincial Park, Alberta, have been sculpted into interesting shapes by erosion. M. FINKELSTEIN, COPYRIGHT PARKS CANADA.

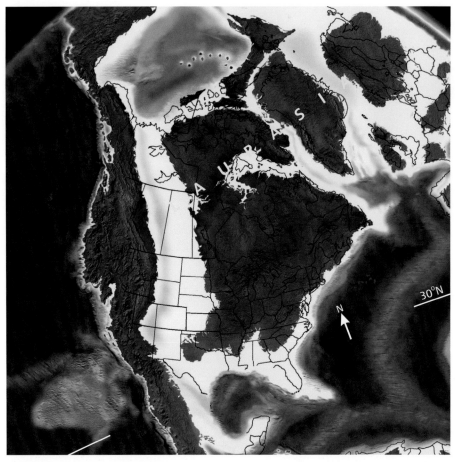

Paleogeography of what was to become North America and adjacent regions in the late Cretaceous, 85 million years ago. Land is shown in brown, with shading showing topography. The lighter blue areas represent possible coastal or nearshore areas, darker blue represents deeper ocean waters, and black indicates trenches. Aspects of modern geography are shown for orientation.

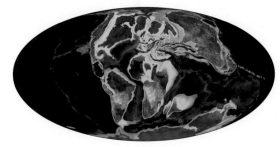

Global paleogeography 85 million years ago, during the late Cretaceous. Colours as for previous figure.

Today, alternating units of marine and non-marine Cretaceous sediments are widespread in the Prairie Provinces. One of the best places to see these sediments is along the Red Deer Valley from the city of Red Deer to the Alberta-Saskatchewan border. Here, a 100-metre-thick marine shale deposited in the Bearpaw Sea is sandwiched between non-marine strata.

Down East

In the early Cretaceous, then, the Canadian scene involved oceans growing to the east and north, mountains rising to the west, and a seaway intermittently covering the area that would become the Western Interior Plains. Between the Western Interior Seaway and the Atlantic, Cretaceous sedimentary rocks are restricted to the relatively small Moose River Basin in the Hudson Bay Lowlands, and a few pockets in the Maritimes. All of these deposits are non-marine.

It may come as a surprise that a group of some of Quebec's more familiar landmarks are of Cretaceous age. The Monteregian Hills are made largely of early Cretaceous intrusive igneous rocks that are more resistant to erosion than the surrounding Paleozoic limestones and shales, and so they form prominent hills in the St. Lawrence Lowlands. Mount Royal, which gave its name to the city of Montréal, is the best known of the Monteregian Hills. Others include Mont Saint-Hilaire, Mont Shefford, Mont Mégantic, and Oka. The hills are not suitable for agriculture, but their rocks are important as aggregate resources (Chapter 14), and some have been developed as ski resorts and parks. Mont Saint-Hilaire is world famous among mineral collectors due to the unusual chemical composition of the intrusive bodies. More than 300 different minerals have been found there, including the first discoveries of monteregionite, and the tonguetwisters fluorbritholite and manganokomyakovite. The Monteregian Hills form a broadly east to west linear trend and roughly align to the southeast with the White Mountains intrusions of New Hampshire and, ultimately, the offshore New England Seamounts, some of which are as young as 80 million years. Some geologists interpret this line of magmatic remnants as recording the motion of the North American continent over a hot spot.

Around 140 million years ago, the direction of sea-floor spreading in the nascent Atlantic Ocean basin gradually switched

Small pockets of early Cretaceous deposits, rarely exposed at the surface, occur in the Maritimes. Some of the white fluvial sand and gravel from this quarry, south of Sussex, New Brunswick, is used for sand traps on golf courses. ROB FENSOME.

Fall colours on Mont Saint-Hilaire, Quebec, one of the Monteregian Hills. PIERRE BÉDARD.

Aerial view of Mont Rougemont (the forested area at bottom left), one of the Monteregian Hills of Quebec. ROB FENSOME.

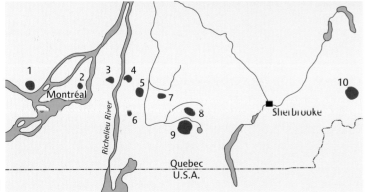

A shows parts of eastern North America and the western North Atlantic including the location of the offshore New England Seamounts, the White Mountain intrusions in New England, the Monteregian Hills in Quebec, and the Moose River Basin in northern Ontario. B, represented by the box in A, shows the distribution in southern Quebec of the Monteregian Hills: 1 = Oka, 2 = Mount Royal, 3 = Mont Saint-Bruno, 4 = Mont Saint-Hilaire, 5 = Mont Rougemont, 6 = Mont Saint-Grégoire, 7 = Mont Yamaska, 8 = Mont Shefford, 9 = Mont Brome, 10 = Mont Mégantic. ADAPTED FROM VARIOUS SOURCES.

from roughly northwest-southeast to east-west, leading to the separation of the Grand Banks from Iberia. The widening ocean had already severed eastern Laurasia's land links with the African part of Gondwana, but a land route between future North America and Eurasia via Greenland would endure into the Cenozoic. As the opening of the Atlantic Ocean shifted orientation and expanded northward, sedimentary basins such as the Jeanne d'Arc on the Grand Banks of Newfoundland, in which Hibernia and other offshore oil fields are located, continued to develop on the new continental margin.

Off Nova Scotia, the Sable and other deltas continued to fringe the Atlantic Ocean in the early Cretaceous. But as the Cretaceous progressed, continuing erosion onshore led to lower relief on land and more sluggish rivers. Because of this and global sea-level rise, the delta systems shrank. During the later Cretaceous, between about 90 and 70 million years ago, the sea off Atlantic Canada was shallow, warm and clear—conditions under which a broad blanket of chalk accumulated. Chalk is composed almost entirely of coccoliths (Chapter 4). Coccolithophores, the protists whose shells are covered by coccolith scales, live mostly in warm water today, so the widespread distribution of chalk deposits in mid- to northerly latitudes indicates that much of the Earth's late Cretaceous climate was comparable to that of the modern tropics and subtropics. Most chalk deposition ended before the close of the Cretaceous and the sea temporarily retreated as delta systems again built out. No late Cretaceous chalks occur on the Labrador Shelf because of cooler conditions compared to those farther south. Instead, the late Cretaceous sediments there are marine sandstones and shales.

The igneous rocks that form the Monteregian Hills are a source of rare minerals, such as this specimen of orange serandite from Mont Saint-Hilaire, Quebec. HELEN TYSON, FROM THE COLLECTION OF HELEN AND ROD TYSON.

Greenland's temporary separation from North America began in the Cretaceous. Between Canada and Greenland today are, from south to north, the Labrador Sea, Davis Strait, Baffin Bay (collectively the Labrador-Baffin Seaway), and the narrow Nares Strait. It was clear to many early twentieth-century explorers that the two facing coastlines match up like pieces of a jigsaw puzzle. Even before the concept of continental drift had been generally accepted, several geologists made the case that Greenland on the one side and Baffin Island and Labrador on the other had drifted apart. They further suggested that Nares Strait was underlain by a transform fault with a left-lateral displacement of about 500 kilometres. Recent research has shown that this scenario is partly correct. About 100 million years ago, a series of rift basins developed, and these ultimately joined to form a late Cretaceous seaway from the Labrador Sea to Baffin Bay. Once formed, the Greenland Plate rotated away from North America as the Labrador-Baffin Seaway widened. So far then, early speculations were right. Where they went astray is a topic for the next chapter.

Cretaceous Park

Late Cretaceous rocks of the southern Western Interior Plains preserve some of the richest dinosaur assemblages in the world. So many fossils are found there mainly because of the wide range of environments in which the rocks were deposited, including rivers, lakes, swamps, and coastal settings. Much of Alberta and southwestern Saskatchewan was then a subtropical coastal plain between the rising Rocky Mountains to the west and the Western Interior Seaway to the east. Over time, sea level fluctuated and the shoreline migrated east and west, so particular types of dinosaurs and their habitats also changed. Floods, created by storms or by rivers breaking their banks, covered the landscape with sediment and buried the remains of animals and plants. Some of these remains became the wonderful diversity of fossils that continue to be discovered in the region. Today, much of southern Alberta and southwestern Saskatchewan is a semi-arid grassland with deep river valleys in which poorly vegetated steep slopes and soft bedrock are easily eroded to reveal abundant fossils.

Over 80 dinosaur species are known from the region, from rocks ranging from 80 to 65.5 million years old. Half of these come from an 80-square-kilometre expanse of badlands near Brooks, Alberta, designated as Dinosaur Provincial Park, a UNESCO World Heritage Site. There, the great diversity of dinosaurs comes from an even shorter interval that covers only about 1 percent of the roughly 160 million years that dinosaurs existed, but represents about 10 percent of known species. Hundreds of dinosaur specimens have been recovered from the Park, and many of these are displayed in museums around the world. Some of the best are on display at Alberta's Royal Tyrrell Museum of Palaeontology in Drumheller, the largest museum in the world dedicated to paleontology.

An excavated bone bed in Dinosaur Provincial Park, Alberta. DAVID EBERTH.

A skull of the late Cretaceous ceratopsian dinosaur *Anchiceratops* found between Morrin Bridge and Tolman Bridge in the Red Deer Valley and now displayed at the Royal Tyrrell Museum of Palaeontology. COURTESY OF THE ROYAL TYRRELL MUSEUM OF PALAEONTOLOGY.

One of the most remarkable features within Dinosaur Park is the *Centrosaurus* bone bed. *Centrosaurus* is a herbivorous horned dinosaur, one of the ceratopsian group, the best known of which is *Triceratops*. In an area the size of a football field, so many ceratopsian bones litter the ground that it is hard to put your foot down without stepping on one. It is estimated that hundreds of *Centrosaurus* are represented in the bone bed. Research has shown that the dinosaurs in the bed may have died simultaneously, probably during a flood. The carcasses were feasted upon by meat-eaters such as *Albertosaurus*, and then the bones were washed around by subsequent floods so that the skeletons were broken up into a dense scattering of bones. This find suggests that centrosaurs may have gathered, at least from time to time, in large herds.

Dinosaur Park's story doesn't end with dinosaurs. It is home to one of the most diverse ancient ecosystems known: aside from 44 dinosaur species, 31 mammal species have been extracted, along with 9 lizards, 13 turtles, 5 amphibians, 2 crocodiles, 2 pterosaurs, 6 birds, and 35 fish. And Dinosaur Park, as remarkable as it is, is by no means the whole paleontological legacy of the Western Interior Plains. For example, at Devil's Coulee near Warner, Alberta, discoveries from rocks of about the same age as those in Dinosaur Park include several nests containing the eggs of duck-billed dinosaurs (hadrosaurs) and small meat-eaters (theropods), as well as the bones of unhatched hadrosaur embryos and babies. To the south in Writing-on-Stone Provincial Park, famous for its rock art and magnificent hoodoos, the rocks contain some of the oldest dinosaur body fossils in the province. They include bones and teeth of small and large theropods, ankylosaurs (armoured dinosaurs), hadrosaurs, pachycephalosaurs (bipedal, dome-headed dinosaurs), and ceratopsians.

Dinosaur fossils have also been found in southwestern Saskatchewan, the most striking among them a more-than-half-complete *Tyrannosaurus rex* skeleton, found in the Frenchman Valley. Locally known as Scotty, this dinosaur can be viewed at the T. rex Discovery Centre in Eastend.

The dinosaur fossil treasury of the Western Interior Plains also includes trackways. For example, near the town of Grande Cache in west-central Alberta, folded rocks preserve thousands of tracks left by dinosaurs as they walked on a muddy shore about 105 million years ago. The footprints tell us that small to large dinosaurs, birds, and small mammals were all thriving together.

Because the sea flooded the Western Interior Basin from time to time, many marine fossils are also known. Marine vertebrate remains have been discovered in oil-sand exposures near Fort McMurray in Alberta, including sharks' teeth and

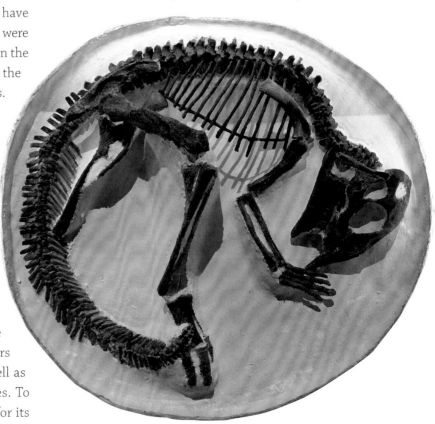

Fossil embryo of the hadrosaur *Hypacrosaurus* from Devil's Coulee, Alberta. COURTESY OF THE ROYAL TYRRELL MUSEUM OF PALAEONTOLOGY.

Skeleton of the short-necked late Cretaceous plesiosaur *Trinacromerum kirki* from rocks in the Treherne area of Manitoba. Short-necked plesiosaurs (called pliosaurs) were fish-eating marine reptiles that thrived in the Mesozoic seas. HANS THATER, COURTESY OF THE MANITOBA MUSEUM.

plesiosaur and mosasaur skeletons. Similar fossils are also found in Saskatchewan and Manitoba. Plesiosaurs were streamlined swimming reptiles, typically with long necks, broad bodies, flippers, and short tails. Mosasaurs, closely related to monitor lizards and snakes, were highly streamlined animals up to 17 metres long that preyed on fish and shelled invertebrates in the shallow sea. Gigantic mosasaurs have been found in Cretaceous shales, most notably near Morden, Manitoba. These fossils are associated with plesiosaurs, sea turtles, and huge fish, such as *Xiphactinus*, which could grow to 6 metres in length. Another impressive creature of the Western Interior Seaway was the long-snouted crocodile *Terminonaris*, the most complete skeleton of which was found in a quarry at Carrot River, Saskatchewan. This 6-metre-long creature would have competed with plesiosaurs and sharks for the fish that it ate. Also found in the Carrot River quarry were fossils of a variety of fish, other marine reptiles, and toothed birds.

These seas also teemed with invertebrates. Remains of clams, snails, ammonites, and decapods (crayfish) occur in the brown to grey, marine and brackish-water mudstones throughout the Western Interior Basin. Some of the ammonites are exquisite, with iridescent shells that are so vivid they have spawned a lucrative local jewelry industry in Alberta. Colourful pieces of ammonite shell are made into "ammolite", used to make rings, pendants, and other jewelry. Highly valued pieces of ammolite exhibit rich red, green, and blue colours.

Around the margins of the interior seaway the land was densely vegetated. The terrestrial deposits contain a rich bounty of plant fossils, including flattened conifer needles, stems, fern fronds, and cycad and gingko leaves. More rarely, coal deposits contain three-dimensional fragments preserved in silica, sometimes including stumps of cypress trees complete with roots and

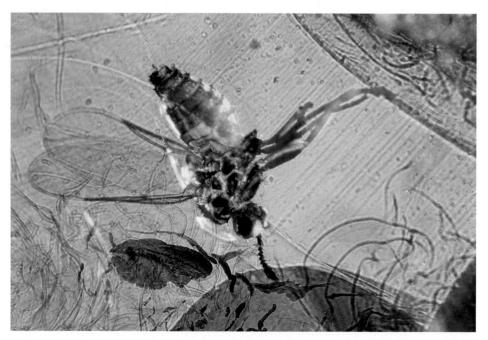

Scatopsia fly in late Cretaceous amber from Grassy Lake, Alberta. COURTESY OF THE ROYAL TYRRELL MUSEUM OF PALAEONTOLOGY.

tree rings. Fossil forests with multiple in-situ cypress stumps about 70 million years old are known from Willow Creek near East Coulee, Alberta. Amber (preserved tree resin), sometimes containing beautifully preserved insects, is also found in Cretaceous coal beds, most notably at Grassy Lake in Alberta and Cedar Lake in Manitoba (in the latter case probably transported from farther west by modern rivers). One intriguing recent discovery in amber from Grassy Lake is helping to resolve how feathers evolved. The find is of so-called protofeathers that appear to be from dinosaurs.

So the fossil treasures provide evidence of a rich ecosystem in the late Cretaceous of western Canada. But this seemingly idyllic Mesozoic world was not to last forever.

The Asteroid Strikes

Toward the end of the Cretaceous, about 69 million years ago, seas that had intermittently formed the Western Interior Seaway began to retreat from most of the region for the last time. Vast areas of forest and bush blanketed newly emergent areas, and the climate was evidently cooling. But temperature change was not the main challenge to life around this time. The end of the Cretaceous was marked by one of the most famous events in Earth's history—the asteroid impact that killed off all the dinosaurs (or all the non-avian dinosaurs, if we accept birds as dinosaurs). This mass-extinction event is also commonly known as the K/T boundary event: K is geological shorthand for Cretaceous and comes from the German word Kreide for the period, and T stands for Tertiary. During the K/T event, about 70 percent of known species disappeared. The extinction event resulted in the annihilation of several animal groups besides the dinosaurs. Among vertebrates, mosasaurs, plesiosaurs, and pterosaurs all disappeared; among invertebrates, ammonites and their cephalopod relatives, the belemnites, met their end. In addition, many other animal and some plant groups suffered devastating losses.

The event was likely caused by the impact of an asteroid that hit our planet near what is today the town of Chicxulub on Mexico's Yucatán Peninsula. Evidence for the impact sparked the imagination of the public and scientists alike during the 1980s and 1990s. Scientists proposed many drastic scenarios, including dust clouds that brought darkness and turned off photosynthesis, acidification of terrestrial and oceanic environments, and global wildfires. We now realize that no single process explains all of the changes that occurred in the asteroid's wake. And some wilder speculations, such as worldwide wildfires, have been discounted. Based on the 180-kilometre diameter of the buried Chicxulub crater and the extent and nature of material ejected in the crash, it is possible to determine the energy of the impact and the extent of changes to the atmosphere. The selectivity among organisms, both marine and terrestrial, that became extinct, also provides clues. And the abrupt changes in the fossil record, especially of spores and pollen from land plants across North America, confirm the sudden nature of the event.

So what is the evidence we see in the rocks? About fifty localities with a complete K/T boundary interval have been found in mid-continental North America, including fifteen from western Canada. A good place to see a boundary section is at Zahursky Point in Grasslands National Park of Canada, south-western Saskatchewan. Well-developed boundary claystone layers also occur along the Red Deer River north of Drumheller, Alberta. These Western Interior Basin localities provide a precise record of the event and how it affected non-marine plant and animal communities. The impact is recorded in the Basin by a thin, discontinuous claystone sequence that, where complete, consists of three distinct layers. Forming the lowest layer are beads of rock melted at or near the site of the impact and hurled into the atmosphere before falling to the ground. The beads,

Cretaceous-Tertiary boundary layers at the base of a coal bed, Wood Mountain Creek, south-central Saskatchewan. The thin white layer, which marks the boundary, was originally a layer of glass beads derived from rock that melted during the asteroid impact event but is now weathered to kaolin clay, a white aluminum-rich clay. The overlying rusty-brown to dark grey impact layers (1 to 1.5 centimetres thick) are enriched in iridium and contain shocked quartz (Box 11). ART SWEET.

Ferns proliferated within weeks to months following the asteroid strike at the end of the Cretaceous, much as they do today after a forest fire. This scene is near Halifax, Nova Scotia. ANDREW MACRAE.

each of which is 1 to 2 millimetres across, are known as tektites, and the bed they form is known as an ejecta layer. They would have reached western Canada, some 3,500 kilometres from the impact site, about an hour after the asteroid struck. The ejecta layer in Alberta and Saskatchewan is 1 to 2 centimetres thick and pale grey to pinkish-white. Above the ejecta layer is a dark-grey to black layer of mudstone 0.5 to 1.5 centimetres thick. This layer was derived from fine material that fell out of the impact-generated dust plume, which would have soared into, and possibly above, the atmosphere. Together with a thin laminated layer above, it contains abundant crystals of shocked quartz and unusually high abundances of the element iridium, both signatures of an extraterrestrial body. The thin laminated layer is thought to have formed from yet finer material falling to the ground during the weeks and years that followed the impact. Microdiamonds also occur in these boundary layers.

Where present, the ejecta layer looks surprisingly innocuous considering that it represents the decimation of vast gymnosperm forests with their understoreys of flowers and ferns. In the sediments above, a huge increase, or spike, in spores indicates the rapid regeneration of a single opportunistic fern species, revealing that a fern prairie extended from Alberta to central Saskatchewan and surely beyond. With the spread of coal swamps, starting about a decade after the impact, pollen and spores indicate that shrubs and herbs progressively migrated in from isolated refuges to re-establish forested swamps.

The amazingly abrupt shift to a fern prairie immediately above the ejecta layer is puzzling. Evidently, within weeks to months following the loss of the forest cover, ferns had reached reproductive maturity during a time when other evidence points to near-total darkness, which would have switched off all photosynthesis. This is truly an enigma.

George Dawson would have been fascinated by our current understanding of the demise of the dinosaurs and the rise of the Rockies. He would surely have agreed that the interval from 251 to 65.5 million years ago was an exciting time in the evolution of the land that was to become Canada. But what happened after the dust from the asteroid had settled was equally engrossing, as we will see in the next chapter.

A Permian scene from Prince Edward Island, showing a group of *Bathygnathus* (possibly equivalent to the more famous *Dimetrodon*, or at least a close cousin) prowling among cycads, with copses of small horsetail trees in the background. PAINTING BY JUDI PENNANEN, COURTESY OF THE NEW BRUNSWICK MUSEUM.

BOX 8 · GROWING FUR

Signing in with Synapsids

Both birds and mammals descended from reptiles, though from two very different groups. Whereas birds evolved from diapsid reptiles, more specifically dinosaurs, mammals evolved from the synapsid, or mammal-like reptiles, which had their heyday in the Permian and included such favourites as the sail-backed *Dimetrodon*. The diversity of mammal-like reptiles was severely reduced in the extinction event at the end of the Permian. But they regrouped somewhat in the early Mesozoic and may even have survived into the Cenozoic—a single contested fossil from the Paleocene, found in Alberta, could be the last known mammal-like reptile. Fortunately for us, by the latest Triassic, these distinctive vertebrates had given rise to mammals.

Mammals have several distinctive features: live birth (though this also occurs in some reptiles, including ichthyosaurs; and monotreme mammals lay eggs); hair; lactation; teeth differentiated into types like canines and molars, the latter commonly promoting a complex chewing action; advanced parental care; warm blood (shared with some reptiles and birds); and double circulation, in which blood flows through the heart twice in a cycle, once oxygenated and once deoxygenated. Most of these features are invisible in the fossil record because soft parts are not generally preserved. For fossils, the distinction between a mammal-like reptile and a mammal usually relies on the structure of the lower jaw. In mammal-like reptiles, each half of the lower jaw is made up of several bones including the dentary, which holds the teeth, and multiple bones behind the dentary. In mammals, each half of the lower jaw is formed only by the dentary, and the other "reptilian" jaw bones are reduced to serve as ear ossicles. This development allowed mammals to evolve a more sophisticated hearing apparatus.

Modern mammals represent three groups: monotremes, marsupials, and placentals. The only major group of mammals now extinct is the multituberculates. Although current molecular evidence places the origin of mammals at 180 million years ago, fossil evidence takes the group back to about 220 million years, just a bit later than the earliest dinosaurs. Early mammals were mostly small, up to a few centimetres long, and probably nocturnal, factors that may have been invaluable for surviving the age of dinosaurs. Only after dinosaurs became extinct did mammals fully explode onto the scene, with most modern mammal families becoming recognizable in the Eocene.

Thanks to recent fossil finds, molecular research, and plate tectonic studies, the evolution of the mammals is now much better understood than it was even just a few decades ago. Perhaps the biggest surprise has been the crucial role played by plate tectonics, starting with the splitting of Pangea into Laurasia and Gondwana, followed by Gondwana's fragmentation into Australia, Africa, and South America, a process that gave rise to isolated evolutionary "continental nurseries" for different groups of mammals. Within these nurseries, mammal evolution tended to follow parallel courses, but the end results were very different. Such differences, substantiated by molecular studies, are the basis for the modern classification of mammals.

Monotremes, Cone Teeth, and Marsupials

The most primitive living mammals are the monotremes, the only ones that lay eggs. Today, monotremes are restricted to Australia and New Guinea, and include the lovable-looking duck-billed platypus. Monotremes are the only survivors of the Gondwanan nursery. Although they have proper mammalian jaw- and ear-bone structures, their anatomy is otherwise a mosaic of reptilian and mammalian features. Other than egg-laying, their reptilian features include the single posterior hole (which serves as the anus, urinary tract, and reproductive tract) and the absence of nipples. Their mammalian features, aside from jaw and ear structures, are fur and lactation. Monotreme fossils

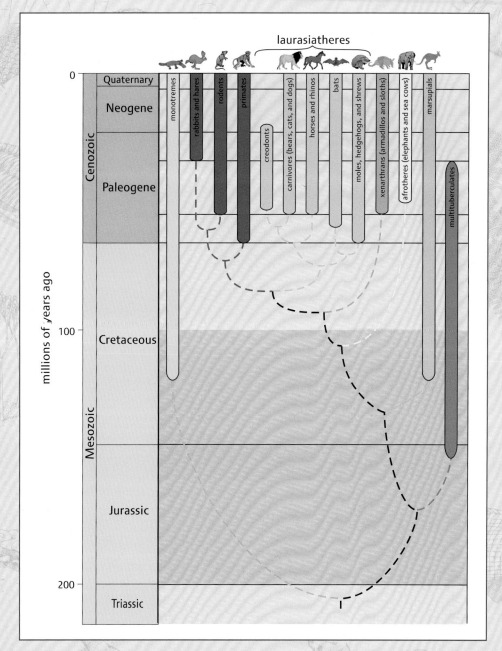

Evolutionary relationships and the timing of first appearances and ranges of the main groups of mammals. Probable inter-relationships are shown by the dashed lines.

teeth. They may have been the first mammals to live in trees, probably to avoid being crushed or eaten by dinosaurs. Multituberculates are common in 85- to 35-million-year-old rocks of Alberta and Saskatchewan.

Marsupials (pouch bearers) represent a second branch of the Laurasian nursery. They give birth to tiny young, which have to make a perilous journey to the mother's pouch before hooking on to their milk supply. Marsupials appeared about 140 million years ago in China and were common members of faunas as far north as Saskatchewan until about 20 million years ago. Today, marsupials are mostly restricted to Australia and New Guinea where (at least before European settlement) they filled many of the ecological niches that placentals fill on other continents. They have evolved into carnivores, herbivores, and omnivores; tree-dwellers, ground-dwellers, and burrowers; and include animals such as wombats, kangaroos, and koalas. Marsupials were also successful for many millions of years in South America. As in Australia, South American marsupials filled many ecological niches from 65 to 5 million years ago. The reason South America today does not have dominant marsupial fauna is that the continent became permanently connected to North America about 3 million years ago: the new link allowed a few marsupials, such as opossums, to move into North America, but it also let many placental mammals move south and outcompete, if not directly eat, South American marsupials.

The Placenta Rules

A key feature of placentals is, of course, the placenta, a magnificent two-way thoroughfare providing food to the young while taking away waste. Molecular studies show that the first placentals to appear, about 105 million years ago, were the afrotheres, which originated in Africa. Living members of the group include the hyraxes, elephants, manatees, dugongs, and aardvarks. Although elephants and their kin originated in Africa, they migrated successfully to other continents—hence the many records of mastodons and mammoths in Eurasia and North America until they became extinct about 11,000 years ago.

About 95 million years ago, another group of mammals, the xenarthrans, diverged from the main placental stock. This group originated in a South American nursery, isolated for millions of years and for a long time was protected from the invasion of some of the more aggressive placentals that evolved elsewhere. Sloths, anteaters, and armadillos are all xenarthrans. During their history, the group made

are mostly from Australasia, where jaw bones and molar teeth have been found in early Cretaceous rocks; none are known from Canada. Some studies suggest that the group originated in the Triassic.

Multituberculates are an extinct group that gets its name from the shape of their molar teeth, which have many cusps. (Many extinct mammals bear names related to teeth because these hardest of skeletal parts are the most common mammal fossils and are characteristic of different groups.) The nursery for multituberculates appears to have been northern Laurasia. No Gondwanan multituberculates are known. Multituberculates appeared about 160 million years ago, perhaps reflecting the early separation of Laurasia and Gondwana, and died out about 35 million years ago. From the narrow space between the pelvic bones, it appears that multituberculate young were born at an early stage of development, as in marsupials. Similar to modern rodents, multituberculates had a pair of lower incisors, but no canine

Scene from Beringia during the last interglaciation (Chapter 11) depicting a faceoff between a peccary (a laurasiathere, with Eurasian ancestry) and giant ground sloth (a xenarthran, with South American ancestry). In the background is a mammoth (an afrothere, with African antecedants) and two camels (laurasiatheres). BERINGIA INTERGLACIAL PERIOD, COPYRIGHT GOVERNMENT OF YUKON / ARTIST GEORGE "RINALDINO" TEICHMANN 2001.

several successful incursions into North America. Xenarthran fossils such as the giant ground sloth are known from sediments deposited during the last 1.8 million years in western and northern Canada, including Yukon.

The laurasiatheres split off about 85 million years ago and, as the name implies, originated in Laurasia in the northern hemisphere. Laurasiatheres are a diverse crew, with shrews, moles, hedgehogs, bats, cetartiodactyls (camels, pigs, deer, sheep, hippos, and whales), perissodactyls (horses, tapirs, and rhinos), carnivores (cats, dogs, bears, weasels, hyenas, seals, and walruses), and pangolins. Laurasiatheres filled all ecological niches: they became carnivores, herbivores, and scavengers; and developed into flyers, aquatic acrobats, and burrowers as well as ground dwellers. Fossils of these groups are prolific in the Cenozoic rocks of Alberta and Saskatchewan and the Quaternary rocks of much of Canada, onshore and offshore.

Rodents make up a fourth placental group, also originating in Laurasia. More than 40 percent of all mammal species are rodents. Thus it is no surprise that their bones and teeth are commonly found in the Cenozoic rocks of Saskatchewan, Alberta, and British Columbia.

Humans are in the group that, with great vanity, we refer to as primates, meaning first ones. Primates are distinctive in that they have retained five fingers (the "primitive" condition), possess fingernails, developed binocular colour vision through forward-facing eyes, and have an opposable thumb. Well-known primates are the lemurs, tarsiers, New World and Old World monkeys, and apes, to which we humans belong. The fossil record of primates also highlights the influence of plate tectonics on evolution. Unfortunately, there are few Canadian fossil finds.

Mammal fossils, unlike dinosaur remains, are commonly small. Here expedition members sieve sediment in their search for tiny vertebrate remains in lake deposits within the Haughton impact structurer, Devon Island, Nunavut (Chapter 10). MARTIN LIPMAN.

10 · FINAL APPROACH
Canada 65 million years ago to today

CHAPTER SUPPORTED BY A DONATION FROM JOHN 'T HART, TALISMAN ENERGY (RETIRED)

Volcanoes have been active in the western Cordillera throughout the Cenozoic. Eve Cone is a 150-metre-high cinder cone on the northern flank of Mount Edziza, one of three large shield volcanoes in the Stikine Volcanic Belt of northern British Columbia. Eve Cone was active around the year 700 CE. CAROL EVENCHICK.

North of Sixty

Our knowledge of Canadian Arctic geology is intimately linked to the colourful history of the region, not least with the ill-fated Franklin Expedition. John Franklin set sail from England in 1845 to seek—like Martin Frobisher, Henry Hudson, and others before him—a northwest passage between the Atlantic and Pacific oceans. Neither Franklin nor the 128 men aboard his two ships, *Erebus* and *Terror*, survived to tell their stories. Many searches were made for the lost expedition and, during some of these, rocks and fossils were recovered and taken back to England. However, these earliest geological discoveries in the Canadian Arctic caused only ripples of interest in academic circles. Exploration was revitalized with the Second Norwegian Expedition from 1898 to 1902 led by Otto Sverdrup in his ship,

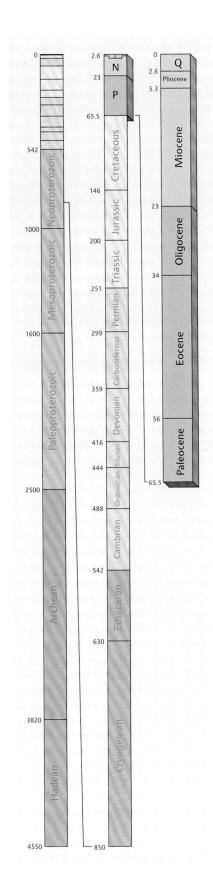

Geological time scale, showing the interval covered in this chapter. Numbers indicate millions of years ago. P = Paleogene, N = Neogene, Q = Quaternary.

Aerial view of Eureka on Ellesmere Island, Nunavut, which was one of a network of weather and radar stations set up by Canada and the United States in 1946. KELLY BENTHAM.

the *Fram*. This new exploration led to the discovery and naming of islands such as Axel Heiberg, Ellef Ringnes, Amund Ringnes, and King Christian. The expedition's brilliant young geologist, Per Schei, carried out some of the earliest geological studies of what later became known as the Sverdrup Basin, in the process losing several toes to frostbite. Canadian expeditions to the Arctic in the early twentieth century were deployed in part to counter Norwegian claims to sovereignty, which were eventually withdrawn in 1930.

The next milestone was inspired by political events immediately following World War II. In 1946, Canada and the United States set up a network of weather and radar stations, such as Eureka, Alert, and Cambridge Bay (now Iqaluktuttiaq), across the Arctic Islands. George Hansen, then Chief Geologist of the Geological Survey of Canada, immediately recognized the stations' potential as bases for geological exploration. He also realized that such exploration meshed perfectly with the aerial photography program of the Arctic Islands carried out by the Royal Canadian Air Force between 1948 and 1953. So it was that, during the 1950s, several young Survey geologists found themselves building careers as pioneers in Arctic geological research, exploring on foot and by canoe, dogsled, fixed-wing aircraft and, during an extensive survey known as Operation Franklin.

Since the 1950s, expeditions have explored to the most remote parts of the Canadian Arctic. One such venture in 1985 led to an extraordinary discovery—a fossil forest exposed on a windswept ridge on Axel Heiberg Island. Although the region was already renowned for Cenozoic plant fossils, the nature and preservation of the new find was exceptional. Among the plant litter between stumps and logs were cones that looked as if they had fallen off the tree that very day. The only trees now growing in this remote and hostile setting are tiny willows no taller than a dandelion.

Tree stump, probably of the dawn redwood, from the Eocene fossil forest on Axel Heiberg Island, Nunavut. JIM BASINGER.

Dawn redwood cones from the Eocene fossil forest on Axel Heiberg Island, Nunavut. HANS DOMMASCH.

Dawn redwood foliage from the Eocene fossil forest on Axel Heiberg Island, Nunavut. JIM BASINGER.

Walnut seeds from the Eocene fossil forest on Axel Heiberg Island, Nunavut. JIM BASINGER.

During further investigations, the site yielded remnants of the once-lush forest that nurtured trees up to 45 metres tall and 2 to 3 metres in diameter. Tree rings show that some individuals lived for as long as a thousand years. Some of the trees are dawn redwoods, among the fastest-growing conifers known. Other conifer fossils found in the leaf litter include beautifully preserved leaves of Chinese water pine and leaves and cones of larches. The angiosperm flora included members of the walnut, oak, birch, and sycamore families. Also found were vertebrate fossils, including remains of *Coryphodon* and brontothere teeth. Brontotheres, which became extinct at the end of the Eocene, were rhinoceros-like browsing creatures related to horses. *Coryphodon* was a large semi-aquatic herbivore with a tiny brain.

The fossil assemblage shows that the forest thrived in the middle Eocene, about 45 million years ago. Based on comparison with modern floras, the climate may have been similar to that of present-day North Carolina. Mean annual temperature was probably about 7 to 15°C (compared to about minus 20°C on Axel Heiberg today), with plenty of rain. Does this mean that Axel Heiberg drifted north in the past 45 million years? The answer is an emphatic no. As we discovered in the previous chapter, paleomagnetic evidence suggests that the Canadian Arctic has actually moved a few degrees southward since the Eocene. Even though the fossils tell us that climate was mild, the environment would have been challenging because of the long periods of continuous daylight in summer and of continuous darkness in winter. However, many of the plants—such as the dawn redwood, larches, and Chinese water pine—were deciduous, a major advantage for surviving winters at such high latitudes. The trees must have grown like weeds during the long summer daylight hours and become dormant in fall, as months of darkness descended. Many of the ancient Arctic

A scene from the Eocene Axel Heiberg forest. Holding centre stage is the four-legged herbivore *Coryphodon*, at about 1 metre tall and 2 metres long, the largest mammal of its time. *Coryphodon* probably wallowed in swamps and marshes, a major plus when the climate was so warm, even in the high latitudes. Hiding in the undergrowth and branches of trees are smaller mammals. FROM A DIORAMA AT THE AMERICAN MUSEUM OF NATURAL HISTORY, PHOTO BY DENIS FINNIN, COPYRIGHT AMERICAN MUSEUM OF NATURAL HISTORY.

dawn redwoods grew to great heights and formed dense forests in conditions very different from those in the region today. Such climatic fluctuations played an important role during the Cenozoic, the interval covered in this chapter and the next.

Setting the Scene

The Cenozoic Era is divided into three periods, the Paleogene, the Neogene, and the Quaternary, with the first two commonly being referred to collectively as the Tertiary. The Paleogene is subdivided into three epochs: the Paleocene (65.5 to 55.8 million years ago), the Eocene (55.8 to 33.9 million years ago), and the Oligocene (33.9 to 23.0 million years ago). The Neogene consists of two epochs: the Miocene (23.0 to 5.3 million years ago) and the Pliocene (5.3 to 2.6 million years ago). And the Quaternary has two epochs: the Pleistocene (2.6 million to 11,700 years ago) and the Holocene (11,700 years ago to the present). (Except for Quaternary dates, from here on we will round off Cenozoic boundary ages to the nearest million years.) In this chapter we will cover all aspects of the Paleocene through Pliocene and follow the tectonic history to the present. The Quaternary Ice Age has had such a fundamental influence on Canada that it deserves its own chapter.

At the beginning of the Cenozoic, North America was still technically part of Laurasia. Even though the tip of the North Atlantic Ocean was unzipping northward during the Paleogene, eastern North America and western Eurasia remained connected well into the Eocene. Two land routes evidently existed: a more northerly one that linked North America with Greenland and Scandinavia until it was drowned about 46 million years ago; and a more southerly route via the Faroe Islands, which existed from 55 to 50 million years ago. No plate boundary separates Alaska from northeasternmost Siberia, which have been connected for most of the past 100 million years. The present, not-so-obvious North American-Eurasian plate boundary lies within eastern Siberia. All three land routes played roles in shaping the Cenozoic mammal faunas found in the Arctic and western Canada. In contrast to these northern links, North America and South America were separated by sea throughout most of the Cenozoic, and became connected only about 3 million years ago. In spite of these complications, for simplicity we will treat North America as if it were a separate continent and plate during the Cenozoic. Over the past 65 million years, North America has moved generally in a southwesterly direction. And most of the continental part of the North American Plate has been above sea level during this time.

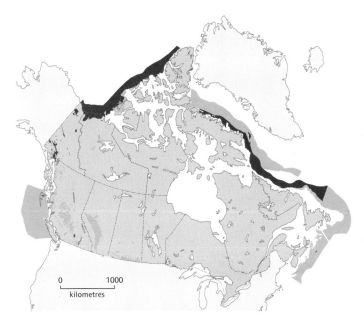

General extent of Cenozoic rocks at the surface (beneath glacial deposits), onshore and offshore. The lighter shaded areas denote either uncertainty or areas where rocks of the particular age have been confirmed but are intimately associated with rocks of other ages and the scale of the map doesn't allow us to show them separately. ADAPTED FROM WHEELER ET AL. (1996).

Paleogeography of North America and adjacent regions in the Paleocene, 60 million years ago. Land is shown in brown, with shading showing topography. The lighter blue areas represent possible coastal or nearshore areas, darker blue represents deeper ocean waters, and black indicates trenches. Aspects of modern geography are shown for orientation.

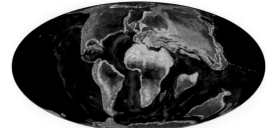

Global paleogeography 60 million years ago, during the Paleocene. Colours as for previous figure.

Climate Swings

Fossil evidence suggests that for the 3 million years following the end-Cretaceous crisis, Paleocene landscapes on the Western Platform consisted of vast stands of conifers and ferns. This vegetation thrived in coal-forming swamps, promoted by a humid climate and preserved in part because of the loss of large herbivorous dinosaurs. The Paleocene climate, though humid, was initially cool relative to the Cretaceous. But temperatures rose toward the end of the epoch, reaching a Cenozoic peak around 56 million years ago. A geologically brief interlude at that time, known as the Paleocene-Eocene Thermal Maximum, was one of the hottest spells of the past 100 million years. We know this through analysis of oxygen isotopes in deep-sea cores. Average sea-surface temperatures, even in the Arctic, were evidently above 20°C—substantially warmer than today. They may even have been briefly closer to 30°C. Poleward migration of warmth-loving plankton species, now preserved as fossils, as well as remnants of palm trees and alligators found in the Arctic Islands and Greenland, add more evidence. We don't know what triggered this dramatic temperature spike, but it may have been caused by the release of methane gas from frozen gas hydrates (Chapter 13).

After the heat wave 56 million years ago, temperatures stayed generally moderate through the Eocene, with cooling episodes around 50 and 38 million years ago. Temperatures dropped markedly about 34 million years ago and remained cool through the Oligocene, then warmed up in the Miocene. Cooling at the end of the Miocene, about 5 million years ago, was accompanied by a drop in sea level and widespread drought, which resulted, for example, in the drying up of the Mediterranean Sea. Temperatures continued to decline, plummeting to a minimum at certain times over the past 2.6 million years, the duration of the Quaternary Ice Age.

The causes of Cenozoic climate fluctuations are slowly being unravelled. For example, the Oligocene cooling that heralded the onset of extensive Antarctic glaciation was probably due in part to the split between Antarctica and Australia in the earliest Oligocene, about 33 million years ago. This separation and the opening of the Drake Passage between Antarctica and South America about 30 million years ago led to the birth of the Antarctic Circumpolar Current. Cooler dense waters flowing around Antarctica prevented warmer equatorial waters from migrating into high southern latitudes, and the descent of these colder, denser, and saltier waters to the depths generally chilled the world's oceans. Australia's northward drift may have had another chilling effect on climate—it constricted the flow of warm tropical water between the Indian and Pacific oceans, reinforcing the global cooling trend.

Other tectonic events, such as the rise of extensive lofty mountain belts in southern Asia during the late Cenozoic, also had a significant impact on atmospheric circulation, forcing greater regional differences in climate and the distributions of plants and animals. The rise of mountain belts would expose more rocks to erosion and weathering, which perhaps accentuated the cooling trend (Box 4).

One puzzle has been the coexistence during parts of the Cenozoic of fossil plants typical of modern temperate regimes

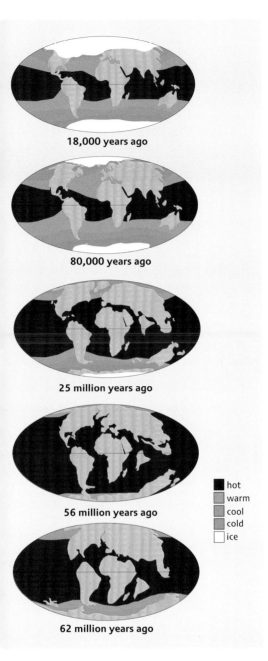

Changing general ocean temperatures through the Cenozoic, showing the trend from a "greenhouse" world 62 million years ago and the Paleocene-Eocene Thermal Maximum 56 million years ago, to the "icehouse" world of today. JONATHAN BUJAK.

with those of much warmer climatic affinities. Perhaps the climate, especially in the Paleogene, was much more equable than today with little or no frost, allowing warmth-loving species to survive year round at higher latitudes. Or perhaps some Cenozoic plants (and animals) had broader tolerances to temperature and other factors.

Shuffling Plates

During the early Cenozoic, the Atlantic Ocean continued to widen and the North American Plate kept on moving southwestward. The plate's western margin was still converging and colliding with oceanic plates of the eastern Pacific Ocean, uplifting mountains in the Cordillera in the process. As the Atlantic widened and the Cordillera rose, vast areas of oceanic lithosphere that once lay east of the main spreading ridge in the eastern Pacific Ocean were subducted into the mantle and overridden by the advancing continent. Today, all that remains of the big oceanic plates that were between the spreading ridge and western North America are small remnants, including the Juan de Fuca Plate off British Columbia. The oldest rocks in these small plates date from about 9 million years ago, and so their magnetic anomalies record only a short interval. But a record of the mirror image of these young magnetic anomalies, continuous with older ones extending back at least 140 million years, is preserved

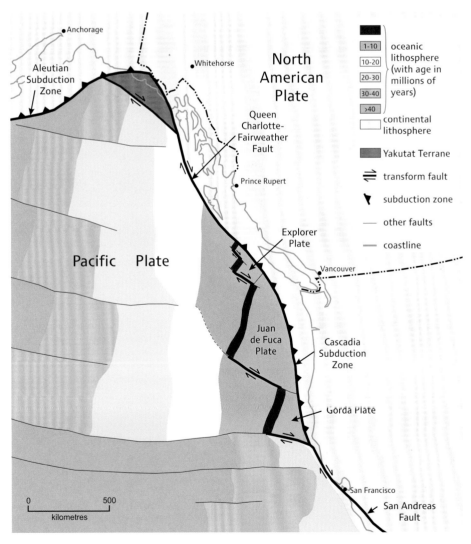

Ages of oceanic lithosphere in the northeastern Pacific Ocean and the structures bounding the western margin of the North American Plate. Young oceanic lithosphere, coloured red, reflects the position of spreading ridges. ADAPTED FROM VARIOUS SOURCES.

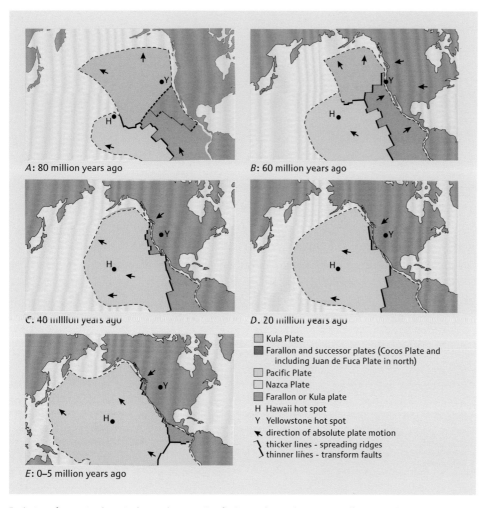

Evolution of tectonic plates in the northeastern Pacific Ocean during the past 80 million years, based on a modern map. ADAPTED FROM ENGEBRETSON ET AL. (1985).

Between 45 and 40 million years ago, the North American Plate overrode an extension of the Pacific Plate that projected eastward into the Farallon Plate, and the latter was consequently split in two. The northern part became the ancestral Juan de Fuca Plate, and the southern part the Cocos Plate. Also in the middle Eocene, the Kula Plate disappeared beneath Alaska (the word kula is Athapaskan for all gone), and the northward-moving Pacific Plate took over from the Kula Plate at the subduction zone. About 40 million years ago, as a result of these changing plate geometries, the western boundary of the North American Plate changed from one dominated by plate convergence and subduction to one where subduction zones alternated with transform faults, a pattern that continues today (Chapter 2). California's San Andreas Fault is today linked at its northern end to the offshore Cascadia Subduction Zone. This in turn is connected off the northern end of Vancouver Island to the Queen Charlotte-Fairweather Fault, which in turn connects to the north with the Aleutian Subduction Zone off southern Alaska.

Landward of the Cascadia Subduction Zone is the Cascade Magmatic Arc, which originated about 40 million years ago and is still active. In Canada, it includes Oligocene volcanics in Queen Elizabeth Park in Vancouver, as well as isolated volcanic centres in the Coast Mountains for about 300 kilometres north of Vancouver.

on the Pacific Plate west of the Juan de Fuca Ridge. From these anomaly patterns, we can get an idea of the configuration of the ocean floor that was subducted under western North America. We also gain insights into past plate interactions from the changing patterns of arc magmatism and tectonic activity that are preserved on the western margin of the North American Plate.

Such evidence suggests that from about 80 to 50 million years ago two large oceanic plates, the Kula to the north and the Farallon to the south, were subducting beneath the North American Plate. The Kula and Farallon plates had different trajectories: the Kula Plate was moving northward and subducting obliquely beneath the North American Plate, whereas the Farallon Plate was approaching the continent almost head-on. Until about 45 million years ago, subduction of the Kula Plate beneath the North American continent gave rise to a magmatic arc that extended from northwestern Washington to southwestern Yukon. Widespread Eocene volcanic rocks from Princeton to Smithers in the interior of British Columbia are surface remnants, and granitic plutons of the eastern Coast Mountains are the roots of the arc, laid bare by subsequent uplift and erosion.

Rolling up the Rim

In the previous chapter, we learned how erosion of the rising Coast Mountains supplied sediment into basins on former Wrangellia and beyond, onto what was to become the Pacific Rim Terrane. This terrane was carried on the Farallon Plate, which was subducting beneath the western edge of North America. In the process, about 55 million years ago the Pacific Rim Terrane was scraped off and accreted to the continent. Remnants of this terrane are now found in coastal areas of southern and western Vancouver Island.

About 40 million years ago, as subduction continued, Eocene basalts and clastics deposited on the ocean floor began to accrete to the continental margin of northern Washington and southern British Columbia. These rocks form the Olympic

Pebbly mudstone of the Pacific Rim Terrane, at Chesterman Beach, Pacific Rim National Park Reserve of Canada, British Columbia. The Pacific Rim Terrane is composed of generally coarse clastic rocks and igneous rocks of Jurassic to Cretaceous age. JIM MONGER.

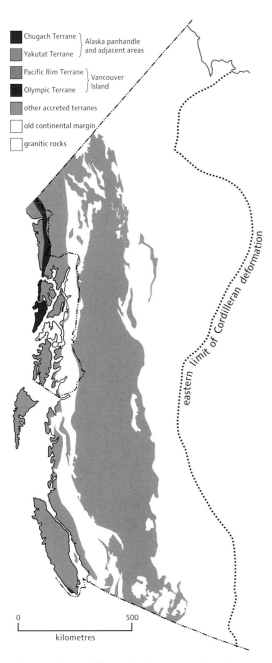

The Canadian Cordillera and adjacent parts of southeastern Alaska showing the westernmost terranes, the latest additions to the continent. ADAPTED FROM MONGER AND BERG (1987), COURTESY OF THE US GEOLOGICAL SURVEY.

Terrane (locally called the Crescent Terrane), which is the youngest fragment to become welded to the Canadian Pacific margin. Remnants of the Olympic Terrane today lie south and west of the Pacific Rim Terrane, separated from it by major faults. Late Paleocene to early Eocene parts of the Olympic Terrane include an ophiolite suite (Chapter 2) that originated as a series of seamounts or an ocean island similar to present-day Iceland. The assemblage includes layered gabbro, sheeted mafic dykes, pillow basalts, subaerial lava flows, and associated sediments. Its rocks underlie the communities of Colwood, Metchosin, and Sooke west of Victoria, where exposures of this ancient sea floor can be seen at many localities, particularly at Sooke Potholes and Witty's Lagoon regional parks.

Docking of the Olympic Terrane was probably responsible for the development of westerly dipping thrust faults on southern Vancouver Island. The accretion of the Pacific Rim and Olympic terranes caused the Wrangellian margin of North America to be uplifted, resulting, for example, in the mountains east of Pacific Rim National Park Reserve of Canada. The accretion also caused the late Cretaceous clastic sediments of the Gulf Islands to be folded and faulted, resulting in alternating hard and soft rocks today forming a series of ridges and valleys inland and promontories and bays along coastlines.

During the Oligocene, shallow marine sandstones and conglomerates accumulated across both the Pacific Rim and Olympic terranes along western and southern Vancouver Island. But material eroded from the uplifted regions accumulated mostly in submerged basins on the continental shelf, including the Tofino and Queen Charlotte basins. The Tofino Basin, which underlies the continental shelf west of Vancouver Island, contains a thick succession of mainly deep-marine sediments on top of the Olympic Terrane. The Queen Charlotte Basin contains a thick sequence of non-marine clastics, as well as widespread volcanic rocks produced during an episode of rifting in Queen Charlotte Sound. During the 1960s and earlier, several petroleum exploration wells were drilled in both basins without commercial success. Since then, a moratorium has prevented further offshore drilling.

Slipping and Stretching

In the previous chapter we learned that since about 110 million years ago the continent has been moving generally in a southwestward direction relative to oceanic lithosphere west of it. This movement has continued to be accommodated in part by right lateral strike-slip faults within the Cordillera through the Cenozoic. Most of the large strike-slip faults active during the Cenozoic are marked by great valleys that have been cut along belts of easily eroded, fractured rock ground up during fault movements. The majority of these valleys parallel the trend of the Cordillera. Examples are the Northern Rocky Mountain Trench in northeastern British Columbia and its continuation, the Tintina Trench, in southern and central Yukon. (The term trench here refers to big modern valleys and should not be confused with the oceanic trenches associated with subduction.) The total combined length of the Northern Rocky Mountain and Tintina trenches is nearly 2,000 kilometres. Another large fault in southwestern British Columbia, the Fraser Fault, differs in that it cuts acutely northward across the Cordilleran grain. Mapping along the Fraser Fault shows that between 45 and 35 million years ago rocks on the west side moved northward by about 140 kilometres relative to those on the east side. Before this displacement, the sites of Hope and Lillooet would have been adjacent to one another. When the amounts of displacement of all the late Cretaceous and Cenozoic strike-slip faults in the Canadian Cordillera are added up, it appears that during the past 100 million years westernmost parts of the continent have moved nearly 1,000 kilometres to the north relative to the continental interior.

While large parts of the Canadian Cordillera have been jostling northward along strike-slip faults during the Cenozoic, the crust beneath the orogen has been thinning. During the early Cenozoic, the tectonic situation of the Canadian Cordillera was probably similar to that of the central Andes in South America today. The modern central Andes contains a large plateau, the Altiplano, about 4,000 metres high, as well as some peaks approaching 7,000 metres. Beneath the Altiplano, the crust is locally up to 60 kilometres thick. In contrast, few peaks in the present Canadian Cordillera exceed 4,000 metres. Much of the region is less than 2,000 metres above sea level, and the crust is generally thinner than 35 kilometres. What has caused the Canadian Cordilleran crust to thin and the mountains to become lower over the past 55 million years? At least part of the answer lies with a suite of normal faults, which developed from about 55 to 45 million years ago in central and southern parts of the Canadian Cordillera. These faults may have resulted from changing plate trajectories, which caused the crust to be stretched rather than compressed, and the Cordillera to become wider and lower.

In southern British Columbia, the normal faults have a roughly north-south orientation, so the crust must have stretched in an east-west direction. Many of the large valleys that extend from the southern Canadian Cordillera into the northwestern United States follow the traces of major Cenozoic normal faults. They include the Okanagan Valley, the Purcell Trench containing the southern part of Kootenay Lake, the southernmost part of the Southern Rocky Mountain Trench, and the Flathead Valley. Most of the normal faults are steeply dipping in the cold, strong rocks near the surface, but they curve and flatten with depth. At depths greater than about 20 kilometres, the rocks are hotter and more ductile and thus stretch more like toffee, with thinner pinches and thicker swells. Small-scale versions of pinches and swells, collectively called boudins, can be seen in

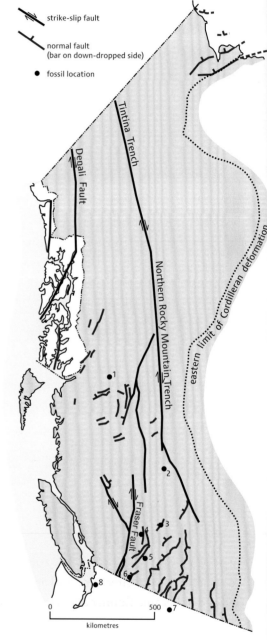

Locations of the major strike-slip faults and prominent normal faults active in the Cordillera during the Cenozoic. The numbered locations show fossil sites: Driftwood Canyon (1), Horsefly (2), Chu Chua (3), McAbee (4), Quilchena (5), Princeton (6), Republic (7), and Chuckanut (8). The last two localities are in the United States.

The Okanagan Valley Fault southeast of Okanagan Falls, British Columbia. Bluffs above the lake at left consist of Jurassic to early Cenozoic gneiss. The surface that gently slopes away from the top of the bluffs marks the fault plane. Above this normal fault are weakly metamorphosed Paleozoic to Cenozoic rocks. JIM MONGER.

metamorphic rocks exposed in road cuts along the Trans-Canada Highway at Three Valley Gap, southwest of Revelstoke.

Movement on some of the normal faults has been so great that in places unmetamorphosed and faulted Eocene sedimentary and volcanic rocks are now found alongside gneiss that formed at the same time but at depths of 25 kilometres or more. This happens on the Okanagan Valley Fault, where slip on the fault surface has been estimated to be as much as 80 kilometres. The fault surface dips westward at an angle of 10 to 15 degrees, low for a normal fault, and has been traced seismically down to depths of about 20 kilometres. The cumulative amount of stretching on all the Cenozoic normal faults in southern British Columbia may be at least 200 kilometres in an east-west direction. And as the crust stretched, it became thinner and unable to support the load of lofty mountains or high plateaus like the modern Altiplano in the Andes.

In this exposure at Three Valley Gap, southwest of Revelstoke, British Columbia, the darker, dog-bone-shaped body is a boudin, which was originally part of a mafic intrusion but has been metamorphosed to amphibolite. This boudin formed during the Eocene at the same time as brittle normal faulting in the upper crust. RAY PRICE.

Faulting has left the Cordillera with small Cenozoic basins, many of which contain sediments with well-preserved fossils. The fossils are in strata generally deposited in lake or swamp settings, as at Quilchena, Horsefly, and McAbee in southern British Columbia and in the Driftwood Canyon area near Smithers in central British Columbia. The deposits are of early Eocene age, and thus contemporaneous with the main phase of crustal stretching in the Cordillera. Early Eocene fossil assemblages in these deposits include superbly preserved insects, fish, rare feathers, and mammalian bone fragments. Many of the plants are familiar today: preserved leaves, seeds, fruits, and flowers represent some of the earliest-known members of the rose family, including apples, cherries, plums, serviceberry (or Saskatoon berry), and raspberry. Maples, elms, and birches are also present. These plants thrived alongside others long gone from natural habitats in western Canada, such as dawn redwood, Chinese water pine, *Ginkgo*, bald cypress, Chinese golden larch, and Chinese rubber tree. Other plants that first appear in these early Eocene rocks, such as *Florissantia*, are now extinct. Although we're not sure if *Florissantia* was a tree, shrub, or vine, its well-preserved flowers suggest that it was probably

Paleogeography of North America and adjacent regions in the late Eocene, 35 million years ago. Land is shown in brown, with shading showing topography. The lighter blue areas represent possible coastal or nearshore areas, darker blue represents deeper ocean waters, and black indicates trenches. Aspects of modern geography are shown for orientation.

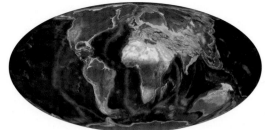

Global paleogeography 35 million years ago, during the late Eocene. Colours as for previous figure.

a member of the mallow family, which includes plants that produce chocolate, cotton, and kapok.

Coal deposits in the Eocene basins, as at Quesnel and Merritt, testify to the former presence of swamps. The lush vegetation needed to produce coal, as well as the diverse plants, suggests a warm, humid climate with little or no winter frost. Modern analogs might include the diverse Appalachian forests and bald cypress swamps in the southeastern United States.

A Pivoting Battering Ram

Let's now shift our attention to Canada's northeast, where the inception of the Labrador-Baffin Seaway was heralded by an episode of volcanic activity 62 to 56 million years ago. This episode, which involved a vast outpouring of flood basalt in west-central Greenland and southeastern Baffin Island, is associated with widespread uplift and consequent erosion around the middle Paleocene. Subsequently, rifting evolved into drifting and sea-floor spreading in the Labrador-Baffin Seaway, which continued to widen until the late Eocene.

As we saw in the previous chapter, early ideas that Greenland had separated from North America through continental drift were correct. But recent research has shown that no transform fault exists along Nares Strait, the narrow linear seaway between Ellesmere Island and northwestern Greenland. Rather, it seems that as Greenland pulled away, the plate boundary developed not along Nares Strait but as a series of faults running from northern Baffin Bay into the Arctic Islands via Lancaster Sound at the eastern end of the Northwest Passage. Hence, the southeastern part of Ellesmere Island was then part of the Greenland-Eurasia Plate.

About 55 million years ago, a new axis of sea-floor spreading formed to the east of Greenland in what is now the northern North Atlantic between Greenland and Scandinavia. The axis extended into the Arctic as the Gakkel Ridge, which separated the Lomonosov Ridge from the Barents Shelf. Greenland became a distinct plate, separated from North America by the Labrador-Baffin Seaway to the west, and from Scandinavia by the new spreading ridge. Changes in plate movements meant that Greenland changed course and acted as a giant battering ram, colliding with Arctic Canada and pushing up parts of Axel Heiberg and Ellesmere islands. This late Paleocene to early Eocene episode of deformation is known as the Eurekan Orogeny. It produced thrust faults in central and northeastern Ellesmere Island, some of which are exposed along northern Nares Strait. These faults carried Neoproterozoic sedimentary rocks over Paleocene conglomerates shed from the new Eurekan Mountains. Farther-travelled sediment shed from these new highlands came to rest as non-marine and shallow-marine deposits in Baffin Bay, along western Ellesmere Island, and among the salt diapirs of Axel Heiberg, Ellef Ringnes, and adjacent islands.

As with all continent-to-continent collisions, this Greenland-North American impact couldn't last, and so the pivotal motion of Greenland away from the Arctic Islands stalled—with broad implications for the evolution of the North Atlantic Ocean. Sea-

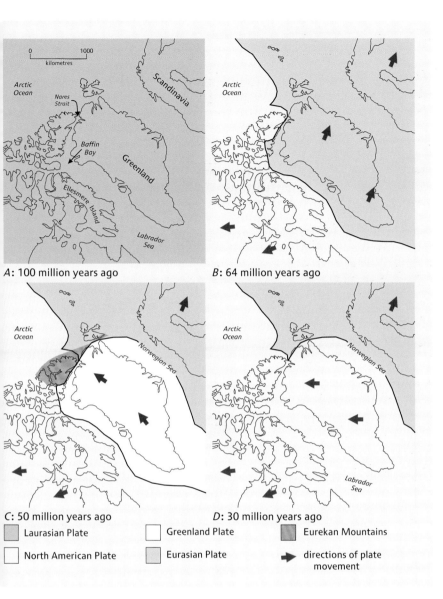

The changing plate-tectonic relationships of Laurasia, Greenland, North America, and Eurasia for middle Cretaceous (A), early Paleocene (B), early Eocene (C), and Oligocene and later times (D). The modern shoreline is shown for orientation, and Eocene plate positions are shown on all figures for simplicity. ADAPTED FROM VARIOUS SOURCES.

floor spreading finally ceased in the Labrador-Baffin Seaway in the latest Eocene, about 34 million years ago. For the remainder of the Cenozoic, Greenland would once more become part of the North American Plate. Had sea-floor spreading in the Labrador-Baffin Seaway not stalled, we might now have a Labrador-Baffin Ocean. Instead, sea-floor spreading jumped eastward, separating Scandinavia from Greenland and severing what had been important land routes for animal migrations.

While Canada and Greenland drifted apart, sedimentary basins on the continental shelves of the Labrador-Baffin Seaway became dumping grounds for large volumes of later Cretaceous and Cenozoic clastic sediment shed from the surrounding landmasses. An intriguing discovery from oil-exploration wells drilled in the Labrador-Baffin Seaway is the abundance of *Azolla* in some early Eocene sediments. *Azolla* is a small, moss-like, freshwater fern used by farmers in tropical Asia as a nitrogen fixer for increasing rice production. Because of its low tolerance for salt, the presence of *Azolla* in these Eocene sediments suggests not only warmer temperatures than today but also the proximity of freshwater lakes. *Azolla* is also abundant in Eocene sediments drilled on the Lomonosov Ridge in the Arctic Ocean, leading to speculation that parts of this ocean basin were freshwater lakes during the Eocene, with an outflow through the Labrador-Baffin Seaway.

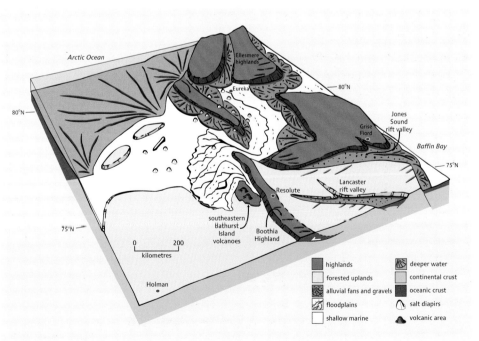

Paleogeography of what was to become the Canadian Arctic in the late Paleocene to Eocene, 62 to 35 million years ago. The approximate location of Holman, Resolute, Grise Fiord, and Eureka are included to provide a guide for orientation, as well as paleolatitudes.

CHAPTER 10 · FINAL APPROACH

An early Eocene summer lakeside scene based on fossils, about 50 million years old, found near Quilchena, British Columbia. On the right is a large swamp cypress tree with flowering "water willow" (*Decodon*) shoots at its left base. Leaves of tea-family plants can be seen at far and lower right, with a fruiting branch of ash at upper right. The shallow waters around the cypress support floating mats of algae, the fern *Azolla*, and leaves of water lily. At left, an alder branch with cones hangs over a fertile shoot of dawn redwood. The small blade-leafed plants at the middle water's edge, although not true grasses, may be early relatives of that group or other grass-like plants. Animal fossils at Quilchena include insects, which likely provided food for the swallow-like swifts (upper left), fossils of which have been found recently. ARTWORK COPYRIGHT ROLF MATHEWES.

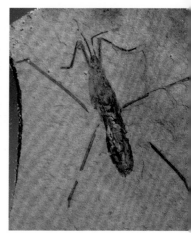

Fossils found in the Eocene lake deposits at Quilchena, British Columbia. The photo at centre left shows the wing of a lacewing insect. The photo in the centre of the page shows dawn redwood foliage at left and a cypress branch crossing the bottom-right of the image. The larger photo at centre right shows a flower calyx *Florissantia quichenensis*. At bottom left is a pine branch with needles. The centre-bottom photo shows a flight feather. And at bottom right is a waterstrider. RON LONG (CENTRE) AND ROLF MATHEWES (ALL OTHERS).

Flat-lying Paleocene basalt flows resting on brown volcanic breccia, on the coast of Baffin Island, northwest of Cape Dyer. BARRIE CLARKE.

The Princess Margaret Range, Axel Heiberg Island, Nunavut, reflects in part the deformation caused by the Eurekan Orogeny. ALAN MORGAN.

Farther south, off Nova Scotia and Newfoundland, depressions such as the Scotian and Jeanne d'Arc basins, which had formed during the Mesozoic, continued to receive continent-derived sediment during the Paleogene. Some of this material was transported into deeper water by turbidity currents. Seismic surveys of the continental margin have identified a series of buried canyons, similar to The Gully, a modern canyon off Nova Scotia, down which many of these turbidity currents flowed. Carbonates accumulated at times in the Eocene in the Scotian and Jeanne d'Arc basins, reflecting interludes with a warm climate and clear seas.

On the Plains

Apart from a limited Paleocene invasion of the so-called Cannonball Sea in southern Manitoba, seas had retreated from what is now Canada's Western Interior Plains in the late Cretaceous. The marine deposits in the foreland basin were succeeded by Paleocene non-marine clastic rocks. Later, in Eocene through Miocene times, extensive sheets of gravel and sand were deposited by streams and rivers. Currents in these watercourses were sometimes very strong, as shown by the presence of boulders the size of small garbage cans, many bearing percussion marks formed by collisions with other boulders. The source areas of these deposits were the Cordillera, as well as areas east of the mountain front in Montana, such as the Sweetgrass Hills, that had been pushed up by local igneous intrusions.

Cenozoic conglomerate and sand deposits that once blanketed broad areas of the Plains are now preserved as isolated remnants, as in the Cypress Hills, which straddle the southern Alberta and Saskatchewan border, and Wood Mountain in southern Saskatchewan. Similar isolated patches of pre-Quaternary Cenozoic sediments extend all the way up to the Arctic, and include the Hand, Swan, and Nose hills in central Alberta, the Grandview Hills along the western side of the Mackenzie River, the Storm and Caribou hills along the east side of the Mackenzie Delta, and the Smoking Hills on the Arctic coast. Detailed studies of the Cenozoic sediments of the Western Interior Plains indicate that once out of the mountains, rivers made their way generally eastward across the Hudson Bay area, ultimately emptying into the Labrador Sea. The major component of this proposed drainage, the Bell River System, would have been much larger than the modern Amazon Basin.

The Cenozoic fluvial deposits of the Western Interior Plains contain a treasure trove of vertebrate fossils. Mammals dominate, but fish, amphibians, reptiles, and birds are also present. Deposits near Swift Current, Saskatchewan, vividly illustrate changing faunas and environments. In the middle Eocene strata there, multituberculates are still represented, but most mammal fossils are marsupials, especially opossums, and placentals. Members of modern placental groups found in these strata include primates, carnivores, perissodactyls, cetartiodactyls, and early North American rabbits, a group that had migrated from Asia. Also present was the condylarth *Hyopsodus*, an extinct small, hoofed animal with a long body, as well as an extinct group of tree-living mammals that superficially resembled flying lemurs.

In later Eocene deposits, representatives of modern mammalian families increased. Fewer opossums are present in the late Eocene mammal assemblages from the Cypress Hills. But entelodonts—monstrous pig-like animals with massive teeth and jaws up to 1 metre long, presumably with an appetite

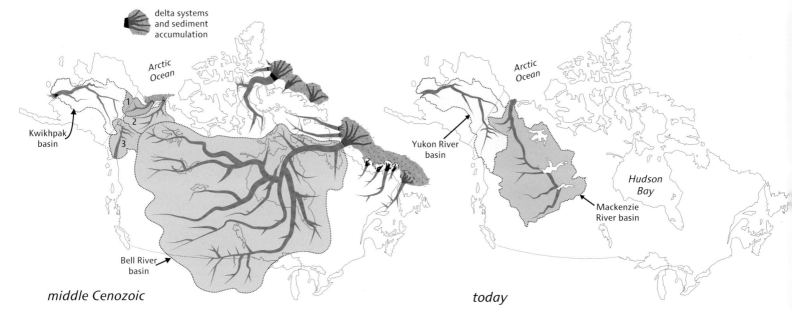

A comparison of middle Cenozoic (left) and modern (right) river drainage basins of northern parts of North America. For much of the Cenozoic two small river systems drained into the Arctic Ocean, and most of the interior was drained by the vast Bell River System. Today, the Mackenzie River basin covers a large area of northwestern Canada. Note too how the river basins of Alaska and Yukon have changed (see pages 210–211). In the map at left, 1 = ancestral Porcupine River Basin, 2 = ancestral Peel River basin, 3 = ancestral Yukon River basin.

to match—made a dramatic appearance. Also present were brontotheres, some as big as elephants, and the last-known multituberculates.

Vertebrate, pollen, and other evidence indicate that toward the end of the Eocene the swampy landscape that had prevailed on the Western Interior Plains since the late Cretaceous, gave way to open, dry, woodland savannah. The widespread advance of grass is reflected in the increasing proportion of grazers, which greatly outnumbered browsers by the end of the Oligocene. Among the mammals that thrived in this new grassy world were ancestors of the horse, such as the small *Mesohippus*, which stood about 65 centimetres tall and had three toes, a striking contrast to the single hoof of modern horses. Horses originated in North America and migrated to Eurasia perhaps four times during the Cenozoic—underlining the importance of land routes.

Remarkable Paleogene fossil mammal assemblages have also been found in the Arctic, including the brontotheres associated with the middle Eocene Axel Heiberg forest. Slightly older mammal remains have been found in western Ellesmere Island and eastern Axel Heiberg Island. The fossils include a multituberculate, creodonts (an extinct group that were the dominant carnivores in the Eocene), forms related to modern carnivores, rodents, pantodonts (bulky cow-sized herbivores with tusks), horses, brontotheres, and early tapirs. Most of these mammals had relatives living in Eurasia.

Tertiary conglomerate, originally fluvial gravels, Cypress Hills, southwestern Saskatchewan. FLOYD WIST.

An angry-looking pair of entelodonts, pig-like animals whose remains have been found in Eocene to Miocene deposits of the Prairies. PAINTING COPYRIGHT MARIANNE COLLINS, ARTIST.

Paleogeography of North America and adjacent regions in the Miocene, 15 million years ago. Land is shown in brown, with shading showing topography. The lighter blue areas represent possible coastal or nearshore areas, darker blue represents deeper ocean waters, and black indicates trenches. Aspects of modern geography are shown for orientation.

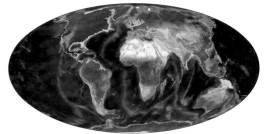

Global paleogeography 15 million years ago, during the Miocene. Colours as for previous figure.

Latest Eruptions

With climate generally cooling and grasslands increasingly dominating North America's heartland, we move from the Paleogene into the Neogene and Quaternary, the past 23 million years, focussing first on the far west. During this interval, the Pacific Plate continued to slide northward past central and northern British Columbia along the Queen Charlotte-Fairweather Fault. Sometime in the past 10 million years, the already-small ancestral Juan de Fuca Plate broke into three pieces: the Gorda Plate off northernmost California; the present-day Juan de Fuca Plate off Oregon, Washington, and southern Vancouver Island; and the Explorer Plate off northern Vancouver Island. Today, the three microplates are being overridden by the North American Plate along the Cascadia Subduction Zone, which continues to generate the Cascade Magmatic Arc. The youth and consequent buoyancy of the Juan de Fuca Plate make it stick to the overriding continental plate as it tries to subduct. Eventually, over several hundred years, enough strain builds up between the converging plates to force them to suddenly slip past each other. The sudden slips generate great earthquakes, as we'll discuss in Chapter 17.

Neogene and Quaternary rocks formed in the Cascade Magmatic Arc include Miocene granitic intrusions near Hope and Miocene volcanic rocks on Coquihalla Mountain. Today, mounts Garibaldi, Cayley, and Meager, which rise to between 2,000 and 3,000 metres, are all stratovolcanoes formed mostly over the past 2 million years. Although these volcanoes appear dormant, a few postglacial lava flows and widespread volcanic ash as young as 2,400 years attest to relatively recent activity. One potentially lethal stratovolcano is the magnificent, 3,200-metre-high, snow-capped cone of Mount Baker, only about 100 kilometres southeast of Vancouver in northernmost Washington. Its last major eruption was in 1843.

Neogene and Quaternary volcanism elsewhere in the western Canadian Cordillera reflects a variety of tectonic settings. Most extensive are the Chilcotin and Cariboo basalts, which cover a large area of south-central British Columbia between 20 and 2 million years ago, forming plateaus.

Mount Garibaldi, a stratovolcano of the Cascade Magmatic Arc northeast of Squamish, British Columbia, was last active about 13,000 years ago. This view is from the northwest.
PAUL ADAM.

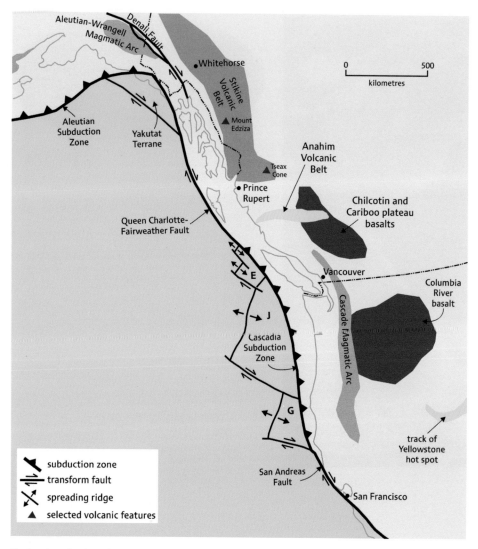

The location of active volcanic areas in western North America and the distribution of plates and plate boundaries associated with the continent's western margin and the northeastern Pacific Ocean. The Columbia River basalts form a huge area of volcanics in the northwestern United States akin to the Chilcotin and Cariboo basalts of British Columbia. E = Explorer Plate, J = Juan de Fuca Plate, G = Gorda Plate.

Other, generally younger, basaltic flows—some only 500,000 years old—occupy valley floors.

Crossing west-central British Columbia from the western Coast Mountains to near the Fraser River is the east-west-trending Anahim Volcanic Belt, which includes the shield volcanoes of the Rainbow, Itcha, and Ilgachuz ranges. Volcanic centres in this belt generally become younger from west to east: one explanation for their origin has been that the Belt marks the westward passage of the North American Plate over a hot spot in the mantle.

Farther north, a group of scattered flows, shield volcanoes, and cones extends from the Coast Mountains near Prince Rupert into central Yukon. This group of volcanoes, which is landward of the Queen Charlotte-Fairweather Fault, is the Stikine Volcanic Belt. The magma that fed these volcanoes may have originated in the mantle and leaked upwards along fractures that developed within the North American Plate margin in response to movements along the Queen Charlotte-Fairweather Fault to the west. The Stikine Volcanic Belt includes the Tseax Cone, the scene of an eruption that took place in the late eighteenth century, probably 1775. Tseax is on the land of the Nisga'a people, and its eruption produced a dramatically explosive pyrotechnics show remembered vividly in Nisga'a oral history. Lava also poured out of nearby fissures and flowed down a tributary of the Nass River for over 20 kilometres. It changed the course of the tributary and its damming action created Lava Lake, which still exists. The Tseax eruption destroyed a productive fishery and two large villages. Nisga'a tradition records that the eruption killed more than 2,000 people in one of the villages, making this perhaps Canada's deadliest-known disaster of geological origin. Evidence and stories suggest that the fatalities were due to buildups of poisonous gases, but recent fieldwork points to a more complex story, not yet fully unravelled. Regardless of the details, the eruption and the lives it destroyed in an instant of time have left a permanent imprint on Nisga'a culture.

Finally, volcanoes of the Aleutian-Wrangell Magmatic Arc have continued to perk above the subducting Pacific Plate throughout the Neogene and Quaternary. For example, about 1,500 years ago, explosive eruptions from a volcano in the St. Elias Mountains spread an ash layer known as the White River Ash over southern and central Yukon.

Close-up of the base of a lava flow over till at Mushbowl Hill, Wells Gray Provincial Park, British Columbia. The volcanic rocks at Wells Gray are possibly related to the Anahim Volcanic Belt. DALE GREGORY.

Modern Mountains and Rivers

Today, the Canadian Cordillera consists of eastern and western mountain belts, reflecting the zones of greatest deformation during the Mesozoic, separated by a central belt of generally more subdued topography, including the Interior Plateaus.

Canada's tallest peaks are in the Icefield Ranges of the St. Elias Mountains of southwestern Yukon and northwesternmost British Columbia, culminating in Mount Logan, named after pioneering Canadian geologist William Logan, at just under 6,000 metres high. To the northwest, in the Alaska Range, is Denali (or Mount McKinley), North America's tallest peak at almost 6,200 metres. The St. Elias Mountains and the Alaska Range continue to be raised by convergence and collision of the northwest-moving Pacific Plate with the North American Plate. As part of this process, a great wedge of sediment that includes the Yakutat Terrane is being subducted, jacking up the overriding North American Plate to form the the Saint Elias Mountains. We know that uplift of the mountains in this region began around 5 to 6 million years ago because earliest Pliocene fossils of plants that thrive under wet conditions have been found north of the Alaska Range, in areas that today are in the rain shadow of the mountains. Also, fission track studies (Chapter 3) suggest that some rocks in the mountains have risen between 2 and 9 kilometres over the past 5 to 6 million years, although erosion over the same time has been wearing them down.

British Columbia's rugged, ice-sculpted mainland Coast Mountains are underlain by hard granitic rocks. Mount Waddington, at just over 4,000 metres, is the highest peak. Although the ancestral Coast Mountains first rose in the middle Cretaceous, various lines of evidence show that the present topography is geologically recent and that parts of the Coast Mountains have risen by more than 3 kilometres since the late Miocene, driven at least in part by ongoing convergence between the Juan de Fuca and North American plates. For example, plant fossils show that 10 million years ago the vegetation of British Columbia's now semi-arid interior was much like that of the coast today. It seems that the mountains in the late Miocene were not then high enough to block the paths of moisture-laden westerlies from the Pacific Ocean.

At 5,951 metres, Mount Logan in the St. Elias Mountains of Yukon is the highest peak in Canada. Mount Logan consists mainly of Jurassic and Cretaceous granitic rocks. The lower southwestern slopes are underlain by early to late Cretaceous sedimentary strata of the Chugach Terrane. A fault, clearly visible here as the abrupt change from grey to dark brown, separates the granitic rocks from the sedimentary strata. CHRIS YORATH.

Mount Waddington, formed of late Cretaceous gneiss, is the highest peak in the predominantly granitic Coast Mountains. JOHN SCURLOCK.

The highest peak in the Rocky Mountains is Mount Robson, at almost 4,000 metres. As we discovered earlier, the folding and thrusting of rocks that now form the Rocky Mountains ceased about 60 million years ago. Yet the Rocky Mountains still stand high. That the Rocky Mountains have not been worn away by erosion since their formation suggests that they must have been elevated subsequently. But by what, and when, are still questions without clear answers.

The flow patterns of two of the largest rivers in the North America Cordillera—the Fraser and the Yukon—were radically reorganized in the late Pliocene. Today, the Fraser River starts high in the Rocky Mountains, then flows west and drops steeply into the Southern Rocky Mountain Trench. The river then runs north along the Trench to near Prince George, where it turns sharply westward, then southward, passing the towns of Quesnel and Williams Lake on its way to the sea at Vancouver.

In the 1930s, geologists realized that some tributaries of the upper Fraser River between Prince George and Williams Lake flowed northward, a pattern that made more sense if the river had also previously flowed in that direction. That this was the case was confirmed by the discovery of sediments deposited by a northward-flowing Fraser River in the Miocene. Also, maps and satellite images reveal an extensive eastward-draining river system in the Nechako Basin of central British Columbia that was joined by the northward flowing upper ancestral Fraser River. Given this evidence, it seems likely that what are now the upstream reaches of the Fraser River flowed through the Rocky Mountains along the present course of the Peace River. The reversal seems to have occurred about 2 to 3 million years ago.

A similar story has emerged from studies of Yukon River, which today starts in the northern Coast Mountains not far from the Gulf of Alaska. It flows north, gathering waters from

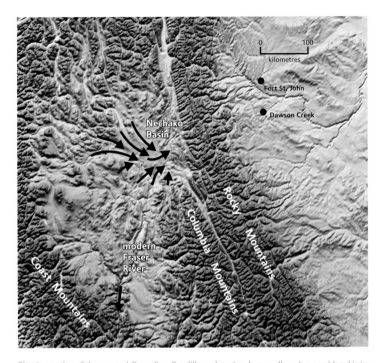

Physiography of the central Canadian Cordillera showing how valleys (arrows) lead into the Nechako Basin at angles that suggest the original flow of the headwaters of the Fraser River was to the east. The geologically recent rise of the Rocky Mountains blocked the route and flow was diverted southwestward down the modern Fraser River. BASE MAP FROM AMANTE AND EAKINS (2009).

The modern Fraser River near Big Bar Creek, British Columbia, flows in a deep inner Holocene valley incised into a broader pre-glacial valley. Terraces mark former levels of the valley floor. The Fraser River now flows south and southwest to the Pacific, but during the Miocene and Pliocene it flowed northward, then eastward. JOHN CLAGUE.

southern and central Yukon, and then turns west along the north side of the Alaska Range, eventually reaching the Bering Sea, almost 3,200 kilometres from its source. As was the case for the Fraser, geologists long ago noted odd directions of Yukon tributaries, suggesting at one time that the Yukon flowed southward, opposite its present direction (see maps on page 206). Subsequently it was discovered that Pliocene bedrock terraces, cut by the Yukon River near Dawson City, slope to the south. The combined evidence strongly suggests that as recently as 3 million years ago the Yukon River flowed southward into British Columbia, although exactly where it met the sea is uncertain.

Ice sheets probably played a significant role in both the Fraser River and Yukon River stories. Before the Ice Age the ancestral lower Fraser River had a small drainage basin in southwestern British Columbia. Subsequent waxing and waning of the ice, combined with rapid headward erosion along the crushed and broken rock marking the Fraser Fault, led to capture of the Nechako drainage and a reversal of its flow. Similarly, 2.6 million years ago, the Yukon River was blocked and diverted as glaciers advanced across the Yukon Plateaus. The changes wrought by the ice were so great that the Yukon River never regained its old southward flow.

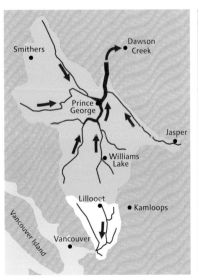
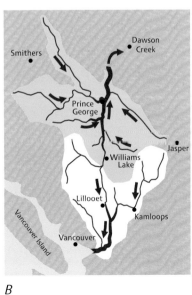
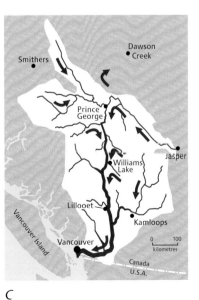

A   B   C

☐ Pacific drainage   ☐ northward (leading to eastward) drainage

The evolution of the Fraser watershed in British Columbia. A shows the situation prior to the Ice Age, when the Fraser was a small coastal river. B shows an expanded Fraser system as the river worked its headwaters inland and the watershed captured the upper reaches of other streams. C shows the modern drainage. ADAPTED FROM A GRAPHIC BY RICHARD FRANKLIN FOR CLAGUE AND TURNER (2003), USED WITH PERMISSION FROM TRICOUNI PRESS.

Reconstruction of *Puijila darwini* swimming. This creature was a "walking seal", intermediate between terrestrial mammals and pinnipeds. PAINTING BY ALEX TIRABASSO, CANADIAN MUSEUM OF NATURE.

Up North

During much of the Neogene, as in the Paleogene, the Arctic was surprisingly mild and nurtured abundant vegetation and wildlife. We know this from fossils preserved in sediments within the Haughton impact structure on Devon Island, the result of a collision between an extraterrestrial body and the Earth during the late Eocene. Early Miocene lacustrine deposits in the impact structure contain the remains of fish, swans, rabbits, and shrews, as well as rhinoceros- and deer-like species known only from this northern outpost. One intriguing find from the lake deposits is a new carnivore, *Puijila darwini*, that resembled an otter but is related to pinnipeds (seals, sea lions, and walruses). Modern pinnipeds are fully flippered, streamlined forms with reduced tails. The Haughton fossil appears to be a "missing link", a walking seal intermediate between terrestrial forms and today's pinnipeds, suggesting that the group may have evolved in the Arctic.

About 15 million years ago, a warming phase led to a poleward advance of mixed hardwood and softwood forests to 75°N. We know this from fossil pollen—including those of hardwoods such as alder, birch, hazel, hickory, linden, mulberry, tulip tree, walnut, and willow—preserved in fluvial deposits and peat on Banks Island. Also present are two climbing vines, *Epipremnum* and *Phyllanthus*, now found only in the tropics and subtropics. Conifers include dawn redwood, fir, hemlock, larch, pine, spruce, and swamp cypress.

Climates deteriorated markedly in polar regions after 7 million years ago. Convincing evidence from offshore drilling indicates that major glaciers were flowing into the North Atlantic from mountains in southern Greenland during the late Miocene and early Pliocene, between 6 and 3 million years ago. Although no evidence has yet been found, it seems likely that glaciers were also forming on higher parts of Baffin and Ellesmere islands during this interval.

The last significant warm spell in the Arctic was in the middle Pliocene, when sediments were deposited widely on a coastal plain that stretched from the Mackenzie Delta to Axel Heiberg Island. Fossils from sediments of the same age, now about 400 metres above sea level in the Strathcona Fiord area of west-central Ellesmere Island, provide evidence of a boreal forest, with alder, pine, larch, spruce, and willow. The latitude, then as now, was about 80°N. One peat deposit contains larch trunks up to 2 metres long, beaver-chewed sticks, and bone fragments of a small species of beaver—together suggesting remnants of a beaver pond. Most surprising from these middle Pliocene sediments was the discovery of camel bones, the size of which suggests that the animal was about a third larger than the average modern camel.

The Arctic Islands are the most distinctive geographical feature of the Canadian Arctic today. However, this geography of islands and intervening channels is a geologically recent development. The channels between islands reflect the courses of late Cenozoic rivers on the broad coastal plain that were widened and deepened by glaciers over the past 2.6 million years, and invaded by the sea when the permanent ice bodies receded (Chapter 11).

Beaver-chewed stick from Pliocene deposits near Strathcona Fiord on Ellesmere Island, Nunavut. MARTIN LIPMAN.

Bones of a large camel have been found in deposits from about 3.5 million years ago on Ellesmere Island. The camels lived in a boreal-like forest.
PAINTING BY JULIUS T. CSOTONYI, COURTESY OF THE CANADIAN MUSEUM OF NATURE.

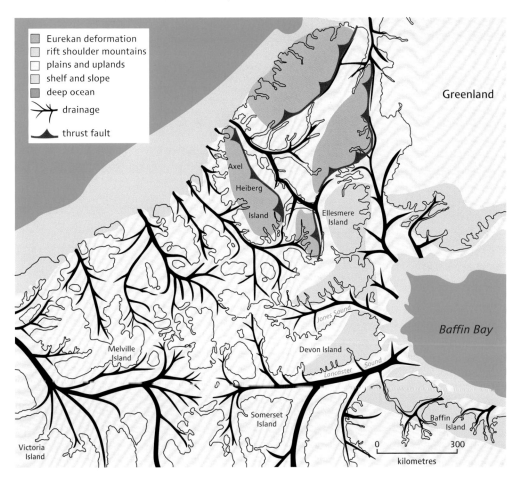

Late Cenozoic paleogeography of the region that was to become the Arctic Islands, showing the courses of major rivers. The valleys of these rivers were broadened and deepened by glaciers over the past 3 million years, so that when the ice receded after the latest glacial advance, the valleys became sea channels and the higher areas between them became the modern islands. Also shown is the distribution of rift-shoulder mountains flanking Baffin Bay and the mountains produced by the Eurekan Orogeny. ADAPTED FROM TRETTIN (1991), REPRODUCED WITH THE PERMISSION OF NATURAL RESOURCES CANADA, COURTESY OF THE GEOLOGICAL SURVEY OF CANADA.

Paleogeography of North America and adjacent regions in the Pliocene, 3 million years ago. Land is shown in brown, with shading showing topography. The lighter blue areas represent possible coastal or nearshore areas, darker blue represents deeper ocean waters, and black indicates trenches. Aspects of modern geography are shown for orientation.

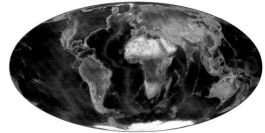

Global paleogeography of the Pliocene, 3 million years ago. Colours as for previous figure.

North American Savannah

Global cooling and widespread drying at the end of the Miocene meant that in North America the trend of increasing grasslands and declining forests persisted. Grazing mammals continued to diversify and woodland dwellers continued to decline. Miocene faunas of the Western Interior Plains were in some ways similar to the wildlife of the modern African savannah because similar climates and environments prevailed.

About 20 million years ago, Asian mammals migrating via Beringia (the land connecting Siberia and North America) replaced several North American groups. Among the new arrivals were oreodonts (sheep-sized cetartiodactyls with four toes), long-legged running bears, beavers, pikas, and entelodonts. Although older relatives of some of the newcomers had previously lived in North America, these new invaders, seasoned in the tough environments of the larger Eurasian landmass, outcompeted many of the native species.

The most abundant large mammals in a 14-million-year-old Miocene fossil assemblage from the Wood Mountain area of southern Saskatchewan, are *Merychippus,* a three-toed horse, and *Merycodus,* an extinct pronghorn. Elephants had arrived from Asia about 2 million years earlier, and remains of an ancestor of the American mastodon and a gomphothere (a four-tusked elephant) occur in sediments from the Wood Mountain area. The youngest known Neogene mammal assemblage on the Canadian Plains comes from a 5-million-year-old Pliocene deposit in the Hand Hills, northeast of Drumheller in Alberta. Among the discoveries in this deposit are rodents, horses, and camels.

Beringia was not the only invasion route during the Neogene. Starting in the late Cretaceous, organisms began to move between the two American continents by island-hopping through the ancestral West Indies. Opossums were among the earlier northward migrants, first appearing in North America toward the end of the Cretaceous. But massive faunal exchanges did not happen until about 3 million years ago, in the later Pliocene, when the two continents finally became directly connected by the Isthmus of Panama. This event changed the distribution and evolution of mammals, insects, and flowering plants on the two continents.

This new link had an impact far beyond faunal distributions, however. The union of North and South America radically altered ocean circulation. Restriction of circulation in both the Pacific and Atlantic oceans triggered more climatic extremes between high and low latitudes. And it contributed to the first extensive growth of ice sheets in the Northern Hemisphere 2.6 million years ago, as Chapter 11 will explain.

A view reminiscent of the Wood Mountain area of Saskatchewan during the Miocene. The foreground is dominated by *Merychippus*, a three-toed horse, while a group of *Merycodus*, an extinct pronghorn, look on in the middle distance at right. PAINTING COPYRIGHT MARIANNE COLLINS, ARTIST.

11 • THE ICE AGE

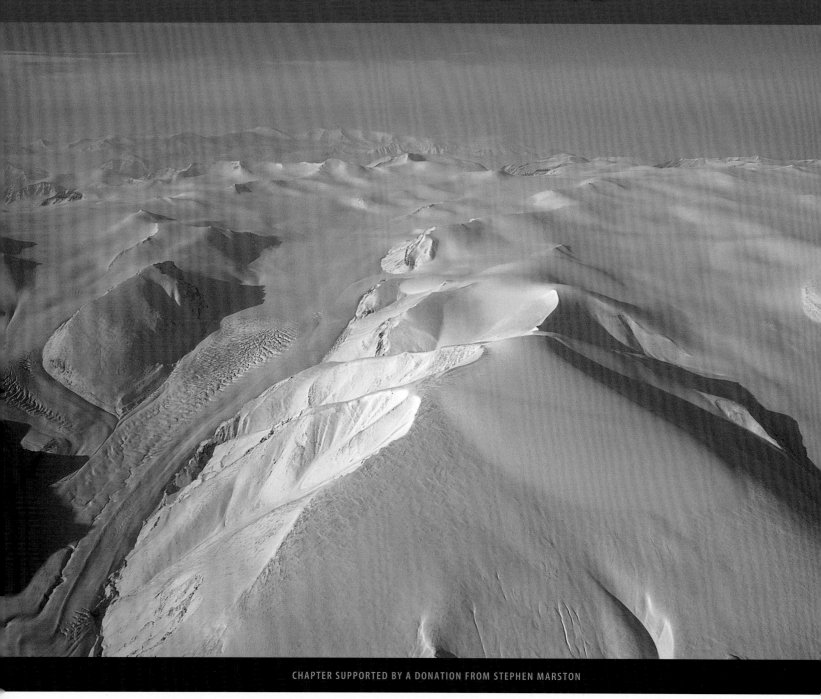

CHAPTER SUPPORTED BY A DONATION FROM STEPHEN MARSTON

Aerial view of Axel Heiberg Island, Nunavut, an icy scene reminiscent of what much larger areas of Canada would have looked like during the last glaciation. KELLY BENTHAM, FROM THE LORITA (LOMONOSOV RIDGE TEST OF APPURTENANCE) FIELD PROJECT, GEOLOGICAL SURVEY OF CANADA UNCLOS PROGRAM.

A Mountaineer, a Brickyard, and a Bluff

Arthur Coleman was a mountaineer, artist, and geologist with a passion for the Ice Age. He travelled across Canada and around the world in pursuit of knowledge and adventure. But one of his greatest discoveries was made in his own backyard, in Toronto's Don Valley Brick Works. Coleman, a professor at the University of Toronto, first visited the Brick Works in 1891 after he heard of workmen there taking home fossils. The Don Valley Brick Works began operations in 1889, producing bricks from Ordovician shale. Many famous Toronto buildings, including Casa Loma, Osgoode Hall, and Massey Hall, were built from its bricks. To reach the shale, workmen had to dig through about 35 metres of much more recent "overburden", but it was this softer top material that drew

Coleman's interest. His research, and that of others who followed him, has revealed an incredible story of climate and life stretching back 130,000 years.

Directly on top of the Ordovician rocks at the Brick Works, Coleman found a dense grey till—a mixed sediment deposited directly from glaciers (Chapter 6). Today geologists call this deposit the York Till. Above the York Till, he found layers of blue mud and fine sand, which he named the Don Beds. These beds are rich in fossils, including tree limbs, well-preserved leaves of deciduous plants, charcoal, mammal and fish bones, shells of mollusks (bivalves and gastropods), insects, and pollen. Collectively the fossils suggested a

Arthur Coleman—mountaineer, artist, and geologist—in his later years. COURTESY OF VICTORIA UNIVERSITY LIBRARY (TORONTO).

climate about 2 to 3°C warmer than today. Reflecting on his finds, Coleman wrote, "With a little imagination one can see the ancient forest of maples and oaks and many other trees on the river shore with deer coming down to drink, bears tearing open a rotten log for its small inhabitants, and at some creek mouth the giant beaver fells a tree with a splash to feed on its branches; while openings in the forest show buffalo grazing." Subsequent research has shown that the Don Beds were deposited in a lake during a warm period between 130,000 and 80,000 years ago.

Don Valley Brick Works, Toronto, Ontario, in 1970. The lower grey rocks are the Ordovician strata used to make bricks. Above these are cream-coloured late Quaternary sediments. ALAN MORGAN.

Geological time scale, showing the interval covered in this chapter. Numbers indicate millions of years ago. P = Paleogene (Paleocene to Oligocene), N = Neogene (Miocene and Pliocene), and Q = Quaternary.

Remains of giant beaver have been found in Quaternary sediments at various locations across Canada, including the Toronto area. For scale, the inset shows an imagined confrontation between a giant beaver, which could be up to 2.5 metres long, and a black bear. GIANT BEAVER, COPYRIGHT GOVERNMENT OF YUKON / ARTIST GEORGE "RINALDINO" TEICHMANN, 1997.

Higher in the succession at the Brick Works, and also exposed in the Scarborough Bluffs on the modern shoreline of Lake Ontario a few kilometres away, was a sequence of mud and sand that Coleman called the Scarborough Beds. These beds contain abundant pollen of pine, spruce, fir, and other plants indicative of a mixed forest and tundra environment like that of northern Ontario today. From this we know that the climate had cooled significantly since the time the Don Beds were deposited. In the upper half of the Scarborough Bluffs, Coleman found thin-layered mud, the basal part of the Sunnybrook Drift, that we now know to have been deposited in an ice-dammed glacial lake some 50,000 years ago. Fossils indicate that the lake was cold and deep. The mud contains isolated pebbles and boulders that were dropped from icebergs floating in the lake.

At the top of the Scarborough Bluffs section are the Newmarket and Halton tills. The older Newmarket Till was deposited by an ice sheet that covered the Toronto area from about 30,000 to 16,000 years ago and then retreated briefly to a position north of Toronto. But ice soon advanced again and deposited the Halton Till. About 14,000 years ago, ice retreated from the Toronto area for the last time,

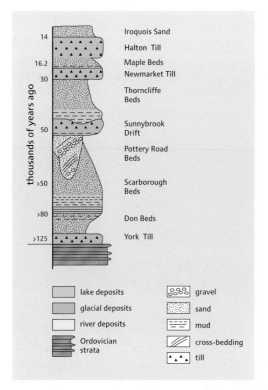

Composite column of the succession of sediments at the Don Valley Brick Works and Scarborough Bluffs, near Toronto, Ontario, and their ages. ADAPTED FROM KARROW ET AL. (2001).

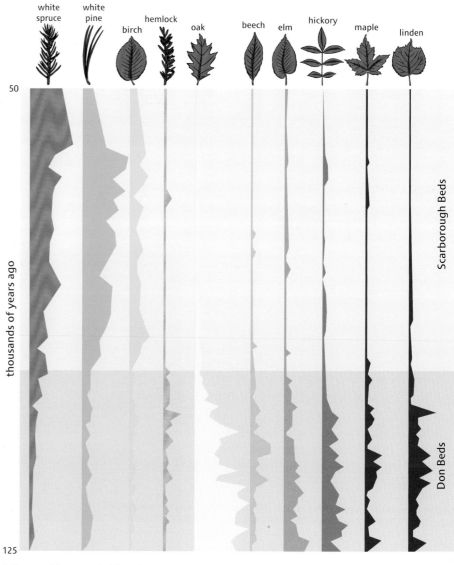

Pollen assemblages recorded from the Don and Scarborough beds, largely reflecting changing climatic conditions. The coloured graphs represent the proportions of particular trees in the pollen assemblage, the broader the coloured band the greater the proportion. The dominant pollen in the interglacial Don Beds are from deciduous trees (oak to linden on the right side of the figure) that flourished during the last interglaciation, whereas the relative abundance of conifer pollen (spruce and pine) in the Scarborough Beds reflects a cooler climate as the last glaciation began to gain momentum. ADAPTED FROM TERASMAE (1960).

leaving the landscape of rolling plains that now form southern Ontario's fertile farmland. The Halton Till is overlain by lacustrine sediment, the Iroquois Sand, which Coleman attributed to a precursor of Lake Ontario, larger and deeper than its modern counterpart.

The Ice Age

The work of Arthur Coleman and other geologists of his time revealed extreme climatic change in our planet's recent past. Evidence collected over the past century has shown that during much of the Earth's existence, ice and snow have been absent or restricted to polar or high-altitude regions. But at times when global temperatures cooled significantly, ice sheets extended into the mid-latitudes. Such chilly times are called ice ages. The most recent of these episodes, referred to as "the Ice Age", began about 2.6 million years ago in the Northern Hemisphere. But as we will see, rather than involving just a single event, ice advanced and retreated over parts of North America and Eurasia many times throughout the Ice Age.

People who live in mountainous regions have long been aware that glaciers advance and retreat. In historical time, farms and even small settlements in Switzerland disappeared under advancing ice. Shepherds noticed that some large boulders (now called erratics) were different from the bedrock beneath them, and that stony ridges distant from glaciers resembled those forming today at glacier margins. However, many scholars who did not live near modern glaciers found it hard to imagine extensive lowland landscapes as the work of ice. Until the middle of the nineteenth century, erratic boulders and till deposits were generally thought to have been swept across the land by the catastrophic biblical flood or dropped from icebergs and ice floes adrift in a former cold sea. Such sediments thus came to be called drift, a term still used for glacial deposits. The ice-floe idea sprang in part from reports by whalers and polar explorers, who saw boulders and mud dropping from pack ice. Geologists had already found marine fossils in sedimentary rocks now well inland, suggesting former inundations of the sea. So they imagined a similar younger episode to explain drift deposits. Prominent Victorian adherents to the drift theory included Charles Lyell and Swiss naturalist Louis Agassiz. Both were later to change their minds—Agassiz enthusiastically and influentially and Lyell reluctantly. Indeed, Agassiz is generally considered the founder of glacial geology. He revised his earlier ideas after a trip to the Alps in 1836. The following year, he presented a paper proposing widespread expansion of alpine glaciers during a great "Eiszeit" (Ice Age).

In the mid- to late-nineteenth century, fieldwork by William Logan, Joseph Tyrrell, George Dawson, and others under the auspices of the Geological Survey of Canada, provided the framework for interpreting the glacial geology of Canada. It was Dawson who conceived the idea of a major Cordilleran glacier that had its crest on the plateau east of the Coast Mountains. He visualized ice flowing down valleys from the ice centre and out through the Strait of Georgia. Dawson also named the ice sheet that by then was known to have covered much of central and eastern Canada—the Laurentide glacier. Today we call these two great former bodies of ice the Cordilleran and Laurentide ice sheets.

By the turn of the twentieth century, Canadian geologists had assembled a detailed picture of the last glaciation in North America. Through the mapping of glacial grooves and striations (the abrasion marks left on bedrock by rock fragments embedded in moving ice) and by determining the bedrock sources of erratic boulders, they had recognized that the Laurentide Ice Sheet originated not primarily in the mountains or at the poles, but rather on the broad uplands of the Canadian Shield. Moreover, they had realized that this vast ice sheet had multiple centres of outflow that shifted in position over time.

Two significant events near the middle of the last century greatly improved our understanding of glaciation in Canada. The first, in 1949, was the discovery of radiocarbon dating, which allowed geologists to date carbon-bearing plant and animal fossils as old as 50,000 years (Chapter 3). The second significant development was the publication in 1958 of the first glacial map of Canada. This map showed the three-dimensional extent of glaciation and many features of glacial erosion and deposition on a subcontinental scale. Before delving further into Canada's Ice Age history, let's look at features that make landscapes associated with glaciation so distinctive.

Flowing Ice

A glacier is a thick mass of ice that flows under its own weight. Enough snow must fall for a glacier to form, and winter accumulation must be greater than summer melt. As snow builds up, it first recrystallizes into small grains of ice. Once ice accumulates to a thickness of 20 to 30 metres, the pressure at depth causes the ice crystals to interlock, and so to flow, albeit slowly. Most glaciers also move by sliding along their beds, on top of a thin layer of meltwater produced mainly by friction and pressure from the glacier's own weight, much as a thin film of water forms beneath an ice-skater's blades. Ice in the upper 40 metres or so of a glacier is brittle and so develops fractures and crevasses, especially where the underlying land surface is irregular or steep.

Valley glaciers originate in mountains and flow down valleys at speeds of 20 metres to about 10 kilometres per year. Larger masses of ice can form ice fields such as the Columbia Icefield in

Crevasses can have concentric and radial patterns, as on this tumbling glacier on Baffin Island, Nunavut. CRYSTAL HUSCROFT.

Glacial grooves and striations on a rock surface near Clam Harbour Beach, Nova Scotia. The rock surface was scratched and grooved by stones and grit frozen into the bed of the glacier. The grooves show various orientations, evidence for changes in the direction of glacier flow. ROB FENSOME.

A characteristic feature of a formerly glaciated rocky landscape is a roche moutonnée, a glacially smoothed rocky prominence with a gentler slope on the up-ice side and a steeper slope on the down-ice side. This example, at Prospect, Nova Scotia, occurs in Devonian granite of the South Mountain Batholith. ROB FENSOME.

the Rocky Mountains, or ice caps such as the Barnes Ice Cap on Ellesmere Island. Ice sheets are glaciers up to several kilometres thick that cover areas of tens of thousands of square kilometres; examples are today's Greenland Ice Sheet and the former Laurentide and Cordilleran ice sheets. Glaciers of all types thin and their outer boundaries retreat when ice is lost faster than it is replenished. But even in retreat, a glacier continues to flow toward its receding margin. Rates of retreat depend on climate, glacier size, and topography. For example, the Laurentide Ice Sheet retreated at an average rate of 150 metres per year, whereas the Athabasca and Saskatchewan glaciers in Jasper National Park of Canada have retreated at rates of 10 to 25 metres per year since 1950. Glacier retreat can be interrupted by pauses and advances at times of cooler or wetter climate.

The modern Canadian landscape has been fundamentally shaped by Ice Age glaciers. They have scraped, pushed, carried, and dumped rock and sediment, acting like great conveyor belts that have moved material from high to low ground, tending to smooth out the landscape in the process. Glaciers erode rock through plucking and abrasion. Fractured bedrock and frozen sediment can be plucked from the ground below a flowing glacier to become frozen into the ice. Abrasion occurs when debris-laden ice grinds over bedrock like sandpaper, smoothing, scratching, and grooving the rock surface. Striations (scratches) formed in the abrasive process leave a record of the direction of ice flow. If flow direction changes, the succession of striations with different orientations provides a record of the switches. Flowing ice shapes the underlying land surface, smoothing angular protuberances into landforms that become streamlined on the upstream side and angular, due to ice-plucking, on the downstream side.

The glacial origin of the valley containing Ten Mile Pond in Gros Morne National Park of Canada, Newfoundland, is evident from its U-shaped cross-section. MICHAEL BURZYNSKI.

Mount Assiniboine, on the Alberta-British Columbia border, is a Matterhorn-like peak eroded on all sides by glaciers. JOHN SCURLOCK.

The erosive action of ice sheets tends to create rounded hills and small depressions, the bottoms of the latter filling with lakes after the ice has retreated. Such a moderately bumpy landscape is typical of many parts of the Canadian Shield today. In contrast, valley glaciers tend to form broad, steep-walled valleys with U-shaped cross-sections. Ice-plucking at the head of a valley glacier hollows out an amphitheatre-shaped bowl called a cirque. Plucking in adjacent cirques can sculpt intervening areas into narrow ridges (arêtes), and ultimately into pointed peaks (horns) such as Mount Assiniboine in the Rocky Mountains.

The Dirt from Glaciers

Most of Canada has been covered by ice one or more times during the Ice Age. Sediment deposited from the ice covers much of the landscape, generally in the order of 10 to 200 metres thick; but it can be more than 300 metres thick where it has infilled pre-glacial valleys. Glaciers are able to carry the materials they pick up, from microscopic particles to large blocks of rock. This debris is deposited when it lodges against objects on the glacier bed or when the ice melts. Glacial deposits are mainly till.

Boulders can be transported by a glacier for hundreds of kilometres, ultimately to be deposited as erratics. Erratics are useful indicators of the direction and distance of ice flow and, along with striations and streamlined glacial landforms, can be used to reconstruct the flow paths of glaciers. An example is the Foothills Erratics Train in southwestern Alberta, which extends for more than 700 kilometres from the Mount Edith Cavell area in Jasper National Park of Canada to northern Montana. The distinctive quartz-rich sandstone boulders that make up the Train were carried down the Athabasca Valley on a tongue of the Cordilleran Ice Sheet to the eastern margin of the Rocky Mountains where it met the Laurentide Ice Sheet. From there, the boulders were carried southward along the zone of contact of the two ice sheets until they were released from the ice. The largest of the erratics—the Okotoks Erratic or Big Rock—is about 40 by 18 metres, with an estimated weight of 16,500 tonnes. It is actually composed of several large blocks and many smaller ones.

A sediment-laden glacier resting on early Permian sandstone, northwestern Ellesmere Island, Nunavut. BENOIT BEAUCHAMP.

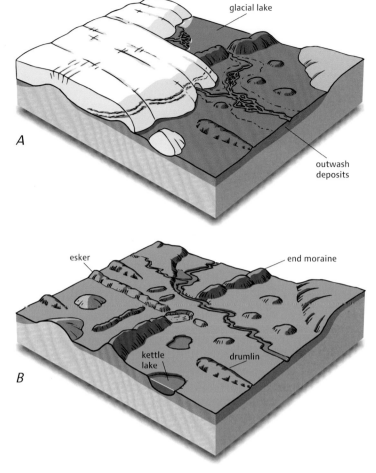

Features produced by glacial deposition or glacier-related water-lain deposits. *A* shows a landscape still partly glaciated; *B* shows the same landscape after deglaciation.

Till deposited during the last glaciation in the Chibougamau area of Quebec. The orientation of elongated boulders and pebbles provides evidence of the flow direction of the last ice to cover the area; in this case it was to the southwest. GILBERT PRICHONNET.

Valley glaciers and ice sheets deposit till directly from their bases. Sheets of till, called ground moraine, form much of Canada's prime agricultural land. As they flow, ice sheets also shape till into drumlins, elongated streamlined features that are 15 to 40 metres high and hundreds of metres long. Drumlins resemble inverted spoons aligned in the direction of ice flow, with the up-glacier end relatively steep and the down-glacier end more gently sloping.

A ridge of till tens of metres high, called an end moraine, may form at the margin of a glacier through the slow release of rock debris from the ice. Related landforms associated with valley glaciers include lateral moraines, which develop along glacier sidewalls, and medial moraines, which mark the contact between two valley glaciers that have merged. Where the lobes of an ice sheet converge, huge interlobate moraines develop. Such moraines are broad areas of elevated hummocky ground that are formed of both till and meltwater-deposited sand and gravel. Many interlobate moraines, such as the Oak Ridges Moraine north of Toronto, are important groundwater reservoirs.

Whereas sediments deposited directly from glaciers are mixtures of coarse and fine debris, those of meltwater streams are layered and consist primarily of sand and gravel. Meltwater streams may flow on the surface of a glacier or in tunnels at the glacier bed. After the glacier has retreated, sand and gravel deposited in such tunnels are left as sinuous ridges called eskers that can be up to 50 metres high, 150 metres wide, and hundreds of kilometres long. Swarms of eskers reveal the "plumbing system" within the former Laurentide Ice Sheet. When meltwaters emerge from a glacier, they lose velocity and so drop sediment, forming sheets of sand and gravel known as outwash.

Large blocks of ice left behind during ice retreat may be buried by outwash material. When this ice finally melts, pits known as kettle holes may be left behind. Many ponds and small lakes on the Western Interior Plains occupy kettle holes. Kettle holes are commonly associated with small hills composed of sand and gravel that was transported by flowing water on top of a glacier before being deposited in a depression in the ice. When the ice melts the sand and gravel form a hill known as a kame. Kettle

"Big Rock" at Okotoks, Alberta, is the largest of a string of erratics collectively known as the Foothills tics Train. The rock is a quartz-rich sandstone (commonly referred to as a quartzite, but it is from a mentary sequence) of probable early Cambrian age. QUATERNARY ENVIRONMENTS, ROYAL ALBERTA MUSEUM.

These snow-covered drumlins near Nicola Lake, south of Kamloops, British Columbia, were formed beneath the Cordilleran Ice Sheet. From the shape of the drumlins, we can infer that the ice flowed toward the lower right. DARREL LONG.

Aerial view of a glacier on Baffin Island, Nunavut, surrounded by moraines. TED LITTLE.

Outwash sand and gravel, Hawkesville Kame, near Waterloo, Ontario. ALAN MORGAN.

A valley glacier with two medial moraines descends toward a fiord on eastern Baffin Island, Nunavut. A meltwater stream flowing along the margin of the glacier to the right has eroded a lateral moraine. Sediment carried down-valley by this stream has been deposited on a delta that is building into the fiord. ALAN MORGAN.

CHAPTER 11 · THE ICE AGE

An esker snakes across the landscape in the Northwest Territories. IAN WARD / AIRSCAPES.CA.

Kettle holes punctuate the farmland in this aerial view taken east of Saskatoon, Saskatchewan. The blue-flowering crop is flax. ROB FENSOME

Moraine-dammed lake at the foot of Bishop Glacier in the southern Coast Mountains of British Columbia. JOHN CLAGUE.

and kame topography consists of a series of ponds and small lakes interspersed with isolated low hills.

Sediment from glaciers also finds its way into glacial lakes, formed where meltwater is trapped between the ice margin and higher ground. Such lakes change in size and depth over time as the glacier margin shifts and new outlets open and close. A variety of sediment is deposited in glacial lakes. Streams flowing into the lake drop their coarse sediment load in deltas near the shoreline, whereas fine material such as mud and clay can be carried for several kilometres before settling out onto the lake floor. Thin cyclic layers of finer and coarser sediment, known as varves, are common in glacial lakes. Varves form because sediment washed into the lake during spring, when currents tend to be stronger, are coarser than those deposited later in the year. The distinctive rhythmic couplets of sediment so formed thus represent annual cycles, and their variation in thickness provides clues to climate change. Large areas of excellent agricultural land on the Prairies are underlain by clays deposited in glacial lakes. Other features associated with glacial lakes include sandy beach ridges and wave-cut terraces, far from any modern lakeshore. We will discuss some impressive examples of former glacial lakes later in the chapter.

Frozen Ground

Throughout most of southern Canada the ground is frozen hard for several months each winter, but readily melts come spring. In contrast, most of northern Canada, almost two-thirds of the country, is underlain by permafrost—layers of dry or ice-bonded soil, sediment, or rock that maintain a temperature below freezing for two or more years. Clearly the permafrost zone was much farther south when ice sheets extended to lower latitudes. In very cold locales, usually farther north, permafrost will be continuous over an entire region. However, where average annual temperatures are slightly below zero, only patches of ground may be permanently frozen, and so the permafrost is discontinuous.

These cyclic layers of mud deposited in a former glacial lake and now exposed along the shore of Lac Témiscamingue, Quebec, represent annual accumulations and are known as varves. JEAN VEILLETTE, REPRODUCED WITH THE PERMISSION OF NATURAL RESOURCES CANADA 2013, COURTESY OF THE GEOLOGICAL SURVEY OF CANADA.

Ice lens in permafrost in the Katherine River valley, Torngat Mountains, Labrador. M. J. VAN KRANENDONK, REPRODUCED WITH THE PERMISSION OF NATURAL RESOURCES CANADA 2013, COURTESY OF THE GEOLOGICAL SURVEY OF CANADA.

During the warmer summer months, especially in more southerly parts of the permafrost zone, surface layers melt and become saturated because water cannot penetrate the frozen permafrost below. This seasonally melted top is known as the active layer. Large areas previously or currently underlain by discontinuous permafrost are now occupied by the infamous muskeg of the Canadian north, in which roads, trucks, and even animals such as moose can sink into a quagmire of mud and peat.

A distinguishing feature of continuous permafrost is polygonal patterned ground, the surface expression of intersecting ice wedges. Ice wedges form as the ground contracts and cracks from the cold, forming a void that penetrates the frozen substrate beneath. Snow falls into the crack, melts the following spring, and then refreezes, gradually wedging the ground apart over hundreds to thousands of years. Small ice-cored hills known as pingos also occur in the continuous permafrost zone. They result from the upward bulging of near-surface frozen ground above the ice core. Another feature involves exposed bedrock that becomes shattered into blocks, resulting in a loose, rocky, more or less level ground surface known as a felsenmeer ("sea of rocks" in German).

As climate warms, the melting of areas of permafrost with a lot of ice can produce a thermokarst landscape, so called because it is distinguished by features that resemble the sink holes developed in carbonate rocks. Thermokarst lakes can be seen in the Mackenzie Delta region and in areas of northern and eastern Nunavut. During the warmer summer months, surface layers can melt and the resulting active layer may move downslope in characteristic lobes.

Palsas and string bogs are features mainly of discontinuous permafrost in boggy wetland areas, and many different ideas for their origins have been suggested. Palsas comprise mounds, irregular patches or linear ridges of peat that contain ice lenses

Aerial view of polygonal patterned ground, produced by ice wedges, on Ellef Ringnes Island, Nunavut. CAROL EVENCHICK.

and may be 1 to 7 metres high; mounds and patches may be more than 30 metres across and ridges may be 150 metres or more long. Some palsas occur in complexes with dimensions of several hundred metres. String bogs cover large areas of Quebec and Labrador, their presence seeming to indicate that permafrost is melting in a region. Generally found on gentle slopes, string bogs consist of distinctive ridges of peat, commonly ice-cored, that may be 2 metres or more high and extend for tens of metres.

Relict permafrost features can be seen in the landscape across Canada, where they echo the colder times of the last glaciation. For example, vestiges of patterned ground reflecting the presence of ancient ice-wedge polygons can be seen from the air over farmlands in many parts of southwestern Ontario, and relict ice-wedge casts occur in the same region.

An ice wedge, near Arundel, Quebec, formed about 11,500 years ago. After ice had opened the wedge and melted, probably several times, gravel partially filled the hole, followed later by windblown dust. More recent gravel has been deposited over the top of the structure. GILBERT PRICHONNET.

Pingo near Erly Lake, Tuktut Nogait National Park of Canada, Northwest Territories. I. K. MACNEIL, COPYRIGHT PARKS CANADA.

Felsenmeer on a small plateau high in the Torngat Mountains, close to the Quebec-Labrador border. The frost-shattered bedrock is composed of Archean gneisses from the eastern margin of the Superior Craton. JAMES GRAY.

Aerial view of thermokarst, Wapusk National Park of Canada, Manitoba. N. ROSING, COPYRIGHT PARKS CANADA.

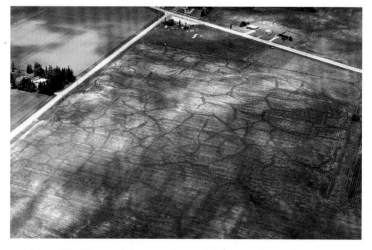

Aerial view of "fossil" polygonal patterned ground dating from the last glaciation at Muir, near Woodstock, Ontario. ALAN MORGAN.

String bogs in discontinuous permafrost in the James Bay Lowlands near the Quebec-Ontario border. The "strings" are ribbons of vegetation several metres wide and about one metre high, aligned perpendicular to water flow. ALAN MORGAN.

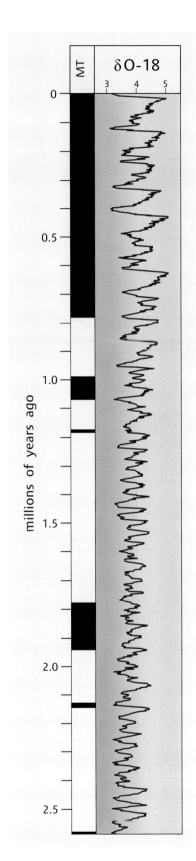

Oxygen isotope (δ O-18) curve for the past 2.6 million years (column at right) against the magnetic timescale (black represents normal polarity, as at the present day, and white represents reverse polarity—see Chapter 3). In the right hand column, the colour gradation reflects relatively warm (orange) to relatively cold (blue) temperatures. Note the cyclic nature of the curve, representing warmer (interglacial) versus colder (glacial) intervals, the changing shape of these cycles, and the overall gradual cooling trend over 2.6 million years. MT = magnetic time scale. ADAPTED FROM LISIECKI AND RAYMO (2005).

One Glaciation Becomes Many

The presence of more than one layer of till at the Don Valley Brick Works and Scarborough Bluffs showed that multiple glaciations had occurred during the Ice Age in southern Ontario with intervening warmer periods, a pattern that had already been established for the Alps. Similar evidence came from tills, interglacial sediments, and fossils across Canada and elsewhere, leading to an early interpretation that four major glacial advances had occurred during the Ice Age in both North America and Europe. In North America, the advances were named, from earliest to most recent, the Nebraskan, Kansan, Illinoian, and Wisconsinan glaciations. Evidence from deep-sea cores in recent decades, however, has revolutionized ideas about the Ice Age. The oxygen isotope composition of foraminiferal shells from deep-sea cores provides a more or less continuous record of temperature fluctuations over the past several million years. And a parallel record for the past 800,000 years has come from trapped air bubbles in ice cores from Greenland and Antarctica. Such studies reveal not just four glaciations during the Ice Age, but at least twenty-two.

According to the new data, each glaciation began slowly and culminated with glaciers much more extensive than those of today. Major glaciations ended with an abrupt change to a warm interval, termed an interglaciation. Oxygen isotope studies indicate that, from about 2.6 million until about 800,000 years ago, glacial and interglacial intervals occurred about every 41,000 years. More recent swings have occurred about every 100,000 years. What causes the cyclicity, and why does it vary? An answer came in the 1920s from Serbian geophysicist Milutin Milankovitch, building on earlier work. He proposed that cyclical variations in the shape of the Earth's orbit around the Sun, as well as in the tilt and wobble of our planet's axis of rotation, lead to differences in the distribution of radiation reaching the planet's surface. These variations have become known as the Milankovitch cycles (see illustration on the next page).

The Earth's orbit around the Sun is elliptical rather than circular, and so the distance between the two changes slightly through the year. This means that the energy reaching the Earth from the Sun also changes through the year. More importantly, the Earth's orbital path varies from being only a little elliptical (as at present) to being mildly elliptical and back on a cycle of about 100,000 years. This variation provides us with the first Milankovitch cycle, known as the eccentricity cycle. When the Earth's orbit is most elliptical, seasons in one hemisphere are intensified relative to those in the other. At times of least eccentricity, the seasons in the two hemispheres are most similar.

The Earth's rotational axis, around which it revolves every 24 hours, is tilted relative to the ecliptic axis, which is an imaginary axis perpendicular to the orbital plane. Because of this tilt, the poles receive 24 hours of daylight in midsummer and 24 hours of darkness in midwinter. If the rotational axis were not tilted, uniform amounts of solar radiation would be received at each pole throughout the year, hours of daylight would be of constant length everywhere, and we would not have seasons. That the tilt varies from 22.1° to 24.5° and back again every 40,000 years or so gives us the second Milankovitch cycle, known as the obliquity cycle. As the tilt or obliquity of the rotational axis increases, seasonal cycles intensify: more radiation reaches both hemispheres in their respective summers, which thus tend to be warmer; winters, with less radiation, tend to be cooler. At times of low obliquity (as is the case today), the opposite is true: both hemispheres tend to be

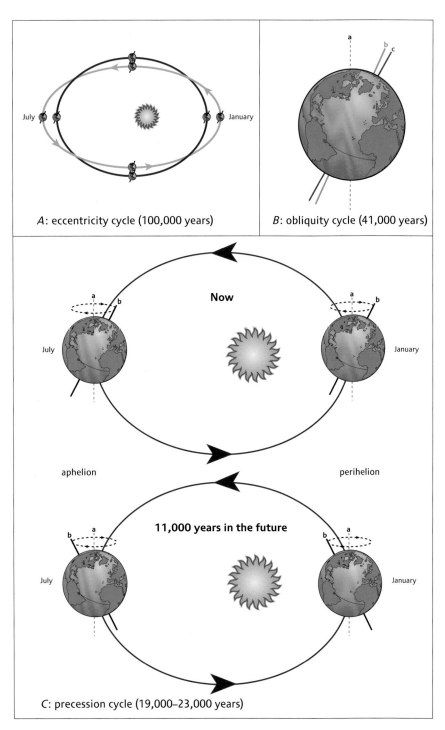

A. The orbit of the Earth around the Sun changes from strongly ellipsoidal to almost circular and back to strongly ellipsoidal every 100,000 years; this is the eccentricity cycle. B. The ecliptic axis (a) is an imaginary line perpendicular to the Earth's orbit around the Sun. The angle between the Earth's rotational axis (b and c) and the ecliptic axis varies between 22.1° (b) and 24.5° (c) and back every 41,000 years. This is the obliquity cycle. C. Earth's rotational axis (b) completes a precession cycle by circling the ecliptic axis (a) every 19,000 to 23,000 years. The upper diagram represents today, with the Northern Hemisphere tilted towards the Sun in July and away from it in January. The bottom diagram shows the halfway stage in the cycle, about 11,000 years in the future, when the Northern Hemisphere will be tilted away from the Sun in July and toward it in January. To avoid confusion, the shape of the Earth's orbit and the position of the Sun are shown unchanged in this diagram.

cooler in summer when they receive relatively less radiation, and warmer in winter when there is more. Cooler summers are thought to be favourable for the growth of glaciers because less of the previous winter's snow melts.

Milankovitch's third cycle, the precession cycle (or precession of the equinoxes), has to do with perihelion (the point in the Earth's annual orbit when the planet is nearest to the Sun) and aphelion (the point in the Earth's annual orbit when the planet is farthest from the Sun). This cycle reflects the circling of the planet's rotational axis around the ecliptic axis and can perhaps best be visualized as the Earth wobbling like a spinning top. The precession cycle has a frequency of about 20,000 years and determines whether the Northern Hemisphere is tilted toward or away from the Sun at the times of perihelion and aphelion. At present, the Northern Hemisphere is tilted toward the Sun at the aphelion, which is in July, and away from the Sun during the perihelion, which is in January.

By combining the three Milankovitch cycles, net changes in radiation can be determined for any point on the Earth back through time and to any time in the future. The combination of the Milankovitch cycles affects the distribution of this radiation between hemispheres and between high and low latitudes. Over the course of about 100,000 years, the amount of solar energy that each hemisphere receives varies by 20 percent. In the Northern Hemisphere, ice sheets may form during intervals when summers are cool and winters are mild. The net changes in the past correlate well with the tempo of glaciations and interglaciations, although the exact relationship of glacial cycles and Milankovitch cycles is complex. Evidence indicates that different Milankovitch cycles can dominate at different times: the 41,000-year cyclicity in glacial cycles prior to 800,000 years ago suggests that the obliquity cycle ruled then, whereas the 100,000-year cyclicity since 800,000 years ago suggests a dominance of the eccentricity cycle.

Milankovitch cycles generate only small changes in seasonality. They can act as a pacemaker of glacial cycles only when other factors are conducive to glaciation. However, Milankovitch or similar cycles have surely affected climate throughout our planet's history. Indeed, their signature cyclicity has been identified in many sedimentary rock sequences (see for example the photo on page 44), some possibly as old as Precambrian. The cycles may have influenced aspects of climate other than cold and warm intervals, such as wet and dry periods. Within the late Cenozoic, glaciation-interglaciation cycles started only when global temperatures cooled below a threshold level around the Pliocene-Pleistocene transition.

The Last Interglaciation

The most recent, or Sangamon, interglaciation lasted from its abrupt beginning about 125,000 years ago to about 75,000 years ago. Soon after its start, annual temperatures rose to a peak of up to 3 to 4°C warmer than at present, enough for sea level to rise about 6 metres higher than today. Sangamon deposits consist mainly of fluvial, lacustrine, and marine sediments, and are generally covered by till.

The bench at the top of the jutting cliff of Silurian strata at Arisaig, Nova Scotia, marks a raised beach cut by the sea during the Sangamon interglaciation about 125,000 years ago. Above this beach is till deposited during the last glaciation. RALPH STEA.

Sangamon deposits can be found in sea-cliff exposures and on wave-cut bedrock benches capped by till around the Gulf of St. Lawrence. One of the best sections is at Woody Cove, near Codroy in Newfoundland. There, deposits indicate an initial subarctic climate, which changed rapidly to one warmer than today before deteriorating again to a sub-arctic climate at the beginning of the last glaciation. Sangamon sediments near Green Point on Cape Breton Island contain remains of trees such as oak, white pine, hickory, basswood, and sweetgum, species that live today in the southeastern United States. This flora suggests that summers in Nova Scotia about 125,000 years ago were 4 to 8°C warmer than they are today, and that winters were wetter. Later, climate cooled and the local vegetation was replaced first by conifers (mainly spruce, fir, and jack pine) and later by tundra plants, including dwarf willow, dwarf birch, and alder. In contrast to the warm climates along the Atlantic coast, pollen and beetle remains from the Hudson Bay Lowlands and Yukon suggest that the interglacial climate was similar to or only slightly warmer than that of today.

The ground sloth *Megalonyx*, which lived in Beringia during the Ice Age, could have been up to 3 metres long and weighed about 850 kilograms. GROUND SLOTH BY GEORGE "RINALDINO" TEICHMANN, COPYRIGHT 1999.

Many of the animals that thrived in Canada during the last interglaciation are now extinct. For example, in sediments exposed along the South Saskatchewan River near Medicine Hat, Alberta, rich interglacial faunas include lions, ground sloths, elk, mammoths, horses, camels, pronghorns, and giant bison—a very different world than today's.

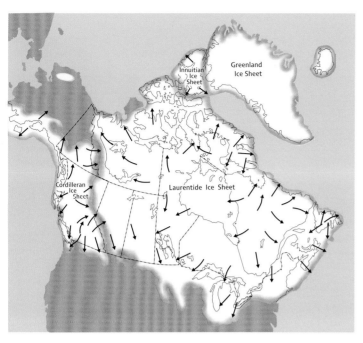

Ice cover over North America and adjacent areas during the last glacial maximum, about 19,000 years ago. The arrows show directions of ice flow from the major ice centres. ADAPTED FROM A GRAPHIC BY RICHARD FRANKLIN FOR CLAGUE AND TURNER (2003), USED WITH PERMISSION FROM TRICOUNI PRESS.

The Big Chill

About 75,000 years ago, mean annual temperatures in the Northern Hemisphere dropped by about 3°C, resulting in increased snow cover in Canada. The greater cover reflected more solar radiation and thus accelerated the decrease in local mean annual temperatures, which plummeted by as much as 10°C. As a consequence, glaciers expanded. Dramatic changes may have happened in just a few centuries. Small ice caps grew and coalesced to form ice sheets that eventually covered most of Canada. Only parts of Yukon and the Mackenzie Valley, the western Arctic, and the southwestern Prairies lay beyond the limits of the ice. As ice sheets advanced over northern North America, they were also growing in northern Europe, Greenland, Antarctica, and parts of Siberia.

In North America, climate warmed about 50,000 years ago, freeing most of the Cordillera, the Prairies, southern Ontario, and parts of the eastern Arctic from ice. But about 35,000 years ago, the ice sheets expanded again and, less than 10,000 years later, most of the northern half of North America was blanketed in thick ice. Three ice sheets were involved: the Laurentide Ice Sheet was the largest; the others were the Cordilleran Ice Sheet in the west, and the Innuitian Ice Sheet over the Arctic Islands. At its maximum, about 19,000 years ago, the Laurentide Ice

Ring Mountain, a tuya northwest of Whistler, British Columbia, was active during the last glaciation. PAUL ADAM.

Sheet reached the edge of the Atlantic continental shelf in the east, merged with the Innuitian Ice Sheet in the north, adjoined the Cordilleran Ice Sheet in the west, and southward extended into the central and eastern United States. The Laurentide Ice Sheet was 3 kilometres thick at its maximum.

The Laurentide Ice Sheet developed from centres over Foxe Basin-Baffin Island, central Quebec-Labrador, mainland Nunavut (Keewatin), the Miramichi Highlands of New Brunswick, Newfoundland, and possibly northern Ontario. Glaciers flowing out from these centres grew until they coalesced to form a huge multi-domed ice sheet. The locations of the centres, or domes, shifted from time to time, resulting in complex flow patterns. For example, the Keewatin ice centre migrated more than 700 kilometres to the east during the last glaciation.

Most of Canada west of the Prairies was covered by the Cordilleran Ice Sheet, which at its maximum about 18,000 to 19,000 years ago was up to 2 kilometres thick. It formed when glaciers sourced in many mountain ranges coalesced over the plateaus of British Columbia and southern Yukon to produce a single ice sheet with several domes. Lobes of the Cordilleran Ice Sheet extended southward into Washington, Idaho, and Montana. To the north, Cordilleran ice flowed over the southern and central parts of the Yukon Plateaus; to the east it eached the Rocky Mountain Foothills, where it briefly coalesced with Laurentide ice; and to the west it covered parts of the continental shelf. Some peaks in the Coast Mountains poked above the ice to form rocky knolls known as nunataks. Some volcanoes in the Stikine Volcanic Belt spewed lavas beneath the ice, creating flat-topped volcanoes known as tuyas; and volcanoes in the Cascade Magmatic Arc, such as Mount Garibaldi, were built partly on or against the ice, making for unstable topography when the supporting ice melted. Northern Yukon and northern Alaska were ice-free; climate in these areas was cold enough to support glaciers, but was too dry for ice to build up.

The Innuitian Ice Sheet covered the Arctic Islands north of Lancaster Sound and extended onto the continental shelf. At its maximum it coalesced with the Greenland and Laurentide ice sheets to the east and south, respectively. Most Innuitian ice flow was westward, down channels separating what were to become the Arctic Islands. Over the islands, ice was about 1 kilometre thick, and thicker over the channels. The high Arctic was so cold that the Innuitian Ice Sheet was frozen to its bed and most movement was by creep within the ice rather than by sliding at the base. Consequently, few erosional and depositional landforms were produced, although raised beaches show that the lithosphere was weighed down by an ice load, and meltwater landforms mark former Innuitian ice margins.

Thawing Out

During late stages of the last glaciation, minor advances of ice lobes at the southern margin of the Laurentide Ice Sheet deepened depressions. It was among these depressions, when the lobes began to retreat again about 18,500 years ago, that the Great Lakes were born. The early Great Lakes changed in extent and level as their water outlets shifted in response to fluctuations of the ice edge.

By 17,000 years ago, ice had retreated from the outer continental shelves of western and southeastern Canada. The Grand Banks of Newfoundland and other parts of the Atlantic and Pacific continental shelves were exposed as land because much water was still locked up in ice sheets and consequently sea level was low. Around the same time, the outer Gulf of St. Lawrence became ice-free and the Atlantic ice-flow centre shifted to mainland Nova Scotia.

An aerial view northwest of Winnipeg, Manitoba, showing iceberg scour marks that formed on the floor of former Glacial Lake Agassiz. ALAN MORGAN.

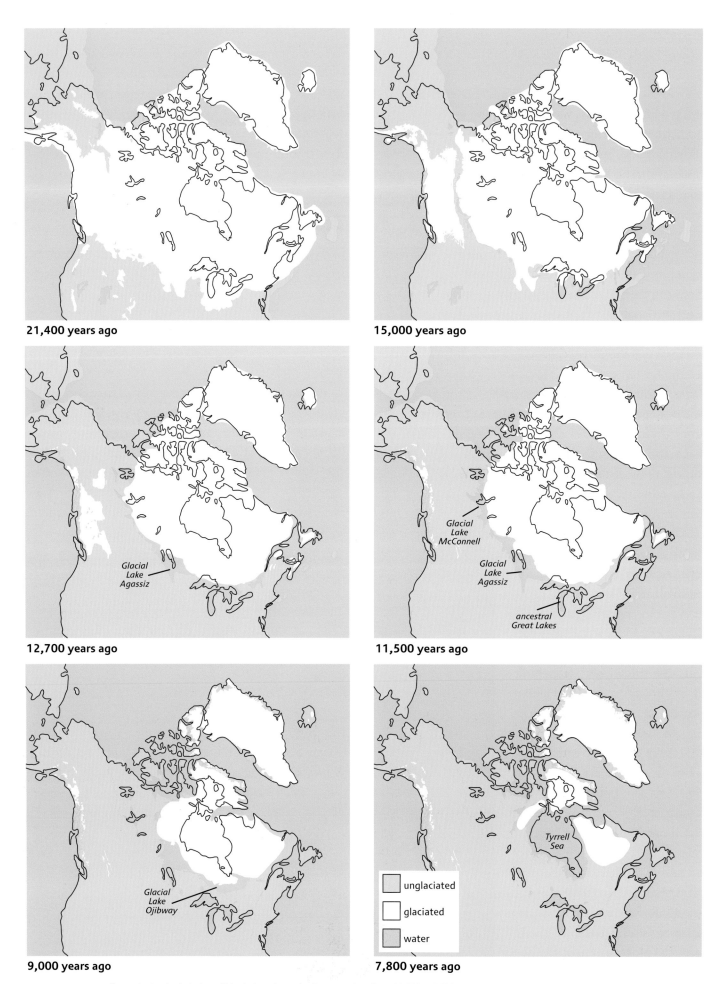

Stages in the deglaciation of North America and adjacent regions from 21,400 to 7,800 years ago. ADAPTED FROM DYKE ET AL. (2003).

The next few thousand years saw huge changes. By 15,600 years ago, the western margin of the Laurentide Ice Sheet had receded, with ice-free areas gradually extending northward as the ice retreated. But even with the ice gone, meltwater, loose sediments, and lack of vegetation led to an unstable landscape. About 14,000 years ago, the huge Glacial Lake Agassiz began to form. As the ice margin retreated, the lake expanded to the north and at one time or another covered more than a million square kilometres, although not all areas were inundated at the same time. During its existence, Glacial Lake Agassiz had four main outlets: southward down the Mississippi River, northwestward to the Arctic Ocean, eastward through the Great Lakes, and northeastward across northern Ontario.

By 13,500 years ago, water levels in the Great Lakes (except Lake Superior, which was still ice covered) were changing rapidly. Southern Ontario was ice-free, with open woodland vegetation and a fauna that included mastodons. Glacial Lake Agassiz was still expanding, and a series of smaller lakes began to develop in the Prairies along the retreating western margin of the Laurentide Ice Sheet.

Climate chilled again shortly after 13,000 years ago. This cold period, known as the Younger Dryas, lasted about 1,000 years and coincided with an expansion of the Laurentide Ice Sheet and with regrowth of ice caps in Atlantic Canada and northern Europe. At the same time, the Cordilleran and Innuitian ice sheets also advanced. The Younger Dryas cooling appears to have resulted from the outpouring of immense quantities of cold fresh water from glacial lakes in North America into the Atlantic Ocean. This occurred when retreat of the Laurentide Ice Sheet suddenly opened new, lower outlets, and rapidly released large volumes of previously impounded meltwater. These waters reduced the salinity (and thus the density) of surface ocean water and prevented surface waters in northern latitudes from sinking during winter cooling as they do today,

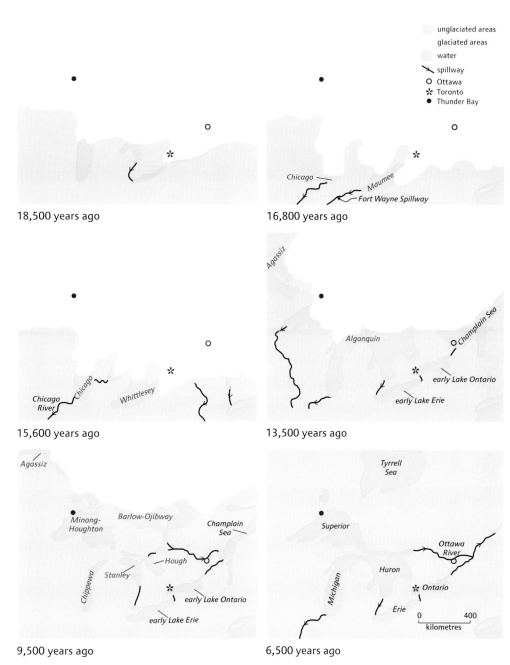

Stages in the evolution of the Great Lakes from 18,500 to 6,500 years ago. Glacial lakes are labelled in red. ADAPTED FROM TRENHAILE (1998) AND DYKE ET AL. (2003).

thus cutting off heat transport via the northward-flowing Gulf Stream and shifting the meteorological polar front southward. Glacial lakes and their outlets continued to develop and shift along the western margin of the Laurentide Ice Sheet from 13,000 to 11,500 years ago.

Except during the Younger Dryas interlude, the general warming trend was rapidly melting the Cordilleran Ice Sheet, so that by about 12,000 years ago, ice cover in the Cordillera was not much more extensive than it is today. In the melting process, glacial lakes developed in many Cordilleran valleys, although they were small compared to some farther east.

This excavated section in sediments near Lismore, Nova Scotia, records rapid climate change at the end of the last glaciation. Reddish sediments at the base were deposited about 14,000 years ago in a glacial lake centred over the present-day Northumberland Strait. The glacial-lake sediments are capped by black peat and pale reddish to cream-coloured clay, which formed during a warmer phase from about 14,000 to 12,500 years ago in a forested landscape. The overlying sequence of reddish sediments was deposited when ice re-advance blocked local drainage and re-established the glacial lake. This occurred during the cooler Younger Dryas. Above the upper lake deposit is the modern soil. RALPH STEA.

Aerial view of a Saskatchewan coulee, with the town of Esterhazy at the bottom. Such valleys on the Prairies, commonly with "misfit" streams too small to have carved them, were cut by meltwater flowing from glacial lakes. ALAN MORGAN.

Gateway

During much of the last glaciation, as for most of the past 100 million years, the area from northwestern Yukon, through Alaska and into eastern Siberia was land—a region known as Beringia. Beringia provided a land connection that allowed plants and animals to migrate both ways between Asia and North America. Toward the end of the Pleistocene, as ice waned farther south, Beringia was unglaciated and home to great herds of game animals—untold numbers of bison, camels, horses, mammoths, musk oxen, caribou, and saiga antelopes. Competing for this prey was a fearsome array of predators such as wolves, lions, and giant short-faced bears. Among these predators was one that was less fearsome-looking but more lethal—*Homo sapiens*.

After leaving Africa between 60,000 and 50,000 years ago, our species had migrated to most corners of Eurasia and to Australia by the late Pleistocene. However, most scholars agree that they had not yet reached the Americas. People apparently reached Siberia 40,000 to 30,000 years ago, though a continued eastward migration was then probably stalled for thousands of years as the last glaciation approached its peak. Archaeological finds show that people eventually reached the coast of northeastern Siberia about 16,000 years ago, by which time they had learned how to design suitable clothing and shelter for cold weather. With such skills, humans could now migrate into the area of the modern Bering Strait and, from there, the Americas.

How did the ancestors of today's Aboriginal peoples enter North America? Most experts favour one of two hypotheses. One is that people from Asia migrated across Beringia and occupied interior Alaska and Yukon before moving into the rest of North America via an interior corridor between the waning Laurentide and Cordilleran ice sheets. The other posits that people occupied and travelled along Pacific coastal areas, from where they moved eastward into the interior of North America and southward to South America. In recent years the balance of opinion has been tipping towards the second hypothesis. The two routes are not necessarily mutually exclusive, but the coastal route would have been much easier to traverse. The next question concerns timing and, again, two scenarios have emerged. Some archaeolo-

Remains of the giant short-faced bear, a laurasiathere that in life may have weighed up to 900 kilograms, have been found in Yukon and on the Prairies. This animal was the largest carnivorous land mammal to have lived in North America. The inset compares the size of a giant short-faced bear (right), a grizzly bear (centre), and a black bear (left).
GIANT SHORT-FACED BEAR BY GEORGE "RINALDINO" TEICHMANN, COPYRIGHT 1999.

This Clovis point from Alberta is about 7 centimetres long and made from a honey-brown, translucent chert from North Dakota that was traded widely across the northern Western Interior Plains. The whitish coating is due to weathering when the point was buried in soil. Clovis points are said to be fluted because chips of rock have been removed from the base to create a channel (or flute) for attaching the point to a spear.
ARCHAEOLOGY, ROYAL ALBERTA MUSEUM.

gists propose that people arrived in North America towards the end of the last glaciation, around 15,000 years ago, following their arrival in far northeastern Siberia. That northeastern Siberia was not occupied until about 16,000 years ago would seem to constrain the timing of migration into North America. Convincing evidence comes from Bluefish Caves in northern Yukon, from where small stone artifacts (burins) have been dated to about 15,000 years ago. At the time that the caves were occupied, eastern Beringia was isolated from the rest of North America by ice sheets, an isolation that lasted until about 13,400 years ago.

Other archaeologists have proposed that humans arrived in North America much earlier—most speculatively and controversially as early as 45,000 years ago. Radiocarbon dates indicate human occupation of a site known as Monte Verde in south-central Chile at about 14,500 years ago. Some argue that if people were living in South America at that time, they had to be living in North America by about 24,000 to 18,000 years ago. However, many experts have found it difficult to understand how people could have lived in the region south of the ice sheets for thousands of years without leaving more evidence than has so far been found. Recently, a few sites, including one in Texas that dates to about 15,500 years ago, offer tantalizing evidence that people may have lived in North America for several thousand years longer than previously thought.

South of the ice sheets, animals followed the retreating ice margins and plants colonized newly deglaciated landscapes. Much of our understanding of postglacial plant and animal communities is from pollen and other microfossils recovered from sediment cores collected from ponds and wetlands. The record shows an early period of tundra vegetation, followed in southern Canada by establishment of open woodlands and finally deciduous forests, with grasslands on the Prairies. By about 13,400 years ago, groups of hunter-gatherers were also spreading into these new environments across North America south of the glaciers, moving northward and eastward as ice melted. Evidence for this migration comes from the distinctive stone spear tips known as Clovis points. From that time on, there is abundant evidence for human occupation in Canada.

Based on faunal remains found at archaeological sites, we know that the main prey animals of the first hunters in what is now Canada were bison. But the earliest populations may also have hunted remnants of the Pleistocene's large mammals such as mammoths and horses. By about 11,500 years ago, many large mammals, including mammoths, mastodons, camels, and horses, had disappeared from North America. Some, such as mastodons, became extinct, whereas others, such as horses and camels, survived elsewhere. Aboriginal people still had other large animals to hunt, such as bison, mountain sheep, caribou, and deer.

Several theories have been advanced to explain the extinctions. As the glaciers receded, the climate was changing dramatically, with a corresponding effect on

Teeth in the lower jaw of a 75,000-year-old mastodon found near Milford, Nova Scotia. HEINZ WIELE, COURTESY OF THE ATLANTIC GEOSCIENCE SOCIETY; SPECIMEN COURTESY OF THE NOVA SCOTIA MUSEUM.

Pleistocene mammoth tooth found on the bank of the Porcupine River about 30 kilometres west of Old Crow, Yukon. BRYAN RUTLEY, SPECIMEN PROVIDED BY ALEJANDRA DUK-RODKIN.

vegetation. These changes would have placed stress on animals. But the same species had previously survived several glaciations in North America by migrating. So the extinctions were probably triggered by other factors, perhaps the most likely being the presence of humans.

Into the Holocene

The official boundary between the Pleistocene and Holocene has been set at 11,700 years ago. Although substantial ice remained in parts of Canada well into the Holocene, by the start of the epoch ice had disappeared from the Prairies, and the southern margin of the Laurentide Ice Sheet lay roughly along the edge of the Canadian Shield. In the Cordillera, glaciers were probably no more extensive than today, and a separate ice cap persisted over Newfoundland. Between 11,700 and 9,000 years ago, the Laurentide Ice Sheet shrank markedly, and the sea entered Hudson Strait. The unstable southern margin of the Laurentide Ice Sheet continued to be flanked by deep lakes. These water bodies included a much-reduced Glacial Lake Agassiz and Glacial Lake Ojibway, the latter flooding parts of northern Ontario as it drained southward via the Ottawa River.

During deglaciation, parts of the ice margin periodically re-advanced or surged into adjacent glacial lakes, forming arc-shaped moraine systems. Immense gravelly interlobate moraines, such as the Harricana Moraine, north of Matagami, Quebec, and the Etawney-North Knife Moraine in northern Manitoba, developed between ice lobes. About 8,600 years ago, the sea entered Hudson Bay, causing the rapid shrinking of the central part of the Laurentide Ice Sheet and the final draining of the remnants of Glacial Lake Agassiz. Seawater flooded into glacially depressed areas surrounding Hudson Bay, creating a salt-water body known as the Tyrrell Sea. About 1,000 years later, the sea invaded Foxe Basin. Final remnants of the Laurentide Ice Sheet disappeared from Keewatin about 6,800 years ago and from northern Quebec about 5,800 years ago. Still farther north, deglaciation of the Innuitian Ice Sheet began in the western Arctic about 14,500 years ago and progressed eastward. By about 9,000 years ago, only small local ice caps on the Arctic Islands remained, and are preserved to this day as the Barnes and Penny ice caps on Baffin Island, the Agassiz Ice Cap on Ellesmere Island, and the Devon Island Ice Cap.

Ups and Downs

At the margins of the continent, sea had been displaced as ice sheets grew beyond what had previously been land. So as the ice sheets retreated, it was natural for salt-water bodies such as the Tyrrell Sea to reinvade. But the story turns out to be far more complicated than simple retreat and advance. For starters, about 35 million cubic kilometres of water were locked up in the Laurentide Ice Sheet at its maximum, contributing to a global sea-level fall of about 130 metres and consequent exposure of vast areas of continental shelves around the world. About half of the subsequent rise in global sea level, which happened between 20,000 and 6,000 years ago, was due to melting of the Laurentide, Cordilleran, and Innuitian ice sheets. Ice is still locked up in the Greenland and Antarctic ice sheets, and if these were to melt, global sea level would rise by about 75 metres, drowning the coastal lowlands where much of the world's population lives.

Ice sheets depressed the lithosphere on which they lay, squeezing out viscous asthenosphere, much as a boat displaces water. (Recall that the lithosphere's ability to depress and rebound in response to the waxing and waning of weighty ice sheets and mountain ranges is known as isostasy.) Those areas depressed the most were beneath the centre of the ice sheets, where ice was thickest. When the ice load was removed during deglaciation, the lithosphere rebounded to its former position, regionally causing the sea to fall relative to the land and shorelines to recede. Early postglacial beaches around Hudson Bay, near the centre of loading, now lie more than 200 metres above sea level: these are classic examples of raised beaches. Tide-gauge records at Churchill in Manitoba indicate that the land continues to rebound there at a rate of 90 centimetres per century.

On the fringes of the ice sheets in regions such as the Maritimes, the lithosphere was not greatly depressed, so the pattern of sea-level change was different. In these areas sea level rose during deglaciation because the rise due to melting of ice sheets was greater than isostatic rebound. The rising sea flooded the once-exposed continental shelves. Sea level continues to rise at a rate of about 20 centimetres per century in these areas, contributing to rapid erosion of coastal bluffs. On Canada's west coast, much of the continental shelf bordering Haida Gwaii was above sea level as recently as 11,000 years ago, is now up to 130 metres below sea level—possibly concealing key archaeological sites that would shed light on the movement of people into North America.

Isostatic rebound is more gradual than the melting of ice sheets, and some interior lowlands can be temporarily inundated by the sea in the interval between melting and sufficient rebound to raise them above sea level. For example, a marine arm known as the Champlain Sea, which existed from 13,000 to 10,000 years ago, extended inland along the St. Lawrence Valley and up the Ottawa Valley as far as Pembroke, Ontario. Fine-grained sediments deposited in the Champlain Sea, known as Leda Clay, form the fertile farmland of the St. Lawrence Lowlands and Ottawa Valley. The deposits contain marine fossils, including mollusk shells and bones of fish, seals, whales, and walruses. Beaches of the former Champlain Sea in the Ottawa area are now 180 metres above sea level.

In Retreat

The retreat of the ice sheets at the end of the last glaciation heralded the present warm interval, a "long summer" during which human civilizations have arisen. Although the continental ice sheets were disappearing, climate continued to fluctuate, and short-term cycles of climatic cooling and warming have occurred throughout the Holocene, confirming that climate change is an ongoing reality with or without human involvement. Cool periods—marked by advancing alpine glaciers, proliferation of icebergs in the North Atlantic, and brief reversions to tundra biomes—alternated with warmer intervals, the warmest being the Hypsithermal, which lasted from about 9,500 to 7,000 years ago. During the Hypsithermal interval, average annual temperatures in North America were about 2°C higher than today, the boreal forest extended up to 100 kilometres north of its present, and some species of mollusks, seals, and whales lived farther north than they do currently.

For most of the postglacial interval, grasslands have prevailed on Canada's Western Interior Plains, the northern part of the vast North American Great Plains. Archaeological sites and artifacts suggest that people have been living on the Plains for more than 13,000 years, and moved into Canadian areas from the south. Hunting tools, including Clovis points, are widespread and denote a similar hunting culture across a vast area, with bison a primary target. Humans were hunting bison in the vicinity of Charlie Lake Cave near Fort St. John in northeastern British Columbia by about 12,600 years ago. Since bison are herd animals, hunters developed various forms of "drives", in which the animals were coaxed into locations from which they couldn't escape. Sites where bison were driven over

Bison were the main prey of the first hunters in what is now Canada. This aerial view shows wood bison in Wood Buffalo National Park of Canada in northern Alberta and southern Northwest Territories. W. LYNCH, COPYRIGHT PARKS CANADA.

Aboriginal hunters used to drive bison over the cliff at Head-Smashed-In Buffalo Jump, Alberta. HANS WIELENS.

a cliff are known as jumps. The best jumps were used again and again over hundreds or even thousands of years. An example is Head-Smashed-In Buffalo Jump near Fort Macleod in southwestern Alberta, a UNESCO World Heritage Site. The oldest bone deposits at this jump date from about 6,400 years ago, and it was last used in the middle of the nineteenth century. Bone piles at many jumps were so extensive that early European settlers based a fertilizer-producing industry on them.

Early archaeological sites have been found in most parts of southern Canada, showing how quickly humans occupied the newly available land. Many camp sites have been found on beaches of former glacial lakes. One example is the Fisher site in southern Ontario, near the southern end of Georgian Bay. This site was occupied about 12,600 years ago and is associated with a former shoreline of Glacial Lake Algonquin. Humans had settled in Quebec by 12,000 years ago in the Lac Mégantic area in the Appalachians, and along the north shore of the Gaspé Peninsula between 11,000 and 9,000 years ago.

In the Maritimes, archaeological evidence of small campsites with stone tools, including fluted points, at Debert in Nova Scotia shows that hunters there pursued caribou about 12,500 years ago. Farther north, evidence exists of human occupation along the southern coast of Labrador from about 10,200 to 3,200 years ago. At L'Anse-Amour on Labrador's Strait of Belle Isle coast in the 1970s, remains were found of a burial mound that contained the red-ochre-stained skeleton of a 12- to 13-year-old youth, interred with several artifacts. Dates obtained from associated charcoal show that the burial is about 8,300 to 7,800 years old. Burial mounds are not usually associated with hunter-gatherer societies, so this site is truly unusual. Although not spectacular to look at, it is the oldest such burial in North America.

Arctic Cultures

Most of Arctic Canada and Greenland were uninhabited by humans until about 4,500 years ago, when a people known as Paleo-Eskimos arrived from Siberia. Their ancestors were probably forest hunters, but by following herds of reindeer north of the tree line each year, some discovered the rich marine mammal life of the Arctic seas. Over time, these people migrated eastwards across the frozen waters of the Bering Sea to Alaska, Arctic Canada, and Greenland. They apparently lived in skin tents, with a central box-hearth and a corridor, both of stone. Remains of such stone structures are common on beaches, some now as much as 50 metres above sea level due to isostatic rebound. Paleo-Eskimo sites are remarkable time capsules and record the presence of small groups of people who, hundreds of years ago, camped together for weeks at a single location. These people were true pioneers in a severe environment.

Paleo-Eskimos occupied Arctic Canada for 3,000 years or more before disappearing 900 to 500 years ago. Around that time, ancestors of modern Inuit migrated eastward, after moving from Alaska into Arctic Canada about 900 years ago. In Alaska, the Inuit had been living off rich marine resources for the previous 1,000 years, using their small hunting kayaks and much larger umiaks. Having umiaks made moving camp easy. Umiaks also provided an ideal platform from which to hunt whales, a food resource that enabled the Inuit to live in relatively large, semi-permanent coastal villages.

The rapid expansion of Inuit across Arctic Canada and into Greenland may be related to the Medieval Warm Period

Thule Inuit archaeological site near Resolute, Cornwallis Island, Nunavut. DAVID ASHE.

(described in the next section). But as the spread of the Inuit occurred several centuries after the onset of warmer conditions, the expansion was more likely due to a search for metal. The Alaskan Inuit had obtained small amounts of iron from Siberian trade, but news of metal in large meteorites may have drawn Inuit populations to northern Ellesmere Island and Greenland. Although these regions were much farther north than the lands of their Alaskan forebears, meat was plentiful for the Thule Inuit, who were the ancestors of all modern Inuit. The Thule Inuit built houses of stone and whalebone, and the remains of their winter villages still dot the landscape over much of Arctic Canada.

Seeds, Sails, and Igloos

Aboriginal peoples, who obtained most of their food by hunting, were familiar with the benefits of plants for food, clothing, household utensils, construction materials, and other purposes. The gathering of plant material is but a small step to the introduction of agricultural techniques. Scholars used to think that agriculture in North America developed by about 8,000 years ago in what is now Mexico, and that it was triggered by the cultivation of maize (corn), which gradually spread northward and eastward. But recent studies indicate that agriculture developed independently in eastern North America and was based on domestication of local plants between 5,800 and 2,600 years ago. Records of corn in southern Ontario date back to 1,600 years ago. Favourable climatic conditions were also a critical factor during the Medieval Warm Period from 1000 to 1300 CE. This relatively warm episode may explain why the Vikings were able to establish settlements in Greenland and at L'Anse aux Meadows in Newfoundland.

The ensuing Little Ice Age lasted until the end of the nineteenth century and was highlighted by periods of cold, wet weather that lasted for decades and had profound effects. Viking settlements were abandoned, glaciers advanced over hamlets in the Swiss Alps, rivers froze, crops withered causing famine in Europe, and winters in Canada were much worse than today. Stories of skaters on the frozen River Thames in England are matched by similar scenes on a frozen Bedford Basin in Halifax Harbour—both places where skaters do not venture these days. During the Little Ice Age, the snow line dropped 100 metres in northern Canada and valley glaciers in the Cordillera expanded beyond their present limits, in some cases by several kilometres.

In the Arctic, the cooler conditions of the Little Ice Age may have led to the abandonment of Thule villages, beginning around the sixteenth century. An increase in sea ice limited whale hunting and the Inuit retreated to areas south of Lancaster and Viscount Melville sounds, where they adapted to a more mobile lifestyle based on increased hunting of seal, caribou, and fish. The new lifestyle and dearth of whalebones created the need for a new form of dwelling, and it was perhaps at this stage that the igloo came into use. The thicker pack ice also thwarted British attempts to find a northwest passage.

Records show that climate has generally warmed since the Little Ice Age. Observations of rapid glacier retreat, melting of permafrost, and sea-level rise are all of concern. But has recent warming responsible for these changes been caused by natural processes, human activities, or a combination of the two? We'll leave a discussion of this hot topic to the following box.

The Little Ice Age was a bad time to go exploring. Watercolour by Samuel Cresswell showing Robert McClure's ship *Investigator* locked (nipped) into heavy sea ice in Mercy Bay, Banks Island, Northwest Territories, from September 1851 until April 1853. The area is presently ice-free in summer. NATIONAL ARCHIVES OF CANADA PHOTO C-016105, PUBLIC DOMAIN.

BOX 9 • BLOWING HOT AND COLD

Storm over the Bessborough Hotel, Saskatoon, Saskatchewan. ROB FENSOME.

Forged by Climate

During Earth's history, one of the few constants has been climate change. So it is no surprise that we have experienced changes in climate over the past 10,000 years. The big question is how much have the recent changes been caused by humans? Most importantly, we must try to predict future climate changes, no matter what their cause, and prepare to adapt. Our survival as a species may depend on it.

What is climate and how does it differ from weather? Climate refers to the characteristic atmospheric conditions of a region over years or decades, and involves parameters such as average temperature and rainfall. For example, we think of the climate of Vancouver as being mild and humid, with lots of rain, especially during fall and winter. Weather describes the atmospheric conditions over much shorter periods, typically hours to a few weeks. We may visit Vancouver even in winter and enjoy bright, sunny, dry weather; conditions such as this in winter are by no means typical, but they are possible. Climate, though, is much more than a story of averages. Two locales may have the same average annual temperature but very different climates. Halifax in Nova Scotia and Kelowna in British Columbia have nearly the same average annual temperature of 7 to 8°C. Yet the annual range of temperatures in Halifax is much less than in Kelowna, whereas Halifax receives much more precipitation.

Canada's national identity has been moulded by climate. The long, cold winters in most parts of the nation have nurtured a passion for skiing, skating, hockey, and, of course, complaining about the cold. Seasonal temperature fluctuations can be startling, particularly inland, where summer highs may be 40 to 60°C warmer than winter lows. Precipitation shows similar extremes. Some areas receive more than a metre of water-equivalent precipitation per year, whereas other regions receive less than one third of that amount.

Over the eons, we have been fortunate that the Earth's climate has varied within a narrow enough range for water to have stayed in a liquid form, allowing animals and plants to evolve. In this sense, our planet is a marvel of self-regulation. Climate is controlled by myriad processes that collectively maintain relatively stable conditions, comparable to the way in which the temperature of the human body is maintained within a narrow range. This climatic stability is dependent on negative feedback processes, which operate to restore a perturbed system (one that has been drastically changed) to its original state. Negative feedback processes contrast with positive feedback processes that amplify perturbations and can cause runaway changes devastating to life. The Earth's self-regulating behaviour led British scientist James Lovelock to conceive of our planet as "Gaia", a living entity.

A Restless Star

In this box we look at some of the factors controlling climate, delving into aspects not dealt with elsewhere in the book, and briefly recalling those already covered. Our Sun is key to the Earth's climate. The surface temperature of the Earth is determined by the amount of solar radiation it receives, modulated by two factors: first, the albedo, or the amount of sunlight that is reflected away from the surface and therefore not absorbed; and second, the degree to which the atmosphere retains heat.

L'Anse aux Meadows, Newfoundland, which was settled by Vikings during the Medieval Warm Period. LYNDA DREDGE.

Most solar radiation that reaches the Earth is in the ultraviolet part of the energy spectrum and has a relatively short wavelength, but some of it can be infrared. About two-thirds of the solar radiation that arrives at the fringe of Earth's atmosphere penetrates to the planet's surface, and much of that is absorbed by water, soil, rocks, and other earth materials. Some of the solar energy that reaches the surface, however, returns to the atmosphere as long-wave infrared radiation, essentially heat. Much of this re-radiated radiation goes back into outer space, but some is absorbed by water vapour, carbon dioxide, methane, and other gases in the atmosphere. The atmosphere emits this radiation in a haphazard fashion, with some of it being scattered back to the Earth and increasing temperatures in the lower atmosphere; this is the so-called greenhouse effect. Without the greenhouse effect, the surface of the Earth would be at least 33°C cooler than it is. Surface water would be frozen, and life would be very limited.

Milankovitch cycles (Chapter 11) show how variations in the tilt of the Earth's rotational axis and its path around the Sun affect how solar radiation is distributed on our planet over thousands to hundreds of thousands of years. These cycles do not change the total amount of solar radiation received by the Earth; rather they change the seasonal distribution of radiation at different places on the planet, which in turn can affect climate.

The Sun is a restless star and its changing energy output may affect our climate. We can estimate changes in solar activity over the past 10,000 years using carbon-14, an isotope we explored in relation to radiometric dating in Chapter 3. Carbon-14 is produced in the atmosphere by collisions between cosmic rays (the intensity of which varies with the Sun's activity) and nitrogen-14 atoms. By measuring the concentration of carbon-14 in tree rings of known ages, we can get an estimate of recent solar output changes.

Records reveal that the Medieval Warm Period (1000 to 1300 CE) was a time of above-average solar radiation. Solar output was at a minimum during the glacial advances of the seventeenth to nineteenth centuries. It appears, therefore, that variability of energy output from the Sun can partially explain climatic variability. But the effect is small, the inferred difference between solar output during the Medieval Warm Period and the Little Ice Age being only about one quarter of 1 percent.

In this aerial view of the southern Coast Mountains of British Columbia, the linear break in vegetation (trimline) delineates the margin of the Icemaker Glacier at the end of the nineteenth century. The glacier has thinned and retreated considerably since then. JOHN CLAGUE.

Terrestrial Controls

In Box 4, we explored the interchange of carbon dioxide between the atmosphere, biosphere, geosphere, and hydrosphere and the critical role of this gas in the carbon cycle. We are aware that a buildup of carbon dioxide, second only to water vapour as a greenhouse gas, leads to global warming. Fortunately, a buildup of carbon dioxide to levels at which life could not survive has been prevented by exchanges among the atmosphere, biosphere, geosphere, and hydrosphere. When carbon dioxide levels are lower, as during many episodes of mountain building, climates are cooler (Box 4). And when mountain ranges are few and continents are less elevated, carbon dioxide concentrations are higher and climates warmer.

Volcanic eruptions affect climates in two ways—by venting of carbon dioxide and through aerosols and particles hurled into the atmosphere. Volcanic particles block incoming sunlight, causing cooling. One of the largest explosive volcanic eruptions in the last 10,000 years was that of Mount Tambora in Indonesia in 1815. The eruption caused months of cool weather, with Europe and North America suffering crop failures and famine during 1816, which came to be remembered as "the year without a summer".

The distribution of continents, oceans, and mountains has a major impact on climates. Especially important is the amount of land area at higher latitudes, where continental glaciers form. If continents cover the poles, such as Antarctica does today, chances are greater that a continental ice sheet will form on them. Mountain ranges and plateaus at mid- and high latitudes are particularly favourable areas

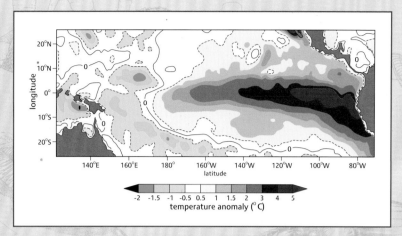

Sea-surface temperature anomalies (in degrees Celsius) in the equatorial Pacific during a strong El Niño event (1997–1998) show how temperatures vary from the long-term average temperature. The anomalies are based on a 7-day average around September 17, 1997. The bar beneath the map shows the temperature differences for the 7-day interval compared to long-term averages. Green and blues represent colder water temperatures than average and yellows through reds to browns represent warmer water temperatures than average. COURTESY OF THE US NATIONAL OCEANIC AND ATMOSPHERIC ADMINISTRATION (NOAA), CLIMATE PREDICTION CENTER/NECP/NWS.

for glaciers to form and grow. If sufficiently large, as for example the Himalayas and Tibetan Plateau are today, they can alter atmospheric circulation. When oceans lie over the poles and the global conveyor belt of oceanic waters is active, heat will be more efficiently transferred from the Equator to the poles. Today, warm ocean currents can neither approach the South Pole, because of Antarctica and the Antarctic Circumpolar Current that encircles it, nor easily enter the mainly landlocked Arctic Ocean. So present-day geography is ideal for glacial conditions.

Climate change over intervals of just a few years is caused by fluctuations in the global conveyor-belt system of ocean-water circulation. A good example is El Niño. (The term El Niño refers to a pattern of ocean circulation in the tropical Pacific Ocean involving the appearance of warm water off the coast of Peru and Ecuador. Its atmospheric component is the Southern Oscillation, which is characterized by a decrease in the strength of the trade winds. For simplicity we'll refer to the entire phenomenon as El Niño.) El Niño dramatically illustrates how changes in atmospheric and ocean circulation can affect the climate of an entire hemisphere over a time scale as short as one year. For example, El Niño is accompanied by drought in parts of Australia, Africa, South America, and southeastern Asia, and wet weather and flooding in parts of western North America. An El Niño event probably begins with a slight drop in the intensity of the trade winds, which flow westward, allowing warm water in the western equatorial Pacific Ocean to flow eastward. The eastward flow further reduces the strength of the trade winds. The slackening of the trade winds in turn causes more warm water to move eastward, so that surface waters across the entire equatorial Pacific from Australia to the west coast of South America become significantly warmer than normal. A strong El Niño event in 1997–1998 contributed to hurricanes in the eastern Pacific, floods and landslides along the California coast, and drought and fires in the southwestern United States. Drought conditions were so bad in Australia that the event is remembered as "The Big Dry".

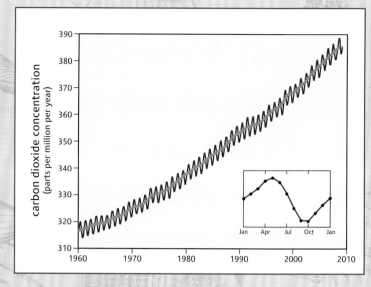

Concentrations of atmospheric carbon dioxide for the years 1960 to 2010 at Mauna Loa in Hawaii. The blue line shows the long-term trend; the red line reflects this trend also, but in addition shows shorter-term annual variations. The insert shows the variability in concentrations of atmospheric carbon dioxide during a single year. IMAGE CREATED BY ROBERT ROHDE FROM DATA PUBLISHED BY THE US NATIONAL OCEANIC AND ATMOSPHERIC ADMINISTRATION (NOAA).

The Human Angle

Although climate change has occurred throughout geological time as a result of many natural processes, most scientists now consider human activity to be an additional, perhaps even dominant, factor. How much are we affecting climate, and what are and will be the effects? These are the questions that generate headlines on a regular basis.

Today, the main concern is the documented increase in greenhouse gases and especially carbon dioxide, which accounts for 60 percent of the anthropogenic greenhouse effect. The increase results from the burning of oil, gas, and coal, and from deforestation and cement production. But what evidence is there for increased carbon dioxide in the atmosphere? The earliest measurable data come from some Antarctic and Greenland ice cores, which contain air bubbles trapped at the time the ice formed during the Ice Age. Measurements of gases in these bubbles indicate that atmospheric concentrations of carbon dioxide have ranged from a little less than 200 to about 300

Storm waves pounding the shore of West Vancouver, British Columbia. BOB TURNER.

Damage at Sambro, Nova Scotia, caused by Hurricane Juan in September 2003. BOB TAYLOR.

parts per million over the past 800,000 years. Ice formed during interglaciations has the highest levels, and the lowest levels are recorded in ice formed during glaciations. The pattern has changed with the widespread burning of fossil fuels over the past several hundred years. Data indicate that at the beginning of the Industrial Revolution in the late eighteenth century, carbon dioxide concentration in our atmosphere was about 280 parts per million. From 1958, when measurements from near the top of Mauna Loa in Hawaii began to be recorded, to 2013, the number rose from 315 to 393 parts per million. Even more alarmingly, values are forecast to be at least 450 parts per million by 2050, almost twice pre-industrial levels.

Computer models of atmospheric processes indicate that such high levels will drive average global temperature up by between 1.4 and 5.8°C by the end of this century. A rise of more than 2°C will likely be devastating in many regions. It will stress plant and animal species by altering their habitats, accelerate desertification of arid lands, increase the frequency of severe droughts and wildfire, and shift the location of agricultural zones. The impact on the Canadian landscape will be dramatic. Trees will invade alpine meadows in the mountains, and boreal forest is likely to replace the southern fringe of tundra plant communities in the Arctic. Grassland may expand at the expense of forest in southern British Columbia and on the Prairies.

Global warming may also increase the frequency and intensity of violent storms, because warmer oceans feed more energy into hurricanes and typhoons. Warmer oceans would increase the frequency of hurricanes that maintain their strength as far north as Atlantic Canada and, perhaps,

Tornadoes are rapidly rotating columns of air that can leave a path of destruction on the ground. This photo shows a category F5 tornado that struck Elie, Manitoba, in June 2007. JUSTIN HOBSON.

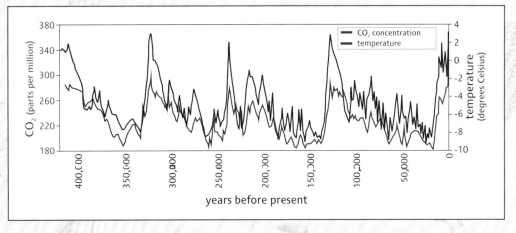

Changes in carbon-dioxide (CO_2) concentrations in air bubbles trapped in the Vostok ice core collected from the Antarctic Ice Sheet. The plot shows the close correlation between carbon-dioxide concentrations and temperature. The latter is inferred from the ratios of two isotopes of oxygen (oxygen-18 and oxygen-16) in the cored ice. Notice how carbon dioxide concentration rises vertically at the right end of the graph. The increase appears vertical because of the long time scale, but it actually occurred over the past 150 years, which corresponds to the modern industrial age. There has not been a corresponding increase in temperature during this period, probably because oceans can store heat and thus delay the increase in atmospheric temperatures. ADAPTED FROM DAVIESAND.COM/CHOICES/PRECAUTIONARY_PLANNING/NEW_DATA/.

Damage by a tornado on August 20, 2005, at Ross Farm, north of Fergus, Ontario. ALAN MORGAN.

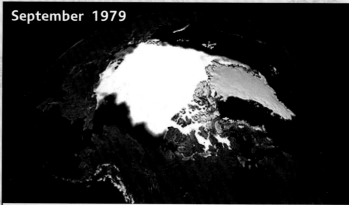

Change in Arctic sea ice cover between 1979 and 2007. Perennial sea ice decreased at a rate of 9 percent over this period. COURTESY OF THE US NATIONAL OCEANIC AND ATMOSPHERIC ADMINISTRATION (NOAA).

trigger severe storms in the north Pacific that would strike the British Columbia coast.

Climate warming will also drastically shrink glaciers. In the mountains of western Canada, glacier cover decreased by one-quarter to one-third during the twentieth century, and similar or larger reductions are likely by the end of this century. Some mountain ranges with only small glaciers will probably become ice-free in the future. During the summer months in watersheds where glaciers disappear, stream discharge will decrease and warmer water will have an adverse effect on fish and other aquatic organisms. Reservoirs may receive less inflow at these times, reducing the amount of storage available for hydroelectric power generation and domestic consumption.

A related issue is the thawing of permafrost. The zone of discontinuous permafrost, which extends through southern Yukon and the northern parts of most provinces, will thaw completely if average air temperatures continue to rise. Loss of permafrost and a seasonal increase in the thickness of the unfrozen surficial layer of soil (the active layer) may damage buildings, roads, and rail lines. It will also increase the amount of methane and carbon dioxide that is released into the atmosphere through plant decay in the Arctic, contributing further to warming.

One of the most serious effects of global climate warming is sea-level rise. Over the past century, sea level around the world has risen at an average rate of 2 millimetres per year, but has now increased to almost 3 millimetres per year. Although this rise may seem small, especially when compared to past changes, it totalled at least 20 centimetres over the twentieth century. And the rate will accelerate if climate continues to warm. Scientists of the Intergovernmental Panel on Climate Change predict that global sea level will be at least several tens of centimetres higher at the end of this century; some argue that this is a conservative estimate because it does not take into account melting of ice in Greenland or Antarctica. Low-lying coastal areas will be flooded by the sea during storms and at high tides. Delta plains, tidal wetlands, and island nations such as the Maldives in the Indian Ocean are especially vulnerable. Low-lying areas in Canada that are at greatest risk from sea-level rise include the Fraser Delta in British Columbia, the Mackenzie Delta in the north, some shores along the St. Lawrence Estuary and eastern Nova Scotia, and shores in southern Hudson Bay. The Fraser Delta is home to a quarter of a million people and will have to be defended against this threat by raising and strengthening the surrounding dykes and by improving the pumping facilities that remove water behind the dykes when tides are high. Water supply in coastal areas will also be at risk, as rising salty water will contaminate groundwater.

On the positive side, global warming will make northern winters milder. Costs of heating homes and workplaces will be lower, although only if the price of energy does not increase dramatically. The winter savings may be offset, however, by the increased demand for air conditioning during hotter summers. Some marginal agricultural areas in Canada and eastern Europe may become more productive. Such shifts do not necessarily mean that prime agricultural zones will also move north. Production of grains and other foods depends as much on soil conditions as on climate. Uncertainty as to how climate change will affect agriculture is a major worry—the disruption of existing grain supplies would be catastrophic.

Although the above observations paint a bleak picture of changing climate, being aware of the link between greenhouse gases and temperature gives us the incentive to foster renewable energy and conservation programs and to reduce carbon dioxide and methane emissions. It also highlights the benefits of research into past climates.

PART 3 • WEALTH AND HEALTH

The flow over Kakabeka Falls in northern Ontario reminds us of the vital role water, in large part a geological resource, plays in our daily lives. ROB FENSOME

12 • FORGING A NATION

CHAPTER SUPPORTED BY A DONATION FROM ROBERT HORN

Aerial view of Flin Flon, Manitoba, a major centre for mining VMS deposits hosted by rocks of the Trans-Hudson Orogen. RON GARNETT / AIRSCAPES.CA.

Klondike Gold

Few events in the annals of mining have seized the public's imagination as much as Yukon's Klondike Gold Rush. The first line of *The Cremation of Sam McGee*, Robert Service's epic poem, captures the grandeur and the struggle for survival in a frozen, gold-strewn land: "There are strange things done in the midnight sun by the men who moil for gold…." It is a story of the lure of the North and of fortunes made and lost in a challenging environment, where winter temperatures hover around minus 30°C for weeks on end. The Klondike Gold Rush and other mining ventures symbolize the struggles and sacrifices that have helped to forge Canada as a nation.

The discovery of gold on Bonanza Creek by George Carmack, Skookum Jim Mason, and Tagish "Dawson" Charlie, following the advice of Canadian prospector Robert Henderson, set off one of the largest gold rushes in history. The year was 1896, but it was not until 1897, when the steamship *Excelsior* arrived in San Francisco carrying more than half-a-million-dollars worth of gold, that news of the discovery spread like wildfire. Thousands

Potential prospectors climb the trail to the Chilkoot Pass at the Alaska-British Columbia border on their way to the Klondike.

In Hot Water

A metal is a chemical element that conducts electricity and heat. Metals come in two types: base metals such as iron, lead, zinc, and copper, which easily corrode (that is, oxidize); and precious metals such as gold, platinum, and silver, which resist corrosion. The durability of precious metals explains why the gold mask of the Egyptian pharaoh Tutankhamun is so well preserved and retains its shiny yellow colour after more than 3,000 years.

In geological settings, metals occur in minerals. As with all minerals, metal-bearing minerals can be made up of a single element, for example native copper, or they can be made up of two or more elements. Minerals composed of two or three elements include many sulphides, halides, and oxides. Sulphates and carbonates contain three or more elements. Concentrations of one or more minerals that can be economically mined to yield metals are termed ores. Ores occur in ore bodies or ore deposits. Such concentrations develop through a great range of geological processes, in which heat and water are critical. Most ores are composed of multi-element minerals and so need to be refined in order to yield their metal resource.

Both heat and water are abundant at plate boundaries. Heat comes from the magmas associated with spreading ridges and magmatic arcs, raising temperatures in the enclosing crust to several hundred degrees. At these temperatures, hot water leaches elements such as copper, zinc, lead, silver, and gold from the

of people from all walks of life sold their possessions, bought provisions including picks and shovels, packed their bags, and booked passage on the many steamers heading north. Although steamers could travel directly to Dawson City by sea and up the Yukon River, this route was expensive and tickets were hard to get. Most "stampeders" booked passage to Skagway in Alaska, then tackled gruelling climbs with heavy loads through the snowy Chilkoot and White Mountain passes to Bennett Lake on the British Columbia-Yukon border. From there it was a sometimes hazardous 800-kilometre boat trip down the Yukon River.

When the exhausted stampeders reached Dawson City, they were greeted with the news that all the most promising claims had been staked two years earlier. Many headed home, but others stayed to become rich from the enterprises supporting the gold miners. By 1898, the population of Yukon soared to over 40,000 and Dawson City was the largest and most cosmopolitan Canadian city west of Winnipeg, boasting the latest Parisian fashions, restaurants serving exotic cuisine and vintage French wines, an opera house, several brothels, and gambling saloons. But by 1899, the Klondike was old news. Most prospectors returned home and the population declined gradually to around a thousand. We'll return to the Klondike later in the chapter to explore why gold occurs there. First, though, we need to explain how nature causes metals to collect in economically attractive concentrations.

A dense billowing plume of mineral-laden water (a black smoker) gushes from a hydrothermal vent on the Juan de Fuca Ridge off the coast of British Columbia. As the hydrothermal fluid meets the near-freezing seawater, minerals precipitate to form chimney structures. The vents are home to an unusual fauna, including tube worms shown in the foreground. VERENA TUNNICLIFFE.

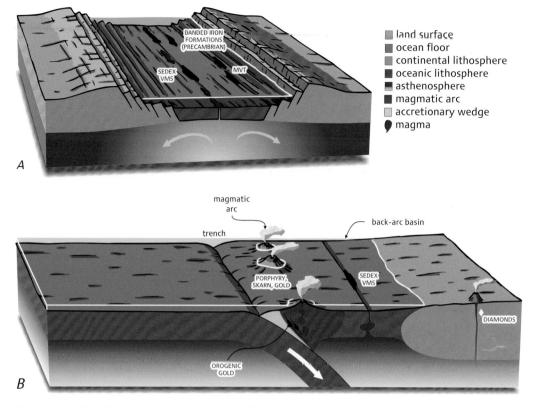

The tectonic settings of some major ore-deposit types broadly associated with divergent (*A*) and convergent (*B*) plate boundaries. VMS (volcanogenic massive sulphide) and SEDEX (sedimentary-exhalative) deposits are commonly associated with spreading-ridge and back-arc-basin settings. Porphyry and skarn deposits, including many gold deposits, are associated with convergent plate-tectonic settings. MVT (Mississippi Valley Type) deposits are associated with carbonate rocks and occur on passive margins and other settings conducive to carbonate deposition. Kimberlites and their contained diamonds are associated with stable and ancient continental crust away from plate boundaries.

VMS deposits are formed from hydrothermal fluids vented from black smokers commonly associated with oceanic spreading ridges. For the deposits to be preserved, the bottom waters must be anoxic.

surrounding rocks, resulting in a hot, metal-rich fluid. At spreading ridges, for example, such hydrothermal fluids rise along fractures toward the sea floor and, to compensate, cold seawater is drawn down to the depths. Thus, a circulation system driven by convection develops and, over millions of years, metals continue to be leached from the enclosing rock as long as temperatures are maintained and any metal remains in the rock.

When hot, metal-rich fluids reach the surface, they issue from hydrothermal vents; there, mixing of the fluids with cold seawater causes the metal sulphides to precipitate and form chimneys. The chimneys collapse from time to time, building up mounds that continue to grow as long as the vent remains active. Some of the hot fluid, still carrying sulphides, escapes from the chimney as a black plume into the sea, giving rise to the term black smoker. These features occur along oceanic spreading ridges such as the Mid-Atlantic Ridge and the Juan de Fuca Ridge off the coast of Vancouver Island.

In modern oceans, over 95 percent of the metal sulphides released at hydrothermal vents are rapidly destroyed by oxidation, and sulphide deposits on the sea floor are partly converted to iron-oxide. This is because the modern global circulation system ensures that bottom waters are well oxygenated. However, at times in the past (and today in the Black Sea), ocean waters lacked such a system. With reduced circulation, seas and oceans became stratified, with an anoxic layer at depth and an oxic layer above. Under anoxic bottom-water conditions, precipitated metal sulphides are preserved. Such deposits are commonly associated with black organic-rich mud on the ocean floor: the black mud reflects the absence or low concentration of bottom-dwelling organisms that consume organic matter raining down from the upper oxic layers.

Many important metal-sulphide ore deposits originated from this combination of leaching, venting, and anoxic ocean conditions. Some ores originated as mounds of sulphide rubble shed from collapsing black-smoker chimneys. Metal sulphide ores that formed largely in association with hydrothermal vents and volcanic rocks are called volcanogenic massive sulphide (VMS) deposits. (Such abbreviations are universally used in metal-mining circles and are arguably less intimidating than the full

Chalcopyrite-rich VMS deposit at a depth of 2,400 metres in the Kidd Creek Mine, Timmins, Ontario. The ore occurs in Archean rocks about 2,700 million years old. JACOB HANLEY.

Folded SEDEX deposit from the Sullivan Mine, Kimberley, British Columbia, originally deposited in the Mesoproterozoic Belt-Purcell Basin. The brown bands and associated thin cream layers are of sphalerite; the mostly thick light-blue-grey bands are galena interbedded with "wrinkled" argillite layers; and the green- to yellow-grey bands toward the top of the specimen are chert. COURTESY OF JOHN LYDON.

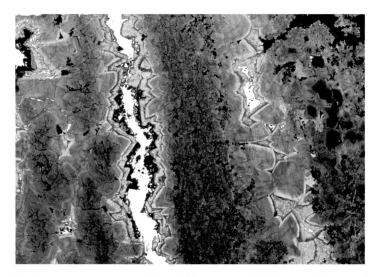

The browny-orange to clear crystals in this thin section are of sphalerite, and form part of a high-grade zinc ore in a middle Ordovician MVT deposit in the former Nanisivik Mine, Baffin Island, Nunavut. GRAHAM WILSON, SPECIMEN COURTESY OF ROSS SHERLOCK.

terms, so we will use them here.) Ancient VMS ores are composed mostly of iron sulphide, mainly pyrite, but other minerals such as chalcopyrite, sphalerite, and galena are also common. Other ores were precipitated from dense hydrothermal fluids that settled as pools on the sea floor, especially in depressions, and formed bedded sulphide deposits. Ore bodies dominated by such accumulations are known as sedimentary-exhalative (or SEDEX) deposits; they are the world's most important sources of lead, zinc, and barite.

Another type of ore deposit produced by hydrothermal fluids is known as the Mississippi Valley Type, or MVT, named after ores of this type discovered in the Mississippi Valley of the United States. MVT deposits form in carbonate rocks from hot, metal-rich fluids generated at depth and which migrate upward along faults until they mix with a sulphur-rich fluid and precipitate sulphide minerals in open spaces, thereby forming potential ore bodies. Ore deposits occur in carbonate rocks in part because carbonates are more chemically reactive than silica-based clastic rocks and tend to be very porous; they thus have open cavities in which ores can precipitate. MVT deposits are not tied to plate boundaries, but can form in areas where thick carbonate successions develop. The main economic metals extracted from MVT deposits are zinc, lead, and silver, all derived principally from the minerals sphalerite and galena.

VMS, SEDEX, and MVT deposits commonly occur in clusters of each type, which reflects the dominant setting at the time of formation. Important VMS clusters in Canada range in age from Archean to Paleozoic. The Abitibi mining district in northern Quebec and Ontario hosts the giant Horne and Kidd Creek deposits, formed within the Abitibi Greenstone Belt of the Archean Superior Craton. Deposits in the Flin Flon area of Manitoba and Saskatchewan are associated with the Paleoproterozoic Manikewan Ocean and Trans-Hudson Orogen. The mining districts around Bathurst in northern New Brunswick and Buchans in central Newfoundland are both associated with Paleozoic back-arc basins related to the demise of the Iapetus Ocean. On Vancouver Island, deposits in the Myra Falls-Buttle Lake district are related to Devonian magmatic arc activity on Wrangellia. Although fewer in number than VMS deposits, SEDEX deposits are generally an order of magnitude larger. Important SEDEX mining areas include the Sullivan district in the Mesoproterozoic Belt-Purcell succession of southeastern British Columbia, and the Selwyn Basin in Yukon and northeastern British Columbia, formed in rocks deposited near the early Paleozoic continental margin of Laurentia/Euramerica. Major MVT deposits in Canada are predominantly of Paleozoic age, and thus formed at a time when what was to become Canada was crossing the tropics and often covered by shallow

Zinc-lead ore from the Pine Point MVT deposit in the Devonian Presqu'ile Barrier reef system, Northwest Territories. This deposit consists of cream and brown sphalerite in a matrix of cream-coloured dolomite. WAYNE GOODFELLOW

continental seas, ideal conditions for carbonate deposition, with the resulting rocks ideal for the development of MVT deposits. Canadian MVT deposits include: the now-inactive Pine Point district on the south side of Great Slave Lake in the Northwest Territories, in Devonian-Carboniferous deposits within the Presqu'ile Barrier reef system; the middle Devonian deposits at Daniel's Harbour in Newfoundland; the late Devonian deposit at the Polaris Mine on Little Cornwallis Island, Nunavut; and the middle Ordovician deposit at Nanisivik on Baffin Island, also in Nunavut.

From Hot Magma

Ores developed directly from magma include so-called porphyry deposits, which are associated with the roots of former magmatic arcs. Their host rocks are intrusive, felsic to intermediate igneous rocks that may have porphyritic textures—typically larger feldspar or other silicate crystals in a fine-grained matrix of quartz and feldspar crystals. As magma rose in the crust, it cooled and started to crystallize. Any remaining liquid became increasingly enriched in particular elements, leading eventually to a residue of hot, metal-rich fluids. Chalcopyrite, molybdenite, and other sulphide minerals precipitated from these late-stage fluids to form potential ores. Porphyry deposits can occur in simple veins that fill fractures, in a complex network of veins referred to as a stockwork, in breccia pipes composed of rocks that have been broken into pieces, and as crystals disseminated throughout the igneous host rocks.

Porphyry deposits include some of the largest and richest copper, molybdenum, and gold deposits. Some porphyry deposits contain mainly copper with variable proportions of gold, others contain mainly molybdenum, and still others contain all three elements in commercial quantities. In Canada, the largest porphyry copper-molybdenum mine is the Highland Valley Mine in south-central British Columbia. The ores mined at Highland Valley formed in a magmatic arc on Quesnellia during the late Triassic, about 210 million years ago. Other deposits of about the same age occur elsewhere in Quesnellia and Stikinia, and in the Jurassic Bonanza Arc in Wrangellia, the latter now on Vancouver Island.

In some cases, the ore-forming fluids derived from a magma may move beyond the igneous intrusion to permeate surrounding rocks. Where the fluids move into carbonate rocks, chemical reactions occur between the fluids and the host material to form potential ores with a distinctive suite of minerals; these are known as skarn deposits. Where the fluids permeate stratified sedimentary rocks, so-called manto deposits may form. The fluids may also discharge from springs at the surface of volcanoes, forming chemical sedimentary rocks called sinters.

Large mafic and ultramafic igneous complexes can host significant deposits of nickel, copper, or platinum-group elements. (The platinum-group elements, usually abbreviated as PGE, are iridium, osmium, palladium, platinum, rhodium, and ruthenium.) These deposits form directly from magma by crystallization during intrusion and cooling. As magma ascends from the mantle through the lithosphere, it cools and leaches sulphur from the surrounding rocks. Droplets of nickel-, copper-, and PGE-rich sulphide liquid form and, because they are relatively dense, collect at the base of the magma body as a potential ore deposit. An example is the Voisey Bay deposit in Labrador. Sulphides of heavy metals, such as lead, zinc, and copper, may also be precipitated and concentrated at bottlenecks along the magma's flow path.

Nickel, copper, and PGE deposits can also occur at or near the base of thick sequences of komatiites, which are low in silica, potassium, and aluminum but high in magnesium. As we discovered in Chapter 5, komatiites formed as volcanic flows, dykes, and sills and are mostly of Archean age. Platinum and palladium are the most abundant PGEs associated with komatiites, but trace amounts of rhodium, as well as of gold, are common. Examples of komatiite-related ore deposits occur at Thompson in Manitoba, at the Raglan Mine in northern Quebec, and in the Muskox Intrusion in Nunavut.

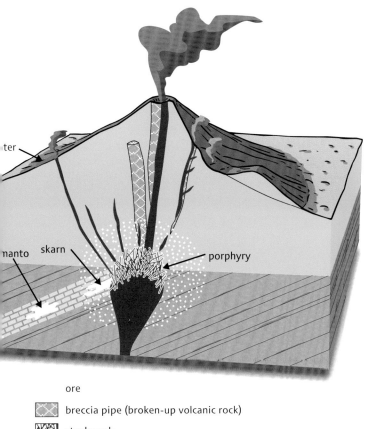

- ore
- breccia pipe (broken-up volcanic rock)
- stockwork
- zone rich in sulphide ore minerals
- intrusion and veins
- volcanic rocks (lavas and pyroclastics)
- carbonate-rich sedimentary rocks
- clastic sedimentary rocks

Porphyry and related ore deposits are generally associated with intermediate to felsic igneous activity. This figure shows how different types of ore bodies are related to such volcanic and intrusive settings.

Rocks within a single porphyry deposit can be varied. In this photo, the first of three on this page from the Mount Pleasant deposit in New Brunswick, granitic rock is cut by veins containing quartz, fluorite, arsenopyrite, and cassiterite, a source of tin. The granitic rock is of Devonian age. DAVID SINCLAIR.

In this example, arsenopyrite- and cassiterite-bearing veins cut fractured and altered granitic host rock. The veins were mined for tin. DAVID SINCLAIR.

This breccia from the Mount Pleasant deposit consists of fragments of granitic rock that have been altered to silica in a dark matrix of sphalerite and cassiterite, a source of zinc and tin. DAVID SINCLAIR.

CHAPTER 12 · FORGING A NATION

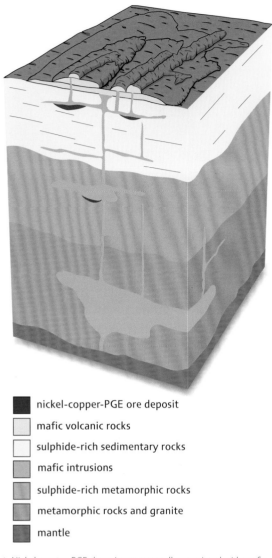

- ■ nickel-copper-PGE ore deposit
- □ mafic volcanic rocks
- □ sulphide-rich sedimentary rocks
- ■ mafic intrusions
- □ sulphide-rich metamorphic rocks
- ■ metamorphic rocks and granite
- ■ mantle

Nickel-copper-PGE deposits are generally associated with mafic igneous activity. This figure shows how different types of ore bodies are related to such volcanic and intrusive settings.

The Impact of Sudbury

It may come as a surprise that the largest and most economically important mining centre in Canada is the result of a single large extraterrestrial impact. The impact occurred about 1,850 million years ago in the vicinity of present-day Sudbury in northern Ontario. Ore deposits related to this impact constitute the largest-known concentration of nickel, copper, and PGE deposits on the planet. The first map of the Sudbury region, published in 1856, was by Alexander Murray, then working for the Geological Survey of Canada. However, the economic significance of the area was not appreciated until a blacksmith employed by the Canadian Pacific Railway discovered nickel sulphides at the current site of the Murray Mine in 1883. This find triggered one of the largest staking rushes in Canadian history.

That the Sudbury deposits are related to a large impact structure was not understood until 1964. Geologists now think that the Sudbury structure was created by an asteroid about 10 kilometres across, which plunged to the Earth at about 25 kilometres per second and on impact excavated a crater 200 to 250 kilometres in diameter (Box 11). Originally the structure would have been circular, but has been squashed to its modern oval shape by later tectonic compression. The heat of the impact formed a pool of molten rock or impact melt up to 2.5 kilometres deep that flooded over 60 square kilometres of the impact structure floor. The melt also intruded shattered host rocks and formed radial dykes that extend as far as 30 kilometres from the impact structure, and concentric dykes up to 14 kilometres around it. Covering the now-solidified melt layer is a bed of impact breccia about 3 kilometres thick, which accumulated soon after the event. The impact occurred on the southern

Breccia from below the ore and impact melt layers of the Sudbury impact structure, Ontario. GEORGE BARDEGGIA, COURTESY OF TOM MUIR, ONTARIO GEOLOGICAL SURVEY.

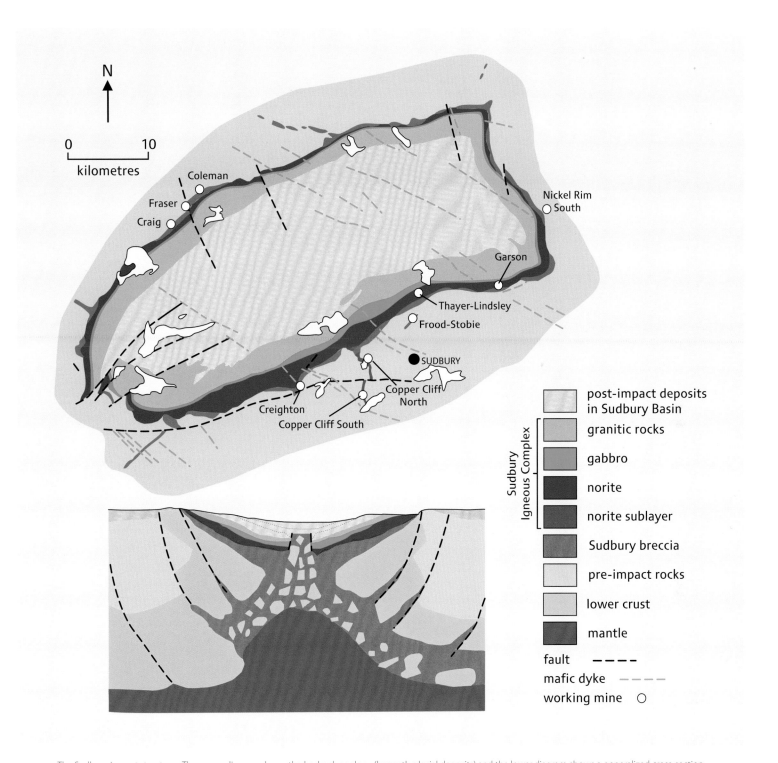

The Sudbury impact structure. The upper diagram shows the bedrock geology (beneath glacial deposits) and the lower diagram shows a generalized cross-section, with glacial deposits outside the impact structure included as post-impact deposits. UPPER DIAGRAM ADAPTED FROM ECKSTRAND AND HULBERT (2007).

flanks of the Superior Craton and left a depression, the Sudbury Basin, into which the sea flooded and began to fill with marine sediments, including muds and turbidites.

The magma generated by the impact was initially about 1,700°C and took more than 300,000 years to solidify and cool. As with many bodies of magma, crystallization of different minerals—and hence the formation of different rocks—occurred at different stages in the cooling process. First to crystallize were the mafic minerals, initially forming the ultramafic rock norite, followed by the mafic rock gabbro. Felsic minerals crystallized during the later cooling stages, producing an upper granitic layer. Collectively, this sequence of rocks is referred to as the Sudbury Igneous Complex.

Ore deposits in the Sudbury structure are of two main types. The first consists of lens-shaped, sulphide-rich, nickel-copper-PGE deposits. The minerals that form these ores settled in depressions at the base of the norite layer because they are heavier. This ore-bearing norite is known as the norite sublayer. Because it was the basal layer of the Sudbury Igneous Complex, the same material filled cracks, forming dykes that radiate out

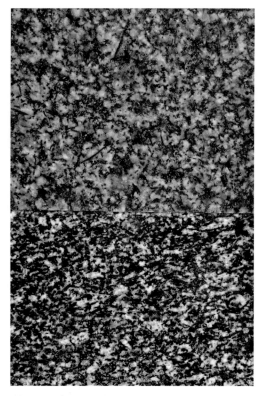

Detail of rocks derived from the upper, granitic melt (top) and from the lower, noritic melt (bottom), of the Sudbury impact structure, Ontario. GEORGE BARDEGGIA, COURTESY OF TOM MUIR, ONTARIO GEOLOGICAL SURVEY.

Ore containing copper and platinum-group elements from the Strathcona Mine in the Sudbury impact structure, Ontario. GEORGE BARDEGGIA, COURTESY OF TOM MUIR, ONTARIO GEOLOGICAL SURVEY.

Aerial view of a dredge used to mine placer gold, Bonanza Creek, Klondike, Yukon. WAYNE GOODFELLOW.

Shatter cones (Box 11) developed in Paleoproterozoic (Huronian) metasandstone, Sudbury impact structure, Ontario. COURTESY OF ONTARIO GEOLOGICAL SURVEY.

from the structure. The second type of ore deposit—mainly sulphides rich in zinc, copper, lead, silver, and gold—occurs in volcanic rocks and marine carbonates that overlie the impact sequence. These ores formed when hydrothermal fluids, their circulation driven by the underlying impact melt, infiltrated along faults and vented upward into the Sudbury Basin. As was probably the case for all seas at the time, the bottom waters were anoxic, so conditions were right for the formation of VMS-type deposits.

Gold and Iron

So far, we've focussed mainly on the processes that concentrate particular types of ore. In this section, we highlight two very different metals: gold and iron. Klondike gold is of a type of deposit known as a placer. Such deposits form when flakes, small grains, and nuggets of a relatively heavy mineral are eroded from bedrock and become concentrated in sand or gravel in rivers and streams. As currents begin to slow down, heavier minerals and rock fragments settle to the stream bottom while lighter materials are carried farther downstream. Gold is commonly concentrated as a placer deposit because it is highly durable and survives extreme weathering. Placer gold is preserved in the Klondike partly because the region was unglaciated during the Ice Age and thus escaped the deep scouring commonly caused by the movement of large ice sheets.

The simplest technique for mining placer gold is by using a bowl, or gold pan, in which water is swirled to separate the gold grains from lighter sediment—a process known as panning. Another technique involves the use of

Quartz vein with gold in the Musselwhite Mine of northern Ontario. This mine is within a greenstone belt in the Superior Craton. The ore occurs as veins and as a quartz-pyrrhotite-rich replacement within folded banded-iron formations. BENOÎT DUBÉ.

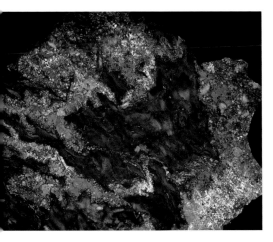

Gold in conglomerate from the Dome Mine, Ontario. The weight of this specimen is 61.8 kilograms, of which about 20 percent is gold. IGOR BILOT, SPECIMEN COURTESY OF NATURAL RESOURCES CANADA (NATIONAL MINERAL COLLECTION NO. 10003).

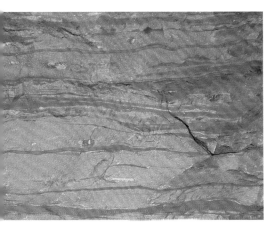

1,880-million-year-old banded iron formation near Schefferville, Quebec. The rock consists of layers of fine-grained hematite and magnetite (bluish grey), jasper (red), and siliceous iron carbonate (brown). ALAIN LECLAIR.

a sluice box in a stream to take advantage of the flowing water to concentrate the gold grains. Large floating dredges have also been used; these remove gravel from a river bed and then concentrate the gold in the same way as a sluice box. The spent gravel is dumped back into the water. Production in the Klondike peaked in 1922, when more than 500,000 ounces of gold were recovered—worth over a billion dollars in today's currency. Placer gold deposits are still being worked in the watersheds of the Klondike River and nearby Indian River.

Gold can also occur as paleoplacers (ancient placers), which were concentrated by fluvial processes like the modern Klondike deposits but later became consolidated by burial and diagenetic processes (Chapter 1), and preserved from erosion. The largest gold deposits in the world, in Precambrian strata at Witwatersrand in South Africa, are most probably paleoplacer deposits.

Gold now found in both placer and paleoplacer deposits originally came from bedrock that contained lodes of gold. The term lode refers to a zone of gold mineralization in which the metal is concentrated in veins, breccias, or pockets, or is finely disseminated within the host rock. Gold can be concentrated in late-stage magmatic fluids, especially those associated with porphyritic granitic rocks. As the fluids cool, gold and other minerals are precipitated in veins or as fine particles disseminated within the igneous rock itself and the adjacent host rocks. Such ores are thus porphyry-type or skarn deposits and generally occur within the top 1 to 3 kilometres of the crust. Gold may also be associated with volcanoes and their plumbing systems, in veins, breccias, and massive sulphide lenses, and as disseminated fine particles.

Finally, gold can be precipitated from hydrothermal fluids. As with VMS and SEDEX deposits, the hot fluids leach the metal from pre-existing rocks. As the fluids rise and cool, the metal may be deposited in veins filling faults and other fractures, usually with quartz and carbonate crystals. Or it may be precipitated as

Disused and flooded open-pit iron mine near Schefferville, Quebec. The pit walls show the multicoloured hues associated with the residual iron ore mined here. This ore is termed residual because groundwater leached silica from the original banded iron formation during the Cretaceous, a process that increased the concentration of iron in the rock. ALAIN LECLAIR.

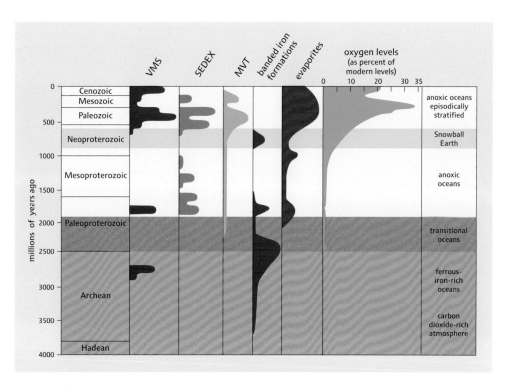

The timing of some major ore-deposit types and evaporites in relation to major events and changes in the Earth's atmosphere and evolution.

disseminated particles, commonly associated with pyrite, within the host rocks. Such deposits are generally formed at depths of 5 to 10 kilometres in association with major faults and convergent plate boundaries, and are thus called orogenic gold deposits. In this setting, tectonic activity generates the pressure and heat that form the gold-bearing hydrothermal fluids. Orogenic gold deposits are second only to paleoplacers in terms of world gold production and are the main source of gold in Canada. Examples include deposits in the Timmins district in Ontario and the Val-d'Or district in Quebec, both within the Superior Craton, and at Yellowknife in the Northwest Territories, within the Slave Craton.

Iron may lack the glamour of gold, but it is fundamental to industry. Historically, iron has been extracted from ores formed under many geological settings. The most important sources of iron today are banded iron formations. Most banded iron formations originated more than 2,000 million years ago, when both the oceans and the atmosphere contained little oxygen, a precondition for their deposition (Chapter 5). Banded iron formations are geographically extensive, commonly several hundred metres thick and consist of layers of the iron oxide minerals magnetite or hematite alternating with bands of iron-poor chert and shale. Banded iron formations are mined today in rocks deposited in the Paleoproterozoic Labrador Trough (Chapter 6).

A New Sparkle

Metallic ores are not the only valuable minerals. Valued for their hardness and lustre, diamonds are a relative newcomer to the Canadian mining scene. They consist entirely of carbon (Box 1), and laboratory experiments show that they form only at temperatures of more than 950°C and at pressures in excess of 45,000 times that at the Earth's surface. Such conditions are found at depths of 140 to 200 kilometres in the mantle beneath continents. Diamonds probably form by direct precipitation from a carbon-bearing melt, or from carbon dioxide, carbonate, or methane-bearing fluids deep in the lithosphere.

Obviously, diamonds cannot be mined at the great depths at which they form. Fortunately, they are carried to the surface in mantle-derived magmas called kimberlites. The very fluid

The brown area marks the presence of a kimberlite diatreme at Elwyn Bay, northeastern Somerset Island, Nunavut. The kimberlite intruded Silurian dolostones (seen in the distant cliffs) during the middle Cretaceous. The brown colour is due to specialized vegetation associated with phosphorus- and potassium-rich soil derived from the kimberlite. BRUCE KJARSGAARD, REPRODUCED WITH THE PERMISSION OF NATURAL RESOURCES CANADA 2013, COURTESY OF THE GEOLOGICAL SURVEY OF CANADA.

Several open pits mark the locations of kimberlite pipes being mined for diamonds at the Ekati Mine, Northwest Territories. WAYNE GOODFELLOW.

kimberlite magmas move up at high velocities through the continental lithosphere as dykes and, as they approach the surface, in pipe-like structures. Near the surface, the magma depressurizes and carbon dioxide and other gases are expelled, as in a freshly opened bottle of champagne. The large increase in volume, possibly in conjunction with contact with ground- or surface water, triggers an explosive eruption that ejects debris hundreds of metres or more into the air. Kimberlite pipes typically have a carrot-shaped breccia zone called a diatreme, and a crater at the surface surrounded by pyroclastic debris ejected during eruptions. Although kimberlites have formed throughout geological history, the major episodes of eruptions were during the Precambrian and Mesozoic. The youngest kimberlites are Pleistocene-Holocene structures in the Igwisi Hills of Tanzania. These African examples constitute the only surface parts of modern kimberlite volcanoes known, though no historic records of eruptions exist.

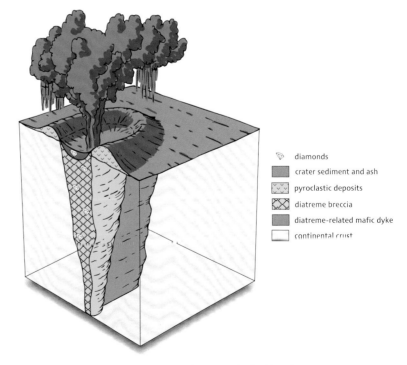

A kimberlite pipe and the related volcanic features at the Earth's surface.

CHAPTER 12 · FORGING A NATION

Economic diamond deposits were unknown in Canada until 1991, when discoveries in a kimberlite pipe in the Lac de Gras region of the Northwest Territories sparked a staking rush. This led to the discovery of hundreds of kimberlite pipes in Canada's north. Today, the Ekati, Diavik, and Snap Lake mines in the Northwest Territories and the Victor Mine in northern Ontario produce gem-quality diamonds, and geologists continue to search for new kimberlite pipes.

Diamonds from the Lac de Gras region of the Northwest Territories. The largest stone in the group is 4.35 carats. The age of the host kimberlite may be late Cretaceous or early Paleocene. COURTESY OF PEREGRINE DIAMONDS LIMITED.

Early Days

Armed with an understanding of how some mineral deposits form, we can now examine their use from a Canadian historical perspective. Long before Europeans arrived in North America, rocks, minerals, and metals played important roles in the lives of Aboriginal peoples. Graphite, hematite, limonite, and agate were used to make tools, as well as paints for decorating skin and clothes. Pyrite and chert provided the sparks to light fires, and soft sedimentary and metamorphic rocks were crafted into cooking utensils and smoking pipes. Archaeological studies suggest that obsidian was being mined in British Columbia 10,000 years ago, that copper was traded in the Lake Superior region 6,000 years ago, and that silver was mined in the Cobalt area in Ontario around 2,000 years ago.

The arrival of Europeans in Canada added another dimension. First to cross the Atlantic were the Vikings, who extracted impure iron from bogs at L'Anse aux Meadows, Newfoundland, about 998 CE. In 1541, the King of France asked Jean-François de la Rocque de Roberval and his subordinate Jacques Cartier to lead an expedition to North America. Cartier and his men arrived ahead of de Roberval at Cap-Rouge, Quebec, in August 1541 and founded the settlement of Charlesbourg-Royal. In his journal, Cartier wrote that they found a good quantity of diamonds and leaves of fine gold. He returned to France in the spring of 1542 with barrels laden with this material, only to learn that what they had collected was quartz and pyrite ("fool's gold"). From this arose the French saying "as false as a Canadian diamond". The English explorer Martin Frobisher had similar luck. During his Arctic expeditions, starting in 1577, Frobisher and his men quarried 182 tonnes of what he thought was gold-bearing ore from a site near present-day Iqaluit on Baffin Island and transported it to England. What Frobisher had brought back, however, was pyrite or gold-coloured mica. To encourage further exploration, in 1603 King Henri IV of France issued a charter to Sieur de Monts to explore for metals and minerals in the New World. This led to the discovery the following year, in what is present-day Nova Scotia, of silver and iron at St. Marys Bay and native copper at Cape d'Or. Fifty years later, in 1654, King Louis XIV of France awarded a concession to Nicholas Denys to mine gold, silver, and copper on Cape Breton Island.

Head frame at the abandoned Enterprise Mine, southeastern British Columbia. The mine produced primarily silver, lead, and zinc intermittently from 1896 to 1977. MIKE PARSONS.

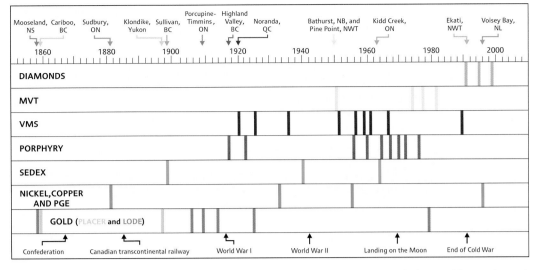

The discovery dates of major Canadian ore deposits and their relationship to some major historical milestones (shown at bottom). BC = British Columbia, NB = New Brunswick, NL = Newfoundland and Labrador, NS = Nova Scotia, NWT = Northwest Territories, ON = Ontario, and QC = Quebec

As Europeans explored farther west, mineral discoveries soared. In 1740, silver was found in galena-rich ores at Anse à la Mine on what is now the Ontario-Quebec border. And in 1771, Aboriginals led Samuel Hearne to native copper occurrences along the Coppermine River near the Arctic Ocean.

But nothing rivalled the allure of gold. In 1835 the first gold nuggets, weighing 1,280 and 1,190 grams, were discovered by chance in the sands of the River Gilbert in the Beauce region of Quebec by Clothilde Gilbert, a local girl. News of this find triggered Canada's first gold rush and by 1847 a placer gold mine had opened there. A few years later, placer gold was discovered along the canyon of the Fraser River, triggering a gold rush that led 20,000 enthusiastic gold seekers to Victoria on Vancouver Island in 1858. They crossed to the mainland, followed the Fraser north, and by 1861 were working the gold fields at Barkerville in the Cariboo district of the Columbia Mountains. This influx of people of European and Chinese ancestry led eventually to the creation of the British colony of British Columbia. But the most spectacular of Canadian gold rushes was the Klondike, as we've already discovered.

One of Canada's first iron mines was opened in the Blairton-Marmora area of Ontario in 1820. Around 1848, mining and smelting of iron and manganese began at Jacksonville near Woodstock, New Brunswick. In 1849, the then-largest iron mine in Canada opened at Londonderry, Nova Scotia. It was at Londonderry that German inventor Friedrich Siemens first tested his open-hearth process for the direct conversion of molten iron into steel in 1874. The first Canadian copper mine, the Eustis Mine, opened in 1865 in Quebec's Eastern Townships. A major asbestos deposit was discovered in 1881 about 20 kilometres northeast of Richmond, Quebec, and mining started there soon after. These events set the stage for Canada to become a global mining giant in the late nineteenth century.

From Rocks to Riches

With more than 200 operating mines, the mineral industry is a major contributor to Canada's wealth. Mining contributed C$63 billion to Canada's gross domestic product in 2011, almost 4 percent of the nation's GDP. In 2012, minerals represented 20.3 percent of the total value of Canada's exports, at over C$92 billion. The Canadian mining industry directly employs over 300,000 people, the majority of them in high paying jobs; and more than 115 mostly remote and rural communities across Canada are dependent on mining. Toronto, Montréal, and Vancouver are home to many of the financial services used by the minerals industry, including stock exchanges, underwriters, brokerage houses, and banks. Over 3,100 Canadian-based companies provide expertise and products to the industry, including hundreds of geological, technical, engineering, environmental, legal, and financial firms operating in Canada and around the world. And Canada was the leading destination for global exploration

Cores of Ordovician rocks from the Bathurst Mining Camp, southwest of Devils Elbow Brook, New Brunswick. STEVE MCCUTCHEON.

spending in 2012, attracting 16 percent of the world's mining exploration budget.

Today, most of the products that we depend on for everyday living—appliances, cars, buses, planes, ships, computers, and communications hardware to mention a few—contain metals or minerals. Metals are important components of industrial plants that supply energy to heat our homes and power our industries. And new alloys have become increasingly vital in the building of new gadgets that facilitate instant global communication and allow us to explore the most remote parts of our planet and Solar System.

The need for minerals and metals is increasing as developing countries raise their living standards. For example, copper is in demand as an excellent conductor, used to transport electricity. Zinc is crucial in the manufacture of galvanized steel and alkaline batteries, and as a white pigment in paint. Gold is not only used in jewellery but, because it is an excellent conductor of heat and electricity and is resistant to corrosion, is also a component of electronic circuits in computers and cellphones. Nickel is an essential ingredient in stainless steel, magnets, and coins. It also forms useful alloys with copper, chromium, aluminum, lead, cobalt, silver, and gold. Platinum and palladium are vital in catalytic converters, and indium is used in light-emitting diodes (LED) and liquid crystal display (LCD) screens.

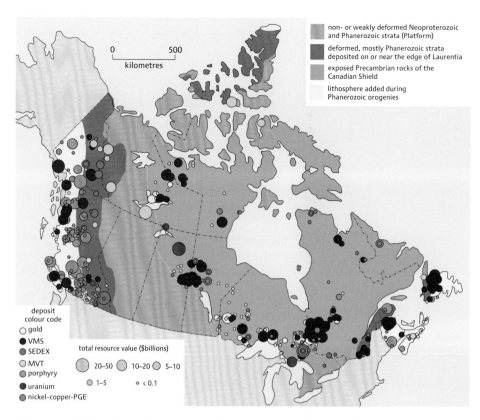

The location and value of Canada's major deposits for 2005 superimposed on a geological map of Canada. Note how almost all mining operations are located on the Canadian Shield (especially its margins, which are closer to potential markets) or in the Appalachian and Cordilleran orogens.

The future of the mining industry in Canada is bright. We have literally only scratched the surface: over 90 percent of operating mines extract ore deposits that are exposed at or very near the Earth's surface, and the potential for discovering deeply buried deposits is high. New exploration tools will help find these hidden deposits. As former Governor General Adrienne Clarkson once noted, it is thanks to the early geologists and mineral explorers that "...the huge territory of this northern land was brought bit-by-bit onto paper and into the minds and imagination of the early political architects of Canada."

The rusty red feature, a gossan, stands out in this aerial view over Baffin Island, Nunavut. Gossans are produced by oxidation of metal sulphide deposits. Because they can be recognized readily on the ground and from aircraft and satellites, gossans are a useful indicator of mineralization. Y. MAURICE, REPRODUCED WITH THE PERMISSION OF NATURAL RESOURCES CANADA 2013, COURTESY OF THE GEOLOGICAL SURVEY OF CANADA.

13 • HAVING THE ENERGY

CHAPTER SUPPORTED BY A DONATION FROM CLAYTON RIDDELL, FOUNDER, PARAMOUNT RESOURCES LTD.

The Mactaquac Dam generates hydroelectricity from the flow of the Saint John River in New Brunswick. It generates about 20 percent of the province's power needs. RON GARNETT/AIRSCAPES.CA

Pioneering Petroleum

People have known about petroleum (which includes gas, oil, and the goo-like substance known as bitumen or asphalt) for thousands of years. Gas bubbling out of the ground was lit by oracles such as those at Delphi in Greece, who persuaded people to pay homage, plus a fee, for prophecies. The first people to drill for petroleum were the Chinese about 3,000 years ago; they recovered gas and piped it through bamboo. In Canada, Aboriginal people have used bitumen for several centuries to make canoes watertight. European pioneers in Canada also valued bitumen because of its perceived medicinal value. A major turning point came in 1846, when Nova Scotian polymath Abraham Gesner discovered the distillation process for producing kerosene. Indeed, Gesner has been acknowledged as the "father" of the petroleum industry. Then, in 1851, Charles Tripp set up the International

Abraham Gesner is acknowledged as the founder of the petroleum industry. Born in Nova Scotia's Annapolis Valley, Gesner became a medical doctor in Parrsboro and, later, provincial geologist for New Brunswick (the first such post in the British Empire). COURTESY OF THE NEW BRUNSWICK MUSEUM.

Mining and Manufacturing Company, the first oil company in North America. Tripp's aim was to exploit the Devonian "gum beds" for bitumen in Enniskillen Township, Ontario, north of Lake Erie. But the company was not a financial success, so in 1856 he sold it to James Williams of Hamilton, Ontario. A year later, Williams founded his own company to extract both bitumen and kerosene from petroleum-rich rocks on the former Tripp properties. To provide a supply of clean drinking water, Williams dug a well 25 kilometres southeast of Sarnia, at Oil Springs, Ontario, in 1858. When the hole yielded oil, not water, it became the first commercial oil well in North America. Williams was interested in marketing the bitumen and in distilling the tarry material into kerosene, which was becoming popular as a less expensive and cleaner fuel than whale oil in lamps, and so he opened a refinery at Oil Springs and went into production. Later, Williams moved his refinery to Hamilton, where he offered refined oil at 16 cents per gallon, as long as the buyer agreed to purchase more than 4,000 gallons.

As the market expanded, so did the search for oil. The first free-flowing oil well was drilled at Oil Springs in 1860. On February 19, 1862, oil soared high above the tree tops when the first gusher, or well with uncontrolled flow, was drilled. Because of the drillers' inexperience, it took a week before the gusher was brought under control. That the first North American oil well was in Ontario has been largely overlooked. That honour is usually bestowed on Edwin Drake, who found oil in a well at Titusville, Pennsylvania, in 1859. Drake's two main legitimate claims to fame are that his well was drilled rather than dug, and that he was first to use pipes to line the hole and prevent collapse, a technique still used today.

But the heyday of petroleum was still to come. In those mid-Victorian times, coal was still "king" in industrialized countries, including Canada, where it was increasingly used to power steam engines.

Black gum, naturally occurring bitumen seeping from the ground at Oil Springs, near Sarnia, Ontario. GRANT WACH.

Old King Coal

Coal, which drove the Industrial Revolution, has been used for heating for many centuries and is still the world's major source of electricity. Heat from the combustion of coal in power plants converts water to steam, which drives turbines that in turn power generators. The United States, China, Russia, and India together hold over 65 percent of the world's coal reserves; Canada has less than 1 percent. Today, Canada is reliant mostly on other energy sources, but in the past, plentiful coal reserves encouraged settlement and growth.

Coal is a sedimentary rock composed mainly of plant material, especially wood, that grew in ancient wetlands (Chapter 1). The earliest trees originated in the Devonian, and so it is no surprise that the earliest coal is about 390 million years old. Most wetlands contain stagnant water

The dark layer in this sequence of rocks from Chimney Corner, Cape Breton Island, Nova Scotia, is a late Carboniferous coal bed. MARTIN GIBLING.

that is low in oxygen, which impedes the decomposition of dead plants by bacteria. Undecayed plant material builds up to form peat, which provides fertile ground on which new vegetation can grow. Indeed, peat is widely used today as a soil conditioner. Burial and accumulation may continue for thousands of years, producing several metres of peat, a prerequisite for coal formation. Historically, peat was used as a fuel, but it generates minimal heat and clouds of acrid smoke. If deeply buried, water is squeezed out and peat changes slowly to coal. Over time, peat transforms into lignite, the lowest grade (or rank) of coal. Further burial and higher temperatures convert lignite to sub-bituminous coal, then to bituminous coal, and finally to coal of the highest rank, anthracite. Increase in rank is accompanied by an increase in carbon content as other compounds, including water, are driven off—the coal becomes harder, changes from brown to black, and has a higher heat-generating potential.

How coal is used depends on its rank. Lignite and sub-bituminous coal are suitable for power plants, and sub-bituminous coal can also be converted to liquid petroleum and gas. Bituminous coal, with its high carbon and low moisture content, is ideal for the production of steel and cement, as well as for electrical power generation. Anthracite has the highest heat value and is used mostly for domestic purposes, for example as a constituent of charcoal briquettes. In Canada, anthracite is found in the latest Jurassic to earliest Cretaceous rocks in the Bowser Basin of northwestern British Columbia, where the remoteness of the deposits and environmental concerns have limited mining. It was once mined in the vicinity of Banff and Canmore, Alberta.

In the early years, coal was extracted mainly from small underground mines, but most is now recovered by surface open-pit or strip-mining, with enormous excavators removing soil and rock covering the coal seams. Strip-mining is much safer than underground mining, but has a far greater impact on the landscape, although attempts are increasingly made to return the land surface as much as possible to its former state.

The earliest record of coal mining in Canada goes back to 1639, when small amounts were recovered from the Minto area of New Brunswick and shipped to Boston. In the early eighteenth century, local coal was mined and used for heating and cooking in the French Fortress of Louisbourg on Cape Breton Island. From the onset of the Industrial Revolution, coal meant wealth, and the Carboniferous coal resources of the Maritimes were a factor leading to Confederation (Chapter 8). During the nineteenth century, numerous coal-mining communities sprang up on Cape Breton Island, as well as in Pictou and Cumberland counties in mainland Nova Scotia. Because of the abundance of coal in the region, Sydney Harbour was at one time designated an emergency base for Britain's steam-powered navy.

Coal production in Nova Scotia peaked in the 1940s, but today all of the province's underground mines are closed, the decline of the industry heralded by disasters at Springhill in 1956 and 1958 and at Westray in 1992; these disasters killed 39, 74,

Cretaceous lignite coal being strip-mined at Estevan, Saskatchewan. ALAN MORGAN.

Open-pit coal mine, Sparwood, British Columbia. Coal in this part of southeastern British Columbia is of Cretaceous age and was deposited in the foreland basin to the east of the advancing Rocky Mountains thrust-and-fold belt (Chapter 9). RON GARNETT / AIRSCAPES.CA.

and 26 miners, respectively. Another tragic aspect of coal mining is silicosis, a deadly, incurable lung disease caused by the breathing in of silica dust, a common by-product of the mining process.

On the Prairies, Aboriginal peoples such as the Blackfoot and Cree knew about the "black-rock-that-burns", but rarely used it. Peter Fidler was the first European to report coal in what is now Alberta in 1792. Commercial mining began in 1872 along the Oldman River in Alberta. In 1884, the same year he came face to skull with the *Albertosaurus*, Joseph Tyrrell found extensive Cretaceous coal deposits in what is now the Drumheller area. These became an invaluable source of power for the construction of the transcontinental Canadian Pacific Railway two years later. Mines sprang up close to the tracks in Alberta, British Columbia, and Saskatchewan, and by 1911 Alberta and British Columbia were major producers of coal. Many towns in the Rocky Mountain Foothills, such as Canmore and Frank, started as coal-mining settlements. Underground mining in these centres lasted from the 1880s until 1980; and open-pit mining started in the late 1960s and continues today, in some cases flattening entire mountains in the process. The coal was deposited in the foreland basin as it migrated eastward in advance of the Rocky Mountains thrust-and-fold belt.

In 1849, Aboriginals informed agents of the Hudson Bay Company that coal was to be found on the west coast. Mining there began on a small scale, but grew rapidly with the discovery in 1850 of coal seams in late Cretaceous rocks of the Georgia Basin at Nanaimo on Vancouver Island; Nanaimo subsequently became a coal-mining centre, with the last mine closing in the late 1960s. Coal was also found in fault-bounded basins in interior British Columbia, formed in the Eocene as the Cordillera became stretched (Chapter 10). These deposits led to the establishment of towns such as Merritt and Princeton as coal-mining centres.

Today, about 99 percent of Canada's coal production comes from Alberta, British Columbia, and Saskatchewan. Alberta has 70 percent of the country's coal reserves, with Highvale near Wabamun Lake being the largest mine by area. Small amounts of coal are also produced in Nova Scotia (from open pits) and New Brunswick. Coal is present in the Northwest Territories, Yukon, and some of the Arctic Islands, but these deposits are too remote to be economically viable. Most modern mines are surface operations, but one exception is the Quinsam Mine near the town of Campbell River on eastern Vancouver Island, where coal is extracted underground. Two-thirds of the coal mined in Canada is used to generate electricity: Alberta produces 43 percent of its power from this resource and Saskatchewan and Nova Scotia 60 percent each. The remainder is transported by rail to the west coast and from there shipped to Asia.

Although coal is the most abundant and cheapest fossil fuel, burning it releases, among other things, carbon dioxide, nitrogen oxides, mercury, and sulphur dioxide (whose emissions cause acid rain). New technologies are reducing many of the pollutants from coal-fired power plants by removing minerals before combustion, and removing sulphur compounds, nitrogen oxide, and particulate matter during combustion. One major goal is to capture carbon dioxide before it leaves the stack, liquefy it, and return it to underground reservoirs. Carbon capture projects, if successful, will turn coal into a much cleaner energy source.

Petroleum's Building Blocks

Over the past century, coal has been gradually overtaken as a primary energy source by petroleum—literally rock oil. Today, petroleum is the world's principal energy source for transportation. It provides fuel and lubrication for cars, trucks, ships, and planes, and is a vital energy source for electricity generation. Petroleum is also the raw material for that ubiquitous modern material known as plastic. But what is this magical elixir? Petroleum consists of hydrocarbons—compounds of hydrogen and carbon. The form petroleum takes, whether solid bitumen, tar, liquid oil, or gas, is determined by the number of carbon and hydrogen atoms in its molecules. As the number of carbon atoms in each molecule increases, petroleum gets heavier and less volatile, progressing along a continuum from gas to condensates, oil, and finally bitumen.

The lightest petroleum is methane, so-called dry gas, which has a simple atomic structure of four hydrogen atoms attached to a single carbon atom (CH_4). Ethane (C_2H_6), propane (C_3H_8) and butane (C_4H_{10}) are the three so-called wet gases. Farther along the continuum is a heavier, less volatile group of hydrocarbons known as condensates, which are gases at higher temperature and pressure at depth but liquids (oils) when brought to the surface. Condensates have molecules with 5 to 7 carbon atoms. An example is hexane (C_6H_{14})

Oils can be light, medium, or heavy. Light oils have from 11 to 26 carbon atoms in a molecule, medium oils have 27 to 39, and heavy oils have 40 or more. The heavy oil in oil sands contains a high proportion of other elements, including sulphur, nitrogen, and oxygen. Of the petroleum products that we use daily, gasoline (distilled from light oil) has molecules with 6 to 12 carbon atoms, diesel has molecules with 12 to 18 carbon atoms, and lubricating oil has molecules with 26 to 40 carbon atoms. Variables other than the number of carbon atoms include shape: for example, a molecule may have straight or branched chains, or single or multiple rings. An example of a ringed molecule is benzene, which has 6 carbon atoms linked to form a circle, with each carbon atom attached to one hydrogen atom.

Liquid petroleum directly from the well is referred to as crude oil and ranges from almost water-clear to yellow, amber, green, brown, and black. Usually crude oil must be processed

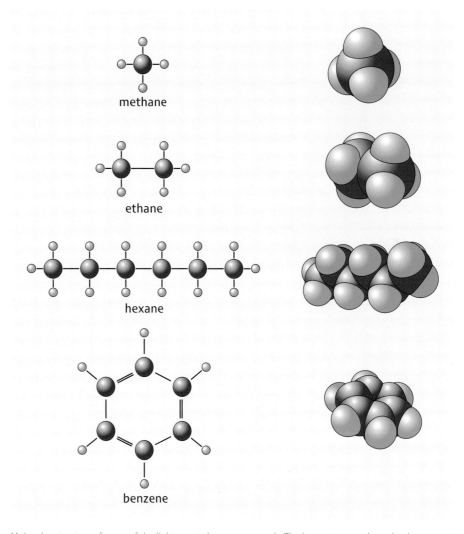

Molecular structure of some of the lighter petroleum compounds. The larger atoms, coloured red, represent carbon and the smaller atoms, coloured green, represent hydrogen. Figures on the left show molecular structure schematically and those on the right, representing the same compounds, show the true relationships of the atoms within each molecule, with the hydrogen and carbon atoms in direct contact. All the molecules shown, except benzene, have a linear structure and single bonds between carbon atoms. Methane is a gas, ethane is a wet gas, and hexane is a condensate. Molecules of benzene are formed of a ring of six atoms and have double bonds between some of these atoms.

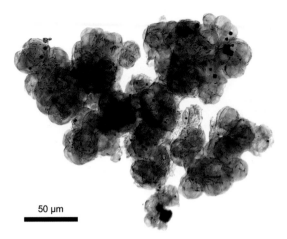

Cluster of organic-walled fossils of *Gloecapsomorpha*, an organism thought to have been a source of petroleum in early Paleozoic rocks. This specimen is from Cambrian rocks in the Saskatchewan subsurface and is about 0.27 millimetres wide. NICK BUTTERFIELD.

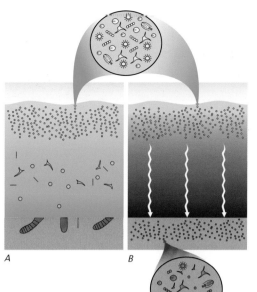

Trillions of microscopic marine organisms can live in a cubic metre of water within the photic zone (where sunlight penetrates). When the organisms die, their bodies sink. If the waters are rich in oxygen, as in A, the remains are not preserved. But if the water has low levels of oxygen, as in B, a large proportion of the organic remains is preserved and buried in the bottom sediments. Within the upper circle are some of the micro-organisms, greatly magnified, that live in the photic zone. Within the lower circle are some of the remains of the micro-organisms preserved in the sediments.

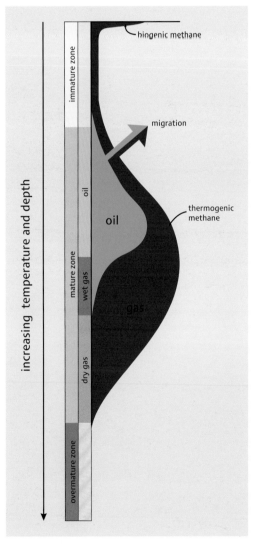

Conversion of organic matter to different kinds of petroleum over time and with increasing temperature and depth. The mature zone denotes where oil and thermogenic gas form. At low temperatures and minimal burial, only biogenic methane is produced. As organic matter matures, oil may be produced along with thermogenic methane and other gases. At still deeper horizons and higher temperatures, all oil has been expelled or is broken down into simpler molecules and only thermogenic methane is produced. The variation in width of the oil (green) and gas (red) curves reflects the relative amounts of each produced at different depths and temperatures.

or refined. One exception was oil from natural seeps at Parson's Pond in western Newfoundland, which provided locals with fuel for their fishing boats during much of the twentieth century. And residents of the eastern Gaspé Peninsula in Quebec were able to use crude oil from surface seeps and old wells to run their cars during the Second World War, when gasoline was rationed.

Whereas most coal and much gas comes from terrestrial plants, liquid petroleum is derived mainly from marine plankton, especially minute algae and bacteria, whose total biomass is enormous. Algae have membranes formed of hydrogen-rich, oxygen-poor molecular chains, which can be readily converted into the long hydrocarbon chains of most petroleum.

When they die, most small marine organisms are consumed by other organisms, but some reach the sea floor. If the bottom waters are oxic (rich in oxygen), their remains quickly decay; scavengers recycle the organic matter and organisms that live in the sediment and feed on the organic matter leave traces of their activity. But if the bottom water is anoxic (with low levels of oxygen), most bacteria that cause decay, as well as scavengers that leave traces, cannot survive. So large amounts of the organic remains are preserved and buried in the sediments. Over time, more and more sediment is deposited on the sea floor, and may eventually form sedimentary rock. Rock containing a lot of organic matter is called a source rock because it has potential to generate oil and gas. In source rocks, smaller organic molecules become linked into long molecular chains or polymers called kerogen. Formation of kerogen is an essential stage in oil and gas generation and also helps to preserve the organic matter. There are three main types of kerogen: that derived from algae living mainly in lakes, that derived from microscopic marine animals and protists, and that derived from terrestrial organic material such as wood that has been carried into lakes or the sea. Bacteria contribute to the formation of all three types.

The generation of petroleum depends on the type of kerogen, the temperature (generally rising with increasing depth of burial), and the length of time of burial. Critical temperatures are between 50 and 120°C, at which kerogen starts to break down into smaller and more numerous molecules of oil and gas. Collectively, these smaller molecules take up more space, thus increasing pressure and forcing the petroleum to migrate out of the source rock, a process that continues as temperatures keep on rising. Above 150°C, the molecules become even smaller and oil is converted to gas, mainly methane. Gas formed through heating is called thermogenic. There is also biogenic gas (also mainly methane), produced by organisms living in wetlands, landfill sites, and sediment at or near the surface.

Trapped

Oil and gas are lighter than water, so when they are squeezed out of a source rock, they tend to move slowly upward along bedding planes, joints, and faults, displacing water in the process. Oil and gas collect in small pores within rocks, and rock bodies that contain significant amounts are called reservoir rocks. The best potential reservoir rocks are carbonates and poorly cemented sandstones that have high porosity, which is the proportion of pore space in a rock. A reservoir rock must also be permeable, which means that the pores must be interconnected. Porosity controls the amount of petroleum that a rock can contain and permeability controls its rate of flow within the rock and into a potential production well. In carbonate rocks, large pores or cavities commonly result from the dissolution of fossil shells and skeletons (such as those of corals), greatly enhancing porosity. Important examples of this type of reservoir rock are the Devonian reefs that host the famous Leduc Oil Field in the subsurface of Alberta. At Grassi Lakes near Canmore, Alberta, cavities in such an ancient reef are the size of small caves that a whole family can climb into.

To remain trapped in a reservoir, the upwardly mobile oil and gas need to be stopped by a seal, a thick, impermeable, fine-grained layer of rock such as a shale, or a thick layer of salt. Otherwise the oil and gas seep to the surface. In the reservoir beneath a seal, the order from bottom to top is water, oil, gas, a sequence that is controlled by relative density. Some of the best traps are anticlines (Chapter 1). As the oil and gas migrate upwards, they are trapped in the crest of the anticline if the reservoir rocks are overlain by impermeable seals. Faults that bring reservoir and seal rocks together are also effective traps. Some of the world's most productive oil and gas fields result from traps beneath and on the flanks of impervious salt structures. As we have seen in the Sverdrup and Scotian basins, salt tends to move toward the surface like toothpaste squeezed from a tube. This upward flow of salt deforms overlying rocks, forming anticlines and other traps. All the traps so far

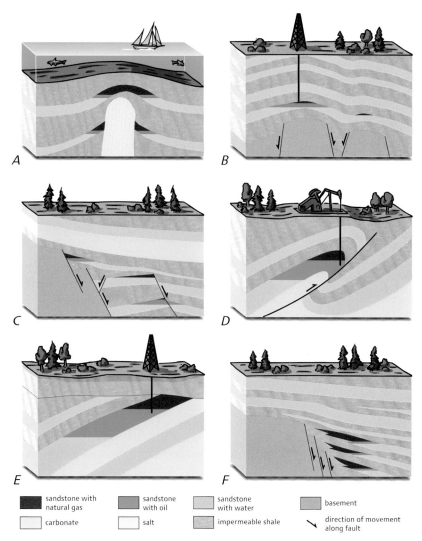

Some types of hydrocarbon traps. Most of the traps shown are structural (A to D) and involve folding and/or faulting of strata. One trap (E) is a combined structural and stratigraphic trap. The last trap (F) is stratigraphic—the rock changes in character laterally. ADAPTED FROM VARIOUS SOURCES.

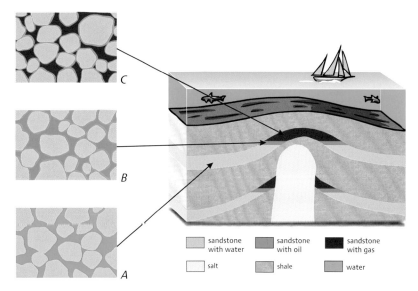

Porosity in sandstone. Liquids and gases occupying the spaces between grains in sandstone are able to migrate upwards until their movement is stopped by an impervious layer, such as shale. In A, pore spaces are filled with water. In B and C, a thin film of water coats each grain, but the rest of the spaces is filled with oil and gas, respectively. ADAPTED FROM ATLANTIC GEOSCIENCE SOCIETY (2001).

CHAPTER 13 • HAVING THE ENERGY 267

Cretaceous deltaic and marine strata tipped up around a diapir of Carboniferous evaporite, Ellef Ringnes Island, Nunavut. Below the surface, petroleum is commonly trapped in structures created by salt domes. CAROL EVENCHICK.

Part of the cairn in front of the Geological Survey of Canada office in Calgary. The rock is from a late Devonian carbonate reef at White Man Gap near Canmore, Alberta, in the Front Ranges of the Rocky Mountains. Its high porosity is typical of many Devonian oil and gas reservoirs beneath Alberta. The holes represent the former positions of stromatoporoids, now dissolved. GODFREY NOWLAN.

mentioned are related to structural deformation—folding or faulting of strata. Other traps, considered stratigraphic rather than structural, occur where reservoir rocks are tilted and wedge out against an unconformity, or grade laterally into finer rocks that act as a trap.

In some petroleum-bearing rocks, especially those near the surface, bacteria that thrive at temperatures below 70°C feed on oil, turning it into a gooey tar. This process is called biodegradation. The most famous example of such biodegraded petroleum is the Athabasca Oil Sands, discussed below. Above 70°C, most bacteria are unable to grow. If temperatures exceed 120°C and gypsum is present, the oil reacts to form hydrogen sulphide or sulphur, as well as bitumen. The addition of hydrogen sulphide makes gas "sour". In some Alberta wells, sulphur is extracted from sour gas, often originating from gas wells penetrating carbonate reservoirs. Sulphur is an added bonus as it is widely used in industry.

The complex sequence of events needed to form petroleum—involving the right kinds of organic remains (the most critical factor), cooking them to just the right temperatures in a source rock, and having the right geological settings for migration and entrapment in a reservoir—explains why the odds of forming an oil and/or gas field are so very low. It also explains why exploration for new resources needs to be technologically sophisticated to find diminishing resources of conventional oil and gas (that is, petroleum reserves in sedimentary basins found in reservoirs beneath traps and produced from vertical wells—we will get to unconventional sources below).

Western Canada's Bonanza

For more than half a century, the Western Canada Sedimentary Basin has been the focus of Canada's petroleum industry, with its hub in Calgary. Alberta's oil industry started in 1902, when the first producing oil well in western Canada was drilled at Oil City, in what is now Waterton Lakes National Park of Canada in the southwestern corner of the province. However, early developments were sporadic until 1914. That year marked the discovery of gas at Turner Valley, south of Calgary, triggering Alberta's first boom. No further significant finds in the province occurred until 1924, when the Royalite No. 4 well penetrated oil-bearing Carboniferous limestone in Turner Valley. The drillers struggled to stop the flow as wellhead pressure rose, but a massive blowout led to a fire that raged for three weeks. Further drilling in 1926 confirmed Turner Valley as Canada's first major oil field, and by the 1930s it was the largest in the British Empire. Massive amounts of gas, then thought worthless, were flared off, thereby leading to the area's designation as "Hell's Half Acre". The flares reduced pressure in the field and changed the oil's viscosity. If this gas had been left in the reservoir, pressure would have been maintained, and 60 percent, rather than 10 percent, of the oil would have been recovered. This waste led the provincial government to introduce conservation measures on oil and gas production and to form the Alberta Energy and Resources Conservation Board to monitor exploration and production and ensure resources were properly managed.

A major milestone in the history of the oil and gas industry in Alberta was Imperial Oil's big oil discovery at Leduc, near Edmonton, in 1947. By then, Imperial had drilled 133 consecutive dry (unsuccessful) wells. A rig boss

Cross-section through the Leduc Oil Field southwest of Edmonton, Alberta. What became the discovery well for this field (Leduc No. 1) was "spudded" (started) in the fall of 1946 and struck oil in a relatively small Devonian coral reef (not shown) in early 1947. Subsequent wells at Leduc that reached slightly older, but much more substantial, reefs that yielded large volumes of oil and gas. ADAPTED AND USED WITH PERMISSION FROM THE EDMONTON GEOLOGICAL SOCIETY.

On March 8, 1948, Imperial Oil's Atlantic No. 3 well near Leduc, Alberta, penetrated a porous, oil-bearing, Devonian limestone reef 1,626 metres below the surface, causing a blowout that caught fire on September 6, 1948. COURTESY OF THE GOVERNMENT OF ALBERTA AND CHRIS YORATH.

CHAPTER 13 · HAVING THE ENERGY   269

Turner Valley was Canada's first major oil field, and by the 1930s was the largest in the British Empire. Massive amounts of natural gas were burned, or "flared off", leading to the field's nickname "Hell's Half Acre". IMAGE COURTESY OF THE GLENBOW ARCHIVES, CALGARY (S-17-23).

named Vern Hunter had been involved in many of them, earning him the nickname "Dry Hole" Hunter. But his luck changed in spectacular fashion when the Leduc No.1 well tapped into oil in a Devonian dolostone. Equally exciting was when Leduc No. 2 penetrated a Devonian reef, destined to become the greatest oil play in Alberta outside the oil sands. Another famous well was Atlantic No. 3, which blew out in an uncontrolled gusher that caught fire on Labour Day in 1948. Success followed success. In 1953, the first well was drilled in the Pembina Oil Field, where the oil is in Cretaceous reservoir rocks. This field was developed so rapidly that by 1956 there were more than 1,500 producing wells.

In western Canada outside of Alberta, still Canada's leading producer of oil and gas, significant discoveries and developments have been made in Saskatchewan, Manitoba, and northeastern British Columbia. Exploration has also been carried out on the continental shelf off British Columbia, where fourteen dry wells were drilled between 1967 and 1969 in the Tofino and Queen Charlotte basins. But this activity was short-lived. A moratorium on drilling was imposed by the Canadian government in 1972 and is still in place there.

Eastern Canada and the Arctic

Oil and gas exploration has also taken place in eastern Canada, with discoveries in Quebec and the Atlantic Provinces as well the groundbreaking finds in Ontario. In western Newfoundland in 1812, a Mr. Parsons collected oil from a seep in the Parson's Pond area and used it as a treatment for rheumatism. The first well in the area, drilled in 1867, encountered small amounts of oil and was followed by at least thirteen other wells over the next 40 years. Drilling for petroleum in mainland Atlantic Canada began in 1859 in Albert County, southeast of Moncton, New Brunswick. This discovery was in the Moncton Sub-basin of the Maritimes Basin (Chapter 8), and in 1909, the Stoney Creek Oil and Gas Field was discovered in the same sub-basin. Oil production peaked in the 1940s, after which it declined rapidly before ending in 1988. In 1875, oil was discovered in the Lake Ainslie area of Cape Breton Island, Nova Scotia. Exploration there continued sporadically up to 1958, without commercial success.

In Quebec, oil seeps were first reported on the Gaspé Peninsula in 1836 and gas was produced from Quaternary sediments in the Trois-Rivières region in 1887, providing power for a short time to some local industries. A gas field was discovered in 1972 near Saint-Flavien, south of Québec City. The field, which was in production from 1980 to 1994, is now used for underground gas storage.

Interest in onshore oil and gas exploration in eastern Canada has been renewed in recent years, and production has resumed in the Stoney Creek Oil and Gas Field. Exploration interest has also reawakened in the Lake Ainslie area. On the Gaspé Peninsula, exploration drilling has led to the discovery of one oil field in Devonian sandstones and one oil and gas field in Devonian carbonates; both fields are now in the development stage. Oil was also discovered in the 1990s on the Port au Port Peninsula in western Newfoundland.

Derelict equipment used when oil was being produced at Parson's Pond, western Newfoundland. Between 1907 and 1909, some 800 to 900 barrels of oil were shipped from there to St. John's. MARTIN FOWLER.

Canada's first offshore well was drilled in 1943, on an artificial island constructed for the purpose, 13 kilometres from shore in Hillsborough Bay, Prince Edward Island. However, the real start of east-coast offshore exploration had to await another two decades. The first exploration wells on the Grand Banks of Newfoundland were drilled in 1966, and the first well in the Hibernia Oil Field in the Jeanne d'Arc Basin was completed in 1979, with production starting in 1997. The Hibernia Oil Field has recoverable oil reserves estimated at 189 million cubic metres (1.19 billion barrels), officially a "giant field" in oil-patch parlance. Since the Hibernia find, three other large fields have been discovered in the Jeanne d'Arc Basin: Terra Nova, White Rose, and Hebron. The reservoir rocks in all four fields are late Jurassic or early Cretaceous in age, and all but Hebron are now in production. Exploration in deeper water to the north of the Grand Banks is also yielding encouraging results. The first exploration well in the Scotian Basin was drilled in 1967, leading to a breakthrough in 1979 with the discovery of the Venture Gas Field in early Cretaceous reservoir rocks. Four other fields have subsequently been discovered in the vicinity of Sable Island, all in Cretaceous reservoir rocks. Drilling off Labrador began in 1971 and has led to several gas discoveries. However, the high costs of building a pipeline and other facilities have prevented development so far north.

In northern Canada, the first major petroleum find was the giant Norman Wells Oil Field in the Northwest Territories, discovered in a Devonian reef in 1920. Oil produced from Norman Wells was considered an important asset during World War II, when it was feared that supplies to Alaska could be disrupted by invasion from Japanese forces. The governments of Canada and the United States collaborated on the Canol Project, which included the construction of a 700-kilometre-long pipeline from Norman Wells to Whitehorse, Yukon, where the oil was refined and then sent on by pipeline to Fairbanks, Alaska. The pipeline was of small diameter (10 centimetres) and, because it was laid on the ground, fractured easily due to repeated freezing and thawing. After the war the pipes from the Canol Project were good only for scrap, but the lessons learned helped in the building of a new pipeline from Norman Wells south to Alberta in 1985.

In 1967 a partnership between government and industry led to the birth of Panarctic Oils, which mounted a big exploration program in the Arctic Islands. Gas was found at Drake Point on

Oil well at Lake Ainslie on Cape Breton Island around 1913. PROVIDED BY GRANT WACH, SOURCE UNKNOWN.

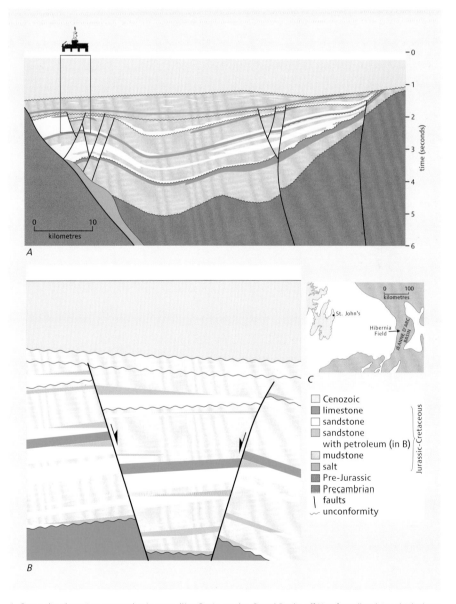

A. Generalized section across the Jeanne d'Arc Basin on the Grand Banks off Newfoundland, in which there are several oil fields, including Hibernia. B. Detail of the Hibernia Oil Field (as approximately shown in red in A; the geology in B is idealized and so does not exactly match that shown in A as the sections are slightly offset. The vertical scale represents the time it takes for seismic waves to travel down to a particular surface, then back up again to the recording instrument. ADAPTED FROM MIALL ET AL. (2008), WITH HELP FROM MICHAEL ENACHESCU.

Semi-submersible rig in the foreground and jack-up rig in the background, Halifax Harbour, Nova Scotia. KEITH VAUGHAN.

The Deep Panuke Platform off Nova Scotia. COURTESY OF ENCANA.

Melville Island in 1969, when the discovery well blew out of control—the gas and salt water created a spectacular cone 61 metres high. Drake Point and Cisco fields near Lougheed Island are respectively the largest gas and oil fields in the Arctic Islands. Estimated reserves for the Drake Point Gas Field are nearly 100 billion cubic metres (3.5 trillion cubic feet) and for the Cisco Oil Field are about 92 million cubic metres (584 million barrels). The Mackenzie Delta-Beaufort Sea region is currently estimated to have reserves of 916 million cubic metres (5.8 billion barrels) of oil and 1.6 trillion cubic metres (58 trillion cubic feet) of gas.

Oil Sands

The existence of the Athabasca Oil Sands in present-day northeastern Alberta was first noted in 1778 by fur trader Peter Pond, and briefly described by Alexander Mackenzie in 1788. The first

Remnants of a 4-inch (10.5-centimetre) diameter pipeline laid during the Canol Project in the 1940s, near Norman Wells, Northwest Territories. HANS WIELENS.

Europeans exploring the area found that local people were using bitumen from the oil sands mixed with tree gum to waterproof their canoes. In 1888, showing remarkable foresight, Robert Bell, then Director of the Geological Survey of Canada, stated, "The evidence...points to the existence in the Athabasca and Mackenzie valleys of the most extensive petroleum field in America, if not the world." We now know that the Athabasca Oil Sands constitute the world's largest accumulation of petroleum. Commercial development of the Athabasca deposits started in 1967, when the Great Canadian Oil Sands Limited opened a strip mine. After processing, the oil sands yielded 4,800 cubic metres of synthetic crude a day, a figure dwarfed by present production rates. The Syncrude consortium began commercial production in 1978 with a new mine, and since then development of the oil sands has been explosive.

Oil sands are a naturally occurring mixture of sand grains coated with a thin film of water (making it "water-wet"), which in turn is coated with bitumen; clay is also present. In the Athabasca deposits, the host rocks are middle Cretaceous marine and fluvial sands. The bitumen that they now contain probably originated as oil from a Devonian source rock, and since migration it has been degraded by bacteria to form bitumen. Recovery of oil from the bitumen is facilitated by the layer of water around each grain. In 1925, Karl Clark (then of the Alberta Research Council) developed a relatively simple procedure to recover commercially valuable oil from the deposits. Clark tested the process, which involves heating the oil sand with water, in large cauldrons on the banks of the Athabasca River.

Along a narrow strip bordering the Athabasca River, the oil sands are at or near the surface and can be mined with enormous power shovels and excavators; these machines load the oil sand

Jackpine Mine oil sands operation, about 110 kilometres north of Fort McMurray, Alberta. RON GARNETT / AIRSCAPES.CA.

into huge dump trucks and take it to processing plants. There, the sands are mixed with steam, hot water, and caustic soda to remove the heavy oil, in a modification of the original process devised by Clark. Deposits that are too deeply buried to be mined economically are extracted in a process that involves the drilling of two horizontal wells, one above the other. Steam is injected into the upper well and adjacent bitumen softens and flows down to the lower well to be pumped out. About 12 cubic metres of water are needed to produce 1 cubic metre of oil, but recycling cuts that amount to 2 to 4 cubic metres of water. The water adds hydrogen that was stripped by bacteria during degradation of the original oil, converting the bitumen back to liquid oil.

In addition to the Athabasca Oil Sands, two other heavy oil accumulations are being developed in northern Alberta: the Peace River and Cold Lake deposits. The combined size of the three deposits is staggering. They underlie a fifth of Alberta, and an area larger than New Brunswick has already been leased to oil sands companies. Recoverable reserves in the Athabasca Oil Sands alone are estimated at about 20.3 billion cubic metres (175 billion barrels) of oil. This makes Canada second only to Saudi Arabia in the size of its oil reserves, although Saudi Arabia's are all conventional. Petroleum reserves such as oil sands, oil shale, and shale gas, are called unconventional.

It's a Gas

Greenhouse-gas emissions from gas combustion are about 30 percent less than those from oil and 45 percent less than those from coal. For this reason, increased use of gas is seen by some as a step toward "greener" energy consumption, until we are able to generate all our power from renewable energy sources. Gas offers the potential to replace other fossil fuels, especially for power generation, heating, and transportation.

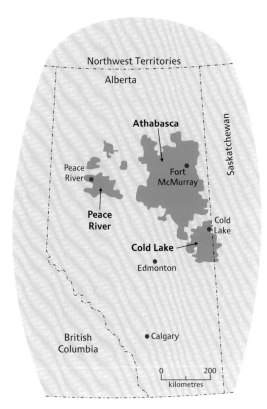

Location of the three major oil-sands deposits: Athabasca, Cold Lake, and Peace River. ADAPTED FROM VARIOUS SOURCES.

Although conventional reserves of gas are abundant, the potential for developing unconventional reserves is huge. As well as having origins in deep source rocks, gas can be formed in young sediments at shallow depths and low temperatures. Such gas results from the activity of microbes known as methanogens, which break down organic matter. Large accumulations of such biogenic gas (in contrast to thermogenic gas from deeper sources) occur in the Medicine Hat area of southern Alberta. This resource was discovered when a crew drilling a water well for the Canadian Pacific Railway near Medicine Hat in 1883 was astounded to see the well burst into flames. The accident led to the development of one of Canada's largest gas fields, inspiring visiting British author Rudyard Kipling to describe Medicine Hat as "the city with all hell for a basement".

Many Canadian coals are also a potential source of gas. At high temperatures, complex organic molecules in buried coal naturally break up, or crack, into simpler molecules such as methane. This methane is stored in pores in the coal, and its buildup in mines can have an explosive impact—as witnessed by the Springhill and Westray tragedies. Such coal-bed methane can now be extracted by drilling wells into seams that are not economic in coal-mining terms, thus providing a potential new resource in coal-mining areas. Unlike strip-mining of coal or oil sands, the impact of coal-bed methane on the landscape is small. But from the subsurface coal seams, the procedure can produce large volumes of impure water—a problem that may be alleviated by re-injecting the water into the ground or collecting it in settling ponds where it can be treated. Methane, gasoline, and diesel can also be produced from coal by a process called gasification. Steam or hot air is used to generate temperatures high enough to crack the complex organic molecules in coal into simpler compounds. The process works as well on other raw materials such as household waste or compost, which can also produce biogenic methane.

Another unconventional reserve involves oil and gas trapped in their original source rocks. One such rock is shale, which now provides more than a third of oil production and about 20 percent of natural-gas production in the United States. In Canada, the only current shale production is from Devonian and Triassic rocks in northeastern British Columbia, where exploration and resource evaluation indicate huge recoverable reserves of gas. But shale gas has potential in other parts of Canada and should considerably add to reserves, provided that environmental concerns over groundwater contamination during fracking—the process of fracturing the source rock to release the gas—can be addressed.

In Canada, economic prospects from shales today are primarily gas. However, an early attempt to recover oil from shale is commemorated at the site of the Craigleith Shale Oil Works in present-day Craigleith Provincial Park near Collingwood, Ontario. There, in 1859, a growing demand for artificial light and lubricating oil led William Pollard to establish a plant to extract oil from Ordovician shales. In a process patented by Pollard, a maximum of 1,137 litres (250 imperial gallons) of crude oil were obtained daily from 30 to 35 tons of oil shale. However, after the discoveries at Oil Springs, the process proved uncompetitive and by 1863 Pollard's venture had failed.

Gas hydrates are another potential energy resource. They are crystalline compounds in which water molecules form a cage that traps molecules of gas. The water cages develop under conditions of low temperature and high pressure, for example, at depths of 200 to 1,000 metres in permafrost, and in sediments on the sea floor at depths of 1,200 to 1,600 metres. Masses of gas hydrates look like clumps of dirty ice, but unlike ice they can burn like a candle. Gas hydrates incorporate huge volumes of methane—about five times more energy than the same volume of conventional gas. It has been suggested that release of gas hydrates was responsible for the Paleocene-Eocene Thermal Maximum about 55 million years ago (Chapter 10), and may be implicated in some of the mass extinction events.

Gas hydrates contain perhaps 6 trillion tonnes of methane, twice the amount of all fossil fuels combined: indeed, they constitute the largest single carbon reservoir on Earth. A Canadian-Japanese team has drilled some experimental wells into a gas-hydrate zone above a natural-gas field at Mallik, Northwest Territories, on the Mackenzie Delta, to test the feasibility of developing gas hydrates at northern and offshore sites. Early results have shown that high flow rates of methane can be maintained over six days, indicating that gas hydrates can be a viable source once the technology for separating and producing the gas is perfected.

Nuclear Power

Uranium is a heavy silvery metal that averages 2 to 10 parts per million in most of our planet's crustal rocks, soils, and oceans. It occurs naturally in the sooty black mineral called uraninite or pitchblende. As we discovered in Chapter 3, isotopes of uranium decay over time to isotopes of other elements. This decay produces energy, which is the basis for nuclear power. Although uranium has several isotopes, almost all of its atoms found in nature are of uranium-238. Nearly all the rest of the atoms are of uranium-235, a less stable isotope that decays naturally to lighter elements, releasing energy in the process. This decay is termed nuclear fission.

Nuclear power plants use the heat generated from the decay of uranium-235 to boil water and create steam. The steam powers a turbine that drives a generator, which then produces electricity. Roughly 15 percent of Canada's electricity, and 50 percent of Ontario's, is derived from nuclear power in Canadian-made CANDU reactors at several nuclear power stations in Ontario and one each in Quebec and New Brunswick. CANDU reactors do not produce material that is suitable for manufacturing nuclear weapons. But they are a source of hydrogen, which could be used as a fuel, and cobalt-60, an isotope with a variety of applications in the medical and food industries.

The first economic uraninite deposit in Canada was discovered in the Northwest Territories by prospector Gilbert LaBine in 1931 at Port Radium on Great Bear Lake. This Paleopro-

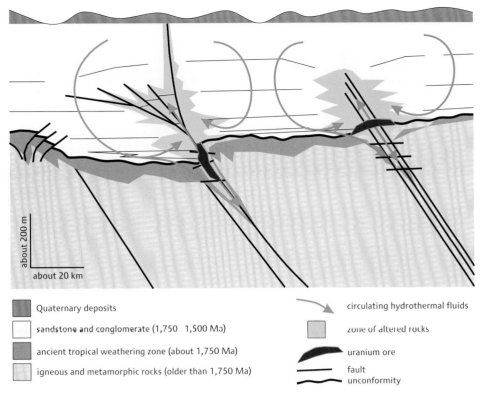

A simplified model of uranium deposits associated with unconformities in the Athabasca Basin of northern Saskatchewan and Alberta. Arrows show the interpreted paths of hydrothermal fluids that dissolved and carried the uranium to the deposit site. The hot waters also altered the surrounding rocks, providing clues to the discovery of these deposits. Hot water flow was driven by temperature differences (convection) and by repeated movements along the faults.

Uranium was discovered in Paleoproterozoic Huronian rocks in the Elliot Lake area of northern Ontario in the early 1950s, with Algoma becoming the world's largest uranium producer in 1953. Twelve mines were developed during the height of the Cold War, and in the late 1950s Elliot Lake became "the uranium capital of the world". During this period, uranium ranked as Canada's fourth largest export, after newsprint, wheat, and lumber.

Mining at Elliot Lake became unprofitable in 1996, mainly because new, large, high-grade uraninite deposits were discovered in the Athabasca Basin in northern Saskatchewan. In Chapter 6, we learned that the Athabasca Basin was a depression within the supercontinent Nuna: the Basin was filled with 1,750 to 1,500-million-year-old fluvial sediments after the erosion of mountains built during amalgamation of the supercontinent. The unconformity between the sandstone filling the Athabasca Basin and older basement rocks represents a late Paleoproterozoic land surface that became deeply weathered while exposed to tropical rain over a long time. Some 50 million years after the Athabasca Basin deposits had accumulated, circulating oxic hydrothermal fluids leached out uranium from the sediments and basement. When the uranium-rich hydrothermal fluids encountered oxygen-poor rocks and fluids at the unconformity, uraninite and coffinite (a uranium silicate) were precipitated. Reactivation of faults that had initially shaped the Basin sustained the flow of hydrothermal fluids, which in turn concentrated the uranium into nearly pure masses of uraninite. Today, uranium mining in Canada is restricted to the Athabasca Basin, where three world-class mines are operating or under development, and some surface mining also occurs. Other potentially economic uranium deposits exist in similar geological settings elsewhere on the Canadian Shield, for example within the Paleoproterozoic Thelon and Hornby Bay basins.

Canada is the second largest producer of uranium ore (11,997 tonnes in 2009), with a 25 percent share of global production, but uses only 20 percent of what it mines, exporting the rest. The estimated Canadian reserves of 1,693 million tonnes of uranium metal are equivalent in energy to about 15.7 billion cubic metres (135 billion barrels) of oil, which is greater than the country's conventional oil and gas resources together. And about 20 kilograms of uranium will produce as much energy as 400,000 kilograms of coal.

Headframes at the Cigar Lake Mine, a uranium mine in northern Saskatchewan. CHARLES JEFFERSON.

terozoic deposit was originally mined for its radium, which is of vital use in medical applications. But during World War II, the United States began to purchase uranium from the mine for the country's first nuclear bombs. In the early years of mining, uraninite was also extracted from vein deposits near Beaverlodge, Saskatchewan, and from granitic pegmatites near Bancroft, Ontario. (Pegmatites are very coarsely crystalline igneous rocks.) All these mines had ceased production by the mid-1980s, because of the discovery of more economical deposits elsewhere.

Sue C open uranium pit at the McClean Lake Mine, just west of Wollaston Lake, northern Saskatchewan. CHARLES JEFFERSON.

Renewing Our Options

Fossil fuels and uranium are non-renewable—once they have been used, they are gone forever—and reserves will run out in the foreseeable future. For this reason, and to avoid adding yet further quantities of carbon dioxide to the atmosphere and to avoid the real, potential, and perceived dangers of the use of nuclear power, we need to continue to pursue conservation measures and to develop renewable energy. At present, renewable energy includes solar power, wind power, tidal power, geothermal energy, and hydroelectricity. Together these sources now supply less than 5 percent of Canada's needs.

The amount of solar energy reaching the Earth in one year exceeds all the non-renewable energy that we have used or will ever need. Solar energy can produce heat, light, hot water, and electricity, and even cool us down. At present, its use in western economies is modest—most of it goes toward heating swimming pools and buildings. However, in countries such as China and India, solar heating of water is common and highly effective. Photovoltaic cells can produce solar energy for myriad uses, such as telecommunication repeater stations, water pumps, and laptop computers.

Wind has powered sailing ships, pumped water, and ground grain for many centuries. The best locations for wind turbines, a modern legacy of windmills, are places where winds are reliable and steady. One such location is Pincher Creek, Alberta, just east of the Rocky Mountains, where turbines are driven by winds flowing east from the mountains. Lakes and offshore areas are potentially favourable sites. A drawback to wind power is that no electricity is generated when the air is still, so that other sources have to fill the gaps in power supply. Also, storage of excess electricity in batteries is still inefficient and costly. Today, wind turbines are widely distributed across Canada and account for 1 percent of the country's electricity use. Ontario produces about one-third of this wind power.

Another alternative is geothermal energy. Measurements taken in boreholes show that the temperature rises, on average, 15 to 50°C for every 1,000 metres of depth—30°C per kilometre as a rule of thumb. This geothermal energy can be harnessed to produce electricity or heat buildings. Geothermal energy can be produced from steam and superheated water that reaches more than 150°C but stays liquid because of pressure. These conditions occur locally in tectonically active areas, and the steam is used to drive turbines. Iceland, astride a spreading ridge, produces about 20 percent of its energy from geothermal sources. Tectonically active areas in the Cordillera of British Columbia and Yukon have great potential for geothermal energy. Wells up to about 3,300 metres deep drilled at Mount Meager, a volcano about 100 kilometres north of Vancouver in the Cascade Magmatic Arc, have encountered temperatures up to 280°C. Development of this clean renewable resource is a distinct future possibility.

Even much cooler waters (50 to 150°C) can be used directly to heat buildings. A well drilled on the campus of the University of Regina, Saskatchewan, found water with a temperature of about 60°C some 2,200 metres below the surface. Energy from this source could satisfy the ongoing heating needs of about 3,000 people. Below 50°C, heat pumps can be used to increase the water temperature. In Springhill, Nova Scotia, water in old coal mines has a constant temperature of about 19°C. This warm water, which is pumped from about 140 metres below the surface, provides heat for some of the town's industrial buildings. In the process, it cools to about 13°C, but is then returned to the depths, where it re-circulates in the mine shafts and heats up again. The system provides an uninterrupted source of heat and reduces heating costs considerably. Even water cooler than 10°C can be used as an energy source if there is enough of it. At Carleton University in Ottawa, Ontario, groundwater with a temperature of 9°C provides energy for heating in winter and cooling in summer.

Canada is the world leader in hydroelectric power, the generation of electricity through the energy of flowing water. The country produces over a half of its electricity needs in this way. Indeed, all territories and provinces except Prince Edward Island now produce some hydroelectricity. A recent development is the generation of electricity from small turbines (micro-hydro facilities) placed in flowing water. Such facilities can be built at a frac-

Powering with wind, the modern way: wind turbines at North Cape, Prince Edward Island. RON GARNETT / AIRSCAPES.CA.

tion of the cost of large dams, but they produce relatively small amounts of energy and only work in rivers that flow year round.

A source of hydroelectric energy with much potential is tidal power, in which incoming and outgoing tidal flow drives turbines. Unlike wind power, tidal power is very predictable. In North America, the only present tidal power operation is the 18-megawatt plant at the mouth of the Annapolis River in Nova Scotia. The Annapolis River discharges into the Bay of Fundy at a site with a tidal range of about 8 to 9 metres. Incoming tides fill a dammed inlet, and the outflow is used to generate electricity. Drawbacks are that the plant operates much less than half the time (when the tide is ebbing), and that mud builds up behind the dam.

Still, tidal sources in the Bay of Fundy region have the potential to supply a large part of the power needs of Nova Scotia. With this in mind, a tidal power demonstration test site has been established in the Minas Passage region, near Parrsboro, to evaluate the design and operation of turbines to produce electricity from the strong flowing currents there. Such newer systems feature turbines that use the tidal current in both directions without a dam. As of 2012 one turbine has been deployed at the site, tested and later removed, and others will soon follow. The plan is to connect the turbines to the power grid with marine cables. Aside from the Fundy region, potential locations for tidal power occur in Ungava Bay in northern Quebec and in some bays in British Columbia.

Energy is a vital aspect of our lives, and many people consider that the future success of our society involves an increase in use of renewable resources and a decrease in reliance on fossil fuels and uranium. Access to more renewable resources would certainly help in our efforts to reduce greenhouse gases, and would lead to a cleaner planet.

The pool at Radium Hot Springs, Kootenay National Park of Canada, British Columbia, is heated to 39°C by geothermal energy; the water moves upward to the surface along a fault. B. WROBLESKI, COPYRIGHT PARKS CANADA.

14 · BUILDING CANADA

CHAPTER SUPPORTED BY A DONATION FROM INSTITUT NATIONAL DE LA RECHERCHE SCIENTIFIQUE, QUÉBEC

The Chateau Laurier Hotel (1908–1912, 1927–1929) in Ottawa, Ontario, is faced mainly with early Carboniferous Indiana Limestone from the United States. ROB FENSOME.

Aftermath of a Fire

June 1877 had been warm and dry in southwestern New Brunswick. At about 2:30 PM on Wednesday the 20th, a spark from a boiler shop or lumber mill—its origin is unclear—drifted into Henry Fairweather's storehouse near present-day Market Square in Saint John. The spark landed in a bundle of hay and within minutes firefighters were faced with a blaze fanned by a strong breeze. The city's buildings were constructed mostly of wood, so before long the fire had spread widely. In just over nine hours much of the city's core had burned down and nearly 2,800 people were homeless. An estimated 1,612 structures were destroyed, including nearly all the public buildings, as well as eight churches, six banks, and fourteen hotels.

Saint John, then as now, had a strong sense of community and enterprise, and within a year new buildings were appearing in the business district along Prince William and King streets. The building boom saw the development of

View along Prince William Street, Saint John, New Brunswick, about 1900. The first structure on the left is the old Bank of New Brunswick, constructed of a white sandstone of unknown, but possibly Nova Scotian, origin. The white stone is more fitting to a classical style than darker New Brunswick sources offered. Next to the Bank is the old Post Office Building, its street front exhibiting darker sandstone from the Dorchester area of New Brunswick. Farther down the street, the buildings are predominantly of brick with sandstone trim. COURTESY OF THE NEW BRUNSWICK MUSEUM.

The Saint John County Courthouse (1826–1829) on Sydney Street, Saint John, New Brunswick, predates the province's domestic dimension stone industry. The back wall, shown here, is a mixture of cheaper stones from a variety of sources, some local and some possibly retrieved from ballast. The nicely finished sandstone blocks to the right, representing the edge of the side wall, are of unknown origin, but may have arrived in New Brunswick as ballast from England. RANDALL MILLER.

Detail of the Pugsley Building (1879) at the corner of Prince William and Princess streets in Saint John, New Brunswick. Red and grey sandstone blocks used in its construction are possibly of late Carboniferous age and from Marys Point, New Brunswick. RANDALL MILLER.

an expanded commercial district along the waterfront, and new homes and churches higher up the hill. The Canadian government erected a new post office and the Custom House on Prince William Street. Architects from Saint John, Boston, New York, and Montréal were engaged to design office buildings and churches to replace those destroyed in the fire.

Many earlier stone buildings in Saint John had been constructed of imported rubble and ballast. These include St. John's Church, also known as the Stone Church, which was built between 1823 and 1828; the Saint John County Courthouse built between 1826 and 1829; and the fire hall, built in 1840. The discerning eye can pick out these earlier buildings based on their construction. By 1877, New Brunswick's building-stone industry had expanded, so fireproof material was readily available for reconstruction. Quarries in Curryville, Marys Point, and Caledonia, all in eastern New Brunswick, were producing fine-grained late Carboniferous sandstone in beautiful shades of brown, yellow, and red. Quarries in the St. George area were selling red and black varieties of so-called St. George Granite, used mainly for decorative work; and the quarries at Hampstead were producing white and pink Devonian granite. Saint John also had its own stone resources. Neoproterozoic marble from the Green Head Island area was used for walls and fences, and was also ground into lime to make mortar and plaster to rebuild Saint John. Indeed, it was in demand throughout the Maritimes and New England. The remains of lime kilns on Green Head Island are still there, overlooking the quarry.

Not all of Saint John's new buildings were made solely of stone. Brick buildings were erected, and many stone buildings had inner walls of brick. The rich variety of materials allowed architects, builders, and stonecarvers to create interesting effects by mixing colours, textures, rock types, and finishes. Sandstone was cut into regular blocks for large buildings and finished with a variety of patterns. Some blocks display simple striations and some have chiselled margins. Others are carved with complex worm-like trails, and still others have a rough, hammered finish. Several buildings combine brick walls with stone trim, columns, and carvings around win-

Carving in yellow sandstone by James McAvity of a head spitting coins. This "grotesque" appears on the Palatine Building (1877–1878) on Prince William Street, Saint John, New Brunswick. RANDALL MILLER.

The City Market (1876) in Saint John, New Brunswick, is the oldest continuously operating farmers' market in Canada. It has decorative sandstone trim, but is largely made of locally manufactured brick. RANDALL MILLER.

Granite columns surround a doorway of the Palatine Building (1877–1878), Saint John, New Brunswick. Features carved in granite are still shiny and fresh, in contrast to sandstone carvings, which are losing their detail. The grey and red pillars shown here are of Silurian-Devonian age and were quarried near St. George, New Brunswick. RANDALL MILLER.

dows and doors. Granite columns and window frames were adorned with sandstone carvings of stone grapes, heads, and animals. Local stonecarver James McAvity did much of this decorative work. His sense of humour shows through his artistry: on the old Bank of Nova Scotia Building (now the Palatine Building), one column displays a head spitting coins, and the Chubb Building across the street is trimmed with faces that McAvity carved to resemble colourful local politicians of the day. Not everyone in Saint John was impressed with McAvity's talents. The *Daily Sun* wrote that the Chubb Building had "been highly disfigured by these meaningless heads, which stand boldly out in all their ugliness."

Visitors to Saint John today will find a city core that owes its charm to a combination of the regional geology and a disaster. Entering the historic Trinity Royal Preservation Area, they will see a collection of small, architecturally intricate buildings made of stone and brick. Many of these buildings provide a level of elegance that is commonly missing in modern steel and glass structures. It's not so surprising, then, that Prince William Street has received a Designation of National Historic Significance as an important late nineteenth-century architectural and commercial landscape.

Setting the Scene

Aside from their practical importance, building stone and construction materials provide insights into a region's geology and history, as is the case for Saint John. Building stones are classified as industrial minerals, which also include gravel and crushed stone (collectively called mineral aggregate) for roads and railways, limestone for cement, shale for brick, clay for dinnerware and paper coatings, feldspar for paint and toothpaste, and quartz for glass and computers. Evaporites—such as the salt we use on food, gypsum for wallboard, and potash for fertilizer—are also considered industrial minerals.

More substantial building materials such as stone are rare in the Prairies, so many of the original farmsteads were made of wattle and daub, basically sticks and soil. This farmstead, photographed in 1979, was near Mendham, Saskatchewan. ROB FENSOME.

Most industrial "minerals" are actually rocks, not minerals. Commercially, "stone" is any type of rock used for building or ornamental purposes. Building stone is valued mainly for its strength and durability, and can range from unfinished "fieldstone" for a rubble wall or fireplace, to cut and polished stone, so-called dimension stone, for a bank or a church. Dimension stone can also be used for items such as curbs and planters. Ornamental stone is used for statues and monuments and is selected for its colour and texture.

Commercial building-stone categories include "granite", "marble", "limestone", "sandstone", and "slate"—the terms are in quotes because the commercial definitions of these categories are commonly much broader than the corresponding geological definitions. Commercial granite is any hard crystalline rock that can be cut and polished, and includes granite and its igneous kin, basalt and gabbro, and even some metamorphic rocks such as gneiss. Dark igneous rock such as gabbro is sold as "black granite". Commercial marble is any crystalline carbonate rock that can take a polish, such as crystalline limestone and dolostone, as well as marble in the strict geological sense. Commercial marble also includes serpentinite, a rock associated with ophiolites and made of silicates rather than carbonates. Commercial limestone and sandstone are closer to their geological definitions, although commercial limestone includes dolostone and commercial sandstone includes conglomerate. Slate, commercially and geologically, is a fine-grained metamorphic rock that splits easily into thin slabs along cleavage planes. Stone companies promote building stones under trade names. One that we've already encountered is the famous Tyndall Stone from Manitoba. Another trade name is Stanstead Grey Granite, an igneous rock from Quebec. The trade names reflect the location of the quarry, the colour of the stone, or simply a distinctive brand.

Fieldstone barn at Indian Head, Saskatchewan, recently refurbished. The stones used in its construction are a variety of rock types from the Canadian Shield, transported to the Indian Head area by ice during the last glaciation. ROB FENSOME.

The University of Saskatchewan Observatory (1929) in Saskatoon is constructed from Ordovician dolostone, locally called greystone. Despite its uniformity, much of Saskatoon's greystone is actually a fieldstone, as explained on page 284. ROB FENSOME.

CHAPTER 14 · BUILDING CANADA

Weathered sandstone gravestone, Clementsport, Nova Scotia. ROB FENSOME.

A gravestone for victims of the sinking of the *Titanic*, Fairview Lawn Cemetery, Halifax, Nova Scotia. Although set up over a hundred years ago, this poignant marker of a family tragedy, made of hard gabbro, still looks fresh. ROB FENSOME.

Most early building stone was fieldstone—any loose stone that could be used for constructing walls or foundations. Most Canadian fieldstone consists of glacial cobbles and boulders, readily available wherever the ice left them. For example, the Little Stone Schoolhouse (1887), now on the University of Saskatchewan campus and Saskatoon's first permanent school, was constructed using representative sampling of Canadian Shield bedrock brought south by the last Laurentide Ice Sheet. Fieldstone continued to be used for convenience, style, or just to keep costs down. And it is not always from local fields. For example, Vancouver's Gertrude Lawson House (1939) is made with granitic and volcanic cobbles brought from New Zealand as ship ballast.

Stone is naturally resistant to fire, which is why it forms the oldest surviving structures in most Canadian communities. Indeed, the story of fire and subsequent rebuilding in stone, as in Saint John, is repeated across Canada: the Québec City Lower Town fire in 1682, the Montréal fire of 1852, the Calgary and Vancouver fires of 1886, the St. John's fire of 1892, and the Toronto fire of 1904. The oldest stone buildings in a community reflect the regional geology. In the early days, use of locally quarried stone minimized hauling costs over rudimentary road networks. The 1800s and early 1900s were the heyday of the Canadian building-stone industry. In the Maritimes, stone architecture blossomed and millions of tonnes of granite and sandstone from the region were used throughout eastern North America. British Columbia sandstone was shipped down the Pacific coast. And in Ontario, limestone and sandstone quarries supplied the demand for massive construction projects such as bridges and canals. Completion of the national transcontinental railway in 1885 meant greater access to more distant stone by architects and builders in central Canada. High-quality Quebec granite and marble crossed the country and Tyndall Stone from Manitoba began its rise to prominence.

Graveyards are excellent places to look for exotic rocks, study local history, and observe the effects of weathering on stone over the decades. The stone used for gravestones, statues, and other monuments should be of the highest quality, have aesthetic appeal, and resist the elements. It is possible to estimate the age of a Maritime graveyard from the type of stones used. Early European settlers lacked the technology and stone for elaborate memorials, so they used the easy-to-work local sandstone or slate. In the 1850s headstones of marble from Italy and Vermont became fashionable. By the 1870s, the ready availability of granite coincided with improvements in granite etching techniques.

By the 1890s, cement and concrete began to displace stone in foundations and other structures. Ultimately steel and reinforced concrete were used instead of stone or brick to form the structural framework of buildings. Even so, the aesthetic appeal and lasting finish of stone have ensured its continued use as exterior cladding and interior finishing.

Building with Sedimentary Rocks

Sedimentary rocks such as sandstone, limestone, and dolostone are easy to quarry and finish. These rocks are widespread and thus are found in many of the oldest surviving structures in Canada, but their use as building stone reflects regional geology. Mesozoic sedimentary rocks are the main building stones in western Canada, early Paleozoic sedimentary rocks are more common in Ontario and Quebec, and late Paleozoic sedimentary rocks dominate in the Maritimes.

Building stones in the Vancouver-Victoria region initially were quarried from late Cretaceous sandstones on the Gulf Islands. As early as 1837, Gulf Islands Sandstone was exported south to San Francisco, where it was used in the construction of the city's mint building. Many of the large commercial blocks, warehouses, and hotels in the Gastown area of Vancouver were built of Gulf Islands Sandstone. This rock was not only used for building stones: a quarry on Newcastle Island in Nanaimo Harbour (handy for overseas export) produced pulp-grinding wheels that were shipped all over the world.

Alberta's best-known building stone is the fluvial Paleocene Paskapoo Sandstone, a soft, easy-to-carve stone with subtle hues of yellow, buff, grey, and golden brown. Its prominence started with the devastating fire of November 7, 1886, that destroyed many of the wooden buildings on Calgary's main street. Calgarians were determined to rebuild with a fireproof material. Outcrops of Paskapoo Sandstone along the Bow River provided the material, and stonemasons from Europe provided the skill. The stone was used in government buildings, schools, and churches, and gave the streets an aura of substance that transformed a pio-

Carved heads peek out from ornate fringes on the lower part of the Greenshields Building on Water Street in Vancouver, British Columbia. The stone for the carvings and surrounds is Gulf Islands or related sandstone, though the location of the quarry is unknown. LINDA CAMPBELL.

Paleocene Paskapoo Sandstone was used for trim and columns, together with brick, in the construction of Southminster United Church (1913) in Lethbridge, Alberta. DIXON EDWARDS.

The Kamenka stone quarry in the Bow Valley, just outside Banff National Park of Canada, Alberta, produces hand-split Rundlestone, a Triassic clastic deposit used regionally as a building stone. DIXON EDWARDS.

The Banff Springs Hotel in Banff, Alberta, is largely constructed of Triassic Rundlestone. In the background the Bow River cascades over Triassic strata at Bow Falls. KEITH VAUGHAN.

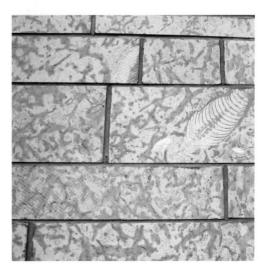

Detail of the TCU Financial Group Building (2003) in Saskatoon, Saskatchewan, which is partly constructed of late Ordovician Tyndall Stone, one block of which shows a section through a nautiloid. KIM MYSYK.

neer settlement into an urban centre. Edmonton also has many buildings of Paskapoo Sandstone, including the Alberta Legislature Building, constructed between 1907 and 1913. Another notable Alberta stone is bluish-grey Rundlestone (or Rundle Rock), a fine-grained Triassic sandstone that originated on the Pangean continental shelf. Rundlestone is a hard, resistant rock that occurs in the thrust-and-fold belt of the Rocky Mountains and is best known for its use in the Banff Springs Hotel (1928).

Historic buildings farther east on the Prairies are made of fieldstone or quarried carbonates. During the Ordovician, much of the region was covered by a continental sea in which the carbonates accumulated, including the dolostone known as greystone, which gave Saskatoon its old nickname, the Greystone City. In Saskatoon, many of the University of Saskatchewan's first stone buildings used greystone. The source of this stone was a large boulder field of erratics north of campus on the south side of the South Saskatchewan River, derived from the outcrop belt 300 kilometres to the northeast.

The Ordovician sea also produced Tyndall Stone, Canada's best-known building stone, which comes from Garson, Manitoba. First used in the construction of Lower Fort Garry, near Winnipeg, in 1832, Tyndall Stone is a light-grey to cream-coloured, highly fossiliferous, mottled carbonate. Its distinctive texture reflects differences in colour between the sediment matrix, made of the mineral calcite, and the network of fossil worm or shrimp burrows, now filled with the mineral dolomite, that penetrate the rock. Tyndall Stone is durable and of high quality. It has been widely used for walls, columns, and decorations, and is familiar to many as the backdrop to television interviews conducted in front of the Parliament Buildings in Ottawa.

The building-stone industry in Ontario started with the construction of fortresses during the War of 1812, and of the Rideau and Welland canals between 1824 and 1831. These large projects employed many European stonemasons, so that by the mid-1830s a skilled labour force was available to quarry and lay stone for buildings and homes. One of the principal stones was Silurian Queenston Limestone, a dolostone from the Niagara Escarpment. This pearl-grey rock weathers brownish-buff and contains distinctive pink calcite crinoid stalks. It was used throughout the Niagara region and was shipped as far away as Halifax, where it is found in the Halifax City Memorial Library (1951), and St. John's, where it was used during renovations of the Colonial Building in the 1960s. Another Silurian rock, the grey and red Whirlpool Sandstone, was used in many buildings in the Toronto and Niagara regions, including the Ontario Legislative Building and the older buildings of the Royal Ontario Museum in Toronto. It is named after the Whirlpool section of the Niagara Gorge, where it outcrops, although it was quarried mainly in the Peel and Halton regions of southern Ontario. Middle Ordovician Black River Limestone quarried around Kingston were

Reinforced concrete of the Whitney Block (1925–1928), Toronto, Ontario, is faced with Silurian Queenston Limestone. ROB FENSOME.

Detail of the Whitney Block, showing the finish of the Queenston Limestone. ROB FENSOME.

Part of the Ontario Legislative Building (1886–1893), Toronto, Ontario. The principal building stone is Silurian Whirlpool Sandstone, quarried near the forks of the Credit and Orangeville rivers. ROB FENSOME.

Detail of the Ontario Parliament Buildings showing an intricately carved detail in a Whirlpool Sandstone block. Note the cross-bedding in the sandstone. ROB FENSOME.

used in so many early buildings that the town became known as the Limestone City. These rocks form the distinctive older buildings of Queen's University. Other limestone building stones from Ontario are the Adair, Silverwater, and Eramosa "marbles", quarried from early Silurian limestone and dolostone on the Bruce Peninsula and Manitoulin Island.

Two limestones, both containing brachiopods and broken bryozoa, were quarried in Quebec and used routinely throughout the nineteenth century. One, the middle Ordovician Chazy Limestone, was used in the construction of Montréal's Windsor Station (1887–1889). The other is the late Ordovician Trenton Limestone, which was used for the Vieux séminaire de Saint-Sulpice (1684–1687), the oldest building still in place in old Montréal.

In the Maritimes, non-marine Carboniferous to Permian sandstones, formed in the Pangean heartland were used extensively, and range from red, brown, or purple to grey, olive, and buff. One of the oldest stone houses in New Brunswick, the MacDonald Farmhouse (1820) at Bartibog, was built from sandstone quarried less than a kilometre away. Sandstones quarried on Prince Edward Island are referred to as Island Stones. The olive-coloured, late Carboniferous Wallace Sandstone from northern Nova Scotia was widely used throughout the Maritimes and in two of the region's most famous buildings: Province House (1811–1818) in Halifax and its namesake, Province House (1843–1847) in Charlottetown, the site of the 1864 conference that led to Confederation. Maritime sandstones were exported to other

Ordovician Chazy Limestone from Quebec was used in the construction of Montréal's Windsor Station (1887–1889). PIERRE BÉDARD.

Government House (1831) in St. John's, Newfoundland, is built mainly of red Ediacaran sandstone quarried from the slopes of nearby Signal Hill. The granite trim is of unknown origin. LAWSON DICKSON.

The MacDonald Farmhouse (1820), Bartibog, New Brunswick, is built of local Carboniferous sandstone. GWEN MARTIN.

The sandstone blocks used to build the Cabot Tower (1897) at Signal Hill in St. John's were in part salvaged from the ruins of a former barracks and hospital built on the site in 1842–1843 and destroyed by fire in 1892. The lower part of the structure is local Ediacaran conglomerate. The upper part is mainly of local Ediacaran grey sandstone, with Nova Scotian sandstone (probably Carboniferous Wallace Sandstone) used around the windows and for trim. LAWSON DICKSON.

parts of Canada as well as to the northeastern United States, where they grace some of New York's famous brownstone buildings and the bridges of Central Park.

In Newfoundland, grey Ediacaran sandstone of deep marine origin from the St. John's area and Cambrian shallow-marine limestone from Conception Bay were used by the English to construct and maintain early fortifications around St. John's. Early non-military stone buildings are rare in the province because many fishermen and officials returned home to Europe for the winter. After 1817, however, colonial governors had to live on the island year round. Government House (1831) required thousands of tonnes of red Ediacaran, shallow-marine to deltaic sandstone quarried from the slopes of Signal Hill in St. John's. Starting in the early 1840s, the Catholic Church constructed stone churches in Torbay, St. John's, and Ferryland. The same grey Ediacaran sandstone used for the fortifications was used extensively for these churches, along with an Ordovician sandstone of shallow marine origin from Kelly's Island in Conception Bay. Limestone from Galway in Ireland was used to face the Basilica of St. John the Baptist in St. John's, but it proved unsuitable for the climate and the outside walls later had to be refaced with the local grey sandstone.

Building with Igneous Rocks

Most igneous rocks are more durable but harder to work than most sedimentary rocks because of their interlocking-crystal or harder-mineral makeup. One of the toughest igneous rocks is granite, which, although expensive, has always been popular because of its beauty and durability. It is an ideal building material for foundations, steps, and curbing. Volcanic rocks are generally less desirable than coarse-grained plutonic rocks because they tend to be less durable, are not as uniform in texture, and are considered less attractive.

Igneous rocks are most common in areas with active or exposed ancient orogens. Most of Canada's igneous building stones come from the Cordilleran Orogen in the west and the Appalachian Orogen in the east. The Canadian Shield is also a source, but only where it is close to consumers—as is the case for parts of the Grenville Orogen in Quebec and Ontario—or where particularly beautiful and thus valuable stones are exposed, such as the Blue Eyes Stone from Labrador, described on page 289.

In British Columbia, building stones were quarried from early Cretaceous granitic intrusions on the islands and along the coast north of Vancouver. Among these stones is Grey Coastal Granite, which has been quarried in the Coast Mountains since the 1890s. Its main use has been in the construction of foundations, including those of the British Columbia Parliament Buildings (1897) in Victoria. It was also used for the breakwaters in Victoria and Vancouver. Above their foundations, the British Columbia Parliament Buildings are made from light-grey Haddington Island Andesite—geologically a dacite—quarried from Miocene lava flows on Haddington Island near the north end of Vancouver Island. Around 1910, this volcanic rock replaced Gulf Islands Sandstone as the principal regional building stone, perhaps because it is more resistant to weathering, is slightly harder, and is especially good for carving.

As no orogens are exposed on the Prairies, igneous building stones are not available locally. However, commercial granites from other parts of Canada were moved there by rail, as with Grey Coastal Granite, which was used to build the foundation and lower walls of the Alberta Legislature Building in Edmonton. Granite has also been brought to the Prairies from abroad. The imposing twin red towers of the Suncor Energy Centre in Calgary, for example, are partly clad with Mesoproterozoic granite from Finland.

Quebec has long been Canada's principal granite-quarrying region. Quebec granites derive from two sources: the Mesoproterozoic Grenville Orogen north of the St. Lawrence River and the Paleozoic Appalachian Orogen to the south, where Devonian intrusions penetrate older sedimentary rocks. The Grenville Orogen provides brownish, red-pink, green, and black granitic rocks. Neoproterozoic Black Granite (geologically an anortho-

Detail of the Hotel Vancouver, on West Georgia Street, Vancouver, British Columbia. Construction on this elegant building began in 1929 but, because of delays caused by the Depression, was not completed until 1939. The exterior is finished with Haddington Island Andesite, shown here. The well-preserved decorations demonstrate this stone's suitability for carving. LINDA CAMPBELL.

Polished anorthosite (a mafic igneous rock) with coarse iridescent plagioclase feldspar crystals and fewer dark pyroxene crystals is used as facing on pillars at Constitution Square (1987), on Albert Street in Ottawa, Ontario. ROB FENSOME.

site) has been quarried at Saint-Nazaire since 1980 and is now one of the most popular monument stones in North America; polishing it brings out the colours of the different minerals. Granite from Rivière-à-Pierre near Québec City is highly prized for paving stones, monument bases, and curbstones.

Appalachian granites of the Stanstead area south of Sherbrooke, Quebec, have been valued by the stone industry since the 1860s. Stanstead Grey Granite—a medium-to-coarse-grained, light-to-silver-grey Devonian granite—was used extensively throughout Canada during the 1920s to 1930s; it forms, for example, the base of Saskatoon's Bessborough Hotel

(1931). In Montréal, Stanstead Grey Granite appears in numerous monuments, foundations, columns, and walls. An example is the Sun Life Building, which at the time of completion in 1918 was reputed to have been the tallest building in the British Commonwealth.

All Maritime granites used as building stones are of Silurian or Devonian age. The St. George Granite, quarried in New Brunswick, ranges from true granite to gabbro and can be red, pink, grey, or black. Many communities in New England and eastern Canada have older buildings or monuments with pillars or columns of this rock, including Halifax City Hall, built in 1841. Granitic rocks of the South Mountain Batholith in Nova Scotia have been quarried at several sites, and have a common grey look. Local granite, probably from Purcell's Cove, was used in the facade and 45-metre spire of St. Mary's Basilica (1820–1829) in Halifax. It has the tallest granite steeple in North America.

In Newfoundland, granite was quarried to construct trestle abutments and piers for the Newfoundland Railway (1881–1897), a narrow-gauge railway extending from St. John's to Port aux Basques. Quarries were developed near the railway at Holyrood, Goobies, Benton, and The Topsails, each with its own distinctive granite. The St. John's Railway Station, now the Railway and Coastal Museum, is constructed with the striking yellow granite from The Topsails, and cobbles of this granite and the Petites Granite were used to pave Water Street in St. John's. The Petites Granite, which was produced from a small quarry at Petites on the southwest coast of Newfoundland, was used also in the Court House in St. John's. The Rose Blanche Lighthouse (1871, restored 1996–1999) was built from local Rose Blanche Granite and is one of the last working granite lighthouses on the east coast. Another Newfoundland quarry near Bishop's Falls, in the central part of the island, produces a striking black Silurian gabbro. Marketed under the commercial name Ebony Black, this rock is used mainly for monuments, but has been also used for black floor tiles and facing stone. The Rooms, a museum constructed in St. John's in 2005, has interior floors

Several buildings on Sainte-Hélène Island, Montréal, Quebec, including the restaurant, tower (shown at top), and fort (now the Stewart Museum), were built using a breccia quarried at the foot of the Jacques-Cartier Bridge. This unusual rock (shown in close-up in the bottom photo) was produced by explosive volcanic eruptions associated with the Cretaceous Monteregian intrusions. It is found on Montréal, Bizard, Cadieux, and Sainte-Hélène islands in the Montréal area. BOTH PHOTOS BY PIERRE BÉDARD.

The granite quarry near the shoreline at Petites, southwestern Newfoundland. DAVE EVANS.

Detail of the Frost Building North (1954), Toronto, Ontario. The exterior is mainly Queenston Limestone (right), with Stanstead Grey Granite (immediately right of the door frame). ROB FENSOME.

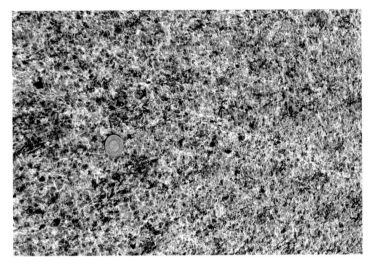

Blue Eyes or Reflect Blue, Mesoproterozoic anorthosite quarried near Nain, Labrador, is one of Canada's most recognizable building stones: the trade names refer to the presence of striking blue labradorite crystals. LAWSON DICKSON.

Details of the Canadian Volunteers Monument (1870) in Toronto, Ontario, commemorating volunteers who perished in the Fenian Raids of 1866. The statue (top) is one of several on the monument that were carved from Italian Carrara Marble. It has passed the test of time better than the motif in the lower photo, carved from soft Carboniferous Nova Scotia sandstone. ROB FENSOME.

and walls that display this black rock in spectacular combination with a Silurian salmon-pink granite from central Newfoundland.

One of Canada's most widely distributed building stones, known in the trade as Blue Eyes or Reflect Blue, is a Mesoproterozoic anorthosite quarried near Nain in northern Labrador (Chapter 6). It is a mafic plutonic rock whose crystals of labradorite (a variety of feldspar) commonly display a beautiful iridescent play of colours. This effect is due to the diffraction of light within the crystal structure. Most of the Nain stone is exported to Italy and China, where it is cut and polished into floor tiles and facing stone. This remarkable material can be seen in the facing stone and floor tiles in the entrance hall of the Natural Resources Building on Elizabeth Avenue in St. John's.

Building with Metamorphic Rocks

Three types of metamorphic rock are used commercially: marble, gneiss, and slate. Marble was first used in ancient times and is still one of the most popular building stones. Geological marble is commonly of uniform texture and is easily worked, and so ideal for sculpting statues. Some marble has subtle colours and veining due to impurities, making it desirable for flooring, steps, and wall facings. Generally, marble is used in interior applications, where it is not subject to salty coastal air, city smog, or acid precipitation. Gneiss, which is commercially called granite, can have an ornamental subtlety similar to marble, but its composition of quartz and feldspar rather than carbonate make it much more durable and so better for exterior cladding. Slate is tough, waterproof, and cleaves into slabs and thin plates. These properties make commercial slate superb for roofing and flooring tiles.

Metamorphic rocks, like igneous rocks, are generally associated with orogens. Hence, among the western provinces, only British Columbia has produced metamorphic building stone. Slate from Jervis Inlet in the Coast Mountains was used as exterior flooring in Government House in Victoria, and deep-red Malaspina Marble from Texada Island was used in the British Columbia Parliament Buildings.

Missisquoi Marble, quarried in the Philipsburg area of Quebec, has been used widely across Canada. Taconic metamorphism of this Cambrian rock in the middle Ordovician has given it a wide range of colours and textures. Missisquoi Marble graces the interiors of the provincial parliament buildings in Alberta, Manitoba, Ontario, and Quebec, and the Centre Block and Peace Tower of the Parliament Buildings in Ottawa. Quebec was the principal supplier of slate in Canada from 1870 to 1880, when about 80 percent of the nation's production came from near Sherbrooke.

A variety of local metamorphic rocks has been quarried and used in Atlantic Canada. Examples include Neoproterozoic marble in the Cathedral of Immaculate Conception (1853) in Saint John and Ordovician-Silurian quartzite in Horton House (1915) at Acadia University in Wolfville. Stone walls and several buildings at Dalhousie University and elsewhere in Halifax were constructed from distinctive "ironstone" quarried at nearby Purcells Cove and from the deep railway cut that runs through the city's south end. The ironstone began as an early Ordovician mudstone from the Meguma Terrane, but was metamorphosed by compression to slate during the Devonian Neoacadian Orogeny, and then heated from nearby magma of the South Mountain Batholith. Away from the Batholith, the equivalent rocks are much more friable and cleaved, and do not make good building stones.

An example of metamorphic building stones from Newfoundland is Cambrian slate from Nut Cove. These rocks were quarried in the late 1880s, producing purple and green slate roofing tiles for export. Today a new quarry at Nut Cove is producing red and green roofing and facing tiles, as well as other slate products. In northern Labrador, builders used local gneiss boulders in the foundation of the eighteenth-century Moravian Hebron Mission.

Summing up Aggregates, Clay, and Concrete

Clay beds, shales, or mudstones can all be a source of commercial clay. Three types of clay are valuable commercially: common clay, consisting mainly of illite; kaolin clay; and smectite clay, especially bentonite, which commonly results from the weathering of volcanic ash. Shale and common clay are present in all provinces and territories and are used to make tiles and bricks. Kaolin clay is

The Prince of Wales Martello Tower (1796–1797) in Halifax, Nova Scotia, was built of local Meguma "ironstone". ROB FENSOME.

a white aluminum-rich clay used in paints, plastics, and ceramics, and also as a filler in the pulp and paper industry. Bentonite is the bonding agent used in oil-well drilling muds. It is used also in the manufacture of impermeable linings for tailings ponds and to form iron ore into pellets prior to heating in a blast furnace. Canada lacks major kaolin and bentonite deposits, although limited amounts of both have been produced from Cretaceous rocks on the Prairies.

Bricks are manufactured from finely ground shale or clay mixed with water, and are then fired in a kiln to make them hard and weatherproof. In the firing process, clay minerals slowly soften and fuse into a tight, solid, water-resistant mass. Brick is an essential construction material, providing fire protection, high insulating value, long life, low maintenance, and aesthetic appeal. Brick manufacturing in Canada can be traced back to the early seventeenth century, when skilled immigrants in Quebec produced bricks for early homes and hearths. Bricks now a century old, made from North Saskatchewan River floodplain clay, were used to construct some of the buildings of the University of Alberta in Edmonton. The Don Valley Brick Works in Toronto, site of Arthur Coleman's pioneering studies of Ice Age deposits (Chapter 11), is the source of bricks for many Victorian buildings in the city, including Convocation Hall at the University of Toronto. The McGill Faculty Club in Montréal was built in 1887 from Trenton Limestone and red brick. In Truro, Nova Scotia, the police station (1884) and many older buildings along Prince Street are constructed of local sandstone and brick.

The term mineral aggregate, or just aggregate, is applied to sand, gravel, and crushed stone. Although perhaps not very glamourous, the aggregate industry is an economic giant. The quantity of aggregate produced in Canada each year is more than twice that of all other Canadian mineral commodities combined. Aggregate is used extensively by the construction industry in concrete, for example to build roads and bridges.

Sand and gravel deposits accumulate through processes that sort and concentrate sediment into different grain sizes. Deposits include those accumulated as sandbars in river systems, beaches, and eskers. Sand and gravel are favoured sources of aggregate because they are easy to excavate and are widely distributed. Pits tend to be small and the products used locally. In British Columbia, gravel has been mined from a late-glacial delta on southern Vancouver Island since the early twentieth century. Similar delta and glacial outwash deposits are quarried for aggregate east of Vancouver and near Bella Coola, Terrace, and Kitimat farther north. Most interior communities in British Columbia use local sources of aggregate of glacial origin.

The local brick and Paskapoo Sandstone used to build Athabasca Hall (1906–1911) at the University of Alberta in Edmonton suffered significant weathering. The building was restored during the 1970s and now looks much as it did when it was first built. DIXON EDWARDS.

The major sources of sand and gravel in Alberta and southern Saskatchewan are glacial outwash and fluvial deposits. Preglacial gravels contain many rock types derived from the Rocky Mountains, including sandstone, limestone, and chert. Quaternary and younger fluvial deposits contain Precambrian granite and gneiss pebbles transported by glaciers from the Canadian Shield. Even in concrete, the difference in pebble compositions is clearly evident. In Saskatchewan, Manitoba, Ontario, and Quebec, outwash deposits (including those forming eskers and kames) are primary sources of sand and gravel for aggregate; all contain igneous and metamorphic Canadian Shield rocks.

Early European settlers in the Maritimes removed sand and gravel from beaches and islands for their aggregate needs. Beach materials are no longer used because their removal tends to destroy the beaches, but the shells visible in some early concrete structures are a reminder of their source.

As unconsolidated deposits become depleted or are made unavailable through competing land use, crushed stone becomes an attractive economic alternative. Crushed stone is manufactured from quarried bedrock. All the crushed pieces are of the same rock type and are angular in shape. Some of the largest crushed-stone operations in Canada are sited where large volumes of quality material can be transported cheaply by ship. In British Columbia, crushed late Triassic Quatsino Limestone from Texada Island in the Strait of Georgia is

Gravel pit developed in an esker (Chapter 11), near McAdam, New Brunswick. ALLEN SEAMAN.

The Decarie Interchange in Montréal, Quebec, illustrates the demand for construction materials. REPRODUCED WITH THE PERMISSION OF NATURAL RESOURCES CANADA 2013, COURTESY OF THE GEOLOGICAL SURVEY OF CANADA.

loaded on bulk carriers and shipped to markets from Alaska to California. Limestone on Manitoulin Island, Ontario, is quarried and crushed, then shipped to ports on the Great Lakes. A quarry on the mainland shore of the Strait of Canso, which separates Cape Breton Island from the rest of Nova Scotia, supplies crushed stone to the Atlantic seaboard. The quarry was originally opened to supply aggregate for construction of the Canso Causeway in the 1950s.

Cement production is another giant industry. A tonne of cement requires about 1.6 tonnes of raw rock, principally limestone. The limestone is burned with smaller amounts of aluminum oxide (from shale, clay, or ash), silica (from sand or sandstone), iron oxide, and a little gypsum. The result is something called clinker, which is then ground to make cement. Sixteen plants in five provinces (British Columbia, Alberta, Ontario, Quebec, and Nova Scotia) manufacture cement. The age of the limestone quarried for cement-making is Permian and Triassic in British Columbia, Devonian in Alberta and southern Ontario, Ordovician in Quebec, and Carboniferous in Nova Scotia.

Reclamation is another important aspect of the industrial minerals industry. Possibly the best-known Canadian examples, both in abandoned quarries, are the Butchart Gardens at Brentwood Bay, near Victoria, British Columbia, and Canada's largest botanical garden, the Royal Botanical Gardens in Burlington, Ontario. The next time you admire the spectacular floral displays in these gardens, you might also reflect on the vast and vital enterprise that created such spaces.

The Butchart Gardens at Brentwood Bay, British Columbia, one of Vancouver Island's prime visitor attractions, is on the site of an abandoned limestone quarry. COPYRIGHT 2013, COURTESY OF THE BUTCHART GARDENS.

In this chapter, our focus has been on building stones, bricks, aggregate, and cement. Other industrial minerals mentioned elsewhere in this book include gemstones (Chapter 12), sulphur (Chapter 13), and evaporites. Many more rocks and minerals—asbestos, barite, magnesite, marl, mica, phosphate, pumice, quartz, soapstone, talc, and zeolite—fall into the industrial-mineral basket. Although lack of space prevents their coverage here, commodities produced from them are all around us and play key roles in our lives.

BOX 10 · QUÉBEC CITY'S EXTRAORDINARY GEOLOGICAL LEGACY

A Geological Crossroads

Many Canadian cities have distinctive geological settings and a variety of natural building materials that make them interesting places to live, each with its own particular personality. However, one city stands alone as a showcase of Canadian geology intertwined with history, culture, and beauty—Québec City. Now a UNESCO World Heritage Site, Québec City is a thriving francophone metropolis with a rich French and British heritage reflected in its architectural treasures. It is the only city in North America to have retained most of its city walls, which bear witness to the region's turbulent history from the seventeenth to nineteenth centuries. The original colony was founded in 1608 at the foot of Cap Diamant, a defensible promontory that towers over the St. Lawrence River, Quebec's freeway to and from France. This mighty river was travelled by Jacques Cartier, Samuel de Champlain, and other early explorers, as well as fur traders and Aboriginal peoples including the Iroquois and Huron. Geographic features such as the narrowing of the St. Lawrence River and the strategic Cap Diamant promontory, fundamental to the very history of Québec City, can be best appreciated through an understanding of the underlying geology.

Québec City stands at the junction of three major geological domains: the Canadian Shield (more specifically the Grenville Orogen), the St. Lawrence Platform, and the Appalachian Orogen. The Grenville Orogen, as we discovered in Chapter 6, resulted from collisions that culminated in the formation of the supercontinent Rodinia about 1,000 million years ago. Rodinia started to break up about 750 million years ago into smaller continents, one of which was Laurentia, the ancestral core of modern North America. During the Cambrian, some 500 million years ago, Laurentia was close to the equator and where

This downstream view of the St. Lawrence River reveals why the Cap Diamant, or Québec, promontory is so strategically important. The promontory consists of resistant rocks of the most northwesterly of the thrust sheets of the Appalachian Orogen. The highlands in the far distance are underlain by rocks of the Grenville Orogen of the Canadian Shield. Most of the lowlands in this view were covered some 12,000 years ago by the Champlain Sea. ROB FENSOME

Québec City stands today was close to the boundary between the continental margin of Laurentia and the Iapetus Ocean. The waters of Iapetus periodically lapped onto Laurentia, depositing layers of sediment that have remained largely undeformed and which underlie the St. Lawrence Platform. From the Ordovician to the Devonian, a succession of magmatic arcs and large and small continents converged and collided with the former Laurentian margin, eventually leading to the formation of the supercontinent Pangea around 310 million years ago. These collisions produced a swath of deformed rocks, the Appalachian Orogen, that extends from Newfoundland through southern Quebec and the Maritimes to Alabama.

As in the Cordillera, the innermost ripples of plate collisions resulted in a thrust-and-fold belt along the former Laurentian margin. The Cap Diamant promontory is located on the edge of this thrust-and-fold belt. Known as Logan's Line after William Logan, this edge is the modern northwestern boundary of Appalachian deformation. Along Logan's Line, we see relatively resistant Ordovician muddy limestones that were thrust over rocks of the St. Lawrence Platform during the Taconic Orogeny (Chapter 7). The escalier Casse-Cou, or Break-Neck Steps, one of the earliest links between the Upper Town and Lower Town, traverses the thrust sheet forming the Cap Diamant promontory. Most of Île d'Orléans and all of the South Shore of the St. Lawrence are also part of the northwestern margin of the Appalachian Orogen. A series of parallel thrust faults and the sheets between them can be traced for great distances along this margin, and typical structures can be seen at chutes-de-la-Chaudière on the south side of the river at Charny.

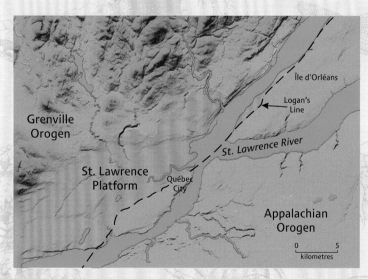

Québec City's geological setting at the crossroads of the Canadian Shield (pink), the St. Lawrence Platform (green), and the Appalachian Orogen (brown). ADAPTED FROM CÔTÉ ET AL. (2001).

Flat-lying late Ordovician limestone, belonging to the St. Lawrence Platform sequence, unconformably overlies hard crystalline Grenvillian gneiss at Montmorency Falls, Beauport, Quebec. A gap of about 550 million years is represented by the unconformity. ROB FENSOME.

This wall along Sous-le-Capin in old Québec City sits on tilted Ordovician calcareous mudstone deposited in the deep waters of the Iapetus Ocean. Ordovician and older rocks were thrust-faulted and folded during the Taconic Orogeny, an event within the Appalachian Orogen. ROB FENSOME.

The Role of Stone

Most of the buildings of the early colony were wooden, but stone was generally used for civic and religious buildings, and increasingly for domestic buildings after the Lower Town fire of 1682. The earliest stone, quarried in the seventeenth century, was the black stone from Cap Diamant, a muddy Appalachian Ordovician limestone. This was not a quality stone, but it came from the heart of the town, and so cost little to transport. Cap Diamant Blackstone was nicknamed "stinking stone" by quarry workers because it smelled of sulphur during cutting. A rare surviving example of a house made of Cap Diamant Blackstone is Maison Cureux at 86 rue Saint-Louis. This house, built in 1729, is reconstructed from a 1709 building, its walls probably made from stone excavated from the house's foundations. Cap Diamant Blackstone splits easily, especially when exposed to air and water, and so most houses made from it, including Maison Cureux until 1968, were encased in wood or plaster.

As the colony grew and prospered, the demand for better-quality stone led to the successive use of Beauport, Château-Richer, and Pointe-aux-Trembles limestones, Ange-Gardien and Sillery-Cap-Rouge sandstones, and finally Saint-Marc-des-Carrières Limestone. Beauport Limestone from the St. Lawrence Platform is a fine-grained sedimen-

Detail of the Maison Cureux (1729 reconstruction of a 1709 building) on rue Saint-Louis, near its intersection with rue d'Auteuil. This is a rare example of a house built of Ordovician Cap Diamant Blackstone. ROB FENSOME.

tary rock formed in middle Ordovician seas that covered the region. It is dark brownish-blue when freshly cut, but weathers to a pleasing light grey. It was transported from Beauport and Côte-de-Beaupré by barge to Québec City and, although primarily used as a building stone, it was also a source of lime for cement. Beauport Limestone was used extensively to build the fortifications, the armoury, and several dwellings around Place Royale. The fossiliferous Ordovician Saint-Marc-des-Carrières Limestone, used for the Quebec Parliament Building (1886) and the rebuilding of the Saint-Louis Gate in 1878, comes from a quarry that has been in production since 1835 at Saint-Marc-des-Carrières, 60 kilometres upriver from Québec City.

Harder, more resistant sandstones were preferred over limestones for military structures, wharves, and cobblestone streets. The French regime used yellowish Ange-Gardien Sandstone from the St. Lawrence Platform. Although the original quarry site is unknown, one quarry was located on the Islets, two small rocky islands near Château-Richer. From about 1740, as the French regime drew to a close, green Sillery-Cap-Rouge Sandstone was quarried at a site between Sillery and Cap-Rouge along boulevard Champlain. This sandstone was formed in the Iapetus Ocean during the Ordovician and was deformed by thrusting and folding during the Taconic Orogeny. A greenish stone that weathers to a brownish colour, it was used extensively by the British for the Martello towers, the Québec Citadel (1819–1832), and other fortifications, as well as for churches and homes of wealthy city residents. And it provided the stone used to rebuild the Saint-Jean Gate in 1938. Together, the Ange-Gardien and Sillery-Cap-Rouge sandstones contribute the greenish and brownish colours that are so characteristic of Québec City's military buildings and fortifications.

Granites provide a contrast to the generally less-resistant sedimentary rocks. The most prestigious monuments in Québec City

The main building stone in Place Royale is Ordovician Beauport Limestone. ROB FENSOME.

Detail of the city wall in Montmorency Park. The wall is composed of Ordovician Ange-Gardien (yellowish) and Cambrian Sillery-Cap Rouge (greenish to brownish) sandstones. ROB FENSOME.

Facades of two houses along the rue des Remparts. The house to the left is constructed of Ordovician Saint-Marc-des-Carrières Limestone, the house to the right of greenish Cambrian Sillery Sandstone. ROB FENSOME.

The Champlain Monument (1898) on terrasse Dufferin. The creamy white limestone of the pedestal is from Château-Landon in France. The steps at the base are Vosges granite, also from France. ROB FENSOME.

Maison Parent (1761), located at the corner of rue Sous-le-Fort and rue Saint-Pierre, is built of a variety of local rocks, many from a former building on the site that was destroyed during the British siege of the city in 1759. The rocks include reddish to brownish calcareous Ange-Gardien Sandstone (its colour due to iron oxide grain coatings), Beauport Limestone, Sillery Sandstone, Rivière-à-Pierre Granite, and Cap Diamant Blackstone. The trim around doors and windows is the Pointe-aux-Trembles Limestone. ROB FENSOME.

have pedestals of Stanstead Grey or Saint-Sébastien granite, which are quarried around Beebe in the Eastern Townships and in the Little Mount Mégantic District, respectively; both are from the Appalachian Orogen. Many churches are also built of these grey granites, including Sainte-Anne-de-Beaupré. The Neoproterozoic Rivière-à-Pierre Granite from the Grenville Orogen is highly valued for paving stones and monument bases. It was carved to form the piers of the Québec Bridge, as well as the pedestal for the Statue of Liberty in New York City.

Although Québec City's earliest building stones were quarried locally, a tour of the modern city reveals building materials from other parts of Canada and the world. For example, Queenston Limestone from Ontario was used for part of the Price Building (1929). It weathers brownish-buff and contains distinctive pink calcite crinoid stems. The Champlain Monument (1898) on the esplanade in front of the Château Frontenac was designed by *Titanic* survivor Paul Chevré and, in commemoration of Québec City's 400th anniversary in 2008, was restored using stone from the original source—the same French limestone that was used in the Arc de Triomphe and Montmartre Basilica in Paris.

Landslides and Seaways

Québec City is thousands of kilometres from active volcanoes, and earthquakes in the region are rare. But another kind of hazard has had a tragic impact on the city: landslides. The sedimentary layers forming the Cap Daimant promontory dip parallel to the slope of the land surface and in places are even vertical. As a result, rockfalls can happen, especially after heavy rain or during periods of alternating freezing and thawing. Landslides at Cap Diamant have caused numerous deaths over the years. On May 17, 1841, part of the promontory above rue Champlain at the west end of terrasse Dufferin collapsed, crushing seven houses and killing twenty-seven people. And on September 19, 1889, forty-five people lost their lives in a similar event at the same site.

The rocks involved in these disasters are hundreds of millions of years old. But our geological exploration of Québec City would be incomplete without mentioning events and deposits of the past 2 million years, during which time glaciers have reshaped most of Canada, including Quebec. About 13,000 years ago, as the Laurentide Ice Sheet was retreating from the St. Lawrence Valley, much of northern North America was still depressed under the weight of the ice. As a result, an arm of the Atlantic Ocean invaded the St. Lawrence Lowlands, creating the Champlain Sea.

The Champlain Sea occupied the St. Lawrence Valley for perhaps 3,000 years. It was gradually infilled by sediments and the region was elevated to its present level due to postglacial isostatic rebound. About 9,000 years ago, the Cap Diamant promontory was an island surrounded by two river channels. The more northerly channel, now the site of Québec City's Lower Town, gradually became dry land as the lithosphere rebounded. Today, the St. Lawrence flows entirely along the southern channel. Many of the lowlands that border the St. Lawrence Estuary, such as those at Cap Tourmente and Montmagny, are underlain by vast accumulations of muddy Champlain Sea sediments. This legacy of the Ice Age provides some of the best farmland in the region and serves the thriving city today.

A cliff of precarious-looking Ordovician calcareous mudstone along a lane in Québec City illustrates the potential danger of rockfalls in the city. ROB FENSOME.

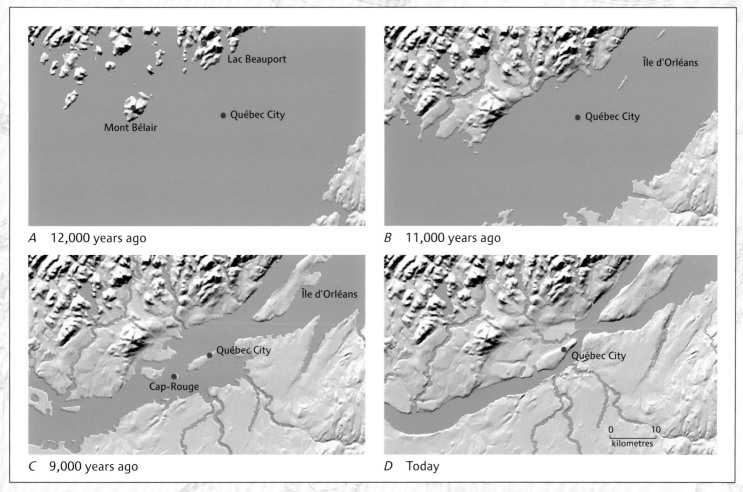

A 12,000 years ago

B 11,000 years ago

C 9,000 years ago

D Today

After the glaciers retreated, the lithosphere took a while to rebound isostatically from their weight, resulting in depressions below sea level. One such depression was the area now occupied by the St. Lawrence Valley, which was flooded by the Champlain Sea. The maximum extent of the Sea's inundation is shown in A and the early stages of its retreat in B. By 9,000 years ago (C) the Cap Diamant promontory was an island in a broad St. Lawrence Estuary. Over the past 9,000 years, sea levels have continued to recede to today (D). ADAPTED FROM CÔTÉ ET AL. (2001) BY ANDRÉE BOLDUC.

15 • WATER: A CLEAR NECESSITY

CHAPTER SUPPORTED BY A DONATION FROM THE CANADIAN CHAPTER OF THE I.A.H., IN MEMORY OF CANADA'S GROUNDWATER PIONEERS

Calm water on Sherbrooke Lake, Nova Scotia. DIANN ROBAR.

A Magic Elixir

Life as we know it would be impossible without water. Although a relatively water-rich country, Canada is not immune to drought, as residents of the southwestern Canadian Prairies know only too well. Between 1857 and 1861, John Palliser, an Irish-born soldier and explorer, was commissioned to lead the British North American Exploring Expedition to then-uncharted lands in the west. In presenting the Expedition's findings to the British Parliament in 1862, he described a large triangular area of what is now southwestern Saskatchewan and southeastern Alberta as "desert, or semi desert in character, which can never be expected to become occupied by settlers." The area became known as Palliser's Triangle. About a decade after Palliser's pessimistic report

A mixed Prairie grassland under good moisture conditions, Grasslands National Park of Canada, Saskatchewan. W. LYNCH, COPYRIGHT PARKS CANADA.

was delivered, Canadian botanist John Macoun predicted that, on the contrary, Palliser's Triangle would be suited for agriculture because rainfall came during the growing season, when it was needed most. So, despite Palliser's warnings and multi-year droughts that occurred in the 1890s and 1910s, the government encouraged settlers. Then came the Dirty Thirties: drought conditions lasted from 1929 to 1937, and the impact was further accentuated by poor agricultural practices. Livelihoods literally blew away in clouds of soil as "dust-bowl" conditions extended from the American South into the Canadian Prairies. Although conditions and practices have generally improved since the 1930s, drought in Palliser's Triangle occurred again during the 1960s, 1980s, and early 2000s. The drought that lasted from 1999 to 2004 was possibly the worst to strike the Prairies in recorded history. For both Saskatoon and Medicine Hat, 2001 was the driest year on record. Crops were ruined and ponds dried up, depriving people and animals of potable water.

The availability of water is a clear necessity to our survival and wellbeing. But what is this magic elixir? Most people are familiar with the chemical formula for water (H_2O), which indicates that each molecule has two hydrogen atoms attached to one oxygen atom. Every molecule has a slightly negative charge at one end and a slightly positive charge at the other. Consequently, water molecules are attracted to one another, giving the substance many of its special properties. Water boils at a much higher temperature than similar chemical compounds, and relative to other liquids it needs a lot of energy to warm up. Another important property of water is that it can readily dissolve many solids such as salt and sugar, as

Farmland suffering drought in southwestern Saskatchewan. ROB FENSOME.

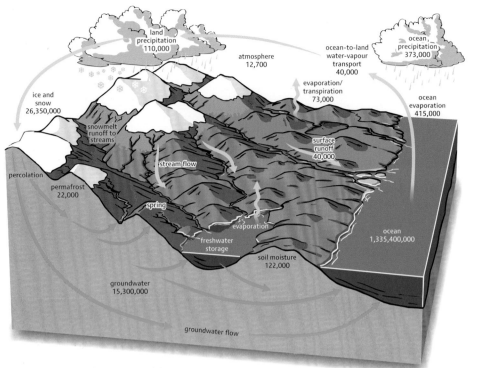

The hydrological cycle begins with the evaporation of seawater to create the world's ongoing supply of fresh water. Moisture-filled air formed over the oceans brings fresh water onto the continents, where it precipitates and eventually returns to the ocean through rivers and groundwater seepage to complete the cycle. Numbers represent in cubic kilometres how much water is either in storage or moving through the various parts of the water cycle on average in one year: for example, 415,000 cubic kilometres refers to the amount of ocean evaporation, of which 373,000 cubic kilometres is returned to the ocean as precipitation, and 15,300 cubic kilometres is stored in groundwater.

well as components of rocks, which explains the development of metal-rich hydrothermal fluids (Chapter 12). The Earth is unique among the known planets in having water in solid, liquid, and gaseous forms at its surface. If our planet were just 5 percent closer to the Sun, the surface would be too hot to retain liquid water or ice. If it were 5 percent farther away, all of its water would be frozen. Because about 70 percent of our bodies are made up of liquid water, the just-right distance between the Earth and the Sun is critical to our existence.

The Earth has about 1.4 billion cubic kilometres of water, the largest single repository being the ocean. Since their formation, oceans have accumulated dissolved material from the land, so that today seawater contains about 35 grams of dissolved salts per litre. The presence of salt—predominantly sodium chloride, but also salts of magnesium, calcium, and potassium—lowers the freezing temperature of seawater to about minus 2°C. Only about 3 percent of our planet's water is fresh, meaning that it contains less than 1 gram per litre of dissolved material, including salts. About 65 percent of fresh water is frozen, locked up in the Antarctic and Greenland ice sheets and the many smaller glaciers around the world. The second largest store of fresh water, nearly 35 percent, is groundwater—water stored in the Earth's upper crust. At the surface, lakes and wetlands hold only 0.3 percent, and rivers contain less than 0.01 percent of the world's fresh water.

Water is constantly moving from the oceans into the atmosphere, onto land, and back to the sea. This circulation is called the water cycle, or hydrological cycle. Driving this cycle is solar radiation, which distils fresh water from the oceans through evaporation. This process adds a remarkable 415,000 cubic kilometres of fresh water to the atmosphere each year. Most of this fresh water falls over the ocean, but about 110,000 cubic kilometres, falls as rain and snow over land. Once it has reached the ground, much of this precipitation is re-evaporated by solar radiation back into the atmosphere. Land plants also absorb large amounts of water through their roots, then return it to the atmosphere through transpiration—the loss of water to the atmosphere through leaves. Around 45,500 cubic kilometres of water remain on land and either infiltrate into the ground or are carried by rivers to the sea to complete the cycle. This annual global runoff, just 0.003 percent of the world's water, represents the renewable freshwater supply that we rely on for our survival.

About 65 percent of the planet's fresh water is locked up in glacier ice, as in this glacier west of McKinley Bay, northern Ellesmere Island, Nunavut. STEVE GRASBY.

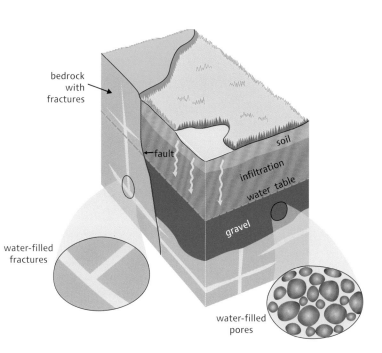

Water that soaks into the ground fills pores and fractures in soil or rock to become groundwater. Pores and fractures below the water table are saturated with groundwater. ADAPTED FROM TURNER ET AL. (2005A).

Tulita Spring, near the confluence of the Mackenzie and Great Bear rivers in the Northwest Territories, brings groundwater to the surface. STEVE GRASBY.

Water in the Rocks

Rainwater is nearly pure, but upon reaching the ground, it moves through soil and rocks as groundwater. If you drill deep enough anywhere in Canada, you will encounter the water table, the level below which rocks are saturated with water. But the distribution of easily accessible potable groundwater is much more limited. Factors such as porosity and permeability, which we explored in Chapter 13 in the context of petroleum, are similarly important in controlling how groundwater occurs and moves. Like petroleum, groundwater may accumulate beneath seals (called aquitards in the case of water) in reservoirs known as aquifers. Most metamorphic and igneous rocks have little or no porosity and hold only small amounts of water in fractures, in contrast to many sedimentary rocks, which have high porosities and can hold abundant groundwater. Unfortunately, most of the groundwater in sedimentary basins is saline. These salty waters originated in ancient restricted seas under a hot tropical sun. Some of the salt came out of solution to form evaporite deposits, but some remained dissolved to form brines twenty times saltier than the original seawater and became trapped in the sediment. Usable groundwater is restricted mainly to sedimentary basins, and then only in the uppermost few hundred metres of rock and sediment. Porous deposits of sand and gravel left as outwash from ice sheets form important aquifers, such as those within the Oak Ridges Moraine, which provides an important groundwater supply in southern Ontario. The Oak Ridges Moraine is also the main source of water for numerous streams and rivers that feed into Lake Ontario, the only water source for the City of Toronto. This has driven extensive debate on limiting development on the Moraine to preserve water supply. There are also concerns about broader impacts, where contaminants in water recharging the moraine may pose a risk of pollution in Lake Ontario.

Groundwater dissolves chemical components of rocks, so its chemistry is partly controlled by the local geology. For example, water moving through carbonate rocks picks up calcium and magnesium, compounds that make it "hard"; such hard water forms deposits on pipes and kettles and makes it more difficult to get soap to lather. Groundwater moving through pyrite-

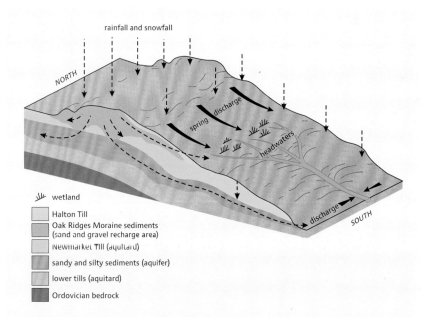

Section through the Oak Ridges Moraine of southern Ontario. Water can move relatively easily into and through porous, permeable material. When saturated by water from rain and melted snow (recharge), the sediments exposed at ground level and extending below the surface become reservoirs of groundwater known as aquifers. Groundwater movement is restricted by impermeable layers, or aquitards. In this example, the groundwater of the Oak Ridges Moraine supports stream and wetland ecosystems. To safeguard this water, areas where recharge occurs must be protected. ADAPTED FROM DOYLE AND STEELE (2003).

CHAPTER 15 • WATER: A CLEAR NECESSITY

bearing shales, which underlie many areas of Canada, causes reactions that form dilute sulphuric acid, in turn causing the water to become acidic and bad-tasting. Acidic groundwater can dissolve trace metals that, in high concentrations, are toxic to aquatic life and even to people. Much of the arsenic in groundwater occurs naturally in this manner. Although passage through certain types of rock sometimes makes water unpalatable or harmful, the vast majority of Canada's shallow groundwater is of good quality and safe for human consumption.

Reactions between water, sunlight, and industrial contaminants in the air can produce acidic rainwater. Limestone neutralizes the acidity and thus mitigates the impact of industrial emissions on surface water or groundwater, a factor that is important in regions such as southern Ontario and the St. Lawrence Lowlands. Metamorphic and igneous rocks have little ability to neutralize acid rain, and so lakes on the Canadian Shield, where such rocks predominate, are commonly acidic. Metals and other elements that are harmful to people can be extracted from water by binding to clay minerals in the rocks. The flow of water through porous rock can also filter microbes, with the rock acting as a natural water purifier.

Staying on Budget

Since water is essential to our survival, it is important to understand how the water cycle functions. In the case of a lake, for example, the absolute volume of water that it contains is less important

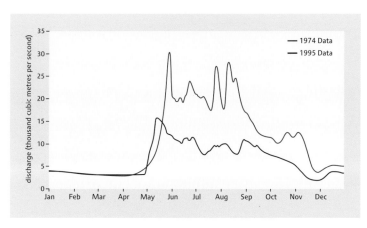

The flow of rivers varies dramatically through the year. The Mackenzie River (shown here) has peak discharge in the late spring when snow is melting and then wanes through the summer and fall. The low-flow period of the winter and early spring is when the water supply is predominantly fed by groundwater discharge. Changes in annual precipitation can have significant impacts on flow from year to year as shown by the pattern of the maximum (1974) and minimum (1995) daily discharge of the Mackenzie into the Beaufort Sea at Arctic Red River. ADAPTED FROM ABORIGINAL AFFAIRS AND NORTHERN DEVELOPMENT CANADA WEBSITE.

than the amount entering and leaving it annually. The water that cycles through a lake each year is the renewable part of the supply, in contrast to the non-renewable part stored in the lake itself. If the latter decreases because more water is taken out than flows in, the lake will shrink. One example of a dramatically shrinking lake is the Aral Sea in Uzbekistan and Kazakhstan. The same principle applies to all water reservoirs, including groundwater, and can be thought of in budgetary terms. Rainfall is a credit and evaporation, transpiration, and consumption are debits. Any positive balance after the debits have been removed is equivalent to your take-home income—what you can plan a budget around. A country's "water income", in other words its annual renewable water supply, is equivalent to the water that flows through its rivers each year. The water stored in a country's lakes, glaciers, and groundwater is like your savings; if you use it all and leave the account empty, you are bankrupt.

About one quarter of Canada is covered by lakes, wetlands, and glaciers. This may seem like a lot, but how do Canada's water resources compare with those of other countries? The amount of fresh water a country or region receives depends on many factors including latitude, distance from the ocean, topography, annual temperature, and air circulation. Of all the precipitation falling on land, South America receives one third, and the Amazon River alone accounts for 20 percent of annual global river discharge. At the other extreme, Australia receives less than 1 percent of global precipitation on land and

Small upland lakes such as those shown here at Hudson Bay Mountain, near Smithers, British Columbia, demonstrate the concept of a water budget. New water inflow from rain, snowmelt, and springs is balanced by downstream water outflow and evaporation. IAN SPOONER.

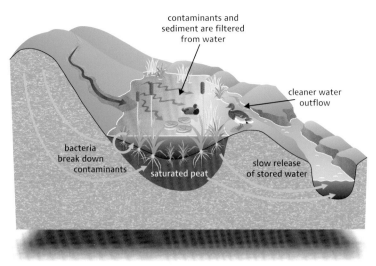

Groundwater and surface water are all part of one interconnected system. Movement of water through ground and wetland environments is a natural process that reduces the contaminants in water. Peat, an especially effective filter, is excellent for storing water, which is then released slowly. The filtered water that leaves the peat has far fewer contaminants and sediment than the water that entered it. ADAPTED FROM TURNER ET AL. (2005B).

accounts for about 1 percent of annual global river discharge. Canada's share of global river discharge is 7 percent, which places its renewable water supply third among countries, with Brazil and Russia first and second, respectively. More relevant than a country's total annual supply of fresh water is its per capita renewable water supply—the amount of water available to each person within a country. Here Canada slides down to seventh place. Even so, each Canadian has access to over 1,700 times more water than someone living in Egypt, a country with an arid climate and dominated by desert.

Canada's large water endowment mostly supports forests and wetlands that produce oxygen and store greenhouse gases. Winter precipitation is mostly snow, which builds up during the season. The annual spring melt and in some regions the greater spring rainfall explain the large river runoff in late spring. Dams can store some of the spring runoff and release it when demand is higher, but even large dams can only store a small proportion of the excess. Most river runoff in Canada is not in areas of greatest water demand. The majority of Canadians live near the country's southern border, whereas over 60 percent of surface water flow is northward into Hudson Bay and the Arctic Ocean. This uneven distribution has prompted contentious plans to divert north-flowing water toward southern population centres and agricultural lands, including a plan to dam James Bay. Other proposals involve transformation of the Southern Rocky Mountain Trench in southeastern British Columbia into a gigantic lake, and diversion of water from the Great Lakes basins. Such major diversions, and even smaller ones, would have serious environmental and societal impacts. And the financial costs would be enormous. Conservation is a much more cost-effective approach to ensuring an adequate supply of fresh water.

Stocking the Warehouses

The movement of water between natural reservoirs such as glaciers, lakes, wetlands, and aquifers regulates the flow of rivers. For example, consider such rivers as the Fraser, Athabasca, and North Saskatchewan, all of which originate from glaciated headwaters within the Cordillera. During spring and summer, meltwater supplements runoff from other sources at a time when downstream water demand is greatest. Over the past century or more, the ice fields have thinned and retreated because more ice is melting than is being added through snowfall. A river that loses icefield input from its headwaters will not dry up, since most of the flow comes from snowmelt, but its discharge will change. In particular, it will experience lower summer and early autumn flows and potentially higher water temperatures, which can harm aquatic ecosystems.

Large lakes also have a regulating effect on the movement of water through a watershed. Lake levels rise during periods of high river inflow, such as in the spring, and drop as the stored water is gradually released or evaporated. Canada has or shares some of the world's largest lakes. The movement of water through these lakes, however, is a small fraction of the total volume stored in them. For example, for the Great Lakes, annual outflow is less than 1 percent of their total volume. Still, these lakes regulate the flow of the St. Lawrence River, the second longest river in Canada.

Sustainable groundwater use requires that the amount of water recharging an aquifer each year is at least equal to that extracted. If extraction exceeds recharge, the aquifer will be drawn down and eventually depleted, causing falling water tables and loss of surface streams. This undesirable situation has already occurred with heavily used aquifers in other countries. Currently, groundwater meets 30 percent of Canada's domestic

A glacier on Mount Brazeau, at the south end of Maligne Lake, Jasper National Park of Canada, Alberta. Glaciers in the Cordillera source and supply many of western and central Canada's great rivers. ROB FENSOME.

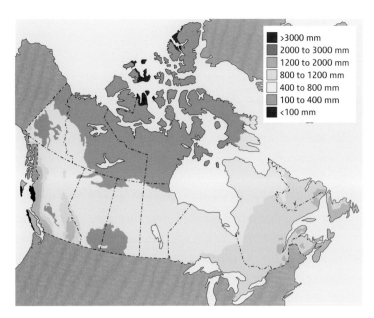

Average annual precipitation across Canada. COURTESY OF ENVIRONMENT CANADA.

Oldman River Dam, Pincher Creek, Alberta, stores water for irrigation. RON GARNETT / AIRSCAPES.CA.

water needs, with most being used in rural areas lacking a main supply, where residents are reliant on wells. However, groundwater availability and use differ from region to region. Among the provinces, groundwater varies from 23 percent of domestic supply in Alberta to 100 percent on Prince Edward Island.

The migration of water to and from aquifers has an influence on the flow of rivers. Precipitation and snowmelt infiltrate the ground, recharging aquifers. But groundwater also escapes to the surface via seeps and springs, augmenting river flow even during periods without rain. Because groundwater maintains a fairly constant temperature throughout the year (7–9 °C), discharge of groundwater maintains river flow through the cold winter months, when precipitation falls as snow and stays frozen at the surface.

Natural water-storage reservoirs, whether icefields, lakes, or aquifers, also play important roles in regulating water over periods of many years. During years of above-average precipitation, glaciers grow, and lake and water-table levels rise. And during drought years, water in these reservoirs supplements stream flow. When prolonged droughts occur, even lakes and groundwater sources can be depleted, with both obvious and unforeseen consequences. A sequence of dry years explains the recent fall in the levels of the Great Lakes, for example. The lower lake levels have reduced discharge through the St. Lawrence Seaway system, so that ships can't carry as much cargo through this section, which has in turn increased transportation costs.

Average annual precipitation in Canada ranges from a high of over 6 metres at Henderson Lake on Vancouver Island to a low of about 0.075 metres at Eureka on northern Ellesmere Island, Nunavut. Variability in precipitation reflects geography and air-circulation patterns. Moist air from the Pacific Ocean tends to cool as it moves inland, releasing rain or snow. The Cordilleran mountain ranges in western Canada enhance this process by forcing air to rise, triggering precipitation along the Pacific coast. Inland ranges wring yet more moisture from Pacific air, so that little is left once it reaches the Prairies. In the east, the Appalachians, lower than the Cordillera, have much less impact on air moisture. The Arctic Islands are the driest part of Canada because cold air cannot hold much moisture. This region is dominated by high barometric pressure.

Water management requires knowledge of how much water moves through all of Canada over specific time intervals. This knowledge is gleaned from a network of gauging stations that measure precipitation and stream flow. From the late nineteenth century to the 1980s, the number of these stations rose from just a few to over 2,500, but since the 1980s the number has declined to about 1,500. Recent research suggests that we need at least 60 years of continuous data to establish a good understanding of a river's natural variation in flow. Although Canada has about 500 monitoring stations with records of this duration, it is not clear that river flow over the past 100 years has been "normal". For example, archaeological and historical data show that over the past few thousand years, the Canadian Prairies have had many prolonged droughts that were more severe than any in the twentieth century.

Using and Losing Water

Most water use in Canada is dedicated to generating electricity (Chapter 13), with Hydro-Québec being the world's largest hydroelectric producer. There are over three hundred large dams in Quebec, including major projects on the Grand River and other tributaries to James Bay. Hydroelectric power generation is an "in-stream use" because it uses the energy of the flowing water, not the water itself. Although commonly viewed as a green form of energy, hydroelectric power is not without impacts. These include the building of dams, the flooding of valleys, and the creation of reservoirs from which water is lost through evaporation. About a thousand large dams and many thousands of smaller ones have been constructed on Canadian rivers over the past century. In reservoirs formed by these dams, about 5 percent of the water is lost annually through evaporation.

More than 63 percent of the water withdrawn from water bodies in Canada is used in the process of generating electricity from the burning of fossil fuels and in the operation of nuclear power plants. Most of this water, which is used mainly for cooling, is not really "consumed" but returns to rivers and lakes, albeit at a higher temperature than when it was extracted. Other water withdrawn from water bodies includes 15 percent used in manufacturing processes, 9.5 percent for municipal and agricultural purposes, and 1 percent in mining operations.

Municipalities are the third largest water users in Canada. On average, Canadians use 335 litres of water per person per day in the home, ranging from 113 litres in Nunavut (where each house has an individual cistern) to 645 litres in Yukon (partly due to the need to run water to prevent pipes in the distribution system from freezing). In comparison, the average European uses about 200 litres. Why is there so much variation within Canada? One reason is that some municipalities charge according to amount of water consumed, which generally persuades residents to reduce their consumption. Another obvious reason is gardening: summer water usage can be double that of winter in some Canadian cities.

Within municipalities, about 57 percent of water usage is residential and 30 percent commercial and industrial, with the rest accounted for by leaks in the system. Surprisingly, about 50 percent of the residential usage is devoted to keeping us and our accoutrements clean and about 30 percent is literally flushed down the toilet; another 10 percent is used for cooking and drinking. Repairing leaks and installing low-flow toilets, faucets, and showerheads are simple and cost-effective ways to reduce consumption.

Most water used in agricultural operations is for irrigation (92 percent) and livestock (5 percent). Only a small fraction of the total used is returned to the rivers, lakes, or aquifers from which it was extracted. The water is not truly consumed, however, since some becomes part of the crops and livestock, and much re-enters the atmosphere through evaporation and transpiration. Even with the best irrigation practices, growing food will always be a water-intensive industry.

Kettle Rapids Generating Station, Gillam, Manitoba. The wide use of flowing river water to generate electricity has led to the word "hydro" being synonymous with electricity in several provinces. RON GARNETT / AIRSCAPES.CA.

Running Low

It is difficult to imagine Canada facing a water shortage. Many regions, including British Columbia, Quebec, and the Atlantic Provinces, generally have an ample supply to meet demands. Nevertheless, some unique challenges exist. For instance, northern communities can face problems with winter supply when rivers and lakes are frozen. Highly populated regions of Canada also face water "stress"—when demand exceeds 40 percent of supply. This has happened in densely populated southwestern Ontario, where the demand from industrial and municipal water users poses significant concerns.

Even British Columbia has localized areas of water stress. The Okanagan Valley, part of the water-rich Columbia River Basin, has many large lakes. But it also has a semi-arid climate and thus a limited renewable water supply. The Gulf Islands, between southern Vancouver Island and the British Columbia mainland, are also facing water shortages. Although located on

Old-fashioned irrigation in the Okanagan Valley, near Oliver, British Columbia. GODFREY NOWLAN.

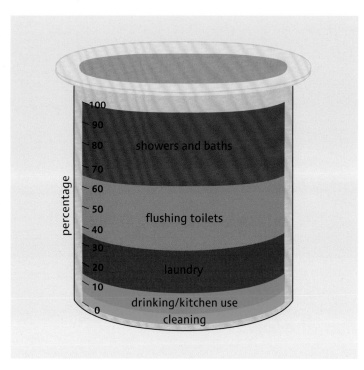

Summary of water consumption in a typical Canadian home. ADAPTED FROM ENVIRONMENT CANADA WEBSITE.

the "wet" coast of Canada, the islands lie in a rain shadow and receive only about 600 millimetres of rain per year. Aquifers that get replenished by winter rains are the only significant storage reservoirs. However, the reservoirs are limited by the low porosity and permeability of the underlying rocks, and recoverable water is largely localized along faults and fractures. Hot, dry summer spells place significant stress on the Gulf Islands' water supply, and wells can temporarily dry up altogether. Similarly, Vancouver relies on several mountain lakes that collect water from the melting of the winter snowpack. Low snowfall over several years can create concerns over the city's summer supply.

Much of the southern Canadian Prairies is also water-stressed. The Bow and Oldman river basins of southern Alberta were declared "fully allocated" by the provincial government, and no new water withdrawal licenses from rivers in those watersheds will be issued. The impact can be seen in Calgary, the fourth largest and one of the fastest growing cities in Canada. Calgary's sole water supply is from the Bow and Elbow rivers. Now that no new water licenses are available, and groundwater supplies are limited, conservation and recycling measures are the only ways to meet the needs of Calgary's growing population. Calgarians have met the challenge by cutting their per-capita water consumption in half since the 1980s, driven largely by municipal efforts to meter usage, mandate low-flow toilets and faucets, and encourage conservation. But the city's population has doubled over the same period, so that total water usage has not changed. To accommodate projected growth, a further 50 percent reduction in per-capita water use will be required over the next 50 years. Because of concerns over water supplies, other communities in the Bow River Basin have opted to restrict growth by, for example, limiting housing construction. For similar reasons, some industries have chosen to locate elsewhere.

Winnipeg, which ironically battles frequent floods, faced challenges in its early days to come up with a good water supply, as water from the Red and Assiniboine rivers was too difficult to clean. The solution was to build a 156-kilometre-long aqueduct to bring water to the city from Shoal Lake in southeastern Manitoba, close to the Ontario border. This aqueduct has been in operation since 1919.

Water is used in the process of drilling for petroleum and to enhance petroleum recovery. In the latter case, water is injected into a well to force oil or gas out of the reservoir rock so that it can be pumped to the surface. The total amount of water used in these petroleum-related processes is small compared to that taken by municipalities and agriculture. But it is an important component of the water budget in Alberta and northeastern British Columbia, where most of the country's petroleum is produced. Originally, most of Canada's oil production was conventional (Chapter 13). Today, unconventional production from the oil sands accounts for over half of the country's output, with this percentage increasing annually. Extracting oil from oil sands requires much more water than conventional production, and that water is derived mainly from the Athabasca River, although other sources, such as saline groundwater, are also used. On average, 2 to 4 litres of water are needed for each litre of oil produced. The total licensed withdrawal of water from the Athabasca River to meet this demand is double that used in Calgary. However, the Athabasca River carries six times more water than the Bow River and withdrawals to support the oil-sands operations are less than 2 percent of the average annual river flow. The main issue is that production requires a constant supply of water throughout the year. Although the annual demand is small, withdrawals from the Athabasca River during winter, when its discharge is low, may exceed 10 percent of the flow. We don't yet know if sustained withdrawals of this magnitude will harm the river's ecosystems. Rightly or wrongly, other rivers in Canada already have much higher licensed water withdrawals. It is not possible to completely eliminate environmental impacts of this usage, but they can be minimized and balanced against Canadians' demand for oil and its by-products.

Today, although Canada is water-rich, some regions are facing water shortages brought on by demands to support Canada's energy and economic needs, and to grow food for domestic use and export. Fortunately, we can learn from the experiences and technology of water-stressed regions and nations, which have dealt with this problem for decades and learned to make the necessary economies.

16 • AT THE BEACH

CHAPTER SUPPORTED BY A DONATION FROM STEPHAN BENEDIKTSON, TAYLOR HILL EXPLORATION LTD.

The coastline near Lower Darnley, Prince Edward Island, consists of alternating headlands of Permian sandstone and sandy beaches backed by cliffs and dunes. RON GARNETT / AIRSCAPES.CA

Harris Misener's Farm by the Sea

Harris Misener was a true Nova Scotian. He was born and spent his early years on Miseners Island, which was named after his ancestors. The island, located off the province's Eastern Shore, was home to three generations of the Misener family, who farmed the land and got their drinking water from a dug water well. All three generations made a living off the land by mixed farming. It was not an easy life, but it was secure. Sadly, the situation was not to last. During Harris's lifetime, Miseners Island was eroded away by the sea, a process exacerbated by the regional rise in sea level of about 30 centimetres during the twentieth century. In an interview, Misener wryly noted, "Where I was born is all out in the ocean…I can show you where Grandfather's well was…there's lots of water in that well now." The disappearance of Miseners Island demonstrates that a coastline, regardless of human intervention, is always changing.

A drumlin island at the mouth of Chezzetcook Inlet on Nova Scotia's Eastern Shore illustrates the vulnerability of these landforms to erosion from the sea, which moves sediment to adjoining spits, beaches, and nearby bays. This 1989 aerial photo shows a spit to the left and a barrier beach to the right (the latter connecting to the mainland and formerly to Miseners Island) trailing from the drumlin. DON FORBES.

To understand why and when Miseners Island disappeared, we need to look at the local geology. Many farms in Nova Scotia, as in other parts of Canada, are located on drumlins. This makes sense because some of the best soil occurs on glacial deposits. Using aerial photographs extending back to 1945, along with earlier historical records, geologists have tracked the destruction of Miseners Island. The drumlin on which Harris Misener lived was relatively stable from 1766, when Captain James Cook mapped the coast, until 1945. Then erosion accelerated, and in the 1970s the island finally disappeared. Between 1954 and 1964, that part of the drumlin most exposed to ocean waves retreated 7 metres each year, and between 1966 and 1974 the rate increased to more than 11 metres annually. At that rate, it doesn't take long to lose a small island.

The boulders, cobbles, and sand eroded from this drumlin and others nearby were moved by waves and currents to form a narrow elongated beach ridge that connected the island to the mainland. This ridge, running from drumlin to drumlin, formed a single beach. Waves washed over the beach during Atlantic storms such as Hurricane Edna in September 1954, pushing it inland. The number and violence of storms increased during the early 1950s, and removal of gravel from the shore for aggregate added to the problem. With the drumlin forming Miseners Island gone, the sediment supply slowed down to the point where it could not support the continued growth of the barrier beach. During one particularly severe storm, the beach was breached, breaking the connection altogether. In one generation, Miseners Island was consumed by the sea and its sediments recycled, leaving the next landward drumlin a target for the waves. Only a rocky shoal now remains where the island once was.

What Harris Misener experienced is the norm along coasts, with land being lost in some places and gained in others. In the high Arctic, for example, remains of house rings and fire pits preserved on former beaches, a testimony to seasonal visits by nomadic hunters going back thousands of years, are now situated inland. The camps become progressively younger towards the present shoreline, due to the slow, continuous emergence of the land since the end of the Ice Age in this area—a phenomenon recalled in Inuit tradition. Similarly, Aboriginal oral traditions on the Pacific coast recall sudden changes to the shoreline caused by great earthquakes and tsunamis that wiped out entire villages. Today too, coastal landowners are aware of changes that happen suddenly, such as during a hurricane. But even though such changes may reflect normal natural processes, loss of land through natural reshaping of the coastline is hard to accept when the land destroyed is yours.

The Ephemeral Coastline

Over longer periods, during which the Earth's plates have rearranged themselves, coastal changes have been profound, with continents continuously forming and reforming and oceans opening and disappearing. Here we will focus on the geologically recent past, during which the advance and retreat of ice sheets and the associated rise and fall of the sea have moved the coastlines around Canada tens to hundreds of kilometres both seaward and landward.

Largely thanks to its many Arctic islands, Canada boasts the longest coastline of any country—a staggering 243,000 kilometres. This dwarfs Russia's 36,000 kilometres, and the 20,000 kilometres that bound the United States. Canada's coast is the country's gateway to and from the outside world. Early explorers arriving at Canada's coast after a long journey must have been glad to see the end of the roiling sea. Although they didn't realize it, these early travellers found a coastline bearing the imprint of

Long Beach, Pacific Rim National Park Reserve of Canada, Vancouver Island, British Columbia. T. W. HALL, COPYRIGHT PARKS CANADA.

Gannet Rock, a shoal off Grand Manan Island, New Brunswick. FRANCIS KELLY, COURTESY OF FISHERIES AND OCEANS CANADA.

glaciation. The most spectacular legacies are the highland coasts indented by magnificent fiords, such as Jervis Inlet in British Columbia, Nachvak Fiord in Labrador, and Pangnirtung Fiord in Baffin Island. Fiords are glacially deepened and sculpted valleys that are open to the sea. They typically have steep rock walls and are hundreds of metres deep; and they may contain tens or even hundreds of metres of sediment, derived in part from the melting of glaciers and subsequently from glacier-fed rivers.

Low-lying coastlines also have a glacial inheritance. This factor explains many of the features of the Atlantic-facing coasts of Nova Scotia and eastern Newfoundland, which have a rocky landscape incised by deeply indented inlets, or rias, that formed as glacial valleys and have subsequently been partly drowned by the sea. This landscape is now draped in glacial deposits. In some Atlantic coastal areas of Nova Scotia, drumlins are so numerous that they comprise drumlin fields, with individual drumlins forming headlands and islands, such as Harris Misener's.

In Arctic Canada, where temperatures are below freezing for much of the year, rocks and sediments near the surface are locked in permafrost (Chapter 11). When exposed in a cliff face, the ice thaws, allowing the sea to attack the sediments. Thaw and

Steep rock walls dominate Tingin Fiord, eastern Baffin Island, Nunavut. DON FORBES.

CHAPTER 16 · AT THE BEACH

This cliff along the tundra shoreline at Stokes Point, Yukon, exposes massive ground ice (permafrost). Although waves undercut and erode such cliffs, warming sea temperatures are also thought to play a significant role. DON FORBES.

Melting of ice wedges and collapse of undercut blocks contributes to rapid coastal retreat along some Arctic coasts, as here at Kay Point, Yukon. DON FORBES.

consequent buildup of excess water also trigger landslides during summer when temperatures are highest.

Constructive and destructive forces are constantly at play along our coastlines. Gravel, sand, and mud are brought to the coast by rivers or glaciers. These sediments are potential materials from which new land can form, adding to the material being eroded at the coast. Waves and tidal currents disperse sediments into the sea and redeposit them. Because most waves and currents are active only in the top few tens to hundreds of metres of the ocean, the best storage space for sediments is in deeper water where, once deposited, it remains less disturbed.

Superimposed on such day-to-day influences on the coastline is sea-level change. If sea level rises, more storage space is created for the transported sediment, but if sea level falls, the space for sediment storage becomes less. Other things being equal, the balance between sediment supply and the ability of the ocean to move and store sediment determines whether the coastline moves seaward or landward. Fighting this natural balancing act may be futile if one is building along the coast. Nature will usually win the battle.

It's All Relative

As we saw in Chapter 11, sea level has moved up and down by many tens of metres as a result of glaciations. During the last glaciation, huge amounts of water were locked up in ice sheets on land, and global sea level was as much as 120 metres lower than today. But about 125,000 years ago, during the last interglaciation, when the climate was warmer than at present, the sea was up to 6 metres higher than now. In the past century, global sea level has risen at a rate of a little less than 2 millimetres per year, due mainly to the warming and expansion of the oceans and to melting of mountain glaciers. Scientists forecast that sea-level rise will increase over the remainder of this century to between

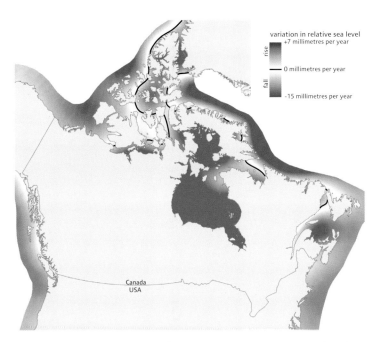

Patterns of relative sea-level rise and fall in coastal and marine areas around northern North America. In this part of the world, relative sea-level rise or fall is controlled mostly by rise and fall of the underlying crust rather than by global sea-level changes. DATA FROM PELTIER (2004); FIGURE COMPILED BY GAVIN MANSON.

A raised beach is a former beach that is higher than today's shoreline because of a fall in relative sea level. At Squally Point, near Apple River, Nova Scotia, sand and gravel deposited on a Holocene beach sit atop cliffs of Devonian and Carboniferous volcanic rocks. Soon after the ice retreated in this area, sea level was about 30 metres higher than at present, but fell rapidly with isostatic rebound to below today's level, leaving this beach high and dry. Today, sea level in the area is slowly rising. RALPH STEA.

5 and 10 millimetres per year or more. If we assume an annual rise of 10 millimetres, that will mean a rise of a metre by the early twenty-second century, a scary scenario since even a few millimetres annually can be devastating. However, global sea-level change is only part of the story. Tectonic factors can cause the land to rise or fall, and this effect can either add to or subtract from the impact of global sea-level changes. This explains why, even though global sea level is on the rise, some parts of Canada's coast have experienced a fall in relative sea level—the key word being relative. Relative sea-level change has a big say in where the coastline is located at any particular time, as well as in the distribution of sediments along it.

Where relative sea level is rising, the sea will usually encroach onto the land and the coastline will retreat. In Nova Scotia, subsidence of the land is adding to the effect of global sea-level rise. In many places, the shoreline is indented by valleys that originated when sea level was lower and have subsequently been drowned to form long, narrow inlets, such as Halifax Harbour. Mapping of the seabed off Nova Scotia has revealed evidence, thousands of years old, of past landscapes, with trees as well as peat that formed in tidal salt marshes since drowned by the sea.

As sea level rises, depositional features such as beaches tend to migrate landward, and may ultimately be drowned or reform elsewhere along the coast as sediment is redistributed. Waves attack cliffs and bluffs more aggressively and to higher levels, accelerating erosion. Increase in erosion at one beach is sometimes enough to tip the sediment balance on adjoining beaches, enabling them to build seaward even as sea level rises, producing a succession of beach ridges. Two to three hundred

The backdrop for the village of Trout River, Newfoundland, includes beach terraces (the flat, ledge-like area in front of the hills) that indicate higher sea levels in the past. MICHAEL BURZYNSKI.

years ago, such extended beach-ridge complexes at Grand Bank and Placentia in Newfoundland were attractive places to sun-dry cod, and this led to the development of communities in both places. But as sea level continues to rise and the sediment supply is exhausted, such settings can slowly drown and shoreline erosion resume.

Although global sea level is rising, in some regions the land is rising more rapidly, causing relative sea level in the region to fall. For example, as we saw in Chapter 10, Vancouver Island sits above the Cascadia Subduction Zone, where the Juan de Fuca

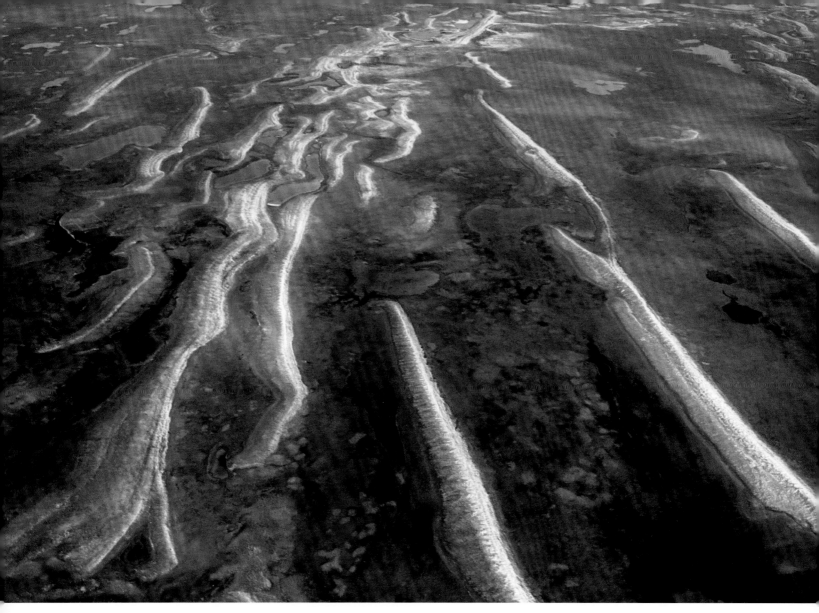

Raised beach ridges, Wapusk National Park of Canada, Manitoba. N. ROSING, COPYRIGHT PARKS CANADA.

Plate descends below the North American Plate. Subduction buckles and raises the crust of the western part of the Island, causing the land to rise relative to the sea. The rate of uplift on western Vancouver Island is at present marginally greater than the current rate of global sea-level rise, so the net effect there is a relative fall in sea level. But an extreme earthquake could immediately cause subsidence of the land, with a corresponding sudden rise in sea level.

In some parts of Canada, the land is still isostatically rebounding after the retreat of the ice sheets. This is the case in northern parts of the Gulf of St. Lawrence region, Labrador, and the central-eastern Arctic. In the Hudson Bay region, relative sea level is falling by as much as 11 millimetres per year due mainly to isostatic rebound. This change has resulted in sets of raised beaches that form steps down to the shoreline. Deltas are also abandoned as the land rises, and this can be seen in many parts of Canada, as far inland as the Ottawa Valley, once occupied by the Champlain Sea. Many fine examples can be seen in Hudson Bay and along the north shore of the St. Lawrence Estuary. Raised deltas are progressively cut into by the rivers that formed them.

The Power of Waves

Most of the sights and sounds of the beach are related to waves, crashing into rocks and rhythmically breaking over sand or gravel. Waves are produced by wind blowing across open water. The stronger the wind and the longer the fetch (the distance over which it blows), the larger the waves. When the wind abates, the waves gradually lose energy and decrease in size. Waves generated by storms in the open ocean can travel for hundreds or thousands of kilometres. Such "swell" waves, the favourites of surfers, can be several metres high and have wavelengths of hundreds of metres.

As a wave approaches shore, its lower part encounters the seabed first and is slowed by friction; thus the wave's upper part moves forward faster and becomes a breaker. When the breaker collapses under its own weight, it sends a mass of water, called the swash, up the beach. The returning water is called the backwash. Swash and backwash are greater during storms, as is their power to erode and move sediments. Storm waves tend to transport sand seaward, sometimes redepositing it as nearshore bars, whereas swell waves have a tendency to carry sediment landward. Gravel usually moves onshore to form storm ridges, which can migrate farther landward as storm waves wash over the crest of the ridge. In the north and around the Gulf of St. Lawrence, winter sea ice protects the shore from wave erosion but can cause severe damage when driven onshore by currents. When this happens, the sea ice sometimes piles up into high ridges, gets pushed over dunes, and shatters wharves and other structures.

The rate of wave erosion depends largely on the type of material forming the shore. Hard sandstones or igneous rocks resist erosion. Cliff erosion takes place by the progressive undercutting of the cliff base by wave action, followed by collapse of blocks along natural joints or bedding planes. Where the coast is formed of relatively soft sedimentary rocks such as shale, erosion can be rapid. The shoreline may consist of loose glacial deposits, including till, gravel, sand, or clay. Shorelines formed in these materials erode much more rapidly than those in hard bedrock. Boulder fields at the base of the gravel and till cliffs armour beaches against wave attack, but as Harris Misener discovered, the relentless battering by storm waves can eventually overcome this resistance.

Beaches are dynamic and ephemeral features and last only as long as sediment supply exceeds erosion. In areas away from the sediment supply provided by rivers, the coastline consists of interspersed eroding and depositional stretches. Erosion of cliffs and bluffs provides sediment for nearby beaches. Sediment is moved from one section of coast to another by a process called longshore drift. Strong currents can form in the surf zone, where waves approach the shoreline obliquely, and mov-

This World War II observation tower (now demolished) near Economy, Nova Scotia, has been undermined by the tides and waves of the Bay of Fundy. BOB TAYLOR.

Pileups of sea ice are created in winter when currents push the thin sea ice onshore, as here at Saint-Siméon, Gaspé Peninsula, Quebec. CHRISTIAN FRASER.

CHAPTER 16 · AT THE BEACH

The features forming this repeating pattern of regular small embayments and protrusions or "horns" are known as beach cusps. These are common and result from the interaction of shoaling waves with beach materials. This example is from the Magdalen Islands, Quebec. PHILIP GILES.

ing sediment not only landward but also parallel to the coast. You may notice the effect of longshore currents when you start swimming in the sea directly offshore from your towel, but after a few minutes find you have drifted along the beach. Shorelines located in the down-drift direction are better supplied with sediment and typically have well-developed beaches.

Long extensions of sand and gravel, called spits, form along irregular coastlines by longshore drift. Sidney Spit in the Gulf Islands of British Columbia is a magnificent example of this. Some spits connect two or more headlands, thus forming a barrier beach such as Melmerby Beach in Nova Scotia. Elsewhere a ribbon of sand and gravel may connect the mainland with an island, creating a tombolo. In other cases, a spit or barrier beach may detach from the land, forming a barrier island, such as the one across the mouth of Malpeque Bay, Prince Edward Island. The New Brunswick coast between Miscou Island and Cape Tormentine is dominated by barrier beaches and spits extending across shallow, drowned estuaries. One spectacular example is the 12-kilometre-long spit at Bouctouche, locally known as the Dune de Bouctouche. Near Stephenville, Newfoundland, the 15-kilometre-long Flat Island spit has formed along the southern shore of St. George's Bay.

The nature of the eroded material is the determining factor in the composition (and popularity) of a beach. Prince Edward Island and the Northumberland Strait shores of New Brunswick and Nova Scotia have some beautiful, wide, sandy beaches and associated dunes, some formed from the continuing erosion of the late Carboniferous and Permian sandstone bedrock and from sandy glacial deposits containing material derived from this bedrock. Where glacial deposits have a wider range of grain size, including gravel, cobbles,

Waves surge onto the shore at Ingonish Beach, Cape Breton Highlands National Park of Canada, Nova Scotia. A. CORNELLIER, COPYRIGHT PARKS CANADA.

Melmerby Beach, Nova Scotia, an example of a baymouth barrier beach formed by a spit extending to meet the opposite side of the bay: this creates a lagoon on the landward side. RON GARNETT / AIRSCAPES.CA.

A mainly granite boulder beach at Kejimkujik National Park of Canada Seaside, Nova Scotia. The boulders are derived form local Devonian granite. ROB FENSOME.

A barrier island, such as this one at Conway Narrows, Prince Edward Island, is similar to a baymouth barrier except that it is separated from the mainland by tidal channels. RON GARNETT / AIRSCAPES.CA.

Coastline with sand dunes, Greenwich Dunes, Prince Edward Island National Park of Canada, Prince Edward Island. J. PLEAU, COPYRIGHT PARKS CANADA.

and boulders, the associated beaches are commonly gravel-dominated, ranging from mixed sand and gravel to beaches almost entirely composed of rounded boulders.

Making Land

Rivers are an important source of sediment supply to the sea. Canada's three largest rivers are the St. Lawrence, Mackenzie, and Fraser. The Mackenzie and Fraser rivers have built large deltas at their mouths over the past 8,000 to 10,000 years, but the St. Lawrence River does not have a delta, in spite of its large discharge. Instead, it empties into a huge estuary extending from Québec City to the Gulf of St. Lawrence. What causes this difference? A key factor is the amount of sediment carried by the rivers. The sediment load of the Mackenzie and Fraser rivers has fed the formation of deltas. In contrast, the St. Lawrence River transports much less sediment because it flows out of the Great Lakes, which act like giant sediment traps, removing nearly all of the silt and sand before they reach the estuary. We will return to estuaries later in the chapter.

Some deltas are among the most densely populated and intensively farmed regions on Earth. For example, 60 million people live on the Ganges-Brahmaputra Delta in southern Asia. In British Columbia, more than 250,000 people live on the Fraser Delta, where the communities of Delta and Richmond are below high-tide level and protected from flooding by a system of river and coastal dykes. Seaward of the coastal dykes are extensive areas of marsh and tidal flats that are habitat for wetland plants, juvenile salmon, and a variety of migratory and resident birds. Twice every year, the Fraser tidal flats host over a million sandpipers on their migrations between South America and Arctic shores.

Over the past century, humans have had a huge impact on the Fraser Delta by altering the natural pattern of sediment dispersal and deposition. They have dredged river channels and built dykes and jetties to fix the position of channels. Old maps

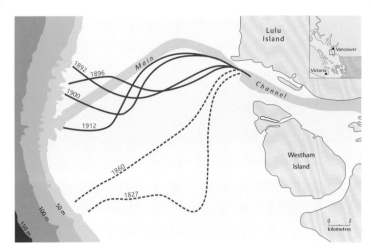

Changes in the Main Channel of the Fraser Delta from 1827 to 1912. Since 1912, advanced engineering has stabilized the Main Channel in the position shown. ADAPTED FROM CLAGUE ET AL. (1983).

of the Delta show that the Main Channel frequently switched its position in the nineteenth and early twentieth centuries, before the river was "trained". The shifting channel spread the river's sediment load across the delta front like an unattended fire hose. Since the early twentieth century, the Main Channel has discharged sediment at only one place, steepening the delta slope there and making it more susceptible to submarine landslides.

The Mackenzie Delta is the second largest delta on the coast of the Arctic Ocean. It lies entirely within the continuous permafrost zone, and the influence of the cold climate can be seen in the myriad lakes that cover its surface. Many of these lakes are connected to channels that meander across the delta plain. In winter, the channels and lakes are covered by nearly 2 metres of ice and the flow in some channels decreases to a trickle.

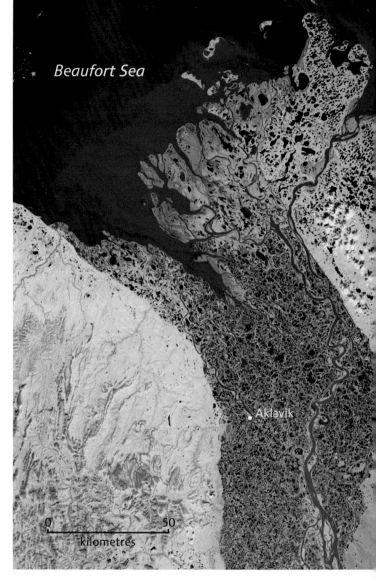

Satellite image of the Mackenzie Delta and adjacent parts of the Northwest Territories and Yukon. The second largest Arctic delta, the Mackenzie covers more than 12,000 square kilometres and has almost 50,000 lakes. The outer part of the delta forms a low floodplain north of the treeline. Its channels are bounded by low levees and its front, facing the Beaufort Sea, is largely retreating as coastal erosion currently outpaces sediment supply across much of the delta. LANDSAT.

Increasing flow in the spring, resulting from earlier snowmelt in the upper river basin to the south, lifts and fractures the ice covering the river—an event referred to as breakup. Broken slabs of ice begin to move down the river, but may jam and cause flooding of the delta surface.

Large amounts of clay, silt, and sand are carried by the Mackenzie River into the Beaufort Sea during late spring and early summer. As the river's flow is checked on reaching the sea, sand is deposited first, and finer silt and clay are transported farther offshore before settling out. When the pack ice melts in summer, waves and currents remobilize these sediments and disperse them along the Beaufort Sea coast.

Tidal marsh on the Fraser Delta south of Vancouver, British Columbia. REPRODUCED WITH THE PERMISSION OF NATURAL RESOURCES CANADA 2013, COURTESY OF THE GEOLOGICAL SURVEY OF CANADA.

Peel Channel on the Mackenzie Delta, with Mount Gifford (or Red Mountain) in the distance. LARRY LANE.

Long Arms of the Sea

In cases where a river doesn't carry enough sediment for a delta to form at its outlet, an estuary can develop. Estuaries represent a transitional zone between the freshwater and marine realms, and so can be fertile areas for a variety of organisms. Some estuaries, such as the St. Lawrence, are large and deep, whereas others are small and shallow. Estuaries can be high-energy environments, with large waves and strong tidal currents, or low-energy ones with small waves and weak currents. The water in an estuary is commonly layered, with less-dense river water floating above denser seawater. This stratification can be disrupted by ocean waves and strong tidal currents.

Canada has spectacular examples of estuaries dominated by tides, among them the Bay of Fundy, Ungava Bay, and Frobisher Bay. The waters of these estuaries are so effectively mixed by tidal currents that stratification never develops. Tides are caused by the gravitational pull of the Sun and the Moon on the surface of the oceans. Where a coastline is deeply indented, tides can be intensified, as in the Bay of Fundy. But not all deep funnel-shaped estuaries have tidal extremes, and even the Bay of Fundy has not always had such a range. At the end of the recent Ice Age, about 12,000 years ago, when the North Atlantic penetrated the Bay, the tidal range was less than 2 metres. The stupendous tides now experienced developed from 7,000 to 4,500 years ago, and reflect the changing shape of the coastline and fluctuating water depths. Today, the time

The St. Lawrence Estuary at Kamouraska, Quebec, is still tidal at this location. That we have a St. Lawrence Estuary rather than a St. Lawrence delta is in part due to the sediment-trapping capacity of the Great Lakes. ROB FENSOME.

Fishing boats at low tide, Hampton, Nova Scotia. The large tidal range is evident from the height of the wharf and the changes in colour on its wall. PHILIP GILES.

Tidal mud and marshes, Shepody Bay on the Bay of Fundy, New Brunswick. RON GARNETT / AIRSCAPES.CA.

interval between tides and the natural lengthwise resonance of the Bay (the sloshing back and forth, such as a bather can produce between two ends of a bathtub) more or less coincide, and the two phenomena reinforce each other, leading to a spectacular tidal range of up to 16 metres. The rush of water can generate tidal currents of several metres per second, capable of eroding and redistributing large volumes of sediment, including coarse gravel. Much of the sediment accumulates in the inner parts of the Bay of Fundy, forming broad areas that are exposed at low tide and flooded at high tide. Within the intertidal zone are sand flats, mud flats, and salt marshes, all dissected by tidal channels. The marshes, many of which have been lost because of human activity, are vital parts of the ecosystem and host large populations of migratory and resident birds.

When the Acadians settled around the Bay of Fundy in the 1600s, they built dykes to prevent the sea invading the salt marshes. The new land was then drained to provide pasture and cropland on fertile soil. Some of the original Acadian dykes are still preserved, for example, on the Tantramar Marshes near Aulac in New Brunswick and in the Grand Pré area of Nova Scotia, where they have been maintained for more than 350 years. The farmland is still in use, but only about 15 percent of the original marsh remains. Furthermore, the dykes prevent deposition of sediments during high tides so that the protected land has not kept pace with rising sea level. This makes it more vulnerable when extreme storms propel seawater over the dykes, as happened in a disastrous tropical storm, the Saxby Gale of 1869 (see Chapter 17).

Lakefront Property

In Chapter 11, we investigated the histories of Canada's glacial lakes at the end of the most recent glaciation. We must bear in mind these dramatic and rapid changes when considering North America's present-day lakes. It is no surprise that the levels of some of the continent's large lakes have continued to change considerably during historic time. Water levels in some of the Great Lakes, for example, have gone up and down by about 1 metre over the past century. These changes have had devastating effects on the lakes' shores, threatening whole ecosystems, as at Point Pelee National Park of Canada on Lake Erie. Lower lake levels have impacted shipping through ports such as Toronto and Chicago, and have reduced the flow of water in the St. Lawrence Seaway. Authorities can manage water levels to some extent using dams and locks, but to maintain the level of a lake, rivers have to supply enough water to keep it topped up (Chapter 15). Studies suggest that the recent trend of falling water levels could persist, with the possibility of even lower lake levels in the future. Scientists are now working to determine how low the Great Lakes might drop with future climate change and water usage. Large lakes are also subject to the effects of different rates of uplift

The delta of the Kaministiquia River near Thunder Bay, Ontario, extends out into Lake Superior, the largest freshwater lake in the world. ALAN MORGAN.

Rock used to armour shorelines against water or ice erosion is known as riprap. The riprap in this example has been introduced to protect summer cottages at Evangeline Beach, Nova Scotia. ROB FENSOME.

Erosion by waves and currents is not restricted to the sea coast. This spit at Limestone Point, Manitoba, is on the northwestern shore of Lake Winnipeg. RON GARNETT / AIRSCAPES.CA.

along their lengths. For example, in Lake Winnipeg, the rate of uplift at its northern outlet to the Nelson River is greater than the rate of uplift at the southern end of the lake, some 400 kilometres to the south. This difference has resulted in a progressive rise of lake level at the south end of the lake, promoting shoreline erosion in areas of shorefront home and cottage development close to Winnipeg.

The shores of Canada's lakes differ from sea coasts in important ways. Lake water is usually fresh, so the fauna and flora are different, and tides are insignificant. Ice forms sooner, typically earlier in the winter. Lakes have beaches, the more accessible ones much loved by summer vacationers, but these are commonly reworked by smaller waves and less powerful currents than those on the sea shore.

Back to Miseners Island

The disappearance of Mr. Misener's island was simply part of the normal state of affairs along an ever-changing coastline. When properties are threatened by the sea, we try to save them, but our actions are commonly expensive, ineffective, or both. Worse, protection of one part of the shoreline often causes erosion of adjacent shores, leading to disputes between neighbours. And erosion is important as a source of sediment to maintain healthy beaches.

With climate change set to raise sea level everywhere, the battle against the sea will become progressively more difficult. Adapting to this inevitable change will be challenging and will require co-operative management of the coast. We need to alter our mindset from a reactive one to one that is proactive and focussed on planning, following the model set by countries such as The Netherlands, which has had centuries of experience in learning how to combat a relentless force. It will be necessary to protect important infrastructure and property against erosion despite the high costs, but choices will have to be made. Good decisions now will pay dividends in the future.

17 • ON DANGEROUS GROUND

CHAPTER SUPPORTED BY A DONATION FROM DARCY MARUD

Practically every year at least one part of Canada is affected by serious flooding. The Red River Valley in southern Manitoba is a particularly vulnerable area. During winter, heavy snow tends to accumulate upstream, leading to the possibility of devastating floods downstream during spring melt. The city of Winnipeg is protected from the worst impact of the "Red Sea" floods by the Red River Floodway, which channels peak flows around the city. RON GARNETT / AIRSCAPES.CA.

The Morning the Mountain Fell

Frank is a small Alberta town in a valley in the Rockies just east of the Crowsnest Pass and south of Calgary. Looming ominously above the town is Turtle Mountain, which the local Aboriginal people called "the mountain that walked", a name that proved to be remarkably prescient. For on the morning of April 29, 1903, tragedy struck. Geologically, Turtle Mountain is part of an anticline. Its upper slopes are of early Carboniferous carbonates, which were folded and thrust eastward over Cretaceous coal-bearing sedimentary rocks during the Paleocene.

In 1903 about six hundred people lived in Frank, mostly workers eking out a hardscrabble existence in the local coal mine. Excavation for the mine began in 1901 when a tunnel was opened at the base of Turtle Mountain. By October 1902 subterranean caverns supported by gigantic rock pillars 12 metres thick extended over 700 metres

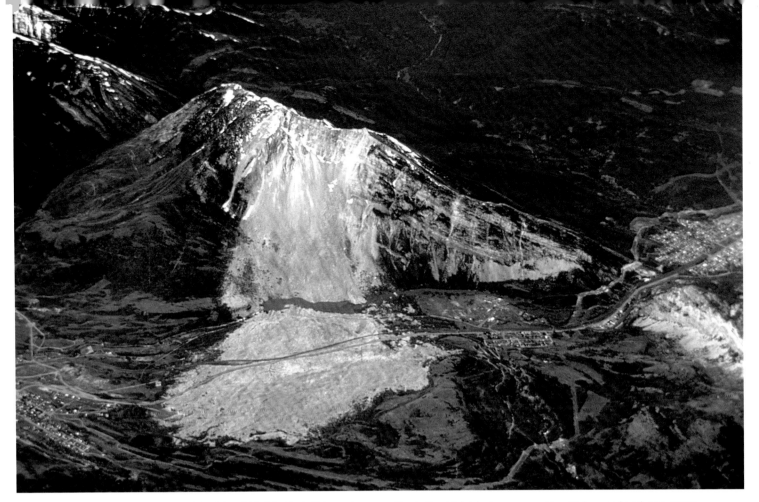

The scar on Turtle Mountain and the debris field of the Frank Slide, southwestern Alberta. The debris lobe is now crossed by a highway and a railway. RAY PRICE.

into the mountain. It was then that warning signs of a coming disaster began to appear. On some days, the ground trembled as Turtle Mountain shifted to accommodate the cavernous openings in the mine below. Although anxious, the miners carried on because they had families to feed and no other work. And extracting the coal was relatively easy because it literally fell from the roof, ready to be shovelled up. The conditions sound grim, but they were far safer than those in most other underground coal mines operating at that time.

Heading underground to start their shift on the night of April 28, 1903, seventeen miners found themselves trapped in the mine by a rockfall. Faced with few options, the miners dug a shaft through the rubble and, after about twelve hours, broke free. But for seventy-six other residents of the town, time had run out. At 4:10 AM on April 29, while the miners were tunnelling their way to freedom, all hell broke loose. About 30 million cubic metres of limestone separated from the summit of Turtle Mountain and hurtled down its eastern slope at speeds up to 140 kilometres per hour. In less than 2 minutes, debris, including some blocks as large as houses, covered about 3 square kilometres of the valley floor. Seven houses on the outskirts of Frank were buried; a small lake formed where the debris dammed Crowsnest River; and railway tracks, roads, and mine buildings were obliterated.

The town proved surprisingly resilient. Most survivors stayed in Frank, and the mine and railway were back in operation in less than a month. Today, Frank is a vibrant community with a population of about two hundred. But Turtle Mountain is treated with respect, and the Alberta government closely monitors its top and east flank for advance signs of another landslide.

What caused the Frank Slide? The mountain is inherently unstable, with hard carbonates overlying softer, younger clastic rocks. It seems that the limestone slid along bedding planes and joints in Turtle Mountain, and hurtled downslope and into the valley. Deforestation and wet weather may have added to the instability. Mining activity at the base of the mountain may have been the trigger. The Frank Slide focussed attention on the importance of warning signs, such as tremors, especially under inherently unstable geological conditions. We know now that the tremors the miners experienced were a warning that Turtle Mountain was about to collapse.

Natural Hazards

Landslides are just one type of destructive earth process or natural hazard. Natural hazards are phenomena that occur naturally at or near the Earth's surface and pose a risk to people and property. They include earthquakes, tsunamis, volcanic eruptions, landslides, snow avalanches, ground subsidence, coastal and river floods, severe storms, dust storms, wildfire, drought, and heat waves. Some hazards are driven by internal planetary forces; volcanic eruptions and earthquakes, for example, are mostly related to movements of tectonic plates. Other hazards result from external forces, such as wind and water.

Since 1995, the world has experienced many natural hazards: deadly earthquakes in Haiti, Japan, China, and elsewhere; devastating tsunamis in the Indian Ocean and Japan; catastrophic flooding in Manitoba, southern Alberta, southern England, central Europe, and Queensland in Australia; hurricanes that have ravaged Central America and New Orleans, the New York City area, and the Philippines; an ice storm that crippled Ontario and Quebec; and globally some of the warmest years of the past century and probably the past thousand years. These natural hazards have been enormously expensive, both in lives and financial costs. The economic cost to Canada alone in the past fifty years has been many tens of billions dollars.

In this chapter we emphasize the more strictly geological hazards, such as earthquakes, tsunamis, volcanic eruptions, landslides, and floods. They have occurred on our planet for billions of years. The word risk is used to define the seriousness of the hazard. It is the chance of injury, death, or damage to property. A hazardous event becomes a disaster or catastrophe. Although the distinction between disaster and catastrophe is somewhat vague, the latter is more massive and affects a larger number of people, generally a broader area, and more infrastructure than the former.

Quakes in Canada

Earthquakes are some of the most deadly natural hazards. During the twentieth century alone, over two million people were killed by earthquakes and related fires, tsunamis, and landslides. With a few exceptions, earthquakes are the result of the movement of tectonic plates. Such motion is usually slow, typically less than 10 centimetres per year. An earthquake happens when stored-up energy from plate movements is suddenly released, causing rocks to break and move rapidly past one another along a fault. The source, or focus, of an earthquake is the location beneath the

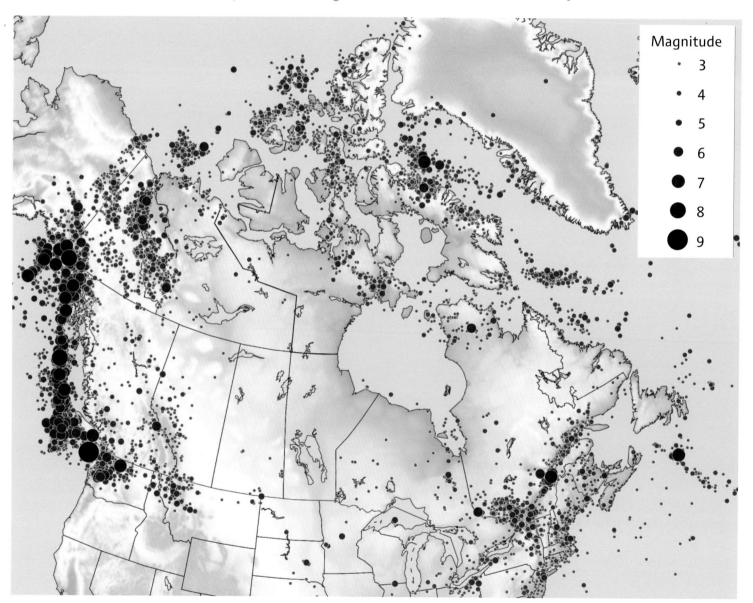

Distribution of earthquakes larger than magnitude 2.5 in Canada. IMAGE COURTESY OF STEPHEN HALCHUK, GEOLOGICAL SURVEY OF CANADA (NATURAL RESOURCES CANADA).

A building that collapsed during the Kobe earthquake in Japan in January 1995. COURTESY OF THE UNITED STATES GEOLOGICAL SURVEY.

The oceanic Juan de Fuca Plate is presently locked to the North American Plate where the temperature of the crust along the Cascadia Subduction Zone is less than 350°C. This locked region is where the next great Cascadia earthquake will be initiated. The 450°C isotherm delineates the landward limit of potential rupture; at higher temperatures the rocks are too ductile to break. Vancouver Island is being flexed upward due to compression and elastic shortening of the North American Plate above the locked interface. This deformation has been detected from satellite measurements of small changes in the relative position of points on the Earth's surface. Sometime within the next several hundred years the plates will suddenly unlock, triggering a great earthquake. FROM CLAGUE ET AL. (2006), USED WITH PERMISSION OF THE AUTHORS AND TRICOUNI PRESS.

surface where the fault initially slips. The point on the Earth's surface immediately above the focus is called the epicentre.

In Canada, the area at greatest risk from earthquakes is the west coast, where the Juan de Fuca, Pacific, and North American plates meet. Fortunately, most big earthquakes on the west coast occur offshore, beneath the Pacific Ocean floor. Although some of these quakes are large, they occur far from densely inhabited areas. Other earthquakes, however, are closer to cities, beneath the southern mainland of British Columbia, Vancouver Island, the Strait of Georgia, and, to the south, Puget Sound. On the west coast, earthquakes are of four types, depending on their relationship to plates and plate boundaries.

The first type occurs on faults within the North American Plate, commonly at depths of 20 kilometres or less below the surface. On average, there is one such quake in British Columbia every day. Nearly all are too small to be felt, but a potentially damaging one occurs about every ten years. The largest earthquake of this type in the past century was a magnitude 7.3 earthquake in 1946 with an epicentre on central Vancouver Island. It caused damage up to about 100 kilometres from the epicentre.

A second type of earthquake takes place within the subducting Juan de Fuca Plate. Such earthquakes are the deepest of the four types, with foci down to about 80 kilometres. Although energy must travel a long way before reaching the surface, several deep quakes have caused severe property damage in the Puget Sound area, including one of magnitude 6.8 near Olympia, Washington, in February 2001, which was widely felt in Vancouver.

Earthquakes of the third type are those occurring on the Queen Charlotte-Fairweather Fault (Chapter 2), which separates the Pacific and North American plates from northwest of Vancouver Island to southeast Alaska. Today, the two plates are locked to one another beneath the ocean floor west, south, and north of Haida Gwaii. However, they occasionally slip, generating earthquakes. Several large earthquakes have occurred on the Queen Charlotte-Fairweather Fault during the past century, including the largest historic quake in Canada, a magnitude 8.1 event in 1949, which damaged communities on Haida Gwaii. Earthquakes of this type, occurring on transform faults, are analogous to those along the San Andreas Fault in California.

The fourth type consists of earthquakes on the Cascadia Subduction Zone, which separates the converging North American and Juan de Fuca plates. Earthquakes on subduction zones are nicknamed "megaquakes" because they are among the most powerful on Earth. The March 2011 quake off northeastern Japan was of this type. During a subduction-related earthquake, huge amounts of energy, accumulated over decades to centuries, are released when the plates slip as much as 15 metres past one another. Along the Cascadia Subduction Zone, such sudden release of energy causes the land to subside along the west coast of Vancouver Island, Washington, and Oregon. No megaquake has been recorded in Canada since the first Europeans settled in the west, but evidence exists for one in January 1700. This quake caused the Pacific coasts of Vancouver Island, Washington, and Oregon to subside by up to 2 metres and produced a tsunami that caused significant damage in Japan. Evidence for older megaquakes includes widespread buried marsh deposits abruptly overlain by intertidal muds in wetlands from Vancouver Island to northern California. Remnants of trees and other plants rooted in the buried marshes are silent witnesses to sudden subsidence that lowered the marshes and drowned the roots of the trees in seawater. Geophysical measurements indicate that energy is accumulating today for another cataclysm. The data show that Vancouver Island is compressing and bulging upwards above the descending Juan de Fuca Plate, supporting the notion that this plate and the North American Plate are currently stuck.

In Canada east of the Cordillera, earthquakes are less common and less damaging, but do occur. As the North American Plate moves westward from the Mid-Atlantic Ridge, stresses at weak places within the plate trigger earthquakes up to about magnitude 7. These quakes are largely related to an ancient rift system initiated during the opening of the Iapetus Ocean (Chapter 7), with most occurring along the St. Lawrence Lowlands from Montréal to Sept-Îles in Quebec. Earthquakes have occurred historically in several clusters in this region, at depths of 5 to 30 kilometres. In southern Ontario and Quebec, earthquakes are more damaging and affect a larger area than similar-sized quakes in western Canada. The reason is that cold, brittle crystalline rocks of the Canadian Shield and the overlying sedimentary rocks more efficiently transmit earthquake waves than the hotter, more ductile, and heavily faulted lithosphere underlying westernmost Canada.

Other areas of Canada with recorded earthquakes include the Mackenzie Mountains, parts of Baffin Island and the Labrador-Baffin Seaway, and some of the Arctic Islands. Earthquakes in the Mackenzie Mountains have been linked to the seismically active region southwest of the St. Elias Mountains, where the Yakutat Terrane is being stuffed beneath the North American Plate (Chapter 10). The Atlantic Provinces have a lower level of seismicity, and earthquakes larger than magnitude 5 are rare there.

Shaken Up

A major effect of an earthquake is ground shaking produced by seismic waves, and generally the larger the earthquake the stronger and longer the shaking and the greater the damage to buildings. Although the ground can crack along the fault generating the quake,

This section dug into a tidal marsh near Tofino, British Columbia, provides evidence for past earthquakes in the region. The tip of the knife is at the top of a buried marsh surface, which is in turn overlain by about 10 centimetres of tsunami-deposited sand. This and other evidence show that the marsh subsided and was inundated by a tsunami during the last great earthquake at the Cascadia Subduction Zone in January 1700. JOHN CLAGUE.

most ground cracks that appear during earthquakes are caused by secondary factors. Strong seismic shaking can liquefy firm but saturated sediments. Artificial fills and unconsolidated sand beneath river plains, deltas, and shorelines are especially prone to this phenomenon. If the sand is capped by cohesive clay, the clay may fracture and slide horizontally on top of the liquefied layer. And sometimes liquefied sediment may flow upward along fractures and erupt at the surface, forming features called sand volcanoes. Large earthquakes can also trigger numerous landslides in mountainous areas, with the potential for major loss of life. In urban areas, fire is a serious hazard associated with earthquakes and is especially dangerous when water pipes or gas lines are broken.

By researching the evidence for, and effects of, past earthquakes, geologists help to reduce damage from future quakes. The history of earthquakes on a fault can be investigated by excavating trenches across the fault at key locations and observing how sediment layers are offset, and by radiocarbon dating those offsets. Geophysicists measure small movements of the ground surface that result from plate interactions and also evaluate how seismic waves will behave as they move through near-surface earth materials. In response to this research, engineers now design buildings and other structures to withstand seismic shaking.

One of the best ways for scientists to inform authorities and the public about earthquakes is to publish maps that show different types of seismic hazards and risk. These maps can show the epicentres of historic earthquakes, the probabilities of future ones of different magnitudes, the amount of shaking likely to occur, and areas vulnerable to liquefaction or landslides. With such information, governments can establish design standards for buildings in earthquake-prone areas.

The two approaches used in anticipating future earthquakes are prediction and forecasting. A prediction states when an

Strong shaking during earthquakes can liquefy loose, water-saturated sediments, as in Christchurch, New Zealand, during the February 2011 earthquake. COURTESY OF NOAA.

earthquake of a given magnitude will occur in a region, commonly within hours or days. A forecast specifies the probability that an area will experience a quake, commonly over many years or decades. Geoscientists are getting better at forecasting earthquakes, but so far they are unable to predict them.

Some attempts at prediction are based on precursor events, such as foreshocks (small quakes preceding a larger one), changes in water level and chemistry in wells, local changes in gravity, and magnetic, electrical, or acoustic properties of the crust. Many of these changes take place when rocks expand and fracture in the hours or days preceding an earthquake. Unfortunately, not every earthquake is preceded by such observable phenomena, and even when such phenomena do occur, they can be difficult to interpret because an earthquake does not necessarily follow. Even if research eventually provides short-term predictions as accurate as weather forecasts, the uncertainty will still put seismologists in a difficult position because of the immense disruption and cost should the quake not happen, as well as the loss of public trust that a false alarm would cause.

Two events in China show the unreliability of earthquake prediction. In late 1974 and early 1975, geoscientists successfully predicted a large earthquake at Haicheng based on a lengthy period of foreshocks and other indicators. On the evening of February 4, 1975, a magnitude 7.3 quake caused extensive damage to houses and buildings in the city. But loss of life was greatly reduced because most inhabitants had been warned and moved outside. However, some 17 months later at Tangshan, a large industrial city about 300 kilometres southwest of Haicheng, a magnitude 7.6 earthquake struck without any warning. More than 250,000 people died in less than 6 minutes.

It is said that "earthquakes don't kill people, buildings kill people", and governments are nowadays better aware of the need for stricter building codes in seismically active regions. Canada has a National Building Code that contains provisions for designing buildings, bridges, pipelines, and other structures to withstand shaking from earthquakes. The seismic provisions in the Code differ locally and regionally in relation to differences in expected earthquake frequency and magnitude across the country. Improving the resistance of existing structures, referred to as seismic retrofitting, is costly but necessary.

Killer Waves

Some earthquakes trigger destructive sea waves called tsunamis. Tsunami is a Japanese word meaning "harbour wave". The commonly used term "tidal wave" is a misnomer because tsunamis have nothing to do with tides. Rather, they are generated by a sudden displacement of large volumes of water in the sea or in a lake, occasionally even in a river. The most common triggers are large earthquakes beneath the ocean floor, but some landslides, volcanic eruptions, and even asteroid impacts cause tsunamis.

Large tsunamis can surge several kilometres inland and reach heights of 30 metres or more, smashing everything in their paths. Surprisingly, a tsunami is imperceptible on the open ocean; passengers on a ship would have no idea that they had crossed one. In the deep ocean, tsunami waves, travelling at speeds up to many hundreds of kilometres per hour, have heights of less than a metre and are kilometres apart. That changes when the waves approach shallow water near a coast. As the water shoals, bottom friction causes the waves to decelerate, become more closely spaced, and grow taller. On reaching the shore they transform into turbulent, landward-surging masses of water that may run many kilometres inland.

One popular misconception is that a tsunami consists of a single immense wave that curls over and crashes on the shore. In fact, tsunamis typically consist of turbulent onrushing surges of debris-laden water. And most tsunamis arrive as a series of

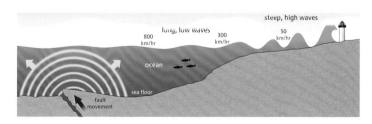

A tsunami can be triggered by rupture of the sea floor along a fault during an earthquake. The sudden upward displacement of the sea floor produces tsunami, which often have a deadly impact when they reach shore. The values in kilometres-per-hour show how much a tsunami loses speed as it moves shoreward. FROM CLAGUE ET AL. (2006), USED WITH PERMISSION OF THE AUTHORS AND TRICOUNI PRESS.

Destruction wrought by the 2010 earthquake and tsunami on the Oshika Peninsula, northeastern Japan. COURTESY OF NOAA.

This house from Port au Bras, Newfoundland, was removed from its foundations by the tsunami that followed the 1929 Grand Banks Earthquake. To prevent further damage, the house was tied up to the Lunenburg schooner *Marian Belle Wolfe*, anchored in Little Burin Harbour. The photograph was taken by Father James Miller on November 19, 1929, the day following the tsunami. PROVIDED BY ALAN RUFFMAN.

waves that can be separated by minutes or even by more than an hour. In many instances, the first wave is preceded by a retreat of the sea. The unusual withdrawal of water may attract the curious, which is one reason why tsunamis can be so deadly.

Earthquakes beneath the Pacific Ocean are the main sources of tsunamis on Canada's west coast. The quakes occur at the subduction zones that surround much of the Pacific. During a major subduction earthquake, a vast tract of ocean floor up to 100 kilometres wide and hundreds of kilometres long moves oceanward, displacing the water above it like a gigantic piston. Waves move outward from the focus, travelling in stealth-like fashion at high speeds toward the surrounding coasts. Many communities on Vancouver Island were damaged by a tsunami triggered by the great Alaska earthquake of March 27, 1964. The tsunami moved outward from the epicentre of the earthquake near the head of Prince William Sound in Alaska. Within a few hours it reached Haida Gwaii and Vancouver Island, causing damage in Port Alberni, Hot Springs Cove, Tofino, Ucluelet, and Zeballos. No one died in British Columbia from this tsunami, but there were over a hundred fatalities in Alaska.

Although tsunamis generated by distant Pacific earthquakes pose a hazard to British Columbia's coastal communities, the greatest threat is at the province's doorstep—the Cascadia Subduction Zone. Should the fault separating the North American and Juan de Fuca plates slip, as it last did in January 1700, the Pacific coast from Vancouver Island to northern California would be struck by tsunami waves far larger than those of March 1964.

Landslides that plunge into the sea or that occur on the ocean floor can also trigger tsunamis. Such an event could happen in one of the fiords indenting the British Columbia coast north of Vancouver. In 1975, a large submarine

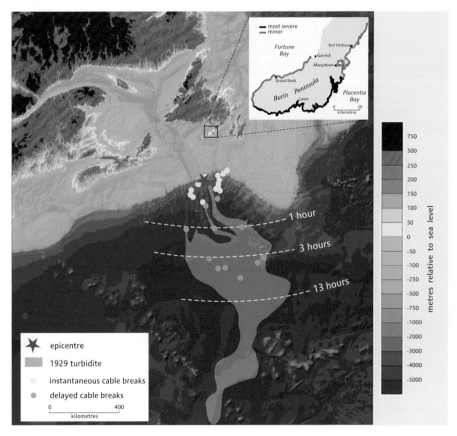

The extent of the turbidity current deposit associated with the Grand Banks Earthquake of November 1929, and the timing of breaks to submarine cables caused by the powerful bottom current. The inset shows the parts of the Newfoundland coastline affected by the tsunami generated by this earthquake. The terms "most severe" and "minor" refer to physical damage to docks, wharves, boats, and buildings. MAIN FIGURE FROM ATLANTIC GEOSCIENCE SOCIETY (2001); INSET ADAPTED FROM CLAGUE ET AL. (2003).

landslide in Douglas Channel, just south of Kitimat, produced waves as high as 8 metres that caused about two million dollars in damage to shore facilities. The only tsunami to have claimed lives in Canada during historic time is the one that struck the Burin Peninsula of Newfoundland on November 18, 1929, killing twenty-seven people. A magnitude 7.2 earthquake beneath the Grand Banks triggered a huge submarine landslide that caused the tsunami and cut vital undersea communications cables. This event was shocking, not just because of its tragic results, but because Canada's eastern provinces are distant from a plate boundary and large tsunami-producing earthquakes are much rarer there than on the Pacific coast.

Warning centres in the Pacific Ocean rely on a network of seismographs to provide real-time estimates of earthquake magnitude and location before issuing a tsunami warning. Data is then retrieved from more than a hundred coastal tidal gauges and sensors connected to floating buoys to verify that a tsunami indeed was produced. The bottom sensors detect small changes in the pressure exerted by the increased volume of water as a tsunami passes over them. These sensors transmit the pressure measurements to buoys at the ocean surface, which then relay the data to a satellite. Finally the satellite transmits the data to the warning centres.

When the source of a tsunami is less than about 100 kilometres away from a populated area, there is generally insufficient time to warn and safely evacuate people. However, those close to the source will probably feel the earthquake and can immediately move to higher ground.

Risk to vulnerable coastal communities can be assessed by determining the frequency and size of past tsunamis from historical records and geological data. Maximum tsunami heights along a reach of the coast are estimated from computer-generated models and the distribution of historic and prehistoric tsunami deposits. Maps can then be produced showing areas likely to be inundated by tsunamis of different sizes. The maps may be used to guide or restrict development in tsunami-prone areas and to educate people living in these areas about the risk they face. Computer models also provide estimates of tsunami arrival times, currents, and impacts on buildings and other infrastructure.

Eruptions

Recent volcanism in Canada is restricted to British Columbia and the southwest corner of Yukon. Although no volcano has erupted in western Canada during the past century, a basalt lava flow in the Unuk River valley near the British Columbia-Alaska border dates from the nineteenth century, and the Tseax Cone north of Terrace, British Columbia, erupted less than 300 years ago. The most immediate threat to a populated area is from Mount Baker,

The crater in the Tseax Cone, which erupted around 1775, producing an explosive pyrotechnics show, is vividly remembered in Nisga'a oral history. The Tseax Cone and its associated lavas are part of the Stikine Volcanic Belt in northern British Columbia. CATHERINE HICKSON.

Seen here from Saturna Island, British Columbia, Mount Baker in Washington is an active volcano, one of many in the Cascade Magmatic Arc. The Arc extends into southwestern British Columbia and includes mounts Garibaldi and Meager. C. CHEADLE, COPYRIGHT PARKS CANADA.

a stratovolcano in Washington, 100 kilometres southeast of Vancouver. Venting of gases and hot fluids from the summit crater in the mid-1970s provided a timely reminder that this volcano is still active and should be considered armed and dangerous.

Active volcanoes come in different shapes and sizes, reflecting the types of eruptions that form them. The most explosive and dangerous are stratovolcanoes (Chapter 1). A chain of active and recently active stratovolcanoes stretches from northern California to southwestern British Columbia, the surface expression of the Cascade Magmatic Arc. Besides Mount Baker, the belt includes such well-known peaks as mounts Lassen, Shasta, Hood, St. Helens, and Rainier. Mounts Garibaldi and Meager are the most important Canadian members. Mount Garibaldi, although a geologically young volcano, last erupted about 11,000 years ago. Mount Meager last erupted about 2,400 years ago and

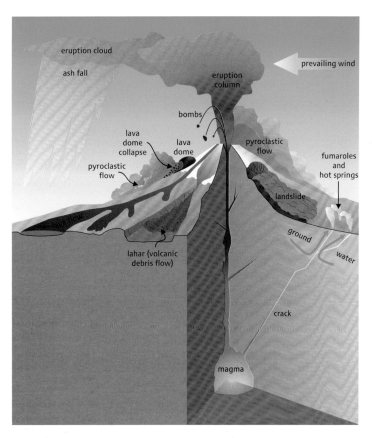

Types of hazards associated with explosive eruptions of stratovolcanoes such as Mount Garibaldi and Mount Meager. Fumaroles are surface openings that emit steam and other gases such as sulphur dioxide and carbon dioxide. ADAPTED FROM CLAGUE AND TURNER (2003); USED WITH PERMISSION OF THE AUTHORS AND TRICOUNI PRESS.

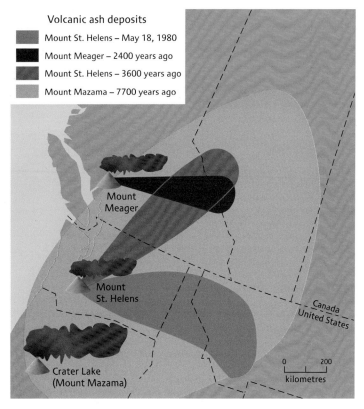

Ash deposits associated with selected volcanic eruptions from Cascade volcanoes. The illustration shows only areas where the ash deposits are thick enough to be visible. ADAPTED FROM CLAGUE AND TURNER (2003); USED WITH PERMISSION OF THE AUTHORS AND TRICOUNI PRESS.

is considered by geologists to be a dormant volcano—meaning that although it is not currently active, it could erupt again.

When a stratovolcano blows, a billowing plume of ash can rise many kilometres into the atmosphere. Ash clouds can extend hundreds of kilometres downwind, depositing ash over large areas. As the eruption of the volcano at Eyjafjallajökull in southern Iceland in the spring of 2010 illustrated, atmospheric volcanic ash can disrupt air travel for days or weeks. Heavy ash fall can collapse buildings and damage crops, and even minor amounts make breathing difficult and damage electronics and machinery.

Gases emitted from a volcano can be deadly. More than 90 percent of the gases are water vapour (steam), but the remaining portion is a toxic brew that may include carbon dioxide, carbon monoxide, sulphur dioxide, hydrogen sulphide, hydrogen, and fluorine. Sulphur dioxide reacts with water droplets in the atmosphere to create acid rain, which is corrosive and harmful to vegetation. Carbon dioxide is denser than air and can be trapped in low-lying areas in fatal concentrations. Fluorine rains out of the atmosphere with volcanic ash particles, poisoning livestock and contaminating domestic water supplies. And even small amounts of carbon monoxide can be fatal.

As we discovered in Chapter 1, pyroclastic flows are avalanches of gas-charged incandescent ash and red-hot blocks that move down the flanks of a volcano during an explosive eruption or when the steep side of a growing dome of lava collapses and breaks apart. Pyroclastic flows are as hot as 800°C and travel at speeds of 150 to 250 kilometres per hour. They tend to follow valleys, flattening and burning everything in their paths. A pyroclastic flow from Mount Vesuvius destroyed the Roman town of Pompeii in the year 79 CE, killing ten thousand to twenty-five thousand people. Today, three million people live within sight of this active volcano. In May 1902 the town of Saint Pierre on the Caribbean island of Martinique was obliterated by a pyroclastic flow from Mount Pelée. Only two of the town's thirty thousand residents

Exposed in a road cut along Highway 3 near Keremeos, British Columbia, the white layer is volcanic ash from the eruption of Mount Mazama in Oregon, 7,700 years ago. DALE GREGORY.

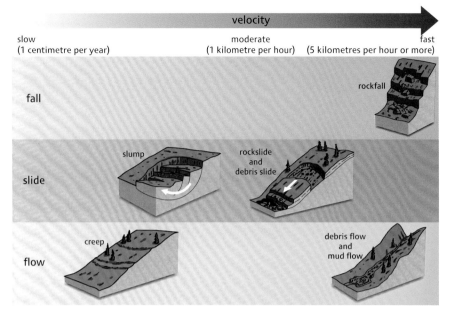

Types of landslides. Landslides are classified according to type of movement (fall, slide, or flow); type of material that fails (i.e., rock or unconsolidated sediment); amount of water or air involved in the movement; and velocity.

Large landslide resulting from the collapse of the south flank of Mount Meager, British Columbia, in August 2010. PAUL ADAM.

survived; one survivor was a prisoner who was captive in a poorly ventilated jail cell below ground level.

Volcanic explosions can produce lower-density hurricane-force winds filled with volcanic debris that overtop ridges and pack a lethal wallop over a broad area. Debris-laden winds that raced north from Mount St. Helens when it exploded in 1980 destroyed over 600 square kilometres of forest.

Some volcanoes have a summit crater lake or are capped by snow and glacier ice. When such a volcano erupts, the hot volcanic debris can melt the snow and ice, forming a lahar, a large fast-moving flow of rocks, mud, and water that can rush more than 100 kilometres down valleys at up to 70 kilometres per hour. Some lahars contain so much rock debris that they resemble fast-moving rivers of wet cement. They rip up and transport trees and huge boulders, and farther downstream they entomb everything in their path in muddy debris.

On a Slippery Slope

Landslides are downslope movements of rock, sediment, or artificial fill under the influence of gravity. They can be triggered by earthquakes or volcanoes, but most, such as the Frank Slide, are unrelated to such tectonic events. Landslides range in volume from tens of cubic metres to tens of cubic kilometres, and move at speeds ranging from millimetres per year to more than 100 metres per second. Larger ones can block rivers, causing upstream flooding, and sudden collapse of such blockages cause devastating downstream floods. Each year landslides kill thousands of people globally and cause tens of billions of dollars in damage.

In Canada, landslides are most common in the Cordillera and in parts of Ontario, Quebec, and New Brunswick. But they do occur elsewhere, for example, along the valleys of the large rivers that cross the Prairies, such as the Athabasca and South Saskatchewan, and along Canada's coasts. Any kind of sediment may slide if the slope is sufficiently steep or saturated with water, but silts and clays deposited in former glacial lakes or seas, Quaternary volcanic rocks, and Cretaceous shales are particularly vulnerable to landsliding.

There are three main groups of landslides—falls, slides, and flows. In a fall, blocks of rock tumble or bounce down a steep slope. A fall that involves rocks that tilt forward and then collapse is called a topple. Piles of rocks that accumulate at the foot of a slope as a result of numerous rockfalls are known as talus cones. Active talus cones lack vegetation; most older cones are covered by plants. Slides involve the downslope movement of rock or sediment along a planar surface; if the surface is concave upward, the slide is called a slump. Flows are water-laden masses

Landslide south of Lytton, British Columbia, in March 1997. The landslide severed the Canadian National Railway line, causing a freight train to derail with the loss of two lives. JOHN CLAGUE.

The 1993 Lemieux Landslide in the valley of the South Nation River, Ontario. Such landslides resulting from the sudden liquefaction of Leda Clay, a deposit of the Champlain Sea. GREG BROOKS, REPRODUCED WITH THE PERMISSION OF NATURAL RESOURCES CANADA 2013, COURTESY OF THE GEOLOGICAL SURVEY OF CANADA.

of sediment or fragmented rock that rush down mountainsides and, more commonly, stream channels. They are often triggered by heavy rainfall. Flows are also common in areas underlain by permafrost, where ice-rich sediments on slopes thaw in summer. A turbidity current (Chapter 1) is a type of flow in which sediment-charged water flows across a sea or lake floor.

Landslides can result from a spreading movement in which blocks of intact land separate and move apart on an underlying weak or soft layer that has liquefied; this was the cause of the failures in Leda Clay in the St. Lawrence Lowland of southern Ontario and Quebec. Over the past century, the sudden liquefaction of Leda Clay at several locations has destroyed many homes and killed over one hundred people. Leda Clay was deposited in the Champlain Sea, which occupied the St. Lawrence and Ottawa river valleys at the close of the last glaciation (Chapter 11). The tendency of Leda Clay to liquefy stems from its unusual structure. The tiny mineral grains that constitute the clay are loosely bound together, with pore water between them. In fact, the sediment may be as much as 60 percent water by weight. When the sediment is disturbed, the mineral grains separate and begin to float in the pore water, causing the clay to flow. Commonly, the area of the landslide rapidly expands, destroying large swaths of flat land.

What steps can be taken to reduce the risk of injury and damage from landslides? We can discourage or prevent development on unstable sites, monitor problem areas, and build corrective and defensive works. For example, in 1973 the British Columbia government denied an application for a residential development on the Rubble Creek fan in southwestern British Columbia. The fan is composed of debris brought down by landslides from a 400-metre-high steep rock face called The Barrier. This cliff formed about 13,000 years ago when a lava flow met a tongue of Pleistocene glacier ice at the head of the Rubble Creek valley. No longer supported by ice and heavily fractured, The Barrier is unstable and has failed repeatedly. The most recent large landslide there occurred during the winter of 1855–1856, spreading a bouldery debris sheet over the area now crossed by the Sea-to-Sky Highway south of Whistler.

Slow movements of unstable slopes may accelerate prior to an impending catastrophic failure. That is why, in recent years, the Alberta government has monitored the fractured crest of Turtle Mountain, source of the 1903 Frank Slide. On a smaller scale, rockfall monitoring is done along many Canadian highways, mainly in the mountainous west. Defensive measures against landslides include reforestation, control works, and protective structures. Control works include berms, rockfall nets, retaining walls, and alteration of surface and subsurface drainage on unstable slopes. Subsurface drainage was what controlled the 1.4-cubic-kilometre Downie Slide in the Columbia River valley north of Revelstoke, British Columbia. The toe of the Downie Slide was flooded when BC Hydro built the Revelstoke Dam. The possibility of catastrophic failure led the company to dewater the landslide by building a drainage system consisting of hundreds of metres of tunnels and thousands of metres of drill holes. Since 1984, when the work was completed, the landslide mass has nearly stopped moving and is considered to be more stable than it was before the dam was built.

Water, Water, Everywhere

Over a hundred people in Canada have been killed by floodwaters since 1900, and nowadays property damage from floods averages well over one hundred million dollars per year. Flooding may be related to the sea or lakes, but mostly it arises from rivers. Many floods occur because rivers, which normally flow within their channels, occasionally overflow and wreak havoc. The time of year during which flooding occurs depends mainly on the size of the watershed and climate. In Canada, many large rivers flood only in late spring, after winters of abnormally heavy snowfall followed by prolonged warm weather. Smaller rivers carry peak flows during late spring, or in the fall during periods of heavy rain. Flooding may result when ice jams form during spring breakup. Small streams can flood at any time of year—in early spring, during mid-winter thaws, or during summer thunderstorms. Very large, short-lived floods result from the sudden draining of glacier-, moraine-, and landslide-dammed lakes. These outburst floods may have peak flows far larger than floods in the same basin triggered by normal rainfall or snowfall. Glacier dams are notoriously unstable and can fail when the lake above the dam fills up enough that the glacier floats or the impounded water escapes through a tunnel at the base of the ice.

Levels reached by Fraser River floods on a barn in the Fraser Valley near Hatzic, British Columbia. COURTESY OF THE BRITISH COLUMBIA MINISTRY OF ENVIRONMENT.

In May and June 1948, a flood in the lower Fraser Valley in British Columbia breached protective dykes in more than a dozen places, inundating some 220 square kilometres of agricultural land, nearly one-third of the valley's flood plain. About 16,000 people were evacuated, 2,000 homes were damaged, and 82 bridges were washed away. Six years later, in October 1954, eastern Canada suffered the ire of Hurricane Hazel, which delivered over 21 centimetres of rain to Toronto, causing the worst flooding in the city's 200-year history. In just two days, streams were turned into swollen torrents that killed 81 people and destroyed or damaged 1,800 homes.

Another Canadian trouble spot is the Red River, which flows through Winnipeg. The river is normally sedate, and the valley is flat and distant from mountains. However, the Red River watershed is large and during some winters heavy snowpacks accumulate in the upper parts of the watershed, leading to devastating floods downstream. In May, 1997, warming temperatures and rain in the headwaters of the drainage basin melted a particularly heavy snowpack. The river rose slowly but relentlessly and eventually spilled across the broad valley floor, flooding it to depths of many metres. The vast temporary lake created by the flooding, dubbed the "Red Sea", was up to 50 kilometres across in parts of North Dakota and southern Manitoba. Many communities were under water or cut off for weeks. Winnipeg stayed dry, however, because the Red River Floodway, which was constructed at great expense after an earlier flood, routed the peak flows around the city (see photo on page 320).

Geology plays an important role in Red River flooding. Most of the valley is underlain by clayey sediment deposited in Glacial Lake Agassiz. This prevents rain and flood waters from easily seeping into the ground, and instead they run off the surface. Furthermore, the gradient of the river is slowly decreasing because the land downstream, to the north, is rising due to isostatic rebound. The decreasing gradient makes it progressively more difficult for the channel to contain flood waters.

A devastating flood hit the Saguenay–Lac Saint-Jean region of Quebec in July 1996. Problems started after two weeks of constant rain, which filled reservoirs, saturated soils, and brought rivers to flood levels. Then, on July 19, about 27 centimetres

The town of Morris, Manitoba, surrounded by floodwaters during the 1997 Red River flood. Although it resembles an island, the ground surface in Morris is lower than the level of the water; a ring dyke around the town prevents it from being inundated. GREG BROOKS.

Flooding in the Lillooet River valley at Pemberton, British Columbia, in October 2003. BRIAN MENOUNOS.

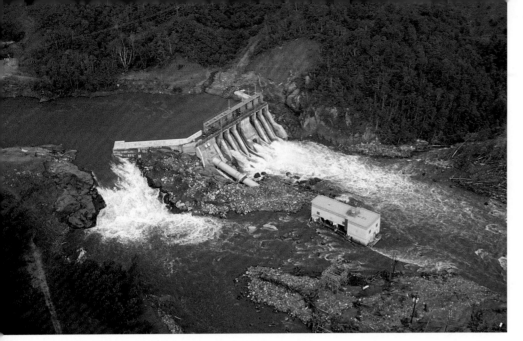

Damage caused by the devastating Saguenay flood in Quebec in July 1996. Waters overflowing the Jonquière Dam on Rivière aux Sables produced a breach about 20 metres wide within the concrete wing of the dam, thereby lowering the reservoir by several metres. Flood waters also severely damaged the powerhouse immediately downstream. GREG BROOKS

of rain fell in a few hours, an amount equivalent to the normal monthly total. Water more than 2 metres deep ran through parts of Chicoutimi and La Baie, completely levelling an entire neighbourhood. Some 16,000 people were evacuated, 10 people died, and more than 2,600 homes and cottages were destroyed. The flood was the worst in Quebec's history.

Torrential rain in June 2013 caused devastating flooding in southern Alberta. Four major rivers overflowed their banks, severely impacting several communities. Calgary, large areas of which are built on the floodplains of the Bow and Elbow rivers, suffered extensive damage. The town of High River was evacuated as the Highwood River inundated the town. And Medicine Hat on the South Saskatchewan River also experienced major flooding. Total damage has been estimated at over five billion dollars, making it by far the worst natural disaster in Alberta's history.

The threat of flooding can be reduced by building dykes, dredging and straightening channels, protecting or armouring riverbanks, and constructing dams. But dykes can cause more problems than they solve, harming river ecosystems and greatly confining the flow of the river during a flood, bringing it to flood stage faster. The benefits of such physical barriers also can be lost because they encourage development on the flood plains they are intended to protect, in effect increasing flood risk by increasing the population at risk. Straightening, deepening, widening, clearing, and lining river channels are common measures to facilitate the passage of flood flows and to protect riverbanks and dykes from erosion. Diversion structures such as the Red River Floodway in Winnipeg can channel peak flows away from communities. Dams built for hydroelectric-power generation regulate flows during periods of high runoff and thus reduce the risk of flooding.

Floodplain management is another strategy for reducing risk. The provincial and Canadian governments operate a joint flood-plain management program aimed at ensuring that all new development on flood plains will be unaffected by floods up to a certain size, for example the "two-hundred-year flood", which is the worst that can usually be expected over a two-hundred-year period. Flood-plain zoning may be a better way of dealing with flooding than the building of dykes and other structures. Restricting flood-prone land to farming and recreation is less expensive than herculean efforts to hold back flood waters.

Coastal areas may also be vulnerable to floods. During what are known as storm surges, strong sustained winds can raise the level of the sea and drive water inland. Flooding and damage are most severe when the storm coincides with a high tide. A famous example is the flooding caused by a tropical storm known as the Saxby Gale, which struck the Maritimes one night in October 1869. The Saxby Gale was named after naval officer and amateur astronomer Stephen Saxby, who had predicted high tides for that date and had warned that flooding would occur if a bad storm happened at the same time—and it did. A two-metre-high storm surge caused extensive destruction along the Maine and Fundy shorelines. Dykes were breached and seawater surged inland, inundating farms and communities and taking more than one hundred lives.

Living on a Restless Planet

Earthquakes, tsunamis, volcanic eruptions, landslides, and floods cannot be prevented, but we can lessen their toll. Although infrequent, natural disasters have high social and economic costs that are rapidly rising. Losses in the 1990s in Canada were seventeen times greater than those in the 1960s. Without action, an even greater impact can be expected in the future as Canada's population grows.

So what can we do to reduce the injury and damage caused by rare disastrous natural events? We need to take action on several fronts, including increased scientific research, improved engineering, better emergency preparedness, and greater individual and public awareness. For example, research in recent decades has greatly improved our understanding of earthquakes on the populated south coast of British Columbia. Better safety standards and construction techniques have resulted, and we are better prepared for an earthquake than ever before. All these measures should ensure that fewer people will be injured or killed during the next large quake.

BOX 11 • EXTRATERRESTRIAL VISITORS

Terms of Impact

Stargazing has always fascinated humans, especially when a "shooting star", fleetingly visible as a streak of light, hurtles through the atmosphere. But what are shooting stars? Space-debris terminology can be confusing. Such pieces of debris in the Solar System are generally termed meteoroids. They comprise material from two main sources: fragments left over after the formation of the Solar System, and small bits of a large planetary body such as the Moon or Mars that have been launched into space as the result of a collision. Meteoroids come in a range of sizes. The smallest are the size of a sand grain, and most are less than 10 metres across. But the largest, known as asteroids, measure about 1 kilometre or more across, and tend to be potato shaped rather than spherical. Ceres, by far the largest asteroid, has a diameter of between 909 and 975 kilometres. Most asteroids orbit the Sun in the main asteroid belt, about halfway between Mars and Jupiter. Meteors (shooting stars) are the visible incandescent traces of meteoroids that enter the Earth's atmosphere and largely burn up because of the friction generated as they pass through the atmosphere at high speed.

Comets are bodies with orbits in the outer limits of the Solar System. They possess a nucleus consisting of a mix of loose ice, dust, and small rocky particles, which is why they are sometimes described as dirty snowballs. Comets also contain carbon dioxide, methane, ammonia, carbon monoxide, and, surprisingly, some carbon compounds. Because of the carbon compounds, some scientists suggest that comets provided the biochemical seeds from which life evolved on Earth, but this notion doesn't explain how such molecules developed in the first place. Comets range in diameter from about 500 metres to 50 kilometres; they differ from meteoroids in being icy, at least in part, and in commonly developing a coma (atmosphere) and a tail as they pass through the inner Solar System. The coma and tail are formed of material ripped off the comet's nucleus by the solar wind (radiation emitted from the Sun) and can be seen through a telescope as visible streamers pointed away from the Sun. Most short-period comets (those with an orbit that takes less than 200 years) are found in the Kuiper Belt, which lies beyond Neptune. Long-period comets (those with an orbit of more than 200 years and in some cases up to several million years) originate even farther out, well beyond Pluto.

Usually, extraterrestrial debris drifts harmlessly in space, but occasionally chunks crash to Earth with consequences that can be

Stony meteorite from Peace River, Alberta. ROB FENSOME, SPECIMEN COURTESY OF NATURAL RESOURCES CANADA.

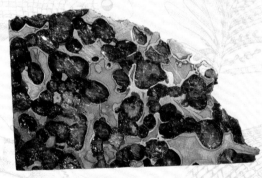

Cut and polished section through a stony-iron meteorite from Springwater, Saskatchewan. ROB FENSOME, SPECIMEN COURTESY OF NATURAL RESOURCES CANADA.

Iron meteorite from Annaheim, Saskatchewan. ROB FENSOME, SPECIMEN COURTESY OF NATURAL RESOURCES CANADA.

Reports of a "fireball" (meteor) near Lone Rock, Saskatchewan, on the evening of November 20, 2008, sent scientists from the University of Calgary on a search that yielded results. A meteorite, shown here with its co-discoverer, was found on a frozen pond in Buzzard Coulee, near Lone Rock. GRADY SEMMENS.

devastating. A meteorite is a portion of a meteoroid or comet that survives the journey through the Earth's atmosphere and is not destroyed when it collides with the ground. There are three basic types of meteorites: stony, iron, and stony-iron. About 93 percent are stony, composed predominantly of silicate minerals like those of most familiar terrestrial rocks, though with some iron and nickel. Roughly 80 percent of stony meteorites are of a variety known as chondrites. These bodies contain chondrules, rounded grains that formed as molten or partly molten droplets in space and which are rich in the minerals olivine and pyroxene. One of the best-known chondrites is the Tagish Lake Meteorite, which fell in northern British Columbia in January 2000. About 6 percent of all meteorites are iron meteorites, which contain iron and nickel alloys in large crystals. Only a minute fraction—about 1 percent—of all meteorites are stony-iron meteorites, which consist of approximately equal amounts of silicate minerals and iron-nickel.

Experts sometimes refer to any unknown large object that has collided with the Earth as a bolide. The effect of the object's impact can be recognized, but whether it originated as an asteroid, meteoroid, or comet is unknown because of the absence of any identifiable original material.

Leaving an Impression

When a comet or an asteroid strikes the Earth, it will blast a crater that is ten to twenty times its own diameter, although the actual size varies according to the speed and angle of impact. A meteorite 50 metres across may generate a crater 1 kilometre in diameter, and a 10-kilometre-wide monster could form a crater perhaps 200 kilometres across. Craters have been recognized on the Moon since before Galileo's time, but their presence and significance on the Earth was not realized until the early twentieth century. Only in the last 50 years have craters been the focus of detailed studies. This oversight was due to the dynamic processes of our planet, including erosion, sedimentation, and plate tectonics, which disguise and eventually obliterate craters. Of those remaining, few resemble the classic craters of the Moon. That is why experts prefer the term impact structure.

About 175 impact structures have been recognized on Earth, with about five new discoveries every year. The shape of these structures can be simple or complex. Simple impact structures have the form of a bowl-shaped depression with a raised rim and are less than 4 kilometres across. Such structures are formed when meteorites up to about 200 metres in diameter penetrate the Earth's surface. Complex impact structures are larger and have a structurally complex and faulted rim, a central peak, and a flat circular trough between the rim and central structure. Some of the largest and most complex impact structures even contain interior or peak rings of upraised structures. An impact structure forms within minutes of collision but it can take hundreds of years or more before stability is reached and the melt crystallizes, as witnessed by the huge Sudbury impact structure (Chapter 12).

Although most terrestrial impact structures have been eroded or completely obliterated, their former existence is commonly indicated by tell-tale evidence in the target rocks. A first clue might be a circular structure on a map or air photo, though such features can be misleading. For example, the shapes of both Hudson Bay and the Gulf of Maine suggest that they might have resulted from impacts, but current evidence suggests that they formed in other ways. Confirmation of an impact origin depends on more tangible

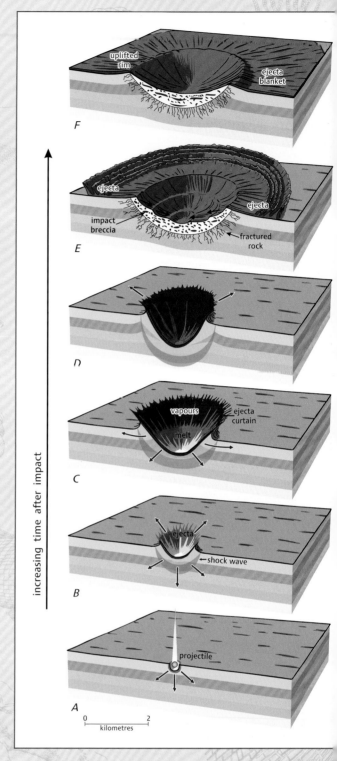

Simple terrestrial impact structures, which can be up to 4 kilometres across, are duced by the collision, or impact, of an extraterrestrial projectile with the Earth's su An impact event can be thought of in a series of stages: impact (A), excavation (and modification (E–F). The collision initiates a shock wave (A). Then excavatio starts to form the impact structure. Surface and projectile materials are shot int air (as ejecta or a curtain of ejecta), vaporized, pulverized, melted, and shocked; s of the ejecta and melted material line the growing cavity (C). At the end of excav (D), which is when the original bowl-shaped impact structure has reached its maxir size, the walls start to collapse inward (E). This collapse transports impact melt, mel material, and shocked material to form an impact breccia lens (the red and blue b represent melt and wall material clasts) over the floor of the impact structure. ejecta curtain distributes material around the crater (ejecta blanket), and some ma may fall back into the crater (F). As the shock wave moves through the rocks, it its intensity but forms a zone of fractured rocks below the crater floor (E, F). Mat blasted out of the impact structure during the impact result in an uplifted rim (F result of the impact is thus a bowl-shaped depression with a depth about twice th the apparent depth and about a third of the diameter. Events shown in A–F occu matter of seconds. ADAPTED FROM GRIEVE (2006), USED WITH PERMISSION OF THE AUTHOR: THE GEOLOGICAL ASSOCIATION OF CANADA.

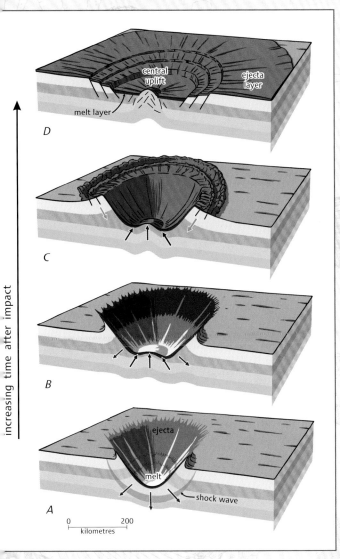

Aerial view of a gorge exposing tilted and faulted Cambrian dolostone in the centre of the Prince Albert impact structure, northwestern Victoria Island, Northwest Territories. The strata were uplifted and tilted immediately after the impact. All the overlying broken and melted rock and the crater itself have been eroded away. BRIAN PRATT.

Complex terrestrial impact structures, which range from 3 to hundreds of kilometres in diameter, take from seconds to minutes to develop, longer than their simple counterparts, although the stages of development are similar: impact (A), excavation (B), and modification (C–D). After the impact, the shock wave propagates into the rocks below (red arrows) and excavation of the impact structure results in surface and projectile material being blown into the air (as ejecta or a curtain of ejecta), vaporized, pulverized, melted, and shocked (A). After excavation ends, an initially deep central portion of the depression starts to rise (blue arrows) while the cavity continues to broaden (B). The central uplift continues to grow, accompanied by downward and inward collapse of the cavity walls (C—green arrows). The impact structure and surrounding area, perhaps tens of kilometres across, are unstable and must adjust via large fractures (C, D). In its final form (D), which may take many years to attain, the impact structure has a flat floor, structurally faulted rim terraces, a moat-like trough, and a central uplift that may stand up as a central peak. Ejecta are distributed on top of and around the structure as a layer. The depth-to-diameter ratio is variable but less than for simple impact structures; it decreases with increasing diameter. Flooring the structure are fractured and shocked rocks, breccia deposits, and melt rocks, which may also radiate outward from the structure as dykes. Depending on its thickness, the impact melt layer may take hundreds to millions of years to fully crystallize. ADAPTED FROM GRIEVE (2006), USED WITH PERMISSION OF THE AUTHORS AND THE GEOLOGICAL ASSOCIATION OF CANADA.

evidence, ideally associated fragments of the impacting body. However, these are not preserved in structures larger than 1 to 2 kilometres across. Fortunately, the target, or host, rocks of large impacts generally show signs of shock metamorphism—changes in minerals and rocks caused by a high-pressure shock wave generated by the impact. Pressures and temperatures reached during such an impact are much higher than those attained during terrestrial metamorphism. The only diagnostic shock-metamorphic features visible to the naked eye are shatter cones, found either

Arrays of shatter cones, produced by the meteorite that formed the Prince Albert impact crater, northwestern Victoria Island, Northwest Territories. BRIAN PRATT.

as individuals or groups. These are striated, horsetail-like, conical fractures ranging from millimetres to metres in length. They are best developed in fine-grained, structurally homogenous rocks such as limestones. Examples can be seen in the Charlevoix region of Quebec.

Microscopically, the best signs of shock metamorphism are defects in mineral crystals, including fracture surfaces and lines of tiny bubbles, or bubble trains. Although these features can develop in many minerals, they are best known in quartz and feldspar, where they are easily recognizable and provide convincing proof of an impact structure.

The debris ejected during an impact is known as ejecta. Ejecta are mostly microscopic, but can range up to gravel-sized particles; such larger pieces composed of green, grey, brown, or black natural glass are called tektites. Perhaps the most famous of all asteroid impacts is the one implicated in the demise of the dinosaurs (Chapter 9). Associated with that impact is a widely distributed (perhaps global) layer of ejecta enriched with the element iridium. The layer contains about 100 times the concentration of iridium normally found in the lithosphere—a concentration level found in

The light-coloured quartz crystal shows features such as fractures and bubble trains produced by the Charlevoix, Quebec, impact event. The field of view is about 2 millimetres across. ANN THERRIAULT.

Pingualuit Crater in northern Quebec is an impact structure almost 3.5 kilometres in diameter, now occupied by a lake over 250 metres deep. Estimated to be about 1.4 million years old, this is the youngest-known impact structure in Canada, and is excavated into 2,700-to-2,800-million-year-old gneisses of the Superior Craton. REPRODUCED WITH THE PERMISSION OF NATURAL RESOURCES CANADA 2013, COURTESY OF THE GEOLOGICAL SURVEY OF CANADA.

asteroids. Other iridium anomalies have been identified, notably in deep-sea cores, and in association with layers containing microscopic ejecta.

Impacts in Canada and Beyond

Because about 70 percent of the Earth's surface is covered by water, most of the comets and asteroids that strike our planet hit the oceans. We know that there are no in-place ocean-floor rocks older than about 180 million years, so evidence for submarine impact structures will be limited. But examples have been identified, including the Montagnais impact structure, 45 kilometres across, on the continental shelf off Nova Scotia, the first offshore impact structure to be recognized.

In Canada, some thirty impact structures have been identified, and most are older than 50 million years. Canadian impact structures range from 2.5 to 250 kilometres in diameter, and are most common in Quebec. The first to be discovered was the Pingualuit (or New Quebec) structure on the Ungava Peninsula in 1950. This structure has a diameter of 3.44 kilometres and is only about 1.4 million years old. Another spectacular example is the Manicouagan structure, some 100 kilometres wide, about 180 kilometres northwest of Sept-Îles, Quebec. When first discovered, the Manicouagan structure contained two arc-shaped lakes. These lakes have now merged into a single ring lake, the Manicouagan Reservoir, the result of a hydroelectric dam and consequent flooding of the natural ring-shaped trough. The reservoir is about 10 kilometres wide and encircles an uplifted inner plateau with a central peak that rises to about 500 metres. Present estimates of the age of the impact place it at about 214 million years.

The Sudbury impact structure is by far the largest in Canada. This huge structure records the collision of an asteroid at least 10 kilometres wide around 1,850 million years ago. Originally, it had an impact diameter of about 250 kilometres, but subsequent tectonic activity has distorted its dimensions and shape. We explored the mineral riches related to the Sudbury structure in Chapter 12. Also in Ontario is the Brent impact structure, which is up to 3 kilometres wide and

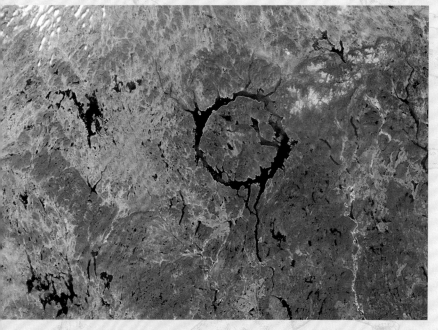

Manicouagan impact structure, northern Quebec. COURTESY OF NASA.

The lighter-coloured area in this aerial view of Haughton impact structure, Devon Island, Nunavut, is composed of impact breccia. MARTIN LIPMAN.

up to 60 metres deep and is located on the northern boundary of Algonquin Provincial Park. This is a classic example of a simple impact structure, formed between 450 and 400 million years ago.

Examples in western Canada include the Carswell impact structure, about 120 kilometres south of Uranium City in northern Saskatchewan. This structure is defined by an almost circular outcrop, 39 kilometres in diameter, of dolostone cliffs. It appears to be of early Cretaceous age. The mining of uranium from the Mesoproterozoic Athabasca Basin (Chapter 13) was made easier by the Carswell impact event, which brought ores closer to the surface through faulting and uplift. The Saint Martin impact structure, between Lake Manitoba and Lake Winnipeg in central Manitoba, was discovered by exploration drilling in the 1960s. This 40-kilometre-wide structure is buried beneath about 30 metres of glacial sediment, but subsurface studies have revealed an uplifted core of Precambrian and early Paleozoic rocks. Local groundwater quality has been adversely affected by the impact melt rocks, which have fluoride levels too high for safe consumption. The Saint Martin structure is estimated to be about 220 million years old. In Arctic Canada, the Haughton impact structure on Devon Island, which contains important fossils (Chapter 10), is one of the most northerly in the world.

A map of impact structures in Canada reveals that most lie on terrain that has been largely unaffected by major tectonic events for hundreds of millions of years. For example, no impact structures have been discovered in British Columbia, likely because any traces of impact have been obliterated in a region so tectonically active in recent geological time. The sparseness of structures in Arctic Canada, however, probably reflects both the need for more surveys and the destructive power of the ice in the past 2.6 million years.

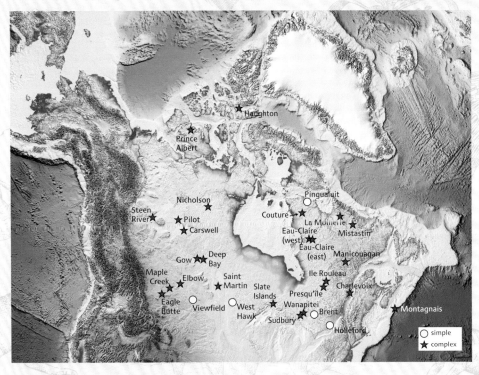

Map showing the distribution of impact structures in Canada. BASE MAP FROM AMANTE AND EAKINS (2009).

18 • ENVIRONMENTAL CHALLENGES

CHAPTER SUPPORTED BY A DONATION FROM THE CANADIAN GEOSCIENCE EDUCATION NETWORK IN MEMORY OF WARD NEALE

The port of Hamilton on Lake Ontario, in Canada's industrial heartland. Industry and related activities are an engine behind Canada's economy and wealth, although they can result in environmental challenges. RON GARNETT / AIRSCAPES.CA.

The Walkerton Tragedy

"We have a terrible tragedy here", said then Ontario Premier Mike Harris as he addressed reporters and residents in the quiet rural community of Walkerton on May 26, 2000. Starting on May 15, residents of the town of about 5,000 people began to experience bloody diarrhea and gastrointestinal infections. By May 21, the number of people hospitalized spurred the region's Medical Officer of Health to issue a boil-water advisory. Walkerton's water supply had become contaminated with a highly dangerous strain of *Escherichia coli* bacteria, more widely known as *E. coli*.

Ultimately 7 people died and more than 2,300 people, almost half of Walkerton's population, became sick from bacterial infection. Contamination originated from groundwater from a shallow well in fractured rock beneath about 5 metres of permeable sand and gravel. Contaminated water from a farm manure pile seeped directly down from the surface into the fractured rock. A subsequent report by the Walkerton Commission included many recommendations for improving the quality of water and public health in Ontario. The recommendations included protection of source water, training and certification of water-supply operators, a quality management system for water suppliers, and better government enforcement. These recommendations have since been incorporated into legislation in Ontario and have influenced policies in other provinces.

The Walkerton tragedy is but one example of the increasing effects that our activities are having on the planet. Humans had minimal impact on the environment until the rise of cities in the Middle East more than 6,000 years ago. In the past 300 years, by burning trees, coal, oil, and gas, we have raised carbon dioxide concentrations in the atmosphere to levels not seen for millions of years (Box 9). We are creating new landscapes by building cities and roads, damming rivers, cutting down forests, and mining resources. The amount of waste rock produced by mining worldwide in the year 2000 was about 18 billion cubic metres, roughly equivalent to the volume of sediment that rivers move to the oceans each year. Urbanization, anthropogenic climate warming, species loss, habitat fragmentation, and ecosystem destruction are now potent human-made forces of global change. The changes are so striking that some have suggested we rename the Holocene the Anthropocene—the epoch of humans—in recognition that surface geological processes are now predominantly influenced by the activities of a single species.

Canada's size and its variety of landscapes and ecosystems multiplies its environmental challenges. Although many environmental concerns are widespread across the country, each region also has specific concerns.

Paving Paradise

During the past 70 years, Canada's population has changed from being predominantly rural to primarily urban, with 80 percent of the country's population now living in urban landscapes that cover less than 0.2 percent of Canada's landmass. Today, more than a third of Canadians live in four emerging super-cities: Montréal, Toronto, Calgary, and Vancouver. Urbanization is a global trend, not just a Canadian one. By 2030, more than half the world's population, about 5 billion people, will live in cities. The exodus from rural areas to urban centres has improved quality of life over the past 150 years and the concentration of populations has led to important economies of scale. However, the same centralizing forces now threaten the global environment.

Cities consume large amounts of resources, especially water, building supplies, food, and energy. Each step in the supply

Satellite image of Canada's "Golden Horseshoe" around the western end of Lake Ontario (north is to the top right), from the Niagara River and the American border (bottom left) around to Oshawa (bottom right). About one third of Canada's people live in this region, which includes Canada's largest city, Toronto. Such extensive urban development leads to great environmental stress. THE NATIONAL AERONAUTICS AND SPACE ADMINISTRATION (NASA).

chain is a source of contaminants to air and water. In 2006, 240 million tonnes of sand and gravel were mined for urban construction in Canada. Some 5 million tonnes of road salt are used across the nation each year to make city roads safe for driving. Cities must dispose of enormous volumes of wastewater and other materials that commonly contaminate water bodies, watersheds, and ecosystems far beyond their borders. People living in suburban communities place additional demands on transport and service infrastructure such as roads and sewers. And increasingly, valuable farmland is being rezoned to accommodate residential growth and associated development. For example, Richmond, part of Greater Vancouver, has been built on the Fraser Delta, occupying some of the richest farmland in Canada.

Emissions from automobiles and power generation from fossil fuels (coal and petroleum) create plumes of contaminants and smog that move downwind, creating health problems and producing acid rain and mercury contamination. Much of central Canada suffers from smog originating in the midwestern United States, augmented by contributions from Ontario and Quebec. Cities around the Great Lakes consume oil, gas, and other materials from half a world away and export contaminants downstream via the St. Lawrence River and in the air across provincial and national boundaries. Although environmental assessments are now routinely required for all urban projects such as new subdivisions and highways, we still tend to ignore the cumulative impact of our activities on natural environments.

Urban development "hardens" watersheds by replacing natural surfaces with impervious roads, parking lots, and buildings. This hardening can greatly increase surface runoff and reduce the amount of water the ground absorbs, thereby limiting the contribution of groundwater to the flow of streams in summer. Consequently, stream flows become more erratic, with abrupt increases in flow and greater runoff alternating with reduced flow in times of fair weather. In summer when water drains off paving material, water temperatures may increase to levels that are lethal for aquatic organisms.

Urban development is also the biggest threat to Canada's water quality. The first flush of waters draining from road surfaces after extended dry spells in summer or winter thaws are loaded with contaminants, as are brackish water runoffs from salted road surfaces in winter and spring.

On the Waterfront

Most Canadian communities once depended on navigable waterways and so are situated on the shores of rivers, large lakes, or the sea. A wide variety of contaminants end up in these water bodies. Many of our cities surround confined harbours, where water quality and fisheries have been badly affected by contaminated storm water and untreated sewage. The Greater Victoria area on Vancouver Island discharges 129 million litres of raw sewage each day into Juan du Fuca Strait: Victoria is planning to have sewage treatment plants operational in the near future. St. John's completed a wastewater treatment plant in 2009, which now treats all sewage entering its harbour. Until 2010, sewage was released into the enclosed waters of Halifax Harbour in Nova Scotia. After considerable expense and some initial setbacks, three sewage treatment plants are now removing all solids and partially treating the liquids before the waste is released into Halifax Harbour. But sewage treatment is not the end of the story, for treatment facilities produce large amounts of solid organic waste that then have to be incinerated, disposed of in landfills, or stabilized (usually by adding lime) and used as agricultural fertilizer.

Enhanced stormwater runoff from hardened watersheds in Toronto, Ontario, led to the extensive use of pipes and culverts to reduce erosion. However, these structures create insurmountable obstacles for fish. PROVIDED BY NICK EYLES.

Frenchman's Bay in east Toronto, Ontario, is a lagoon that has experienced extensive loss of wetland and fish habitat due to stormwater runoff. The lagoon receives several thousand tonnes of road salt every winter from Highway 401 (foreground) and the surrounding urban catchment. PROVIDED BY NICK EYLES.

The Great Lakes are receiving basins for contaminated waters from surrounding urban, industrial, and agricultural lands. Lake Ontario is the most vulnerable because it is at the downstream end of the Great Lakes watershed, home to more than 40 million people on both sides of the border. Levels of organic contaminants in fish and other organisms in Lake Ontario have decreased in recent years, indicating some success in dealing with the pollution. But concentrations of new chemicals are increasing: recent studies have focussed on previously unrecognized environmental risks associated with high levels of pharmaceuticals, personal care products, and pesticides. Nuclear and coal-fired power plants discharge warm water, affecting temperature-sensitive aquatic ecosystems. Outfalls from sewage treatment plants elevate metals in nearshore areas. Every year sewage treatment facilities along the Toronto waterfront dump 350 tonnes of phosphorus and over 212 tonnes of lead into Lake Ontario—the source of drinking water for millions of people.

Contaminants from the Great Lakes are flushed out through the narrow St. Lawrence River, only to be deposited where the river widens and slackens in the vicinity of Montréal. There, metals and organic chemicals extensively contaminate the muddy bottom sediments of Lac Saint-Louis and Lac des Deux Montagnes, both wide spots in the river rather than true lakes. Here, as elsewhere, metals are taken up by organisms, concentrated, and transported higher up the food chain, posing a threat to the entire ecosystem, including us.

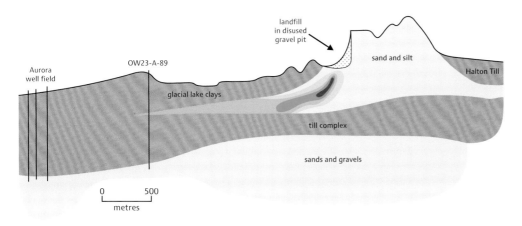

Cross-section through a landfill at Aurora, Ontario, within the Oak Ridges Moraine. Historically, landfills such as this one involved the dumping of waste into an open gravel pit, where it was eventually covered and left. Rainwater and snowmelt pick up contaminants as they flow through the waste, creating a plume of contaminated groundwater that continues to seep slowly, sometimes toward wells. Interceptor wells can be drilled to pump up water from the plume, slowing down the advance.

Fieldwork at a monitoring well that tracks contaminants in groundwater. PROVIDED BY NICK EYLES.

A Lot of Garbage

The huge middens surrounding ancient cities provide intriguing evidence that people and garbage have been companions for a long time. Our modern consumer society, however, creates far more waste than past civilizations. Each Canadian city dweller produces on average some 400 kilograms of garbage each year. Although largely non-toxic, this waste is discarded and becomes part of the environment—a new kind of sediment. We also have to deal with toxic industrial and consumer waste, including computers, cellphones, and mercury-containing fluorescent light bulbs. But we are doing better than in the past, when most garbage dumps were unregulated, and records of disposal sites and types of waste were lost. Such unregulated dumps, often in abandoned gravel pits, still pose potential risks to groundwater and streams.

Today we have better guidelines for the design and construction of landfills, as well as more extensive recycling programs in urban areas. Leachate—a fancy word for waters containing leached contaminants—was once thought to be diluted as it moved through sediments and mixed with groundwater. It was believed that sites underlain by thick, impermeable glacial sediments such as till were ideal locations for waste disposal. We now know that groundwater moves through some tills more quickly than previously thought, resulting in plumes of contaminated groundwater that extend out considerable distances from old landfills. New landfills must have a liner of plastic sheeting and compacted clay, as well as pipes to collect leachate and the biogenic methane produced through decay of organic materials. However, designing and building containment systems that will last until the toxic components of waste degrade to a safe state, potentially several hundred to thousands of years, remains a challenge.

Cleaning Up

In Chapter 12 we explored mining as an integral part of Canada's historical development and it is still vital to the nation's economy. But mining comes with a legacy of related environmental problems. According to one estimate, 27,000 abandoned mines and mine waste dumps exist in Canada, covering a total area of 150 square kilometres. Mine waste includes both waste rock and tailings. Waste rock consists of unprocessed rock removed during mining, usually to gain access to the ore body. The term tailings refers to materials that remain after processing and separation of the valuable fraction from the ore. The largest single area of mine waste is at Sudbury in Ontario; other large mine areas are in southern and central British Columbia and in the Abitibi region of Quebec. At the Tilt Cove Mine on Notre Dame Bay in western Newfoundland, a source of copper ore, almost 8 million tonnes of tailings were released directly into the ocean in the 1950s and 1960s. Direct discharge of tailings to the ocean is no longer permitted in Canada. In recent years, tailings associated

Recycling toilets at a landfill at Waterloo, Ontario. ALAN MORGAN.

Reclaimed land at the Cardinal River Mine, north of Cadomin, Alberta. Waste rock removed from the area prior to coal mining has been spread in the foreground as a rough surface and then re-seeded. GODFREY NOWLAN.

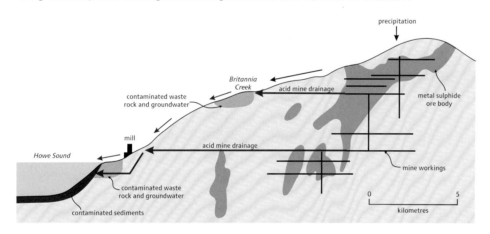

The Britannia Mine, near Squamish in British Columbia, was active from 1902 to 1974, and at one time was the largest copper mine in the British Empire. Some 44 million tonnes of mine tailings were dumped directly into the sea, and the disused mine was a major source of acidic mine drainage containing dissolved metals that flowed into Howe Sound. This source of contamination had a significant impact on the marine biota until a major cleanup was conducted prior to the Vancouver Winter Olympics in 2010.

of low hills and depressions, but the tailings dump is now up to 30 metres high and dominates the landscape. Acidic groundwater, rich in iron and other metals, flows outward from the dump. Attempts to deal with the problem include covering the tailings with sediment, wood chips, and sludge to limit their exposure to oxygen, adding lime to neutralize acidity and limit the movement of dissolved metals, and building dams to impound tailings and sludge. Neutralization of acidic mine waters with lime creates its own problem: a voluminous sludge rich in metal hydroxides that precipitate out of the water. This material then requires long-term treatment and storage. And if dams of holding ponds fail, the discharge of mining waste into the surrounding rivers, lakes, or seas would be a serious concern.

Disposal of radioactive waste, primarily uranium, along with its derived materials, is especially difficult and is not restricted to mining activity alone. Radioactive waste materials are classified as low level, intermediate level, and high level. Low-level radioactive waste includes clothing, rags, contaminated soil, decommissioned nuclear facilities, and isotopes used in research, medicine, and industry. Intermediate-level waste includes resins and chemical sludges. High-level waste is primarily spent fuel rods from nuclear reactors. Every year Canadian reactors produce about 3,000 tonnes of high-level radioactive waste. Although this amount is not comparatively huge, it takes about 100,000 years for radioactivity to decay to the level of the original uranium ore. Elliot Lake in northern Ontario, once the "uranium capital of the world", is now left with 160 million tonnes of tailings that will remain radioactive for thousands of years.

with oil-sands mining in Alberta have become a major environmental issue.

Many metal-mine tailings dumps in Canada are rich in sulphide minerals that react with water and oxygen to produce sulphuric acid. Waters that move through such tailings can carry high metal loads that are toxic to aquatic life, a problem known as acid-rock drainage. The problem has been dealt with successfully in some cases either by covering tailings with water, mixing them with lime, encouraging vegetation, or intercepting and treating acidic runoff.

Tailings have been dumped at Copper Cliff near Sudbury since 1936, and every day some 35,000 additional tonnes are produced. The original natural terrain around Sudbury was one

Spent nuclear fuel rods are initially stored in deep-water pools to allow them to cool and to prevent radiation leakage. After a few years, the waste is placed in concrete canisters. How to dispose of uranium waste permanently and safely is a question still being resolved. One suggestion being considered is to store spent fuel 500 to 1,000 metres below the surface in a stable-rock repository on the Canadian Shield. As an extra precaution, the waste would be packed in ceramic containers filled with bentonite clay and encased in concrete.

Oxidized gold-mine tailings at the historic Second Relief Mine near Salmo, British Columbia. MIKE PARSONS.

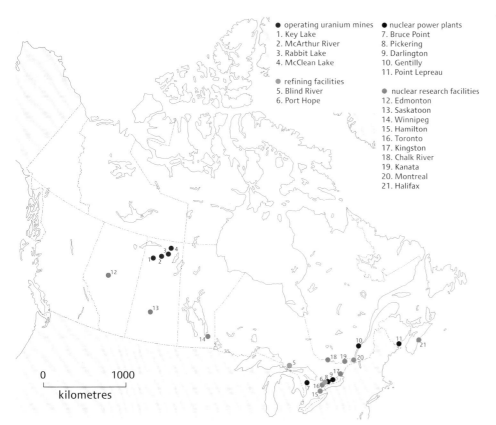

- operating uranium mines
1. Key Lake
2. McArthur River
3. Rabbit Lake
4. McClean Lake

- refining facilities
5. Blind River
6. Port Hope

- nuclear power plants
7. Bruce Point
8. Pickering
9. Darlington
10. Gentilly
11. Point Lepreau

- nuclear research facilities
12. Edmonton
13. Saskatoon
14. Winnipeg
15. Hamilton
16. Toronto
17. Kingston
18. Chalk River
19. Kanata
20. Montreal
21. Halifax

Nuclear-related localities in Canada. ADAPTED FROM VARIOUS SOURCES.

Underground storage of the much larger volumes of low- and intermediate-level wastes has also been considered. Low-level waste disposal sites, such as an underground repository proposed near the town of Deep River in Ontario, have been identified but shelved because of public concerns. Consequently, much low-level waste is now stored in temporary locations throughout Ontario, especially around Port Hope. The safe disposal of nuclear waste needs to be resolved if Canada is to fully utilize its vast uranium resources, either internally or through exports.

Indeed, although cleanup is costly, former mining areas of all types must be reclaimed. An example of successful cleanup comes from the former copper-mining community of Murdochville in the Gaspé Peninsula of Quebec, which is now home to Canada's largest wind-power project.

The Right Atmosphere

Governments and businesses spend more money on preventing air pollution than on any other form of environmental protection. Airborne pollutants are widely dispersed and know no national boundaries. Canada and the United States had considerable success dealing with the acid-rain problems of the 1980s, but trans-Pacific transport of industrial pollutants from eastern Asia and agricultural pollutants from Mexico pose new problems. These pollutants are now found in increasing quantities in Arctic lakes and northern and alpine ecosystems.

Acid rain is produced by the natural conversion of atmospheric nitrous oxides and sulphur dioxide into nitric acid, ammonium nitrate, and sulphuric acid. The main sources of atmospheric nitrous oxides and sulphur dioxide are the burning of gasoline and coal. Ontario, Quebec, and the Atlantic Provinces have been particularly hard hit by acid rain, because they are downwind of the heavily industrialized midwestern United States and because the rocks and soils throughout much of these provinces have little or no carbonate to neutralize acidity.

Airborne mercury is another worrisome pollutant. Coal-burning power plants in the United States are the source of 48.5 tonnes of airborne mercury every year. Cement kilns, smelters, and steel works, on which our urban development is pinned, are also significant mercury emitters. Another problem is airborne arsenic. The Giant Mine near Yellowknife in the Northwest Territories produced gold between 1948 and 1999. The pyrite-rich ore was roasted to remove the

Abandoned buildings, Stirling Mine, Nova Scotia. MIKE PARSONS.

gold, in the process producing large volumes of arsenic trioxide dust. More than 7 tonnes of arsenic were produced each year and stored underground in mined-out caverns, which are within naturally frozen ground. The frozen surrounds continue to prevent migration of contaminants, and the freezing is aided by the use of drill holes filled with carbon dioxide.

In Box 9, we discussed carbon dioxide's role as one of the greenhouse gases contributing to global warming. Canadians release only 2 percent of the world's human-sourced carbon dioxide, but their contribution on a per-capita basis is the second highest among industrialized countries, exceeded only by Australia. Canada's emissions stem from its cold winters, large size, and resource-based industries.

How can we dispose of this carbon dioxide or perhaps put it to good use? The most promising strategy put forward so far is to retrieve carbon dioxide at sources such as power plants and store it underground. In this process, the gas is captured and then purified and compressed, before being stored below ground in rocks where it is isolated from the atmosphere. Ideal storage areas are deep sedimentary basins, oil and gas reservoirs, coal seams, salt domes, and highly saline waters that are millions of years old. In the United States and Canada, carbon dioxide has been injected into oil reservoirs for more than 25 years to increase oil recovery. But the additional environmental benefits of sequestering (separating and secluding) the gas underground are only now being recognized. One Canadian oil company is presently using carbon dioxide generated by a coal-gasification plant in North Dakota to enhance oil recovery at its Weyburn Oil Field in southern Saskatchewan. Over a million tonnes of the gas are being sequestered each year in the underground reservoir. In Ontario, studies are being carried out to see if Cambrian sandstones beneath Lake Erie and Lake Huron might be feasible storage sites for more than 700 million tonnes of carbon dioxide, which would offset the emissions from the coal-fired Nanticoke Generating Station for nearly 30 years.

Alberta is furthest ahead of all Canadian provinces in its plans for carbon sequestration. In early 2008, the Alberta government announced that 70 percent of the province's planned 200 million tonne cut in emissions over the next 42 years would be accomplished by carbon sequestration. But it is unclear what the impact of increased oil sands production will have on this target. The plan clearly illustrates the long lead times required to significantly cut emissions.

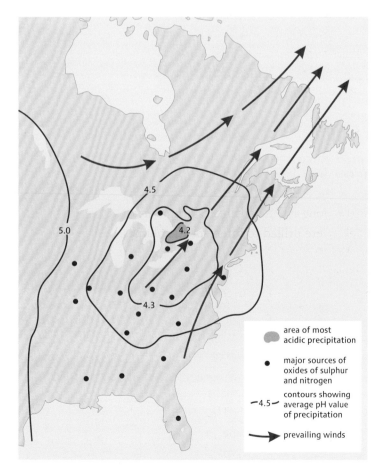

The distribution of acid rain in eastern North America and the impact of prevailing winds.

Recognition and Remediation

In 2005, the Canadian government identified 500 contaminated sites across Canada for detailed environmental assessment, and targeted 97 priority sites for accelerated cleanup. Many of the priority sites are harbours, former metal mines, or plants where coke and gas were made from coal. Some are ex-military installations such as abandoned Distant Early Warning (DEW) Line radar stations in the Arctic. The 400-square-kilometre Suffield Experimental Proving Ground in Alberta, which contains residues of chemical and biological weapons, is also on the priority list. Many contaminated industrial sites have a long history of use, so the potential presence of uncharted buried storage tanks and pipes may make drilling and trenching at these sites hazardous. To locate these hazards, a variety of relatively inexpensive geophysical techniques are used to map the shallow subsurface (Box 3).

The next phase in an environmental assessment is an investigation of the chemistry of the wastes. At many sites, investigators may have to deal with a complex group of organic contaminants called DNAPLs (pronounced D-napples), an acronym for dense non-aqueous phase liquids. DNAPLs are hazardous synthetic compounds that are heavier than and largely insoluble in water. These compounds are used as solvents in many industrial and commercial processes, including dry cleaning and metal degreasing, as well as in paints and adhesives. When spilled or leaked, DNAPLs migrate into groundwater and collect on impervious layers. Remediation is costly and technically challenging. One of the largest cleanups involving DNAPLs in Canada is at a former outboard-engine factory in Peterborough, Ontario, which used degreasing agents and solvents between 1913 and 1990. Contamination in sediments below the site affected two plumes of groundwater, each up to 600 metres long and underlying residential areas. Wells were drilled to intercept the plumes and contaminated water was pumped up and treated. A deep trench was also excavated around the main source of the DNAPLs and filled with bentonite clay to create an impermeable seal.

Synthetic compounds that float on water and thus pool at the top of the water table are termed LNAPLs (L-napples), light non-aqueous phase liquids. These compounds include gasoline, benzene, and other petroleum-based products typical of spills from gas stations. Because they are lighter than water and are largely insoluble, LNAPLs commonly can be pumped to the surface and removed.

Remediation methods may involve dealing with subsurface chemical contaminants in place, or removing the contaminated sediment and treating it elsewhere. Techniques include incineration, adding reagents to solidify wastes, washing with water or solvents, planting trees and letting root microbes biodegrade contaminants, using vacuum systems to draw out gas, pumping and treating liquids, forcing air through the sediments, and isolating the site with barrier walls and a cover.

Huge gaps exist in our knowledge of the impact of the rapid environmental changes that are happening in Canada. Filling these gaps is important for creating credible report cards on the vital signs of the health of our environment. There are critical questions to answer still. Are our activities sustainable? What are the long-term environmental impacts of climate change or of current and planned land use?

More effort is needed to monitor geological processes and understand local geological conditions in order to better guide future resource use and planning. There are encouraging signs: the high levels of contaminants found in many organisms in the 1970s and 1980s have since been greatly reduced and urban air quality has improved. But pressures from a growing population and new emerging issues and contaminants loom large. Of particular concern is the impact of urban areas on water quality in lakes and rivers. New technologies will let us measure and understand environmental change much better than in the past.

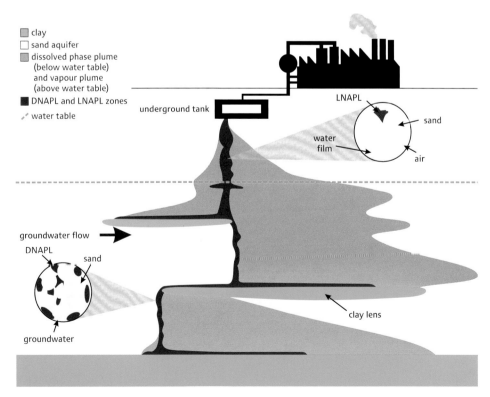

Contrasting subsurface migration of light non-aqueous phase liquids (LNAPLs) and dense non-aqueous phase liquids (DNAPLs). Being dense, the latter move down below the water table and are much more difficult to clean up.

19 · TOXINS IN THE ROCKS

CHAPTER SUPPORTED BY A DONATION FROM THE TAMARATT TEACHING CHAIR IN GEOSCIENCE, UNIVERSITY OF CALGARY

Selenium is one of the elements that we humans require in our diet, but problems can arise if we get too much of it. Selenium from Cretaceous shales in the Canadian Prairies enters the food chain by way of grasses eaten by cattle. Animals need trace amounts of selenium, but not too much. MARK DUFFY: WWW.WORLDOFSTOCK.COM (SEARCH NAN3057).

Too Much of a Good Thing?

There was no explanation for the pain being suffered by a farmer from the Maria area on the Gaspé Peninsula of Quebec. For over a year, every joint in his body was stiff and he had trouble breathing. The local doctor couldn't find a cause, and finally sent the farmer to a hospital in Québec City. There, analyses of his bones and urine revealed that he had skeletal fluorosis, a disease caused by high concentrations of fluoride. The hospital rheumatologist recommended analysis of the drinking water from the farmer's well. The results showed that the water contained 25 milligrams of fluoride per litre, over 12 times the safe daily maximum. Fluoride improves the strength of dental enamel and is beneficial for oral hygiene, but only at concentrations between 0.5 and 2.0 milligrams per litre. At higher concentrations, as in the farmer's well, fluoride is a toxin.

After hearing this story over twenty years ago, Dan Boyle, a geologist with the Geological Survey of Canada, visited the Maria area and found that 15 to 20 percent of the rural population was drinking water with naturally elevated fluoride, which originated from the town's deep Carboniferous sedimentary bedrock. In contrast, local shallow wells had negligible fluoride, significantly below the recommended range for prevention of tooth decay, and posed no health risk to residents. As Boyle pointed out, the old adage of many well drillers that "deep is good" did not apply in this area.

The Maria fluoride study is an example of the relationship between geology and human health. Concerns can cut two ways: sometimes we don't have enough of a beneficial mineral, but occasionally we have too much. This is especially true of elements and compounds that occur in minute (trace) amounts. In this chapter, we explore some natural geochemical hazards, including metals such as mercury and cadmium; metalloids (elements with properties characteristic of both metals and non-metals) such as arsenic; metal-organic compounds such as methylmercury; and organic compounds such as polycyclic aromatic hydrocarbons. We discuss selenium and fluoride, which in the right dosages are essential to life but whose concentrations vary widely, depending on the geological setting. Finally, we highlight some airborne hazards with geological connections, such as radon, asbestos, and particulate matter.

Cadmium, Caribou, and the Ring Cycle

Caribou have been an important part of Aboriginal peoples' diets in Yukon and the Northwest Territories for thousands of years. Even today, caribou and moose meat are the main sources of protein, iron, zinc, copper, and magnesium in Dene and Métis communities along the Mackenzie River. Generally this meat is not a health risk. But occasionally the organs of some caribou become contaminated with high levels of cadmium, a soft, bluish-white metal that occurs widely but in small amounts in rocks, windblown dust, forest fire residues, and volcanic ash. Cadmium is also spread by human activities such as base-metal refining and fuel burning. Acidic conditions, large soil particles, and high soil moisture can increase the amounts of cadmium taken up by vegetation.

Cadmium in caribou comes from the lichens they eat. Lichens contribute to the weathering of rock and soil, and absorb cadmium from air, soil, and water. Concentrations of cadmium in most caribou meat are very low, but it accumulates in the kidneys and, to a lesser extent, in the liver and bones. Contaminant levels tend to increase slowly as animals age and continue to ingest the substance. This process is called bioconcentration. In humans, kidney damage is the most common problem associated with long-term exposure to high levels of

Lichens, which absorb cadmium from rocks, soil, and windblown dust, are consumed by grazing caribou, an important food source for northern communities. WWW.INDEXSTOCK.COM

cadmium. However, you can be exposed to more cadmium from smoking cigarettes than from eating one caribou kidney a week.

Persistent organic pollutants (POPs) are a large group of potentially hazardous organic compounds. Some of them are known or suspected to cause cancer, and others disrupt the endocrine system by mimicking natural hormones. POPs include a group of organic molecules called polycyclic aromatic hydrocarbons or PAHs. Polycyclic means that the atoms in the molecules form multiple cycles, or rings; and as you may have guessed from the term aromatic, these compounds are smelly. As we discovered in Chapter 13, hydrocarbons are molecules that always contain hydrogen and carbon. An example is anthracene, a solid PAH that contains three benzene rings and is derived from coal tar.

PAHs in the environment are derived mainly from incomplete burning of organic material. Although they commonly come from non-geological sources such as forest fires, industrial effluents, smoking tobacco, and charbroiling meat, they also enter the environment from volcanic eruptions, as diesel-engine exhaust, and during combustion of coal, oil, and gas. They are natural components of peat, lignite, coal, crude oil, and bitumin, and can be leached from soft coal seams, ultimately ending up in lakes. Once in water, PAHs tend to concentrate in the organic

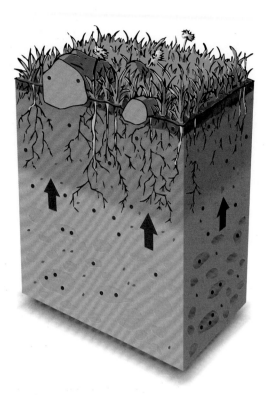

Profile of a soil derived from till. Natural weathering processes release metals such as cadmium from rocks and minerals into the soil. The released metals (represented by red dots) may dissolve in water, or they may attach to organic particles in the soil. Metals are taken up by plants in varying amounts (a process represented by the red arrows) depending on soil conditions and plant type. The transfer of metals from geological sources to the food chain occurs when the plants are eaten by animals that in turn may be eaten by humans. ADAPTED FROM VARIOUS SOURCES.

fraction of soils and sediments. Unlike cadmium, organic contaminants build up in the fatty tissues of marine mammals such as beluga whale and seal.

Mercury Rises

At room temperature and in its pure form, mercury is a liquid. As temperature rises, however, mercury expands and vaporises into a gas. In the last century, liquid mercury was an important component of thermometers, barometers, electrical switches, and fluorescent lights. But the manufacture of these products led to unacceptable levels of mercury in the environment; so by the mid-twentieth century, mercury was no longer mined or widely used in Canada.

Smelter stacks, coal-fired power plants, and mercury-rich mine tailings continue to contribute mercury to the atmosphere, as do volcanoes and some rocks during weathering. For example, Paleozoic black shales in southeastern Yukon produce diffuse emissions of mercury gas over an area of 1,500 square kilometres. More localized emissions come from sulphide-bearing fault zones and numerous small Proterozoic black shale outcrops and quarries near Thunder Bay in Ontario.

Mercury is commonly associated with sulphide minerals in rocks, sediments, and soils. Plants accumulate mercury compounds produced by weathering of these materials. When herbivores eat the plants, the mercury accumulates in their tissues and organs. The herbivores are, in turn, consumed by carnivores, and the mercury becomes increasingly concentrated in their tissues. This progressive concentration up the food chain is called biomagnification. Earlier we discussed bioconcentration, which occurs within an organism, as in the example of cadmium in caribou. Biomagnification is different because it occurs across trophic (food-chain) levels. Humans are exposed to mercury mainly through the food they eat, with shellfish, predator fish such as lake trout and smallmouth bass, and sea mammals being the greatest contributors of this element to our diet. Biomagnification causes higher mercury levels in these animals because they are at the top of the aquatic food chain.

In the environment, mercury is transformed from its inorganic form into methylmercury, which accumulates in the tissues of fish and sea mammals by adhering to proteins in muscles. Concentrations in predator fish at the top of the aquatic food chain may exceed recommended consumption limits. By the time humans consume such fish, methylmercury is considerably biomagnified, sometimes to levels that can cause neurological and developmental disorders, although the degree of toxicity depends on the amount and length of exposure. Aboriginal peoples living a traditional lifestyle are more exposed to mercury because fish from lakes, as well as sea mammals, are a significant part of their diet.

Geologists have identified areas where natural sources of mercury may lead to elevated levels of methylmercury in fish. For example, in 1969 and 1970, geologists discovered elevated levels of mercury in the Kaminak Lake region of Nunavut, attributable to sulphide minerals in black shales and volcanic rocks of Archean age. A commercial fishery that had been established in Kaminak Lake had to be closed in 1972 due to unacceptable mercury concentrations in fish (0.6 to 2.0 parts per million, exceeding the national consumption guideline of

Scientists use chambers to measure natural mercury vapour emissions. At this site on Paleozoic black shale in the Macmillan Pass area, Yukon, mercury vapour is released to the atmosphere at a rate of about 9 nanograms per square metre per hour during the summer. ALEXANDRA STEFFEN.

The Rove Shale, a Paleoproterozoic rock extracted from numerous small quarries west of Thunder Bay, Ontario, is used extensively as aggregate for constructing rural roads and as fill. Emissions of mercury from the Rove Shale enter the atmosphere at a daytime rate of about 32 nanograms per square metre per hour in the summer. PAT RASMUSSEN.

Testing for water quality in a stream flowing into Sherbrooke Lake, Lunenburg County, Nova Scotia. ROB FENSOME.

0.5 parts per million). The fishery was relocated to a nearby lake where mercury levels were lower.

Another example relates to certain lakes around Huntsville, Ontario, near the southern margin of the Canadian Shield. In these lakes, several types of predatory fish, such as smallmouth bass and lake trout, were found to have mercury concentrations as high as 2.5 parts per million, which is more than five times the acceptable limit. These anomalous values could not be explained by long-range atmospheric transport or industrial sources. And only small amounts of mercury are present in sulphide minerals and graphite deposits found in this area. The source of the mercury was finally discovered when samples of vegetation, soil, and lacustrine sediments were collected and analyzed. The analyses showed huge differences in mercury concentrations in sediments from different lakes, with values ranging from 450 parts per billion to less than 5 parts per billion. Samples collected from fault zones in the area had the highest concentrations of mercury. Researchers concluded that the faults provided a conduit for mercury to rise to the surface from natural sources deep in the Earth's crust.

Tasteless but Toxic

Arsenic is another natural constituent of our planet's crust, commonly found in sulphide minerals in association with copper, iron, nickel, cobalt, lead, silver, and gold. Arsenic can move from the crust to the atmosphere, soil, and water by both natural processes and human activity. The element is released during volcanic eruptions and wildfires, and by weathering of minerals and ores; it is also common in oxygen-free groundwater near coals. Combustion of fossil fuels, especially coal, releases arsenic. Other sources include wood preservatives, smelting of sulphide minerals, and gold processing. The largest atmospheric emissions of the element in Canada used to be from the smelting and refining of copper and nickel, but changes in smelting processes in the early 1980s dramatically reduced these emissions.

Trace amounts of arsenic occur in all living matter. Drinking water is the major concern for people living near an arsenic source. Arsenic enters water through the breakdown of some minerals and rocks; the discharge of industrial wastes; or the deposition of arsenic in dust, rain, or snow. Levels of arsenic in Canadian drinking water are generally less than 5 parts per billion—half the Canadian and American standards. Natural concentrations above the standard, due to local geology, have been found in isolated pockets across Canada.

Arsenic in drinking water is absorbed by the body and distributed by the bloodstream. Cancers of the bladder, liver, lungs, and kidneys have been linked to exposures to high levels of the element in drinking water over periods of many years to decades. Long-term exposures may also cause thickening and discoloration of the skin, nausea, diarrhea, decreased production of blood cells, abnormal heart rhythm, blood vessel damage, and numbness in hands and feet. At high concentrations, arsenic is also toxic to many other organisms. It can reduce photosynthesis and crop yields, cause malformations in amphibians, and lead to cancer, blindness, and fetal malformations in mammals.

In some areas, arsenic contamination of water supplies is caused by metal mining and smelting, coal processing in coke ovens, quarrying, and certain other industrial activities. Areas with elevated arsenic levels include parts of Alberta underlain by oil sands, the Sydney area in Nova Scotia, Yellowknife in the Northwest Territories, and the Moira River system in southern Ontario. Perhaps most unexpected are the high arsenic residues in some soils of the Maritime Provinces, caused by the past use of sodium arsenite as a pesticide in potato fields.

As early as the 1970s, naturally high levels of arsenic were found in water supplies in parts of Nova Scotia, related to elevated bedrock concentrations around Halifax and in Guysborough County. Runoff and seepage from the tailings of gold mines that were operated from the 1860s to the 1930s in the Halifax area have also polluted local water wells with arsenic.

Dosage Makes the Difference

Cadmium, mercury, and arsenic are trace elements that have no known benefits to our health and that have toxic effects at elevated levels. Other trace elements, such as fluorine, selenium, zinc, and copper, are essential to our health, but in specific dose ranges. If our intake of any of these elements is below or above certain levels, we get sick or may even die.

Fluorides are compounds containing the element fluorine. They occur naturally in soils, rocks, water, air, and food; are released into the atmosphere by volcanic eruptions, evaporation, sea spray, and industrial pollution; and return to the Earth in rain or snow. Fluorine primarily enters our bodies in the foods and fluids we consume. The right amount of fluoride makes our teeth more resistant to decay. Because of its proven record in fighting tooth decay, fluoride has been added to toothpaste and mouthwash, as well as to water supplies. In 1945, Brantford, Ontario, became the first Canadian city to fluoridate its water. Today, about 40 percent of Canadians have fluoridated water.

Too much fluoride can cause dental fluorosis, which in its mildest form gives teeth a mottled appearance. In extreme cases, teeth become blackened and weakened and may fall out. In addition, fluoride levels that are too high can harm bones. When fluoride enters our bodies, it quickly dissolves in the blood and concentrates in our bones. Initially, the fluoride increases bone mass, but with prolonged exposure, bones become weak and brittle. The end result is skeletal fluorosis, marked by calcification of ligaments, bone deformities, and neurological symptoms resulting from spinal cord compression.

Drinking-water fluoride surveys were conducted in several provinces and territories in the 1980s. Naturally elevated levels were rare, but some communities in the Maritime Provinces, Quebec, Saskatchewan, and Alberta had concentrations as high as 2.52 to 4.35 milligrams per litre, exceeding the 1.5 milligrams per litre maximum recommended by Health Canada. In particular, fluoride is associated with Carboniferous strata that underlie much of the Maritimes. These rocks are of similar age and type to those which are the source of the high fluoride levels in the Maria area of the Gaspé Peninsula.

Canada is one of the world's major exporters of selenium, which is used in glass manufacturing, chemicals, metallurgy, and electronics. It is a by-product of copper refining and smelting. Selenium has been widely studied because it is both an essential nutrient and a dangerous toxin. The element is necessary for maintaining healthy cells and tissues in people and animals. A beneficial concentration range for selenium—the threshold between too low and too high—is narrow and difficult to determine, because its toxicity is controlled by its chemical form and its interaction with other aspects of our diet. Canadians typically get between 0.05 and 0.2 milligrams per day, largely from food. For adults, these values are within the recommended range. Higher amounts (0.5 milligrams per day or more) for long periods are considered detrimental to our health.

Selenium is associated with sulphide minerals, clay-rich sediments, and organic-rich sedimentary rocks such as black shales and coal. The element enters the air through the burning of coal and petroleum, and the processing of copper, zinc, uranium, and phosphate. In most of Canada, plants and soils contain less than one part per million of selenium. Bedrock geology can influence the selenium content of agricultural soils used for crops and livestock feed. Soil alkalinity, climate, and drainage also have a major impact.

The semi-arid belt from the plains of Canada through the western United States and into Mexico is a zone of abundant selenium. Reports dating back to the 1800s for the region describe deaths of horses and cattle from selenium poisoning. Elevated levels of the element that are potentially toxic to livestock occur in isolated areas of Alberta, Saskatchewan, and Manitoba. Overexposure occurs if livestock consume selenium accumulator plants such as cream milk vetch, which can have selenium levels as high as 15,000 parts per million. Unless they are desperate for food, grazing animals tend to avoid selenium-rich plants, the toxicity of which can cause blind staggers, hoof deformation, lameness, and hair and weight loss.

Soils on silica-rich bedrock, which are common in eastern Canada and in British Columbia, tend to be deficient in selenium. Livestock in these areas can suffer from selenium deficiencies because the element is not very soluble in acidic or poorly aerated soils. In acidic soils, the element binds with iron, aluminum, and manganese oxides, reducing the amount available to plants and thus to grazing animals. In contrast, highly oxidized, alkaline soils favour selenium compounds that remain in solution and are available to organisms. Selenium deficiency in livestock has been linked to white muscle disease, which is characterized by muscular weakness, reduced appetite, embryonic deformities, and stunted growth. Fortunately, this disease is easily diagnosed and is readily treated with supplements of selenium and vitamin E in livestock feed.

What Lies Beneath

Radon, a gas, is produced by radioactive decay of uranium, thorium, and radium, which occur naturally in some rocks and soils. Radon is not normally a health concern, because it quickly dissipates in the atmosphere and poses no health risk. However, radon gas is nine times heavier than air and so can accumulate in depressions such as the basements and subsurface floors of buildings where people live and work. Elevated radon levels can increase over time, particularly in well-insulated, energy-efficient buildings where little air is exchanged with the outdoors. People exposed to high levels over many years have an increased risk of lung cancer.

Five factors control radon levels in a home: the abundance of uranium, thorium, and radium in underlying bedrock, sediment, and soil; the ease with which radon can travel upward along faults and joints in the bedrock or through pores in sediment and soil; the ease with which the gas passes through basement floors or walls; the presence or absence of radon collection areas in the house; and interior ventilation.

How much radon is acceptable? In 2007, the Canadian government lowered the threshold for indoor air from 800 to 200 becquerels per cubic metre (a becquerel corresponds to the disintegration of one radioactive atom per second). It recommends that remedial action be taken to reduce exposure where the radon concentrations in a house exceed 200 becquerels per cubic metre.

We can identify sources of radon by mapping radioactivity using gamma-ray spectrometers, both from the air and on the ground. Spectrometers measure gamma-rays given off by radioactive elements, mainly uranium and thorium. Since 1967, the Geological Survey of Canada has mapped radioactivity in Canada. In Nova Scotia, this program has shown that radon levels in 719 homes correlate strongly with uranium concentrations inferred from spectrometer mapping. High uranium concentrations in the area are especially associated with the South Mountain Batholith (Chapter 8), which extends from Halifax southwestward to near Yarmouth.

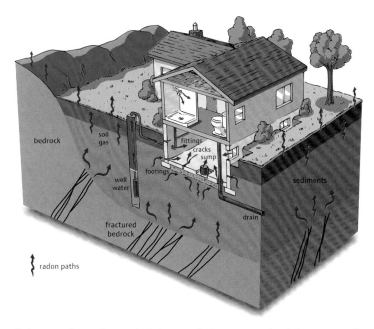

Radon enters homes from geological sources. Radon concentrations in homes are partly influenced by the presence of uranium in bedrock and surficial deposits, and partly by characteristics of the house. Depending upon how easy is it for radon gas to enter, accumulate in, and leave a building, there can be considerable variation in the concentration of radon in homes and offices, even in the same geological setting.

Radon-222 is produced by the radioactive decay of uranium-238. Most of the radon gas that we inhale is exhaled before it decays and is thus harmless. Unfortunately, radon-222 decays to polonium-218, which attaches to particles in the air we inhale. The particles containing polonium can attach themselves to the lungs, increasing the chances of tissue damage through radioactive decay, and potentially resulting in the formation of cancer cells. Population studies suggest that this may be responsible for 10 percent of all cases of lung cancer.

Dust and Volcanoes

Earth's atmosphere contains a variety of fine particles from both natural and industrial sources. Natural airborne particles include those of biological origin, such as viruses, bacteria, pollen and spores, animal and human skin and hair particles, leaf and needle fragments, and forest-fire ash. Geological contributions, such as dust from volcanoes and windblown clays, are also significant. Once airborne, fine mineral dust, ash, and biological particles can travel long distances, as shown by their presence in Arctic snow and ice cores thousands of kilometres from where they originated. Fine airborne particles can penetrate deep in the lungs, where they can cause damage. High concentrations of fine particulate matter are associated with many health problems, including asthma, bronchitis, shortness of breath, painful breathing, and may cause premature death.

Some health hazards are associated with the release of ash and trace elements during volcanic eruptions. Volcanic ash is

Testing for radon. REPRODUCED WITH THE PERMISSION OF NATURAL RESOURCES CANADA 2013, COURTESY OF THE GEOLOGICAL SURVEY OF CANADA.

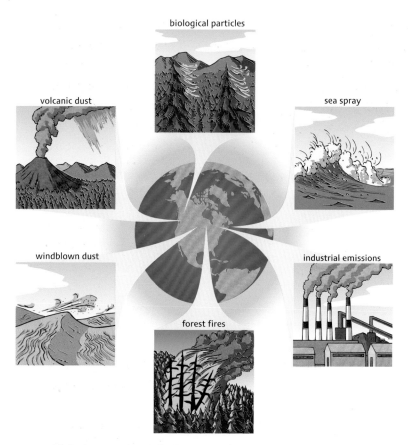

Globally, the major sources of airborne particles are industrial, sea spray, windblown dust, volcanic dust, forest-fire debris, and biological particles (spores, pollen, waxes). Particle emission inventories indicate that industrial emissions contribute about 5 to 7 percent of total annual global emissions.

A specimen of chrysotile asbestos from Quebec. D. MARUSKA, REPRODUCED WITH THE PERMISSION OF NATURAL RESOURCES CANADA 2013, COURTESY OF THE GEOLOGICAL SURVEY OF CANADA.

Ash cloud from the 2010 eruption of Eyjafjallajökull, Iceland. ALAN MORGAN.

composed of small jagged pieces of rocks, minerals, and glass varying in size from less than 0.001 to 2 millimetres. The smallest ash particles can be carried thousands of kilometres by prevailing winds. They can pollute water supplies and disrupt transportation. Thick accumulations of heavy ash can cause buildings or other structures to collapse. Inhaled ash can aggravate asthma and bronchitis. Coarser particles can lodge in the nose, causing irritation, or in the eyes, where they cause corneal abrasions. Silicosis, also known as Grinder's disease or Potter's rot, is caused by inhaling very fine silica particles, and in some cases has been attributed to long-term exposure to volcanic ash. Silicosis contracted specifically from breathing volcanic dust is called pneumonoultramicroscopicsilicovolcanoconiosis, which at forty-five letters is the longest word in the English language.

Resistant Fibres

Asbestos is a general name for many fibrous magnesium silicate minerals that occur naturally in rocks. These fibrous minerals are resistant to acid, abrasion, and fire, and have been used extensively in insulation and other construction materials, brake linings, textiles, industrial ovens, smelting ovens, and electrical appliances. Asbestos minerals belong to two mineral groups: serpentine and amphibole. The serpentine group includes chrysotile, which has a fibrous (asbestiform) habit. Chrysotile is curly, each fibre being a bundle of smaller fibrils; it is more flexible than other forms of asbestos and is essentially the only asbestos used commercially today. The amphibole group includes three commercial asbestos fibres: amosite, crocidolite, and anthophyllite. All of these fibres are made from needle-like crystals and are more resistant to chemicals than chrysotile. From the 1930s to the 1980s, amphibole asbestos minerals were used extensively, after which they were banned in many countries due to health concerns.

Canada supplied more than half of the world's chrysotile until the mid-1950s and remained the leading producer until the mid-1970s, when production dropped drastically due to the growing concerns about asbestos-related health risks. Although there have been chrysotile mines in other provinces, the Canadian asbestos industry has always been concentrated in Quebec's Eastern Townships and is now limited to this area.

Asbestos poses a health risk when microscopic fibres are present in the air people breathe. Health risks are greatest for workers in industries that produce and use asbestos, such as mining and milling, and for staff exposed to old and crumbly asbestos products in buildings. In the past, workers in these environments were 100 to 1,000 times more exposed to asbestos than today's workers. Strict standards now limit exposure, and the ban on amphibole asbestos has reduced industrial risk. During renovations and repairs to older buildings, however, workers and staff may be exposed to very high concentrations of asbestos fibres. Diseases associated with asbestos exposure include a scarring of the lungs that makes breathing difficult (asbestosis), a cancer of the linings of the chest or abdominal cavity (mesothelioma), lung cancer, and probably laryngeal cancer. Smoking in combination with asbestos exposure greatly increases the risk of lung cancer, compared to exposure to asbestos alone.

Amphibole asbestos fibres reside much longer in the lungs than chrysotile fibres and are more likely to inflict damage and cause disease, particularly mesothelioma. The incidence of mesothelioma increased rapidly in most industrialized countries from 1970 to 2000. In Canada mesothelioma rates are highest in Quebec, the province where the asbestos industry has been most active and where asbestos workers were most numerous and most highly exposed until the 1970s. Amphibole asbestos has been replaced by chrysotile in new asbestos products, as chrysotile is much less persistent in human tissues and is less carcinogenic than amphiboles. Nevertheless, chrysotile is a carcinogen and is generally replaced if equivalent and safer materials are available.

Urban Exposure

Atmospheric particles are a major part of urban air pollution. The main sources of these particles are vehicle emissions, industrial pollution, and other combustion processes. In sunlight and at high temperatures, vehicle emissions produce smog, the most visible form of air pollution. Fine particles are also formed in the atmosphere when gases such as sulphur dioxide, nitrogen oxides, and volatile organic compounds are transformed by chemical reactions.

Urban residential dirt (soil, street dust, and house dust) can contain some unpleasant substances such as lead and smelly organic compounds. DAVID GARDNER.

Windblown dust and emissions from vehicles and industrial sources can transport metals and organic compounds into homes. People and pets also track dirt indoors. Sources of metals and organic contaminants in the house include carpets and other furnishings, paints, and building materials. Fireplaces, cooking, and smoking add to indoor pollution. Studies of urban residential areas where there is little industrial activity, such as Ottawa, show that levels of many metals, such as lead, mercury, arsenic, and cadmium, can be up to ten times higher in indoor dust than in garden soil or street dust.

Residential dirt can be an important source of pollutants for preschool children, who tend to ingest dust and soil through hand-to-mouth activities. A major concern is long-term exposure to dust containing high concentrations of lead, which can cause lead poisoning and neurological damage. The risk is not limited to smelter towns, mining sites, and other industrial areas; it may also exist in pre-1960 homes, built in the era of lead-based paint, lead pipes, and lead solder.

An important factor influencing the biological availability and toxic action of metal compounds in dust and soil particles is their solubility in our digestive tracts. The total metal content of dust or soil provides little information on the potential for metal uptake in our bodies. Mineralogy, metal chemistry, and particle size are all important factors that control the bioaccessibility (the potential for biological uptake) of metals in dust and soil. Fine silt and clay-sized particles adhere better to a child's hand and are thus more likely to be ingested than coarse particles. Once inside our bodies, metals are generally more bioaccessible when they are associated with fine, rather than coarse, particles.

Geoscientists are working with toxicologists to develop tests that mimic our digestive systems and yield realistic assessments of internal exposure. This new research is providing better estimates of the proportions of metals in dirt, household dust, garden soil, and playground substrates that are soluble and might be absorbed into our bodies. Geoscientists are becoming increasingly involved in human health studies aimed at identifying, understanding, and minimizing human exposures to natural sources of hazardous substances in Canada.

20 • CANADA'S GEOLOGICAL HERITAGE

CHAPTER SUPPORTED BY A DONATION FROM SCOTT JOBIN-BEVANS, CARACLE CREEK INTERNATIONAL CONSULTING INC., IN MEMORY OF DOUGLAS ROBERT BOWIE (1942–2008)

Coastal exposure of sandstones on the Magdalen Islands, Quebec. These Permian redbeds were deposited in the interior of the supercontinent Pangea. CATHERINE MOONEY

Ancient Beginnings

Our main goals in writing this book have been to present the geological evolution of Canada to a non-specialist audience and to demonstrate its significance to Canadian society. The story has been woven from the contributions of about a hundred experts, and is based on Canada's exceptional geological record, starting with the oldest-known terrestrial rocks. Let's now reflect on the story "so far".

From the Earth's original basaltic magma ocean, granitic material began to segregate as "rafts" and proto-continents too light to sink back into the interior. These early fragments grew into unsinkable Archean cratons such as the Slave and Superior. In Paleoproterozoic rocks, from 2,500 million years ago, we can identify passive margins, for example in the Huronian succession of northern Ontario and the Labrador Trough in Labrador and eastern Quebec. These margins flanked expansive oceans, until about 2,000 to 1,800 million years ago, when many of the oceans closed to produce the first supercontinent, Nuna. In the process, intervening oceanic crust was squeezed between colliding continental blocks and incorporated into mountain belts. The best preserved of these former mountain belts is the Trans-Hudson Orogen, which today hosts metal-mining districts such as Flin Flon and Thompson in Manitoba. About 1,850 million years ago, a large extraterrestrial body smashed into the margin of the Superior Craton, creating the Sudbury impact structure and the nickel-rich ore deposits that now drive the Sudbury region's economy.

The billion years following the formation of Nuna were less eventful than the previous billion, but convergent plate activity occurred along that margin of Nuna today represented by the southeastern edge of the Canadian Shield. From about 1,800 to 1,200 million years ago, this margin was a boundary between a continent and an ocean. In the continent's interior, Amazon-sized rivers unconstrained by vegetation deposited sediments in scattered sedimentary basins, including the Athabasca Basin in northwestern Saskatchewan, in which huge uranium ore deposits are now found. Though evidence for the breakup of Nuna is not well recorded in Canadian rocks, it may be reflected in the Mesoproterozoic rift sequence of the Belt-Purcell Basin now exposed in the southern Rocky Mountains.

Several continents must have existed during the later Mesoproterozoic because around 1,200 million years ago, the southeastern margin of Nuna changed from the site of a continent-to-ocean convergence to one of continent-to-continent collision. This titanic clash culminated in the Grenvillian Orogeny, which ended 1,000 million years ago; the resulting Grenvillian Mountains were probably comparable in size to today's Himalayas. By the end of the Mesoproterozoic, the planet's landmasses had come together to form the supercontinent Rodinia.

From One Supercontinent to Another

Rodinia stayed intact until about 750 million years ago, when it began to break up into several continents. Although any recognizable hint of modern continental geography was still hundreds of millions of years in the future, one of the new continents, Laurentia, eventually was to become the nucleus for North America. Laurentia had been embedded in the core of Rodinia, probably just south of the paleoequator, and was rotated such that the margin that would become today's Canadian Arctic then faced east. By 520 million years ago, Laurentia was surrounded by oceans. Former Laurentian crust today stretches from the eastern Cordillera to Greenland, Labrador, northwestern Newfoundland, and eastern Quebec. Precambrian rocks such as those exposed today on the Canadian Shield formed the Laurentian land surface. These mostly crystalline rocks were widely covered during the Paleozoic by blankets of sediment, including Saskatchewan's potash deposits and Manitoba's Tyndall Stone, which adorns many prominent buildings across Canada, including the Parliament buildings in Ottawa. The legacy of the Paleozoic continental seas not only includes potash and beautiful building stones, but the reefs so important today as oil and gas reservoirs.

Between about 480 and 390 million years ago, several microcontinents and the larger continent Baltica, which comprised what is now Scandinavia and Russia west of the Ural Mountains, collided with Laurentia's eastern margin to form a new and larger continent known as Euramerica. The microcontinents—Dashwoods, Ganderia, Avalonia, and Meguma—and remnants of some intervening magmatic arcs and back-arc basins now underlie the Maritime Provinces, much of the island of Newfoundland, and a small part of southeastern Quebec. The collisions culminated in a major Appalachian-Caledonian mountain belt that extended from the present-day southeastern United States, through Atlantic Canada, Greenland, and

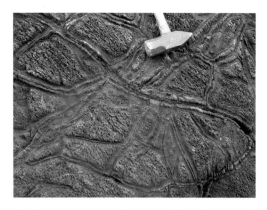

Weathered Archean komatiite, about 2,700 million years old, from the Rae Craton on Baffin Island, Nunavut. The pattern may reflect columnar jointing. MIKE YOUNG.

These well-preserved Paleoproterozoic pillow lavas on central Baffin Island, Nunavut, are 1,970 million years old and likely formed in the Manikewan Ocean prior to the Trans-Hudson Orogeny. The white areas between pillows (not the cross-cutting white veins) represent carbonate that was precipitated in crevices between pillows. DAVID CORRIGAN.

Gneiss and metamorphosed mafic dykes exposed on Franklin Island, off Georgian Bay's east shore. These rocks were deformed at great depth beneath the Grenvillian Mountains over 1,000 million years ago. They reveal what parts of the continental crust might be like today at depths of a few tens of kilometres below the surface. CHRISTOPHER HARRISON.

The strata in the upper left half of this photo, taken from Lake Louise ski hill, Alberta, are Cambrian sandstones. Unconformably beneath these are less well-exposed Neoproterozoic deep-sea deposits of the Windermere Seaway. The grey rocks at centre and immediately beneath the well-bedded Cambrian strata represent a submarine canyon and can be seen to thin toward the distance. BILL ARNOTT.

the northwestern British Isles to Scandinavia. The northwestern edge of the deformation associated with these collisions, known as Logan's Line, runs through Québec City.

By about 310 million years ago, a new supercontinent called Pangea had formed. What is now Atlantic Canada was squeezed obliquely between converging continents. The squeezing formed a complex of structural highs and lows collectively called the Maritimes Basin. Most of the Devonian to Permian sediments that accumulated in the Maritimes Basin are nonmarine and include the coal beds visible, for example, in the mines and along the shores of Cape Breton Island, and the red clastic rocks forming the red cliffs of Prince Edward Island and the Magdalen Islands of Quebec.

During the Paleozoic, the future Arctic and western Canada were still continental margins of Laurentia, Euramerica, and Pangea, successively. Both the Arctic and western margins were initially passive, the former margin facing the ancient Ural Ocean and the latter facing huge Protopanthalassa, the distant ancestor of the present Pacific Ocean. But a convergent plate boundary in the Ural Ocean led to the accretion of the microcontinent Pearya, a remnant of which today forms Ellesmere Island's northern tip. Possibly associated with Pearya for part of its journey was the well-travelled Alexander Terrane, whose remnants now cling to parts of northwestern North America. The Ural Ocean survived until the Permian. Its final closure is marked today by the Ural Mountains in Russia.

Aerial view of early Cambrian sandstones and overlying Carboniferous redbeds east of Hare Fiord on Ellesmere Island, Nunavut. These strata have been intruded by a Cretaceous mafic dyke (the dark band running obliquely down the cliff). CHRISTOPHER HARRISON.

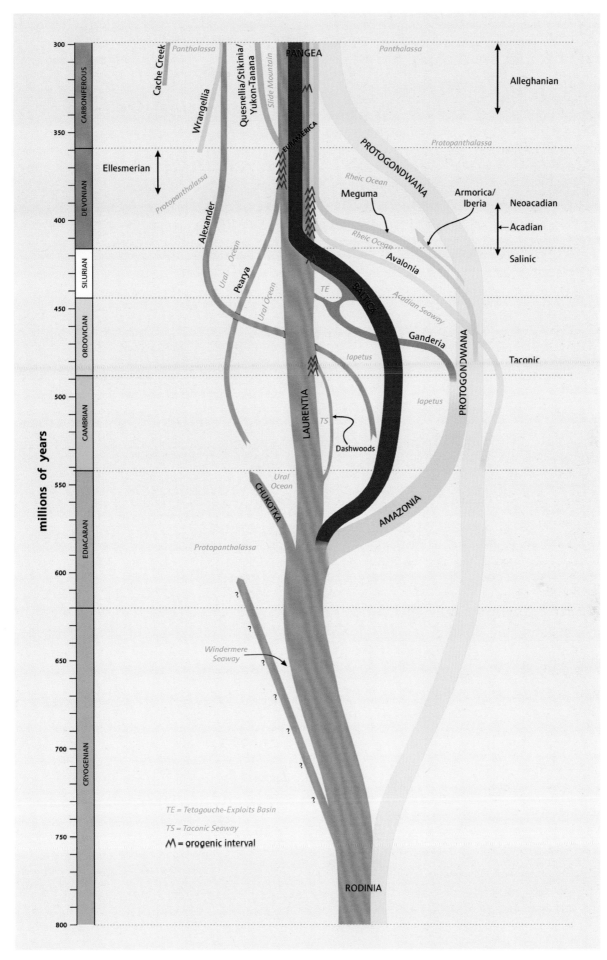

Impressionistic representation of the evolution of continents, oceans, terranes (including microcontinents), orogenies (indicated at right and left), and other major geological features over the interval from 800 to about 300 million years ago. Oceans and seas are in blue lettering; otherwise colours are mainly for clarity and aesthetics. The main focus is on events that relate to the evolution of Canada. It is impossible to accurately represent all events that have occurred during this interval in two dimensions, but this figure provides a sense of how events are related through time and across Canada.

Flat-lying Ordovician limestone beds near the Alexandra Bridge in Ottawa, Ontario. ROB FENSOME.

The brown rocks on these slopes near Trout River Pond, Gros Morne National Park of Canada, Newfoundland, are part of the Bay of Islands ophiolite, which was thrust up from the mantle during the Taconic Orogeny, about 480 million years ago. MICHAEL BURZYNSKI.

The gorge at Grand Falls, New Brunswick, cuts through Ordovician to Silurian limestones deposited within former Ganderia, deformed during the Salinic and Acadian orogenies. The gorge formed between 13,500 and 11,500 years ago as waters from Glacial Lake Madawaska started to drain south into the present-day Saint John River valley instead of northeast into the Restigouche River system. This change gave rise to the modern Saint John River watershed. COURTESY OF TOURISM AND PARKS NEW BRUNSWICK.

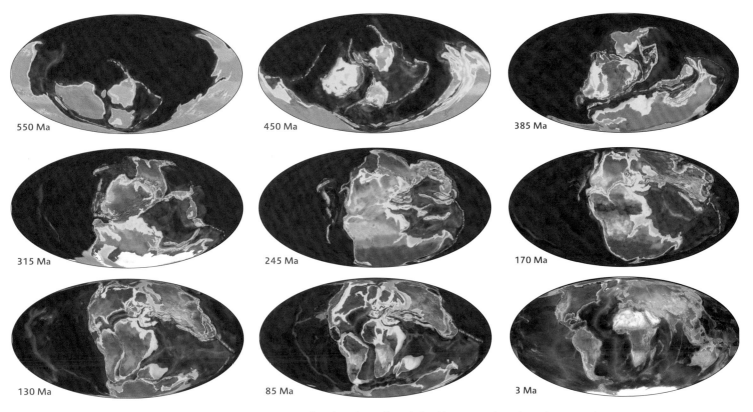

Global paleogeographic reconstructions for selected times from the late Neoproterozoic to almost the present day. Ma = millions of years ago. RON BLAKEY.

The Protopanthalassan margin, in what is today western Canada, remained passive until about 390 million years ago. Then it became a convergent plate margin that presaged, by about 200 million years, the start of Cordilleran mountain building. The onset of convergence in the west roughly coincided with the final closing of oceans in the Appalachian Orogen in the east, suggesting that events on the two sides might be related. The late Paleozoic and Triassic eastern boundary of Protopanthalassa, which became Panthalassa once Pangea had assembled, was more like the modern margin of the western Pacific Ocean, with island arcs and back-arc basins, rather than like today's eastern Pacific margin bordered by continental arcs.

Pangea's Legacy

Most of what is now Canada formed part of Pangea's interior from the late Carboniferous until the early Jurassic, about 320 to 180 million years ago. Some 235 million years ago, extensive rift systems began to develop within Pangea, with the Fundy Basin being the best-exposed example of these rifts in North America. Rifting in the Triassic changed to drifting during the early Jurassic, about 185 million years ago, with new oceanic crust breaching the floor of the nascent North Atlantic, then little more than an extension of the huge ocean known as Tethys. As the Jurassic progressed, Pangea split into two still-huge continents, Laurasia to the north and Gondwana to the south. Laurasia included lands that were to become North America and most of Eurasia; Gondwana included Africa, South America, Australasia, India, and Antarctica. As part of the early breakup between Laurasia and Gondwana, future North America split from future Africa.

Unconformity between Silurian and Carboniferous rocks at Quinn Point, Jacquet River, New Brunswick. The Silurian rocks were tilted during the Acadian Orogeny about 420 to 390 million years ago. ROB FENSOME.

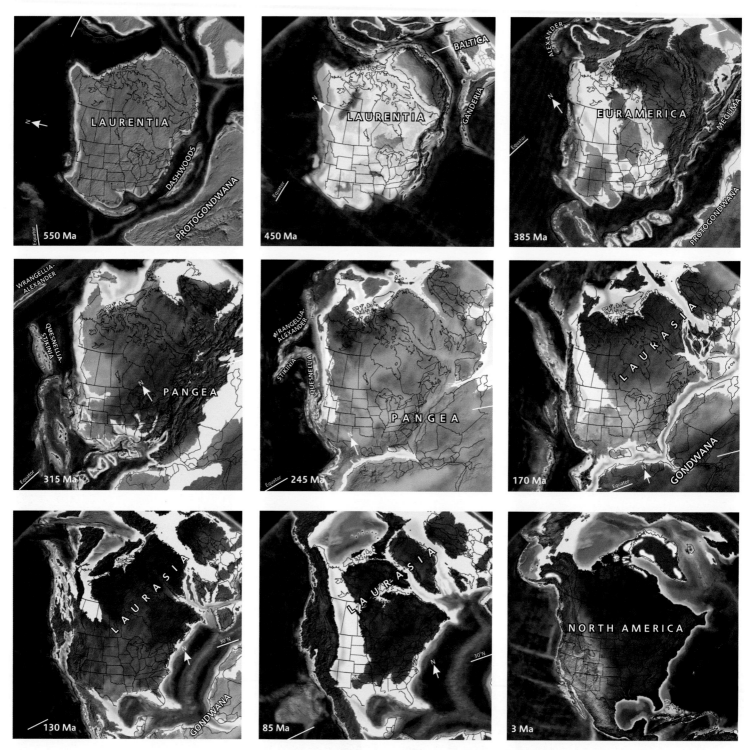

Paleogeographic reconstructions of what was to become North America for selected times (ages in millions of years at top left of each map) from the late Neoproterozoic to almost the present time. Ma = millions of years ago. RON BLAKEY.

However, northeastern North America was to remain connected to northwestern Eurasia for another 100 million years or more.

Between about 185 and 175 million years ago, Stikinia, Quesnellia, Wrangellia, and the Alexander Terrane began to collide with the Pangean continental margin, the collision producing the first Cordilleran mountains. That the opening of the Atlantic and the rise of the first Cordilleran mountains occurred around the same time is probably no coincidence, as the opening of the new ocean pushed the continent westward, where it collided with Pacific crust.

The collision between the Cordilleran terranes and the continent was an oblique one. During the Jurassic and into the Cretaceous, until about 110 million years ago, the continent moved northwestward relative to the terranes and to the Pacific Ocean. However, from the middle Cretaceous to the present, the continent has moved southwestward. The timing of the switch fits well with the start of the opening of the Arctic Ocean and Labrador Sea, as tectonic plates jostled toward a new equilibrium.

The collision between Laurasia and the Pacific was felt hundreds of kilometres inland, where it produced the thrust-

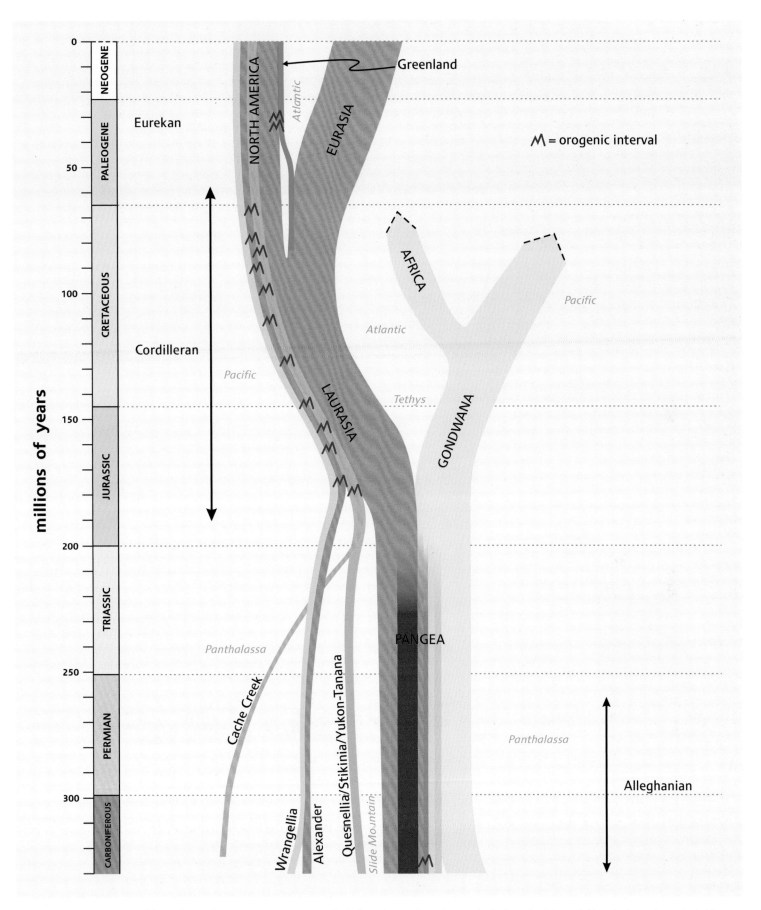

Impressionistic representation of the evolution of continents, oceans, terranes (including microcontinents), orogenies (indicated at right and left), and other major geological features over the interval from 330 million years ago to the present. Oceans and seas are in blue lettering; otherwise colours are mainly for clarity and aesthetics. The main focus is on events that relate to the evolution of Canada. It is impossible to accurately represent all events that have occurred during this interval in two dimensions, but this figure provides a sense of how events are related through time and across Canada.

Folded early Cretaceous sandstone with interbeds of coal at the Cardinal River Mine, north of Cadomin, Alberta. These folds are evidence of the deformation that produced the thrust-and-fold belt of the Canadian Rocky Mountains during later Cretaceous to Paleocene time. The horizontal lines are the old pit roads. GODFREY NOWLAN.

Balancing Rock is a pillar of North Mountain Basalt on Long Island, Nova Scotia. The North Mountain Basalt is part of an episode of late Triassic magmatic activity that presaged the breakup of Pangea and the opening of the Atlantic Ocean. ROB FENSOME.

The black bands in this cliff east of Thompson Glacier on Axel Heiberg Island, Nunavut, are early Cretaceous mafic sills. Here they intrude between late Triassic sandstone and shale layers. The sills are part of the magmatic activity that heralded the early development of the Arctic Ocean. CHRISTOPHER HARRISON.

and-fold belt of the Rocky Mountains. The resulting compression caused rocks to pile up, in part through thrust-faulting, the added weight depressing the crust over a large part of the Western Platform to produce the Western Interior Basin. This in turn filled with sediments eroded from the rising mountains to the west. Processes involved in this series of events and the structures that they produced are responsible, at least in part, for many of the prolific petroleum resources of western Canada. The plate boundary between the North American part of Laurasia and the Pacific Ocean remained one dominated by subduction until about 60 million years ago, when forces related to plate convergence became less intense. By about 40 million years ago the boundary had developed its modern configuration, with subducting segments such as the Cascadia Subduction Zone separated by long transform faults such as the San Andreas and the Queen Charlotte-Fairweather.

Just as northeastern North America remained connected to northwestern Eurasia until about 50 million years ago or later, so northwestern North America has stayed connected to north-

Conglomerate of Tertiary age, Cypress Hills, southwestern Saskatchewan. FLOYD WIST.

eastern Eurasia on and off since the days of Pangea. Indeed, the Bering Sea between Alaska and Siberia today is a shallow continental sea, with the North American-Eurasian plate boundary being marked by a series of faults in eastern Siberia. During the Jurassic, and early Cretaceous, South America was part of Gondwana, but separated during the middle Cretaceous, staying on the western side of the growing South Atlantic Ocean. Its

Glacial erosion produced a lot of sediment, material that came to be deposited widely as till. This landscape in southeastern Alberta is formed of hummocky till deposited by the Laurentide Ice Sheet. ROB FENSOME.

Stromatolites in red dolostone within a small Mesoproterozoic basin on northern Somerset Island, Nunavut. Stromatolites resulted from the activity of microbes and are widespread in Proterozoic sedimentary rocks prior to the evolution of grazing animals. The red colour here is due to oxidation during diagenesis of small amounts of iron in the dolostone. DARREL LONG.

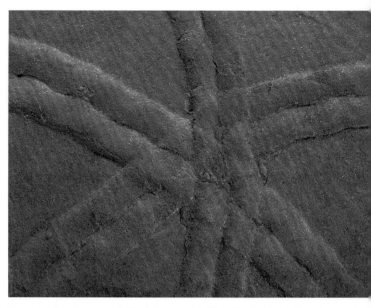
Psammichnites gigas, an early Cambrian trace fossil from Hanford Brook, New Brunswick. The explosion of life from 580 to 530 million years ago, saw not only a spectacular rise in the variety of animal body fossils but also a burgeoning of trace fossils. HEINZ WIELE, COURTESY OF THE ATLANTIC GEOSCIENCE SOCIETY; SPECIMEN COURTESY OF THE NEW BRUNSWICK MUSEUM.

eventual linking with North America occurred with the formation of the Isthmus of Panama only about 3 million years ago; this event reset the pattern of ocean currents and as a consequence may have established conditions for the onset of the Ice Age in the Northern Hemisphere.

Canada's modern landscape has been sculpted primarily by ice during the past 2 to 3 million years, including the most recent glaciation, which ended about 12,000 years ago. Ice sheets, which periodically covered much of the country, moved material around, directly or via meltwater. Ice Age glaciation and its peripheral effects are widely reflected in Canada's modern scenery. Mountains have been sculpted by glaciers, much of the country is blanketed by till, and drumlins and eskers are common in many areas. Soils developed on till and glacial-lake deposits are a major factor in Canadian agriculture. We are living during an interglacial interval, which means that, barring the effects of human influence, ice sheets will sooner or later return to southern Canada.

Most of Canada is more or less tectonically quiescent. However, the west coast is adjacent to an active plate boundary and carries the risk of earthquakes and volcanic eruptions. North America is still moving southwestward relative to the Pacific, with its western boundary a mixture of subduction zones and transform faults. It is no coincidence that the highest peak in Canada, Mount Logan in the St. Elias Mountains, is where the north-trending coast turns westward to directly face the approaching and subducting Pacific Plate.

Early Life

The geology of Canada is not only about rocks and tectonics; it also involves life as revealed by the fossil record. Fossils provide a window into the biosphere and the evolution of life and, along with rocks, tell us much about the history of our planet's atmosphere and hydrosphere. Although Canada cannot claim the oldest fossils in the world, the discovery of prokaryote microfossils from the Gunflint chert in northern Ontario sparked a revolution in the 1960s in the search for Precambrian life. These remains represent organisms that lived 1,900 million years ago in pre-Nuna seas

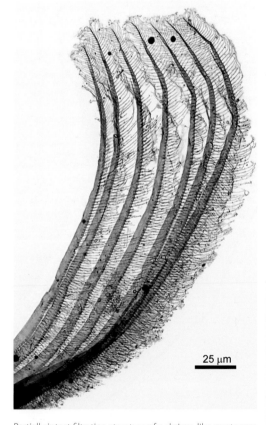
Partially intact filtration structure of a shrimp-like crustacean from early Cambrian rocks in the Northwest Territories. The specimen is about two-tenths of a millimetre long. It reflects the increasing complexity and diversity of life at the time. NICK BUTTERFIELD.

that flanked the Superior Craton. They lived several hundred million years after the Great Oxidation Event, when oxygen had begun to accumulate as a by-product of early life. Oxygen levels were still low, however, and would remain that way for the next billion years or more. Prokaryotes are responsible not only for the oxygen in the atmosphere, but probably also played a role in the development of huge expanses of banded iron formations in anoxic oceans during the Archean and Paleoproterozoic. These rocks are today's primary source of iron ore, mined for example at Fermont in Quebec and Wabush in Labrador.

It was not until the Neoproterozoic that oxygen levels began to climb toward present-day levels. This coincided with big changes in the biosphere. The first diverse assemblages of visible organisms, the Ediacara biota, appeared about 580 million years ago and are found at Mistaken Point in Newfoundland. Between 580 and 520 million years ago, life proliferated with the appearance of almost all the major animal groups living today. Burrows and trails also appeared in the fossil record for the first time, confirming that some animals were beginning to "head forward"; and the presence of trilobites with eyes meant that life was beginning to see the light. The slightly younger Cambrian fossils of the Burgess Shale in southwestern British Columbia shed light on some of the strange early animals that lived in the aftermath of the Cambrian Explosion, such as *Opabinia* with its five eyes. Cambrian animals, apparently, even made brief excursions onto land, as witnessed by trackways found in a quarry near Kingston, Ontario.

In spite of mass extinction events at the end of the Ordovician and within the late Devonian, life generally continued to flourish through the Paleozoic, notably in the shallow seas that for so long covered much of the interior of the continent. Carbonate reefs built by organisms living in these seas are reservoirs for petroleum, as at Leduc in Alberta, and organic-rich muds deposited in the same seas generated the petroleum that migrated to these reservoirs.

Canada has many excellent fish-fossil localities, perhaps the most important being Miguasha in the Gaspé Peninsula of Quebec. At Miguasha, late Devonian lobe-finned fish such as *Eusthenopteron* show us what our remote aquatic ancestors probably looked like, as does *Tiktaalik*, a fossil from slightly older Devonian rocks on Ellesmere Island, Nunavut. Although technically still a lobe-fin fish, *Tiktaalik* could prop itself up under water with its fins and had a flexible neck, just as our earliest tetrapod ancestors must have had.

Some of the earliest-known vascular plants are found in late Silurian and Devonian rocks on Bathurst Island in Nunavut and in Devonian rocks at Dalhousie Junction in New Brunswick. These fossils provide a glimpse of the early evolution of vascular

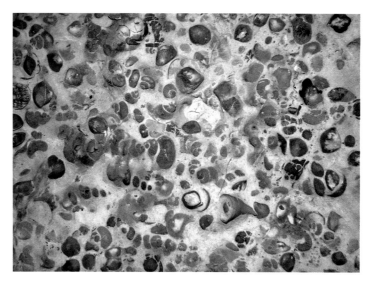

Abundant fossils are exposed on this early Silurian bedding surface from Anticosti Island, Quebec. During the Ordovician and Silurian, invertebrates proliferated in the oceans, where almost all life was then located. ROB FENSOME, SPECIMEN COURTESY OF NATURAL RESOURCES CANADA AND PAUL KOPPER.

A specimen of the arthropod *Belinurus* from middle Carboniferous rocks near Parrsboro, Nova Scotia. ANDREW MACRAE.

In this Devonian estuary scene, the underwater fauna is dominated by fish, based on fossils found at Miguasha, on Chaleur Bay in Quebec. The waters of the estuary have as their backdrop the eroding Acadian Mountains, part of the Appalachian Orogen. FROM CLOUTIER (2001), USED WITH PERMISSION; PAINTING BY FRANÇOIS MIVILLE-DESCHÊNES.

systems and leaves. The proliferating terrestrial plant community tempted animals out of the water, with invertebrates in the vanguard. Indeed, the earliest-known terrestrial invertebrate remains in North America, including those of scorpions and millipedes, come from Dalhousie Junction.

The evolution of wood provided plants the structural support needed to soar above the ground. Among the earliest trees is the late Devonian *Archaeopteris*, found at several Canadian localities. The Carboniferous Period initially was a time of hot and dry conditions in the Maritimes, later changing to wet and tropical. Two internationally renowned sites in Nova Scotia are Blue Beach and Joggins. In the early Carboniferous rocks at Blue Beach, near Hantsport, the earliest-known assemblage of amphibian trackways can be seen, mainly in lakeshore sediments. At Joggins, the remains of a 313-million-year-old tropical forest are preserved, including the lower parts of clubmoss tree trunks. Joggins is also the site of the earliest-known reptiles, sometimes found inside the fossil tree trunks, and land snails.

The flourishing tropical forests of the late Carboniferous were probably responsible for the highest levels of atmospheric oxygen known, stimulating the growth of giant arthropods such as the millipede-like *Arthropleura*, whose trackways have been found at Joggins. Oxygen levels declined during the Permian and Triassic as continental monsoon conditions predominated within Pangea and red clastic sediments stained with iron oxide were widely deposited. The end of the Permian marked the most devastating mass extinction of all time, and cleared the way for survivors to evolve in an array of new evolutionary directions.

Early Carboniferous amphibian footprints at Blue Beach, Nova Scotia. HEINZ WIELE, COURTESY OF THE ATLANTIC GEOSCIENCE SOCIETY; SPECIMEN COURTESY OF GORDON OAKEY.

This strange looking middle Carboniferous rock from Parrsboro, Nova Scotia, may be a paleosol. Before life moved onto land, the Earth's surface was largely either bare rock or loose rubble, as on Mars today. Plants and, later, animals provided the organic component and the mechanisms needed to produce soil and, hence, fertile land. ROB FENSOME.

In this latest Triassic scene based in part on fossil finds around the Bay of Fundy in Nova Scotia, a pair of *Coelophysis* dinosaurs pause by a phytosaur skeleton as the main pack forges ahead. Early pterosaurs soar overhead. The mud-cracked surface and evaporite deposits (white) bordering the lake testify to a hot, dry climate. But the remains of the phytosaur, an animal that lived in wetter environments, testifies to fluctuating climatic conditions in the region during the late Triassic. PAINTING BY JUDI PENNANEN, COURTESY OF THE ATLANTIC GEOSCIENCE SOCIETY.

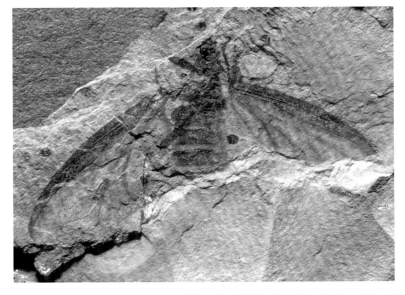

March fly from Eocene lake deposits at Quilchena, British Columbia. Although they originated in the Devonian, insects have achieved incredible diversity over the past 100 million years by co-evolving with flowering plants. RON LONG.

Modern Life

The crisis that ended the Permian also ended the Paleozoic Era and heralded the beginning of the Mesozoic Era. During the Mesozoic, mammals, birds, and flowers evolved; insect life diversified to produce forms that we would be familiar with today; mollusks became an important group; and modern plankton and corals appeared. Life thrived in part because climates remained equitable—warm to hot, and with no ice ages. During the Mesozoic, vertebrate faunas were dominated by diapsid reptiles, including the dinosaurs. In Canada, the first inkling of this dominance comes from marine strata in the Rocky Mountain Foothills of northeastern British Columbia, where the remains of *Shastasaurus* were found. At 21 metres long, this ichthyosaur is the largest marine reptile known.

Canada's oldest-known dinosaur fossils, remains of prosauropods, occur in early Jurassic deposits near Parrsboro, Nova Scotia. But the most stunning Mesozoic fossils of all

A skeleton of *Chasmosaurus*, a ceratopsian dinosaur from Alberta, now on display at the Royal Tyrrell Museum of Palaeontology, Drumheller, Alberta. Some of the finest late Cretaceous dinosaurs have been discovered in rocks of the western Canadian Prairies. COURTESY OF THE ROYAL TYRRELL MUSEUM OF PALAEONTOLOGY.

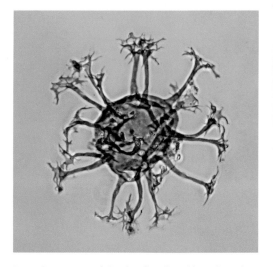

Stained specimen of the dinoflagellate *Oligosphaeridium* from late Cretaceous rocks of a shallow core hole in Baffin Bay. Plankton, such as coccolithophores, diatoms, dinoflagellates, and planktonic foraminifera, evolved during the early Mesozoic. GRAHAM WILLIAMS.

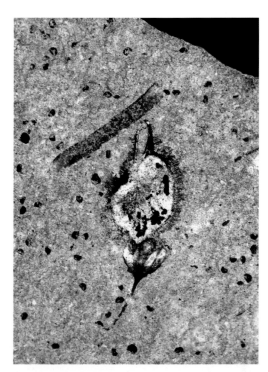

Specimen of the fruit of an elm tree preserved in Eocene lake deposits at Quilchena, British Columbia. Flowering plants (angiosperms) evolved in the early Cretaceous and have since developed a dominant role in land-based ecosystems. ROLF MATHEWES.

come from late Cretaceous strata of the Western Interior Basin. Cretaceous fossils from the region include dinosaurs and other vertebrates, plus a wealth of invertebrates. Plants flourished in the Western Interior Basin between the Jurassic and Paleocene, and their remains accumulated in sufficient quantities in some places to form commercial coal deposits, as at Canmore in Alberta.

Many prominent Mesozoic animal groups, including dinosaurs and ammonites, were killed off by a devastating asteroid impact some 65 million years ago. But this event made possible the ensuing rise of mammals during the Cenozoic. Mammal fossils, especially teeth, occur in deposits on the Prairies, as in the Cypress Hills, and in the Arctic, as in the Haughton impact structure on Devon Island. A remarkable Arctic site is the 45-million-year-old fossil forest on Axel Heiberg Island, with exquisite remains of dawn redwoods and other plants. Surprisingly, what is now Axel Heiberg Island was possibly even farther north at that time than today. Obviously the global Eocene climate was much more equable than at present, although individual species may also have had different tolerances then. Dawn redwoods and diverse flowering plants also occur in early Eocene sediments in the interior of British Columbia.

Global climate remained warm during the Paleocene and Eocene, reaching a peak around the boundary between the two epochs. Since the start of the Oligocene, temperatures have generally cooled, albeit episodically. Grasses increased at the expense of trees, so that during the Miocene the interior of North America may have resembled the present-day African savannah, and was home to an increasing number of grazing animals. Declining temperatures, combined with cycles related to Earth's orbital characteristics (Milankovitch cycles), led to the onset of alternating conditions of glaciation and interglaciation between 3 and 2 million years ago. The Ice Age had arrived, leading to the appearance of cold-adapted animals such as mammoths and mastodons. During the Ice Age, modern Northern Hemisphere biomes such as tundra and boreal forest attained their present form and their boundaries migrated southward and northward in response to the growth and decay of ice sheets.

Canada Today

Today's landscapes and coastlines mark a fleeting moment in the continuing geological evolution of Canada. Precambrian crystalline rocks, long worn down by erosion, are exposed as the Canadian Shield, which forms the northern core of the continent. Much of the periphery of the Shield reflects the latest extent of erosional "back-stripping" of younger rocks that once covered most if not all of it. An exception is the Shield's southeastern margin, where the boundary is marked by a fault, with the resistant rocks of the Grenville Orogen rising above the St. Lawrence Lowlands to the southeast. Crystalline Shield rocks extend as a basement beneath westward- and southward- thickening Phanerozoic strata throughout much of the remainder of North America. The Shield remains important as a source of many of Canada's mineral deposits. Indeed, if you grew up in a town on the Shield, that town was most likely a railroad junction, a forestry centre, or a mining town.

Phanerozoic rocks surrounding the Canadian Shield are generally flat-lying or gently tilted and underlie the platform areas of the Western Interior Plains, the Hudson Bay Lowlands, much of the Arctic Islands, and the Great Lakes and St. Lawrence Lowlands. Where gently tilted rocks that are resistant to erosion reach the surface, they form scarps such as the Niagara Escarpment in southwestern Ontario and the Manitoba Escarpment west of lakes Winnipegosis and Manitoba. Pleistocene ice sheets and flat-lying glacial-lake sediment may have left a cover of till and other deposits, but the basic topography of these areas reflects their pre-Ice Age geology. Soils formed in regions underlain by sedimentary rocks are generally good for agriculture, so regions such as the Prairies, southern Ontario, and the St. Lawrence Lowlands have extensive farmland, essential to stocking our supermarkets and satisfying our appetites.

In this Quaternary Beringian scene, cave lions retreat from a herd of mammoth. SHAMEFUL RETREAT BY GEORGE "RINALDINO" TEICHMANN, COPYRIGHT 1998.

Aerial view of Split Mountain, western Axel Heiberg Island, Nunavut. Resistant, gently folded Cretaceous basalt flows form cliffs several hundred metres high. Beneath the basalts are early Cretaceous mudstone and sandstone. The hills in the foreground are formed of diapirs of yellowish-grey Carboniferous gypsum and salt, whose injection caused the doming up of the Cretaceous rocks. ANDREW MACRAE.

Mount Garibaldi, British Columbia, was built from eruptions over the past 200 thousand years. The volcano last erupted about 12,000 years ago. Canada's only potentially active volcanoes are all in the far west. PAUL ADAM.

Parts of Quebec southeast of the St. Lawrence Lowlands, as well as the Maritime Provinces and Newfoundland, have highland regions that are not unlike those of the northwestern British Isles and Scandinavia. This similarity is not surprising as these regions were once linked by the continuous Appalachian-Caledonian mountain belt prior to the opening of the Atlantic. After the main episodes of accretion, magmatism, and deformation, Carboniferous and Permian sediments blanketed large areas within the former Appalachian Orogen. These undeformed rocks form plateau-like areas on Prince Edward Island and the Gulf of St. Lawrence borderlands in eastern New Brunswick and northern Nova Scotia.

Canada's long, scenic northeastern coast, from Labrador to northern Ellesmere Island, is rimmed by highlands and mountains. The uplifted areas are not associated with an active plate boundary. Instead, they are rift-shoulder highlands originally produced by upwelling in the mantle, which led to the separation of what is now Greenland from Canada during formation of the Labrador-Baffin Seaway from about 100 to 35 million years ago. The seaway is no longer a growing ocean, because Greenland stopped pivoting when it collided with the Arctic Islands. This collision gave rise to the mountains of the northeastern Canadian Arctic.

Perhaps the Canadian Arctic's most characteristic feature is its many islands, which give Canada the longest coastline in the world. A few million years ago, what is today the Arctic Islands was almost all land. During glaciation, ice filled the river valleys and greatly deepened them. The ice sheets also weighed down the land, so that when they melted, the sea flooded the deepened valleys to form the islands we see today.

In the west, the Cordillera remains adjacent to an active plate boundary. This proximity explains, at least in part, why

Farwell Canyon, British Columbia, where glacial outwash and glacial lake sediments have been eroded into cliffs and pillars. PAUL ADAM.

The Rocky Mountains may be Canada's best-known landscape. Part of the range is a UNESCO World Heritage Site. Although they do not reach the heights of the Coast and St. Elias Mountains, the Rocky Mountains attract and inspire many visitors; they were the first recognized and are the best studied of all thrust-and-fold belts. This view is of the Kananaskis Country of Alberta. RON GARNETT / AIRSCAPES.CA.

The Red River meanders across the flat Prairie landscape in this view northward between Letellier and St. Jean Baptiste, Manitoba. GREG BROOKS.

Aerial view of the Canadian Shield at Lefty's Falls on the Grease River, northwest of Stony Rapids, Saskatchewan. RON GARNETT / AIRSCAPES.CA.

western Canada, including the plains east of the Cordillera, is higher than most of central and eastern Canada. Two main belts make up the Cordillera: the Coast and St. Elias mountains to the west and the Rocky Mountains and northern equivalents to the east, with plateau-dominated regions between the two. The height of the St. Elias Mountains reflects a location above an active subduction zone. The majestic beauty of much of the Canadian Cordilleran landscape supports a year-round tourism industry.

How will Canada and the world look in the future. There are so many variables that we can't be sure, but the globe on the next page shows one possible scenario for 50 million years hence, based on an understanding of geological processes.

Final Reflections

Aboriginal peoples were the first to see the land that is now Canada, to appreciate it in ways that we would today think of as "geological", and to utilize its rocks and minerals. We are just beginning to explore what Aboriginal traditions might tell us about their geological insights. Europeans brought with them a sense of commerce and discovery, leading in the later nineteenth and early twentieth centuries to the geological exploration of Canada, especially its resources. Pioneering geological explorers such as William Logan, William and George Dawson, Alexander Murray, and Joseph Tyrrell contributed fundamentally to Canada's growth. Geology witnessed many major advances in the twentieth century, and Canadian scientists made significant contributions. Perhaps foremost among these scientists were John Tuzo Wilson and Lawrence Morley, who belonged

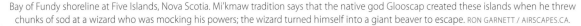

Bay of Fundy shoreline at Five Islands, Nova Scotia. Mi'kmaw tradition says that the native god Glooscap created these islands when he threw chunks of sod at a wizard who was mocking his powers; the wizard turned himself into a giant beaver to escape. RON GARNETT / AIRSCAPES.CA.

Global paleogeography 50 million years into the future. The lighter blue areas represent possible coastal or nearshore areas, darker blue represents deeper ocean waters, and black indicates trenches. RON BLAKEY.

This fall scene of Lake Bouchard in La Mauricie National Park of Canada, Quebec, is typical of much of southern Ontario and Quebec. M. MILLS, COPYRIGHT PARKS CANADA.

to the small coterie of scientists that forged the plate-tectonic revolution in the 1960s, supplying a much-needed unifying theory to geology comparable to evolution in biology.

The success and identity of Canada has been based more than most other nations on its geological resources. The founding of the Geological Survey of Canada in 1842, twenty-five years before Confederation, was aimed at delineating such resources, key to the growth of the colonies. Coal for steam power to drive industry and metals for the fabric of that industry were critical early needs. More recently oil and gas have come to the fore to feed the increasing demand for energy, keeping us warm in winter and cool in summer, for fuelling industries, and sustaining valuable export earnings. Oil sands help to assuage the ever-growing need for energy, but this demand will increasingly be served by new, cleaner sources of energy such as geothermal and solar. Geology is also a principal basis for industries such as agriculture and tourism. Despite its evident importance, Canada's geology is not as well understood or as widely cherished as it should be. And so we have written this book to tell the incredible and important story of the country's four-billion-year geological heritage, a story that continues to evolve with every new scientific paper written about it. Stay tuned. Who knows what fascinating facts will be revealed or critical discoveries brought to light in the coming years.

Aerial view of Signal Hill in St. John's, Newfoundland. The Cabot Tower, at the top of the cliffs at left, is constructed in part from the local Ediacaran rocks that underlie the Avalon Peninsula. The Tower was built in 1898–1900 to mark the 400th anniversary of the landing of John Cabot, so this scene provides a fitting reflection on Canada's intertwined geological and historical heritages. RON GARNETT / AIRSCAPES.CA.

BOX 12 · MILESTONES

THOUSANDS OF YEARS AGO

0.15	Confederation
0.7–0.15	Little Ice Age
0.9	Inuit people move into Arctic Canada
1–0.7	Medieval Warm Period
1.6	agriculture in Ontario
4.5–0.5	Paleo-Eskimo culture in Arctic Islands
5.8–2.6	domestication of plants in eastern North America
6.4–0.15	Head-Smashed-In Buffalo Jump in Alberta in use
8	agriculture begins in North America
8.3–7.8	burials at L'Anse-Amour, Labrador
9.5–7	Hypsithermal interval
11.5	mammoths, mastodons, and other large mammals gone from North America
12	Cordilleran ice cover about same as today; people settle in Quebec
12.6–12.5	people hunting bison in northeastern British Columbia; occupation of Fisher site in Ontario and Debert site in Nova Scotia
13–10	Champlain Sea penetrates along St. Lawrence Lowlands
13–12	Younger Dryas; Laurentide Ice Sheet re-expands and ice caps in Atlantic Canada redevelop
13	Mount Garibaldi, British Columbia, last erupted
13.4	people spread across Prairies
14	ice retreats from Toronto area; Glacial Lake Agassiz begins to form
14.5	human occupation at Monte Verde, Chile
15	people occupy Bluefish Caves in northern Yukon
15.5	people in Texas
16	people reach coast of northeastern Siberia
17	ice retreats from eastern and western continental shelves
19	last glacial maximum
24	Cordilleran Ice Sheet forms
40–30	people reach Siberia
60–50	people migrate from Africa
75	start of latest glaciation
125–75	last (Sangamon) interglaciation; Don Beds deposited in southern Ontario
800 to present	glacial intervals occur every 100,000 years

MILLIONS OF YEARS AGO

2	volcanic mounts Garibaldi, Meager, and Cayley form
2.6–0.8	glacial intervals occur on average every 41,000 years
2.6	Ice Age begins in Northern Hemisphere
3–2	Fraser and Yukon rivers reverse direction
3	Isthmus of Panama forms, fundamentally affecting global climate
3.5	camels live on Ellesmere Island
6–3	glaciers on Greenland and perhaps also on higher parts of Canadian Arctic
6–5	St. Elias Mountains, today Canada's tallest, begin to rise
7	climates start to deteriorate in polar regions
15	warming trend
20–2	Chilcotin and Caribou lavas in British Columbia erupt
30–25	grass becomes abundant and widespread
33–30	development of Antarctic Circumpolar Current leads to glaciation on Antarctica
34	end of sea-floor spreading in Labrador-Baffin Seaway
40	west coast changes from subduction-dominated to alternating subduction and transform plate boundaries
42–40	Olympic Terrane accretes
46	last land link between North America and Eurasia via Greenland severed
50	fossil-rich sediments in fault-bound basins of interior British Columbia deposited
55–50	Eurekan Orogeny
55	Greenland starts to separate from Scandinavia; Pacific Rim Terrane accretes
56	Paleocene-Eocene Thermal Maximum—a short, sharp hot spell
62–56	episode of magmatism heralds separation of Greenland from Canada
65.5	**end-Cretaceous mass-extinction event;** asteroid impact; demise of the dinosaurs
75–70	Bearpaw Sea covers parts of the Western Interior Basin
80–65	age of dinosaurs in Western Interior Basin
90–70	deposits in offshore eastern Canada include chalk
95–85	ancestral Coast Mountains raised and associated magmatic arc begins to move eastward
100	all but marginal parts of Cordillera above sea level; rift basins develop in what was to become the Labrador-Baffin Seaway
105	placental mammals evolve; Wrangellia and Alexander Terrane crunch against continental interior
110–60	thrust-faulting and folding produce the structure of the Rocky Mountains
110	movement of future North America changes from northwesterly to southwesterly
120	seas first invade Western Interior Basin to form Western Interior Seaway
130–95	intermittent magmatism in Canadian Arctic
140	Atlantic Ocean begins to open between Grand Banks and Iberia; earliest angiosperm pollen; earliest marsupials
174	all major Cordilleran terranes now docked somewhere along continental margin; Cache Creek Terrane thrust westward over Stikinia
175	Tethys has largely split Pangea into Laurasia and Gondwana
180	Quesnellia thrust over the continental margin
185	first ash beds on western continental margin since Devonian; sea-floor spreading begins in the central Atlantic
199	Canada's earliest dinosaur bones, in Fundy Basin
200	**end-Triassic mass-extinction event;** volcanic outpourings include the North Mountain Basalt in Maritimes
210–195	magmatic arc on Quesnellia shifts eastward
220	earliest mammals
225	earliest dinosaurs
230–203	Bonanza Arc active on Wrangellia
230–225	Karmutsen Volcanics erupt on Wrangellia

Date	Event
235	rifting presaging formation of Atlantic Ocean begins
250	North America now mostly in Northern Hemisphere
251	**end-Permian mass-extinction event;** Siberian Traps volcanism
270	Ural Ocean closes and Siberia joins Pangea; Sverdrup Seaway becomes Sverdrup Basin
313	Joggins succession in Nova Scotia deposited; earliest-known reptile
315	Euramerica and Protogondwana join to form Pangea
320	Sverdrup Seaway begins to develop
330	rift basins develop on Ellesmere Island
340–260	Alleghanian Orogeny in southern Appalachians as Pangea consolidates
355	earliest tetrapod community on land revealed by trackways at Blue Beach, Nova Scotia
360	Slide Mountain Basin forms, taking slivers off western Euramerica
370	early fish thrived at Miguasha, Quebec
375	**late-Devonian mass-extinction event;** *Tiktaalik* living on Ellesmere Island
380–360	magmatism in Maritimes, for example at Peggys Cove, Nova Scotia
380	earliest trees
385–360	Pearya collides with Euramerica in Ellesmerian Orogeny, producing Franklinian Mountains
390	end of Acadian Orogeny; earliest subduction in Cordilleran Orogen; Meguma begins to dock, setting off the Neoacadian Orogeny; sea transgresses over broad areas of the continent
405	earliest evidence in North America of land animals at Dalhousie Junction, New Brunswick
420	collision of Avalonia with Euramerica; Acadian Orogeny begins; evaporites form in Michigan Basin, including those at Goderich and Windsor, Ontario
425	earliest vascular plants (tracheophytes); earliest lobe-finned fish
440–422	Salinic Orogeny signals amalgamation of Ganderia, Baltica, and Laurentia to form Euramerica
445–444	**end-Ordovician mass-extinction events;** dramatic drop in global temperatures; glaciations in Protogondwana
450	Iapetus closed except for Tetagouche-Exploits Basin and Acadian Seaway; Rheic Ocean growing
460–444	major transgression of the sea over Laurentia; deposits include Tyndall Stone in Manitoba
470	earliest land-plant debris, including spores, appear in sediments
480–445	subduction zone that will lead to docking of Pearya develops in Ural Ocean
480	Taconic Orogeny; Meguma fully detached from Protogondwana
500	Ganderia begins to rift from the Amazonian part of Protogondwana
505	Burgess Shale of British Columbia deposited
510–472	major transgression of the sea over Laurentia
520	Laurentia a discrete continent by this time
521	earliest trilobites
525	earliest fish
540	Avalonia begins to rift from Protogondwana
542	"small shelly" fauna appears
550	Dashwoods Microcontinent separates from Laurentia; earliest animal skeletons (*Cloudina* and *Namacalathus*)
555	earliest trace-fossil evidence of mobile animals (bilaterians)
570	sea-floor spreading begins in Iapetus
580	Ediacara biota at Mistaken Point, Newfoundland
600	sea-floor spreading in the Ural Ocean
720	extensive mafic magmatism, including Franklin Dyke Swarm in Arctic
740–600	tillites indicate intermittent ice ages—Snowball Earth episodes
780–750	Rodinia begins to break up, including rifts that lead to Windermere Seaway and Panthalassa
800	Little Dal fossils, Yukon
810–715	earliest-known biomineralized fossils, Yukon
1000–750	age of the supercontinent Rodinia
1000–900	Amundsen-Mackenzie Mountains Basin succession deposited
1090–1000	Grenvillian Orogeny
1100	development of Midcontinent Rift
1200	fossils from Nunavut testify to oldest-known sexual reproduction
1210	continental collision in Grenville Orogen begins
1360–1290	anorthosites intruded in Labrador
1450	deposition in Belt-Purcell Basin in southern Rocky Mountains
1700	deposits in Nuna's interior supplied from Grenville Orogen by huge rivers
1820	amalgamation of Nuna completed
1850–1000	activity in Grenville Orogen
1850	Sudbury impact event, Ontario
1900	Gunflint chert in Ontario; earliest probable eukaryote fossils in Michigan; Manikewan Ocean (Trans-Hudson Orogen-to-be) separates Superior Craton from lands to west
2000	Labrador Trough succession deposited, including commercial banded iron formations
2450–2000	generally, time of breakup and dispersal of continents
2450–2200	intermittent glaciations
2500–2200	Huronian passive-margin succession of northern Ontario
2500–2400	Great Oxidation Event
2680	amalgamation of Superior Craton completed
2700–2500	acme of banded iron formations
2700	acme of Archean crust formation; plate-tectonic-like processes; controversial geochemical evidence for eukaryotes
2800	stromatolites forming on Superior Craton
3200	earliest microstructures generally acknowledged to be fossil microbes
3500	controversial fossil microbes; geochemical evidence for photosynthesis; earliest stromatolite-like structures
3800	geochemical evidence of life; earliest banded iron formations
3820	end of interval of large impacts on Earth and Moon
4000	"protocontinents" present
4030	age of rock later metamorphosed to form Acasta Gneiss
4400	age of detrital zircons in later Australian sedimentary rock; oceans present
4520	collision of Earth and other planet leads to the origin of the Moon
4525–4475	Iron Catastrophe
4550	origin of rocky planets, including Earth; age of most meteorites
4567	origin of Solar System

Black represents mass-extinction events

Blue represents climatic, oceanic, and atmospheric events

Green represents events in the history of life, including people

Orange represents rock-related events

Red represents astronomical and plate-tectonic events

AUTHORS' AND THEIR AFFILIATIONS

Aïcha Achab, Institut national de la recherche scientifique, Centre Eau Terre Environnement, Québec City, Quebec

Wayne Bamber*, Geological Survey of Canada, Calgary, Alberta

Sandra Barr, Acadia University, Wolfville, Nova Scotia

Alwynne Beaudoin, Royal Alberta Museum, Edmonton, Alberta

Jean Bédard, Geological Survey of Canada, Québec City, Quebec

Andrée Blais-Stephens, Geological Survey of Canada, Ottawa, Ontario

Ron Blakey*, Northern Arizona University, Flagstaff, Arizona

Wouter Bleeker, Geological Survey of Canada, Ottawa, Ontario

Doug Boyce, Geological Survey of Newfoundland and Labrador, St. John's, Newfoundland and Labrador

John Calder, Nova Scotia Department of Natural Resources, Halifax, Nova Scotia

Michel Camus, Health Canada, Ottawa, Ontario

Jean-Bernard Caron, Royal Ontario Museum, Toronto, Ontario

Lesley Chorlton, Geological Survey of Canada, Ottawa, Ontario

John Clague, Simon Fraser University, Burnaby, British Columbia

Thomas Clark*, Ministère des Ressources naturelles du Québec, Québec City, Quebec

Ron Clowes*, University of British Columbia, Vancouver, British Columbia

Maurice Colpron, Yukon Geological Survey, Whitehorse, Yukon

Fabrice Cordey, Université Claude Bernard Lyon, France

David Corrigan, Geological Survey of Canada, Ottawa, Ontario

Pascale Côté, Geological Survey of Canada, Québec City, Quebec

Sonya Dehler Geological Survey of Canada, Dartmouth, Nova Scotia

Keith Dewing, Geological Survey of Canada, Calgary, Alberta

Lawson Dixon*, Geological Survey of Newfoundland and Labrador, St. John's, Newfoundland and Labrador

Lynda Dredge, Geological Survey of Canada (retired), Ottawa, Ontario

Benoît Dubé, Geological Survey of Canada, Ottawa, Ontario

Alejandra Duk-Rodkin, Geological Survey of Canada, Calgary, Alberta

Dave Eberth, Royal Tyrrell Museum, Drumheller, Alberta

Dixon Edwards*, Geological Survey of Alberta, Edmonton, Alberta

Carol Evenchick, Geological Survey of Canada, Vancouver, British Columbia

Nick Eyles, University of Toronto Scarborough, Toronto, Ontario

Gordon Fader*, Geological Survey of Canada, Dartmouth, Nova Scotia

Howard Falcon-Lang, Royal Holloway College, London, U.K.

Rob Fensome, Geological Survey of Canada, Dartmouth, Nova Scotia

Martin Fowler, Applied Petroleum Technology (Canada), Calgary, Alberta

Paul Fraser, Geological Survey of Canada, Dartmouth, Nova Scotia

David Gardner, University of Ottawa, Ottawa, Ontario

Pat Gensel, University of North Carolina, Chapel Hill, North Carolina, U.S.A.

Martin Gibling, Dalhousie University, Halifax, Nova Scotia

Wayne Goodfellow*, Geological Survey of Canada, Ottawa, Ontario

John Gosse, Dalhousie University, Halifax, Nova Scotia

Stephen Grasby, Geological Survey of Canada, Calgary, Alberta

Fran Haidl, Saskatchewan Ministry of the Economy, Regina, Saskatchewan

Christopher Harrison, Geological Survey of Canada, Ottawa, Ontario

Catherine Hickson, University of British Columbia, Vancouver, British Columbia

Phil Hill, Geological Survey of Canada, Pat Bay, British Columbia

Andrew Hynes, McGill University, Montréal, Quebec

Charles Jefferson, Geological Survey of Canada, Ottawa, Ontario

Pierre Jutras, Saint Mary's University, Halifax, Nova Scotia

Ian Knight, Geological Survey of Newfoundland and Labrador, St. John's, Newfoundland and Labrador

Andrew Knoll, Harvard University, Cambridge, Massachusetts, USA

Denis Lavoie, Geological Survey of Canada, Québec City, Quebec

Michael Lazorek, Chevron Canada Resources, Calgary, Alberta

Dale Leckie, Nexen ULC, Calgary, Alberta

Alain Leclair, Geological Survey of Canada, Ottawa, Ontario

Robert Ledoux*, Université Laval, Québec City, Quebec

Bernard Long, Institut national de la recherche scientifique, Centre Eau Terre Environnement, Québec, Quebec

Roger MacQueen*, Geological Survey of Canada, Calgary, Alberta

Andrew MacRae, Saint Mary's University, Halifax, Nova Scotia

Michel Malo, Institut national de la recherche scientifique, Centre Eau Terre Environnement, Québec City, Quebec

Rolf Mathewes, Simon Fraser University, Burnaby, British Columbia

Mike Melchin, St. Francis Xavier University, Antigonish, Nova Scotia

Andrew Miall, University of Toronto, Toronto, Ontario

Randall Miller, New Brunswick Museum, Saint John, New Brunswick

Jim Monger*, Geological Survey of Canada, Vancouver, British Columbia

Alan Morgan*, University of Waterloo, Waterloo, Ontario

Brendan Murphy, St. Francis Xavier University, Antigonish, Nova Scotia

Reuben Murphy, Canadian Nuclear Safety Commission, Ottawa, Ontario

Leo Nadeau, Geological Survey of Canada, Québec City, Quebec

Guy Narbonne, Queen's University, Kingston, Ontario

Walter Nassichuck†, Geological Survey of Canada, Calgary, Alberta

JoAnne Nelson, British Columbia Geological Survey, Victoria, British Columbia

Godfrey Nowlan*, Geological Survey of Canada, Calgary, Alberta

Andrew Okulitch*, Geological Survey of Canada, Vancouver, British Columbia

Paul Olson, Columbia University, New York, New York

John Percival, Geological Survey of Canada, Ottawa, Ontario

Terry Poulton, Geological Survey of Canada, Calgary, Alberta

Brian Pratt, University of Saskatchewan, Saskatoon, Saskatchewan

Raymond Price*, Queen's University, Kingston, Ontario

Gilbert Prichonnet, l'Université du Québec à Montréal, Montréal, Quebec

Rob Rainbird, Geological Survey of Canada, Ottawa, Ontario

Pat Rasmussen, Health Canada, Ottawa, Ontario

David Rudkin, Royal Ontario Museum, Toronto, Ontario

Bruce Ryan, Geological Survey of Newfoundland and Labrador, St. John's, Newfoundland and Labrador

Natalia Rybczynski, Canadian Museum of Nature, Ottawa, Ontario

Bruce Sanford*, Geological Survey of Canada, Ottawa, Ontario

Martine Savard, Geological Survey of Canada, Québec City, Quebec

John Storer, Sechelt, British Columbia

Art Sweet*, Geological Survey of Canada, Calgary, Alberta

Ann Therriault, Natural Resources Canada, Ottawa, Ontario

François Therrien, Royal Tyrrell Museum, Drumheller, Alberta

Pierrette Tremblay, *Elements* Magazine, Québec City, Quebec

Cees van Staal, Geological Survey of Canada, Vancouver, British Columbia

Hans Weilens†, Geological Survey of Canada, Dartmouth, Nova Scotia

Graham Williams, Geological Survey of Canada, Dartmouth, Nova Scotia

Marie-Claude Williamson, Geological Survey of Canada, Ottawa, Ontario

Chris Yorath*, Geological Survey of Canada, Pat Bay, British Columbia

Graham Young, The Manitoba Museum, Winnipeg, Manitoba

Darla Zelenitsky, University of Calgary, Calgary, Alberta

John-Paul Zonneveld, University of Alberta, Edmonton, Alberta

*retired or emeritus, †deceased

ACKNOWLEDGEMENTS

Hundreds of people have been involved in one way or another with the production of Four Billion Years and Counting (FBY), and many have gone above and beyond the call of duty. First and foremost, we must thank the Geological Survey of Canada (Natural Resources Canada) for generously allowing many of those involved in the project to use their time and facilities. Without this fundamental support, FBY would not have been possible. We also sincerely thank other sponsors and supporters for their generosity and encouragement. The idea for the book was conceived by two of us (RAF and GLW) after being inspired by a talk by John Percival at Dalhousie University, Halifax, in March, 2003; and was put on an official footing, as part of Canada's contribution to the International Year of Planet Earth, through the efforts of Richard Grieve, then Chief Scientist at the Geological Survey of Canada, and Irwin Itzkovitch, then an Assistant Deputy Minister at Natural Resources Canada. In addition to members of the current Editorial Board, Simon Hanmer and Bob Turner served on the initial steering committee for the project. In the acknowledgements that follow, we have inevitably but inadvertently omitted some names, for which we apologize in advance.

Many of the key contributors and all sponsors to the project, as well as authors of chapters and boxes are cited at the front of the book. Credits for individual graphics and non-photographic artworks are provided in the individual captions except where the graphic was provided by one of the chapter/box authors. Jennifer Bates, David Frobel, Al Sangster, Scott Swinden, Bob Taylor, and Marcos Zentilli provided additional important advice on graphics. Nelly Koziel carried out multifarious tasks associated with various aspects of the manuscript, and we are very grateful to her; and to Lorraine Thompson for unyielding and invaluable support and encouragement. Credits for individual photos are provided in the individual captions.

Our story has been enhanced by artwork courtesy of several artists and institutions. In chapters 11 and 20 and Box 8 are several oil paintings by George "Rinaldino" Teichmann (Rinaldino Art Studios—www.iceagebeasts.com). These paintings are reproduced with the permission of the artist or the Government of Yukon, Department of Tourism and Culture—Yukon Beringia Interpretive Centre, as specified in the credits to individual paintings. We thank Chris Caldwell, Garnet Muething, and Liz Kosters for facilitating our use of these paintings. Scattered throughout the book are paintings by Judi Pennanen, commissioned variously by the New Brunswick Museum and the Atlantic Geoscience Society; we are grateful to Judi and the two organizations for allowing us to reproduce these paintings. We also appreciative the artworks provided by Marianne Collins; Rolf Matthewes; François Miville-Deschênes and Miguasha National Park; the Royal Tyrrell Museum; and Alex Tirabasso, Julius Csotonyi, and the Canadian Museum of Nature.

The following served as reviewers of various parts of the text or have provided (non-photographic) materials, help, or advice (other than as co-authors): Lyn Anglin, Ken Ashton, Esther Asselin, Jan Aylesworth, Diane Baldwin, Wayne Bamber, Charlie Bank, Jim Basinger, David Batten, Benoit Beauchamp, Jean Bédard, Marc Bélanger, Bob Berner, Peter Bobrowsky, Melissa Bowerman, Jack Brink, Greg Brooks, Michel Caillier, Calvin Campbell, Donald Canfield, Colin Card, Neil Carleton, Bob Cathro, Norm Catto, Joyce Chew, Tom Clark, Barrie Clarke, Barb Cloutier, Louise Corriveau, Pascale Côté, Lynn Dafoe, Eric Damkjar, Steve Daniel, Mark Deptuck, Linda Deschenes, Kevin DesRoches, Jim Dixon, Jon Dudley, Norman Duke, Howard Falcon-Lang, Mark Fenton, Claude Fortin, Norah Foy, David Frobel, Ursula Frobel, Al Galley, Martin Gibling, Susan Goods, John Gosse, Ian Graham, Bob Grantham, Milton Graves, Dale Gregory, Jim Haggart, Simon Hanmer, Jeff Harris, Catherine Hickson, Steve Holysh, Lesley Hymers, Jennifer Jackson, Mike Jackson, Christopher Jass, Chris Jauer, Dick Jennings, Heather Johnson, Kimberley Johnston, Paul Johnston, Paul Karrow, Todd Keith, Russell Knight, Andy Knoll, Elisabeth Kosters, Christine Kulyk, Michelle Landreville, Dale Leckie, Mike Lewis, David Liverman, Darrel Long, Rolf Ludvigsen, Rosemary MacKenzie, Rob MacNaughton, Roger Macqueen, Richard Malcolm, Gavin Manson, Gwen Martin, Brian Matthewes, Charles Maurice, Francine McCarthy, Jock McCracken, Murray Metherall, Andrew Miall, Eldridge Moores, Alan Morgan, Leo Nadeau, Ward Neale, Gordon Oakey, Paul Olsen, Neil Opdyke, Mike Orchard, Georgia Pe-Piper, David Piper, Alain Plouffe, Patrick Potter, Mathilde Renaud, Malcolm Richard, Barry Richards, John Riva, Christine Rivard, Blyth Robertson, Kevin Root, Beverly Ross, Gerry Ross, Pierre-Simon Ross, Barry Ryan, Bruce Ryan, Matt Salisbury, Al Sangster, Hamed Sani, Kevin Seymour, Arnet Sheppard, John Shimeld, Paul Smith, Dan Smythe, Ian Spooner, Ralph Stea, Denis St-Onge, Bert Struik, Scott Swinden, Bob Taylor, Martin Teitz, Jessica Tomkins, Eileen van der Flier-Keller, Mike Villeneuve, Christy Vodden, Miriam Vos-Guenter, Bob Wagner, John Waldron, Dan Walmsley, Tracy Webb, Steve Westrop, James White, Hans Wielens, Jeanette Wielens, Erica Williams, Graham Wilson, Ron Yehia, and Marcos Zentilli.

The following individuals provided photographs (whether included or not) or otherwise helped in respect to photos: Paul Adam, Bill Arnott, David Ashe, Chris Bagley, George Bardeggia, Sandra Barr, Jim Basinger, Jennifer Bates, Benoit Beauchamp, Alwynne Beaudoin, Jean Bédard, Pierre Bedard, Trevor Bell, Philippe Belley, Kelly Bentham, Ruth Bezys, Dave Birrell, Wouter Bleeker, Roberta Bondar, Doug Boyce, Sandy Briggs, Henk Brinkhuis, Greg Brooks, Dave Brown, Paul Budkewitsh, Michael Burzynski, Susan Butorak, Nick Butterfield, John Calder, Linda Campbell, Pat Carey, Neil Carleton, Jean-Bernard Caron, Norm Catto, France Charest, Joyce Chew, Mike Churkin, John Clague, Thomas Clark, Barrie Clarke, Phoebe Cohen, Maurice Colpron, Paul Copper, David Corrigan, Louise Corriveau, Art Cosgrove, Pascale Côté, Janet Couper, Brian Cousens, Isabelle Coutand, Nick Culshaw, Ted Daeschler, Lynn Dafoe, Terry Danyluk, Neil Davies, Sonya Dehler, Steve Deschênes, Jon Devaney, Keith Dewing, Lawson Dickson, Elio Dolente, Hans Dommasch, Al Donaldson, Howard Donohoe, Jean Dougherty, Marten Douma, Stephanie Douma, Lynda Dredge, Benoît Dubé, Thomas Duffett, Mark Duffy, Alejandra Duk-Rodkin, Dave Eberth, Dixon Edwards, Leslie Eliuk, Clive Elson, Ashton Embry, Saskia Erdmann, Carol Evenchick, Nick Eyles, Gordon Fader, Hendrick Falck, Howard Falcon-Lang, Tim Fedak, Alexandra Fensome, Robert Fensome, Laing Ferguson, Fil Ferri, Denis Finnin, Timothy Fisher, Don Forbes, Martin Fowler, Kim Franklin, Christian Fraser, David Frobel, Bob Fulton, Les Fyffe, Andy Fyon, Hu Gabrielse, Ben Gadd, Len Gal, David Gardner, Wayne Garland, Marilyn Garnett, Ron Garnett, Viki Gaul, Patricia Gensel, Eldon George, Martin Gibling, Philip Giles, Wayne Goodfellow, Susan Goods, Steve Gordey, Angela Gordon, John Gosse, Jean Goutier, Bob Grantham, Steve Grasby, Dale Gregory, Melissa Grey, Sandy Grist, Jim Haggart, Fran Haidl, Russell Hall, Terry Halverson, Linda Ham, Warren Hamilton, Phil Hammer, Diane Hanano, Kirk Hancock, Jacob Hanley, Christopher Harrison, Charles Henderson, Catherine Hickson, Phil Hill, Justin Hobson, Hans Hofmann, Zdenek Hora, François Houle, Crystal Huscroft, Steven Ings, Steve Irwin, Fenton Isenor, Clair Israelson, Ruth Jackson, Noel James, Charles Jefferson, Susan Johnson, Rich Jones, Pierre Jutras, David Keighley, Francis Kelly, Bernadette Kennedy, Daniel Kerr, Geoff Kitching, Rod Klassen, Andrew Knoll, Dan Kohlruss, Elisabeth Kosters, Christine Kulyk, Jean-Yves Labbé, Michelle Landreville, Larry Lane, Walter Lanz, Colin Laroque, Denis Lavoie, Alain Leclair, Robert Ledoux, Martin Legault, Dave Leonard, Ben Lepage, Riccardo Levi-Setti, Martin Lipman, Ted Little,

David Liverman, Greg Locke, Darrel Long, Ron Long, Lucas Lourens, Mike MacDonald, Lori MacLean, Bill MacMillan, Robert MacNaughton, Andrew MacRae, Michel Malo, Henrietta Mann, Gavin Manson, Anne Marceau, Christian Marcussen, Gwen Martin, John Mason, Jennifer Mathewes, Rolf Mathewes, Francine McCarthy, Joanne McCloskey, Krista McCuish, Steve McCutcheon, Greg McHone, Margo McMechan, Mike Melchin, Brian Menounos, David Mercer, Andrew Miall, Yves Michaud, Derrick Midwinter, Randall Miller, Angèle Miron, Larry Moncza, Jim Monger, Catherine Mooney, Alan Morgan, David Mosher, Grant Mossop, Tom Muir, Jennifer Mullane, Ron Mussieux, Peter Mustard, Kim Mysyk, Léopold Nadeau, Guy Narbonne, JoAnne Nelson, Godfrey Nowlan, Gordon Oakey, Sean O'Brien, Paul Olsen, Luke Ootes, Gerry Osborn, Miyuki Otomo, Anthony Pace, Etta Parker, Mike Parkhill, Mike Parsons, Jennifer Pell, John Percival, Julie Peressini, Ron Pickerill, Laszlo Podor, Patrick Potter, Terry Poulton, Brian Pratt, Ray Price, Gilbert Prichonnet, Toon Pronk, Rob Raeside, Rob Rainbird, Johanne Ranger, Pat Rasmussen, Nicole Rayner, Barry Richards, Cindy Riediger, Diann Robar, André Rochon, Kevin Root, David Rudkin, Alan Ruffman, Brian Rutley, Bruce Ryan, Natalia Rybczynski, Sue Sabrowski, Osman Salad Hersi, Bruce Sanford, Wayne Sawchuck, Louis Schilder, Donna Schreiner, Steve Scott, John Scurlock, Alan Seaman, Grady Semmens, Ross Sherlock, John Shimeld, Martin Simard, David Sinclair, Deborah Skilliter, Kathleen Smart, Norman Smith, Kermit Smyth, Steve Solomon, Ian Spooner, Terry Spurgeon, Paul Starr, Ralph Stea, Alexandra Steffen, Pete Stelling, Glen Stockmal, Marc St. Onge, John Storer, Arthur Sweet, Jean-Pierre Sylvestre, Bob Taylor, Nicolas Teichrob, Franz Tessensohn, Hans Thater, Ann Therriault, François Therrien, Lorraine Thompson, Owen Thompson, Sheridan Thompson, Vickie Thorn, Brian Todd, Pierrette Tremblay, Hans Trettin, Verena Tunnicliffe, Bob Turner, Suzanne Twelker, Helen Tyson, Rod Tyson, Gilbert van Ryckevorsel, Keith Vaughan, Pierre Verpaelst, Ron Verrall, Mike Villeneuve, Christy Vodden, Miriam Vos-Guenter, Grant Wach, John Waldron, John William Webb, Tracy Webb, Dustin Whalen, Chris White, Heinz Wiele, Hans Wielens, Graham Williams, Nicole Williamson, Alex Wilson, Graham Wilson, Reg Wilson, Floyd Wist, Margaret Wist, Natasha Wodicka, Glenn Woodsworth, Matthew Wright, Jane Wynne, Christine Yankou, Chris Yorath, Graham Young, Mike Young, and Marcos Zentilli. Helen Tyson of Tysons' Fine Minerals generously photographed a selection of minerals for this book; the specimens are from the private collection of Rod and Helen Tyson.

The following organizations provided photographs or otherwise helped with regard to acquisition of photographs (including those not used in the book): Academy of Natural Sciences of Philadelphia; Alberta Parks; Atlantic Geoscience Society; British Columbia Ministry of Environment; Canada Centre for Remote Sensing; Canadian Museum of Nature; Fisheries and Oceans Canada; CIM Magazine; DEVCO; Encana; Fundy Geological Museum; Geological Survey of Canada; Glenbow Archives; Government of Alberta; Hubble Site at the National Aeronautics and Space Administration (NASA); Integrated Ocean Drilling Program; Joggins Fossil Institute; Lithoprobe; Lorita (Lomonosov Ridge Test of Appurtenance) Field Project, Geological Survey of Canada UNCLOS Program; United States National Oceanic and Atmospheric Administration; Manitoba Museum; Miguasha National Park; Mining Association of Nova Scotia; Ministère des ressources naturelles du Québec; Natural Resources Canada; New Brunswick Museum; Nexen Inc.; Nova Scotia Museum; Parks Canada; Peakfinder; Peregrine Diamonds; Potash Corporation of Saskatchewan; Royal Alberta Museum; Royal Ontario Museum; Royal Saskatchewan Museum; Royal Tyrrell Museum; Tourism and Parks New Brunswick; United States Geological Survey; and Victoria University Library, Toronto.

Thanks to all at Nimbus for their help and encouragement, including Patrick Murphy and Whitney Moran.

The French edition was made possible thanks to francophone co-authors, who translated the chapters and boxes with which they were involved. Additional translations were made by Open Text Corporation and Traduction Géosphère. The help of Jean-Luc Charbonneau and Jean-Yves Labbé during the French translation process is much appreciated. We thank the following for reviewing the French texts: Andrea Amortégui, Esther Asselin, Jean Bédard, Jean-Francois Blais, Sébastien Castonguay, Jean-Christophe Caron, Anne Cremazy, Pascale Côté, Claude Fortin, Dominique Gauvreau, Genevieve Girard, Charles Gosselin, Kathleen Lauzière, Richard Martel, Ghismond Martineau, Joanne Nadeau, Didier Perret, Daniele Pinti, Christine Rivard, Pierre-Simon Ross, Robert Thériault, N'Golo Togola, Natasha Wodicka, and Régis Xhardé. Mathilde Renaud exhaustively reviewed and harmonized all French texts for the French edition, for which we are greatly indebted.

The following provided feedback from provincial or territorial perspectives: Diane Baldwin (Northwest Territories), Maurice Colpron (Yukon), Les Fyffe (New Brunswick), Fran Haidl (Saskatchewan), Linda Ham (Nunavut), Jean-Yves Labbé (Quebec), David Liverman (Newfoundland and Labrador), and JoAnne Nelson (British Columbia).

REFERENCES CITED IN CAPTIONS

Amante, C., and Eakins, B. W. 2009. ETOPO1 1 Arc-Minute Global Relief Model: procedures, data sources and analysis. *NOAA Technical Memorandum NESDIS NGDC-24*, 19 p., March 2009. (www.ngdc.noaa.gov/mgg/global/global.html)

Atlantic Geoscience Society. 2001. *The Last Billion Years: A Geological History of the Maritime Provinces of Canada*. Halifax: Nimbus Publishing and Atlantic Geoscience Society, 212 p.

Berner, R. A. 2004. *The Phanerozoic Carbon Cycle*. Oxford: Oxford University Press, 150 p.

Clague, J., and Turner, B. 2003. *Vancouver, City on the Edge: Living with a Dynamic Geological Landscape*. Vancouver: Tricouni Press, 192 p.

Clague, J. J., Luternauer, J. L., and Hebda, R. J. 1983. Sedimentary environments and postglacial history of the Fraser Delta and lower Fraser Valley, British Columbia. *Canadian Journal of Earth Sciences*, 20 (8): 1314–1326.

Clague, J. J., Munro, A., and Murty, T. 2003. Tsunami hazard and risk in Canada. *Natural Hazards*, 28: 433–461.

Clague, J. J., Yorath, C., Franklin, R., and Turner, B. 2006. *At Risk: Earthquakes and Tsunamis on the West Coast*. Vancouver: Tricouni Press, 200 p.

Cloutier, R. 2001. *Miguasha Park: from Water to Land*. Québec: MNH Publications Inc., 141 p.

Colman-Sadd, S., and Scott, S. 2003. *Newfoundland and Labrador: Traveller's Guide to the Geology and Guidebook to Stops of Interest*. Geological Association of Canada and Geological Survey of Newfoundland and Labrador, 97 p. and map.

Coté, P., Achab, A., and Michaud, Y. 2001. Geoscape Québec: an extraordinary geological heritage. *Geological Survey of Canada, Miscellaneous Report* 76, 2 sheets.

Doyle, V. C., and Steele, K. G. (compilers). 2003. Geoscape Toronto. *Geological Survey of Canada, Miscellaneous Report* 83, or *Ontario Geological Survey*, Poster 6.

Dyke, A., Moore, A., and Robertson, L. 2003. Deglaciation of North America. *Geological Survey of Canada Open File* 1574.

Eckstrand, O. R., and Hulbert, L. 2007. Magmatic nickel-copper-platinum group element deposits. In: Mineral deposits of Canada: a synthesis of major deposit-types, district metallogeny, the evolution of geological provinces, and exploration methods. Ed.: W. D Goodfellow. *Special Publication, Mineral Deposits Division, Geological Association of Canada*. 5: 205–222.

Engebretson, D. C., Cox, A., and Gordon, R. G. 1985. Relative motions between continental and oceanic plates in the Pacific Basin. *Geological Society of America Special Paper* 206, 59 p.

Gibson, J., Nedelcu, S., and Budkewitsch, P. 2010. *A Complete Orthorectified Landsat-7 Mosaic of the Canadian Arctic Archipelago*. www.nrcan.gc.ca/earth-sciences/land-surface-vegetation/land-cover/north-american-landcover/9152.

Grieve, R. A. F. 2006. *Impact Structures of Canada*. St. John's, Newfoundland: Geological Association of Canada, 210 p.

Grotzinger, J. P., Watters, W. A., and Knoll, A. H. 2000. Calcified metazoans in thrombolite-stromatolite reefs of the terminal Proterozoic Nama Group, Namibia. *Paleobiology*, 26 (3): 334–359.

Hein, F. J., and Nowlan, G. S. 1998. Regional sedimentology, conodont biostratigraphy and correlation of Middle Cambrian–Lower Ordovician(?) strata of the "Finnegan" and Deadwood formations, Alberta subsurface, Western Canada Sedimentary Basin. *Bulletin of Canadian Petroleum Geology*, 46 (2): 166–188.

Ilyin, A. V. 2009. Neoproterozoic banded iron formations. *Lithology and Mineral Resources*, 44 (1): 87–95.

Karrow, P. F., McAndrews, J. H., Miller, B. B., Morgan, A. V., Seymour, K. L., and White, O. L. 2001. Illinoian to late Wisconsinan stratigraphy at Woodbridge, Ontario. *Canadian Journal of Earth Sciences*, 38: 921–942.

Keller, G. 2005. Impacts, volcanism and mass extinction: random coincidence or cause and effect? *Australian Journal of Earth Sciences*, 52: 725–757.

Kious W. J., and Tilling, R. I. 1996. *This Dynamic Earth: the Story of Plate Tectonics*. Denver: US Department of the Interior, US Geological Survey. pubs.usgs.gov/gip/dynamic/dynamic.pdf

Knoll, A. H. 2003. *Life on a Young Planet: The First Three Billion Years of Evolution on Earth*. Princeton, New Jersey: Princeton University Press, 277 pp.

Leidy, J. 1854. On *Bathygnathus borealis*, an extinct saurian of the New Red Sandstone of Prince Edward Island. *Journal of the Academy of Natural Sciences of Philadelphia, Second Series*, 2 (4): 327–330 (pl. 33).

Li, Z. X., Bogdanova, S. V., Collins, A. S., et al. 2008. Assembly, configuration and break-up history of Rodinia: a synthesis. *Precambrian Research*, 160: 179–210.

Lisiecki, L., and Raymo, M. E. 2005. A Pliocene–Pleistocene stack of 57 globally distributed benthic $\delta^{18}O$ records. *Paleoceanography*, 20, PA1003 .

Lyons, T. W., and Reinhard, C. T. 2009. Early Earth: oxygen for heavy metal fans. *Nature*, 461: 179–181.

Mathewes, W. H. (compiler). 1986. Physiography of the Canadian Cordillera. *Geological Survey of Canada Map* 1701A.

Miall, A. D., Balkwill, H. R., and McCracken, J. 2008. The Atlantic margin basins of North America. In: *The Sedimentary Basins of the United States and Canada*. Ed.: A. D. Miall. Elsevier, p. 474–504.

Monger, J. W. H., and Berg, H. C. 1987. Lithotectonic terrane map of western Canada and southeastern Alaska. *US Geological Survey Miscellaneous Field Studies Map* 1874-B.

Niklas, K. J., Tiffney, B. H., and Knoll, A. H. 1985. Patterns in vascular land plant diversification: an analysis at the species level. In: *Phanerozoic Diversity Patterns*. Ed: J. W. Valentine. Princeton, New Jersey: Princeton University Press, 97–128.

Peltier, W. R. 2004. Global glacial isostasy and the surface of the Ice-Age Earth: the ICE-5G (VM2) Model and GRACE. *Annual Review of Earth and Planetary Sciences*, 32: 111–149.

Poulton, T., Neumar, T., Osborn, G., et al. 2002. Geoscape Calgary. *Geological Survey of Canada, Miscellaneous Report* 72.

Rasmussen, B., Fletcher, I. R., Bekker, A., et al. 2012. Deposition of 1.88-billion-year-old iron formations as a consequence of rapid crustal growth. *Nature*, 484: 498–501.

Read, P. B., Woodsworth, G. J., Greenwood, H. J, et al. 1991. Metamorphic map of the Canadian Cordillera. *Geological Survey of Canada Map* 1714A, scale 1:2,000,000.

Shaw, J., Taylor, R. B., Forbes, D. L., et al. 1998. Sensitivity of the coasts of Canada to sea-level rise. *Geological Survey of Canada Bulletin* 505, 79 p. geoscan.nrcan.gc.ca/starweb/geoscan/servlet.starweb?path=geoscan/downloade.web&search1=R=210075

Terasmae, J. 1960. Contribution to Canadian palynology no. 2. *Geological Survey of Canada, Bulletin* 56, 41 p.

Trenhaile, A. 1998. *Geomorphology: A Canadian Perspective*. Don Mills: Oxford University Press, 340 p.

Trettin, H. P.1991. Chapter 18: Middle and late Tertiary Tectonic and Physiographic Developments. In: Geology of the Innuitian Orogen and Arctic Platform of Canada and Greenland. Ed.: H. P. Trettin. *Geological Survey of Canada, Geology of Canada*, 3: 493–496.

Turner, R. J. W., Franklin, R. G., Grasby, S. E., and Nowlan, G. S. 2005a. Bow River Basin Waterscape: protecting and conserving the shared waters of our Bow River. *Geological Survey of Canada, Miscellaneous Report* 90.

Turner, R. J. W., Franklin, R. G., Journeay, J. M., et al. 2005b, Waterscape Bowen Island: water for our island Community. *Geological Survey of Canada, Miscellaneous Report* 88.

Van Staal, C. R., and Barr, S. M. 2012. Chapter 2: Lithospheric architecture and tectonic evolution of the Canadian Appalachians and associated Atlantic margin. In: Tectonic styles in Canada: the Lithoprobe perspective. Ed.: J. A. Percival, F. A. Cook, and R. M. Clowes. *Geological Association of Canada Special Paper* 49: 41–95.

Walter, M. R., and Heys, G. R. 1985. Links between the rise of metazoa and the decline of stromatolites. *Precambrian Research*, 29: 149–174.

Weissenberger, J. A. W., and Potma, K. 2001. The Devonian of western Canada—aspects of a petroleum system: introduction. *Bulletin of Canadian Petroleum Geology*, 49: 1–6.

Wheeler, J. O, Hoffman, P. F, Card, K. D., et al. 1996. Geological map of Canada/Carte géologique du Canada. *Geological Survey of Canada Map* 1860A.

Wilson, J., and Clowes, R. 2009. *Ghost Mountains and Vanished Oceans: North America from Birth to Middle Age*. Toronto: Key Porter, 248 p.

Yorath , C. J., Dixon, D. F, Gabrielse, H., et al. 1992. Chapter 9: Upper Jurassic to Paleogene assemblages. In: Geology of the Cordilleran Orogen in Canada. Ed.: H. Gabrielse, and C. J. Yorath. *Geological Survey of Canada, Geology of Canada*, 4: 329–371.

Yorath, C. J., and Gadd, B. 1995 *Of Rocks, Mountains and Jasper: A Visitor's Guide to the Geology of Jasper National Park*. Geological Survey of Canada and Dundurn. 170 p.

INDEX

Abbreviations: AB = Alberta, BC = British Columbia, MB = Manitoba, NB = New Brunswick, NL = Newfoundland and Labrador, NPC = National Park of Canada, NPRC = National Park Reserve of Canada, NS = Nova Scotia, NUN = Nunavut, NWT = Northwest Territories, ONT = Ontario, PEI = Prince Edward Island, QC = Quebec, SK = Saskatchewan, YK = Yukon.

Page numbers in bold indicate where a particular term is defined or key aspects are explained.

A

A horizon (soil) **14**
aardvarks 190
abandoned gravel pits 341
abandoned mines 341
Abenaki Bank **171**
Abitibi, QC 341
Abitibi Greenstone Belt 79, **81**, 84, 249
Abitibi mining district 249
Aboriginal hunters 237
Aboriginal languages and customs 161
Aboriginal oral traditions 308
Aboriginal peoples and coal 264
Aboriginal peoples 234, 235, 239, 258, 293, 372
Aboriginal peoples, diet of 348
Aboriginal traditions 372
Aboriginal use of bitumen 272
abrasion **220**
absolute ages **41–42**
Acadian dykes 318
Acadian Mountains 136, 143, 366
Acadian Orogeny ... **133**–134, 141, 357–359, 375
Acadian Seaway109–**110**, 115–116, 133–134, 357, 375
Acadians 318
Acasta Gneiss **77**–**78**, 80, 82, 375
accretion **33**, 36, 370
accretion of island arcs in Cordillera 169
accretionary complexes 157, 165
accretionary wedges 33–34, 248
Acernaspis boltoni 40
Acernaspis copperi 40
Acernaspis orestes 40
acid mine drainage 342
acid mine waters 342
acid rain 104, 265, 339, **343**
acid rain and prevailing winds 344
acid-rock drainage 342
acidic groundwater 302, 342
acidic mine drainage 342
acidic precipitation 344
acidic rainwater 302
acidic runoff 342
acoustic properties of rock 47
acoustic waves **47**
acritarchs **107**
actinolite 16, 79
active layer 225, 244
active plate boundary 370
active subduction zone 372
Adair Marble 285
Adams Sound, NUN 94
Adirondack Highlands, USA 108
aerosols 241
Africa 24–25, 162, 234, 242, 359, 361, 374
African margin of Protogondwana 110
African part of Gondwana 171
African Plate 29
African savannah 369
African savannah, as analog 214
afrotheres 191–190

afrotheres, African origin of 190
Agassiz Ice Cap 236
Agassiz, Louis 219
agate 20–**21**
agate, Aboriginal use 258
aggregate 182, **280**, 290, **291**–292, 349
aggregate, uses of 291
agricultural techniques 239
agricultural zones 243
agriculture 239, 244, 369, 373–374
Aillik Arc, NL 91
air bubbles 227, 242–243
air circulation 65
air guns 47
air pollution 343
airborne arsenic 343
airborne hazards 347
airborne mercury 343
airborne particles, problems caused by ... 351
airborne particles, sources of 352
airborne pollutants 343
Aklavik, NWT 316
alabaster **97**
Alaska .. 37, 111, 115, 139, 155, 195, 231, 234, 238, 247, 363
Alaska earthquake of 1964 326
Alaska Range 209, 211
Alaskan Inuit 239
albedo **240**
Albert County, NB 270
Albert Mines, NB 150
Albert Street, Ottawa, ONT 287
Alberta 10–11, 13–14, 24, 30, 32, 36–37, 51–52, 59, 69–71, 94–95, 103, 113–114, 118, 137–138, 141, 145–148, 159–163, 169–172, 178–182, 184–188, 205, 210, 214, 219–224, 229–231, 233, 235–239, 243, 263–265, 267–273, 276, 282–284, 287, 291–292, 298–299, 303–304, 306, 320–321, 329, 332–333, 339, 342, 344–345, 349–350, 355–356, 359, 361–363, 365, 368–369, 371–372, 374
Alberta Energy and Resources Conservation Board 269
Alberta Legislature Building, Edmonton, AB 284, 287
Alberta Research Council 272
Alberta, flooding in 322
Alberta-Saskatchewan border 182
Albertonia 164
Albertosaurus 51, 185
Albertosaurus libratus 52
alder 204, 212, 229
Alert, NUN 193
Aleutian Arc 166
Aleutian Subduction Zone . 28, 197–198, 208
Aleutian-Wrangell Magmatic Arc 208
Alexander Archipelago, Alaska 139
Alexander Terrane ... 115, 133, 139–140, 142, 152–155, 156, 166, 172–173, 176–177, 356, 360, 360–361, 374
Alexandra Bridge, Ottawa, ONT 358
algae 14, **56**, 98, 126, 204, 266
Algoma 275

Algonquin Arch 119
Algonquin Provincial Park, ONT 96, 337
Alleghanian Orogeny **145**, 357, 375
alligators in Arctic 196
alloys 260
alpha particles **43**, 44
Alpha Ridge 175
alpine ecosystems 343
alpine glaciers 219, 237
alpine meadows 243
Alps 227
alternating subduction and transform plate boundary 374
Altiplano 200–201
aluminum 250
aluminum oxides 19
Amasia 37
Amazon Basin, as analogy 205, 355
Amazon River 33, 302
Amazonia 101, 108, 357
Amazonian margin of Protogondwana ... 110
Amazonian part of Protogondwana 375
amber 186–**187**
American Falls, USA 4
American Great Plains 34
American War of Independence 130
Americas 234
amethyst 20–**21**
Amisk Lake, SK 92
Amiskwia 125
ammolite jewelry 59, **186**
ammonia 333
ammonites .. 39, **59**, 168, 170, 173, 186–187, 368
ammonites, extinction of 187
ammonium nitrate 343
ammonoids 56, **59**, 159, 163, 168,
amniotic egg **61**, 62
amoebas 56
amosite **352**
amphibian footprints, Blue Beach, NS ... 366
amphibian trackways 366
amphibians 2, 54, 56, 60–**61**, 128, 155, 166–167, 185, 205
amphibole asbestos 353
amphibole group asbestos **352**
amphiboles 9, 16, **19**–20
amphibolites 16, 18, 79, 81, 92, 173, 201
Amund Ringnes Island, NUN 71
Amund Ringnes Island, NUN, discovery of . 193
Amundsen-Mackenzie Mountains Basin .. 85, 93, 96–98, 110–111, 375
anaerobic organisms **82**
Anahim Volcanic Belt 208
anapsids 56, 60, **62**
ancestral Great Lakes 232
ancestral Juan de Fuca Plate 198
Anchiceratops 185
anchor marks 50
Anchorage, Alaska 197
ancient flyers 62
Anderson, Michael 101
Andes 7, 22, 33, 128, 169, 200–201
andesite **7**, 9
Ange-Gardien Sandstone 294–296
angiosperm pollen 129, 374
angiosperms **129**, 194, 368
anhydrite **97**
animal activity on land, earliest 107
animal body plans 123

animal exchange between North and South America 214
animal life 67
animal migration routes 203
animal skeletons, earliest 105
animals 54–**55**, 126, 366
animals, early 1, 122
animals, role in weathering 9
ankylosaurs **185**
Annaheim, SK 333
Annapolis River, NS 277
Annapolis Valley, NS 262
annelids 56, **58**
annual average temperature 240
annual cycles 224
annual spring melt 303
anomalocarid 124
Anomalocaris 123–125
Anomalocaris canadensis 123
anorthosite **95**, 287, 375
anorthositic intrusion 95
anoxic conditions ... **14**, 67, 82, 88, 120–121, 125, 248, 254, 266
anoxic layer 248
anoxic oceans 256, 365
Anse à la Mine, QC 259
Antarctic Circumpolar Current .. **196**, 242, 374
Antarctic glaciation 196
Antarctic ice cores 242
Antarctic Ice Sheet 236, 243, 300
Antarctic Plate 29
Antarctica 25, 227, 230, 241–242, 244, 359, 374
anteaters 190
antennae, evolution of 106
anthophyllite **352**
anthracene **347**
anthracite 14, **263**
anthracite, use of 263
Anthropocene **339**
anthropogenic greenhouse effect 242
anthropology 161
anticlines **16**–**17**, 46, 111
anticlines, as petroleum traps **267**
Anticosti Basin 40, 115–116, 119, 135, 149, 365
Antigonish Highlands, NS 148–149
ants 14
apatite 20–21, 43, 94
apes 191
aphelion **228**
Apollo missions 76
Appalachian Basin 115, 119, 135–136
Appalachian basins 118
Appalachian Mountains 70
Appalachian Orogen 102, 112, 114–115, 119, **134**–136, 145, 166, 260, 287, 293–294, 296, 359, 366, 370
Appalachian-Caledonian mountain belt 355, 370
Appalachians 33, 70, 112, 128, 134, 138, 238, 304, 375
appendages 58
appendages, evolution of 106
Apple River, NS 311
apples 201
aquatic ecosystems 302–303, 340
aquatic life, and toxicity 342
aquatic organisms **52**, 340
aqueduct 306
aquifers **301**, 303–304, 306, 345
aquifers, migration of water to and from ... 304

aquitards **301**
Arabia, location in Rodinia 97
Arabian Plate 29
aragonite **19**–20, 84
Aral Sea 302
araucarian trees 167
Arc de Triomphe, Paris, France 296
arc magmatism 198
Archaea 54–**55**
archaeocyathan reefs 114
archaeocyathans **56**, 106–108, 117
archaeological sites 237–238
Archaeopteris **128**, 142–143, 366
Archaeopteris halliana 143
Archean **40**, 67, 74, 78–84, 249–250, 348, 354, 365
Archean cratons . . **78**, 87, 90–91, 93, 112, 354
Archean Eon **78**, 86
Archean gneiss 88–89, 226
Archean komatiite 355
Archean life **83**
Archean microfossils 86
Archean tectonic processes 82
Arctic 180, 238–239, 243–244, 368, 375
Arctic Bay, NUN 94
Arctic coals 174
Arctic Cretaceous faunas 174
Arctic Cretaceous floras 174
Arctic dinosaurs 174
Arctic geology, history of study 192–193
Arctic Islands . . 69, **70–71**, 111–112, 174, 193, 196, 202, 203, 230–231, 236, 272, 304, 308, 324, 369–370, 374
Arctic Islands, coal in 264
Arctic Islands, origin of 212–213
Arctic Islands, petroleum exploration 271
Arctic, Norwegian claims 193
Arctic Ocean 71, 173, **174**–175, 203, 205–206, 210, 233, 242, 303, 316, 360, 362
Arctic Ocean, freshwater lakes in 203
Arctic Ocean, origin of 174, 176
Arctic Ocean, sea-floor spreading in 174
Arctic Plains Platform 112
Arctic salt structures 175
Arctic sea ice cover, September 1979 244
Arctic sea ice cover, September 2007 244
Arctic seas 174
arêtes **221**
Argentinosaurus 168
argillite **15**, 97, 249
argon 66
argon isotopes 42
arid climate and evolution of seeds 128
Arisaig, NS 53, 60, 133–134, 229
Arkona, ONT 10, 59
armadillos 190
Armorica 109, 357
armouring of shorelines 313, 319
arms race 106
arsenic 302, 344, 347, 349–350
arsenic and pesticides 349-350
arsenic contamination 349
arsenic, diseases linked to high level exposure 349
arsenic from bedrock 350
arsenic in groundwater 349
arsenic trioxide 344
arsenopyrite 251
Arthropleura 154–155, 366
arthropleurids 56, **59**, 67, 141
arthropods . . . 56–**58**, 107, 122–123, 133, 135, 141, 154, 159, 365
arthropods, primitive 123
asbestiform habit 352
asbestos 259, 347, **352**
asbestos, as industrial mineral 292
asbestos, health concerns 352
asbestos, health risks 353
asbestos remediation 353
asbestos, uses of 352
asbestosis **353**
ash (tree) 204
ash deposits 328, 374
ash fall 328
ash plume 328
Asia 234, 242
asphalt **261**
Aspidella 100–101, 104–105
Aspidella terranovica 100
Assiniboine River, MB 306
asteroid belt 333
asteroid impact 64, 187–188, 368, 374
asteroid impact as tsunami trigger 325
asteroid impact and dinosaurs 335
asteroids 75, 187, 252, **333**–334, 336
asteroids (starfish) 56, 59, **60**
Asterophyllites 128, 154
asthenosphere . . . **22–23**, 29, 31, 33, 236, 248
astronomical cycles 44
astronomy 38
asymmetrical ripples/dunes **11**
Athabasca Basin 93–95, 275, 337, 355
Athabasca Basin, uranium ores in 275
Athabasca Glacier 220
Athabasca Hall, University of Alberta, Edmonton, AB 291
Athabasca Oil Sands 268, 272–273
Athabasca River, AB 303, 306, 329
Athabasca Valley 221, 272
Atholville, NB 144
Atlantic No. 3 well, Leduc, AB 269–270
Atlantic Ocean . . . 24, 26–27, 31–33, 37, **162**, 170, 172, 176, 182–183, 192, 197, 214, 233, 297, 360–362, 370, 374–375
Atlantic Ocean, earliest sea-floor spreading in 171
Atlantic Ocean floor 32
Atlantic Ocean prehistory 170
Atlantic passive margin, as analogy 111
Atlin, BC 157
atmosphere 44, **65**–68, 88, 241, 364–365
atmospheric carbon dioxide 242
atmospheric circulation **65-66**
atmospheric conditions 240
atmosphere, contamination in 344
atmospheric gases 65
atmospheric nitrous oxides 343
atmospheric oxygen **84**, 366
atmospheric pollution 343
atmospheric processes 243
atolls 33, **156**, 166
atomic mass **41**
atomic nucleus **41**, 44
atomic number **41**
atomic structure **41**
atoms **41**
augite **19**
Aulac, NB 318
Aurora, ONT 341
Aurora well field 341
Australasia 190, 359
Australia . . . 25, 37, 78, 83, 135, 139, 189, 234, 242, 375
Australia (ancient continent) 102
Australia, flooding in 322
Australia, location in Rodinia 97
Australia, northward drift of 196
automobile emissions 339
Auyuittuq NPC, NUN 91
Avalon Peninsula, NL 99–100, 109, 373
Avalon Zone 102
Avalonia 99, 102, 108–110, 115–116, 133–135, 141–142, 355, 357, 375
Avalonian magmatic arcs 109
Avalonian oceanic plateaus 109
average annual precipitation 304
average global temperature 243
average rainfall 240
average temperature 240
Aviat, NUN 90
Axel Heiberg Eocene forest 206
Axel Heiberg Island, NUN 2, 70–71, 129, 151, 175, 194–195, 202, 205–206, 212–213, 216, 362, 368, 370
Axel Heiberg Island, NUN, discovery of . . . 193
Ayles Fiord, NUN 111
Azolla **203**–204
azurite **21**

B

B horizon (soil) **14**
back-arc basin . . 33, **36**, 81, 91, 169, 248–249, 355, 359
backbone 60
Backbone Ranges, NWT 96
back-stripping 369
backwash **312**
Bacon, Francis 24
bacteria 14, **55**, 83–84, 266, 268, 273
bacteria, source of fine atmospheric particles 351
bacterial infection 338
Baffin Basin 170
Baffin Bay 31, 35, 71, 170, 184, 202, 203, 213, 368
Baffin Island NUN . . . 19–20, 35, 71, 79, 91, 94, 97, 110, 202, 205, 213, 220, 222–223, 231, 236, 249–250, 260, 309, 324, 355
Baffin Island, Paleocene volcanism 205
Baffin Island, possible Neogene glaciers . . 212
Baffin Mountains, NUN 34-35
Baja California, Mexico 37
Baker Lake, NUN 90
Baker Lake Basin 93–94
Balancing Rock, Long Island, NS 362
bald cypress 201
Ballantynes Cove, NS 148
ballast, cobbles in 282
ballast, stones from 279
Baltic Sea 23
Baltica **102**, 108, 115–117, 132–133, 139, 355, 357, 360, 375
Baltica, location in Rodinia 97
Bancroft, ONT 16, 20, 94, 275
Banda Sea, as analogy 135
banded iron formations 67, 79, 84, 88–89, 97, 248, 255–256, 365, 375
banded iron formations, leaching in Cretaceous 255
banded ironstones **84**
banded ironstones, earliest 375
banding (in gneiss) **15**
Banff, AB 46, 113, 118, 146, 179–181, 263, 284
Banff NPC, AB 13, 70, 163, 284
Banff Springs Hotel, AB 284
Bangiomorpha 98
Bangiomorpha pubescens 98
Bank of New Brunswick, Saint John, NB 79
Bank of Nova Scotia Building, Saint John, NB 280
Banks Island, NWT 71, 142, 239
Banks Island, Miocene flora 212
bar magnet 25
Barents Sea 139, 151
Barents Shelf 175, 202
Barghoorn, Elso 86
barite 175, 249
barite, as industrial mineral 292
Barkerville, BC 141, 259
Barnes Ice Cap 220, 236
barrier beach 308, **314**–315
Bartibog, NB 285
basalt **7**, 9, 16, 22, 25, 27, 30, 77–78, 165
basalt, columnar **7**
basaltic crust 82
basaltic lava 27, 30, 35, 168
basaltic volcanism 77
base-metal refining 347
base metals **247**
basement **112**, 369
basement detachment zone 179
Basilica of St. John the Baptist, St. John's, NL 286
basins 35–36
basswood 229
batholiths **8**–9
batholiths in Grenville Orogen 94
Bathurst, NB 116, 249, 259
Bathurst Island, NUN 71, 127, 138, 140, 142, 365
Bathurst Island, Paleogene volcanoes 203
Bathurstia 140
Bathygnathus 158–159, 189
bathymetry **69**
bats 190–191
Bay of Fundy . . . 20, 30, 149, 166–168, 171, 277, 313, 318, 367, 372
Bay of Fundy, as estuary 317
Bay of Fundy tides 152, 317–318
Bay of Islands ophiolite, NL 115, 358
BC Hydro 330
beach, composition of 314-315
beach cusps 314
beach ridges 311
beach terraces 311
beaches and sediment supply 313
beaches 10–12, 308, 312–**313**
beaches, as source of sand 291
beaches, landward migration of 311
Bear Lake, ONT 20
Bearpaw Sea 180, 182, 374
bears 190–191, 214, 217
Beauce, QC 259
Beaufort Sea 174–175, 272, 316
Beauport, QC 294–295
Beauport Limestone 294–295
beaver-chewed sticks 212
beaver pond, fossil evidence 212
Beaverlodge, SK 275
beavers 212, 214
Becquerel, Henri 41
bedded sulphide deposits 248–249
bedding 11, 15
Bedford, NS 50
Bedford Basin, NS 50, 239
bedrock 14, **45**, **112**

bedrock benches ... 229
bedrock maps ... 45–46
beds (sedimentary) ... **9**
Beebe, QC ... 296
beech ... 218
beetles ... 229
behaviour ... 54
Belcher Islands, NUN ... 92, 97
belemnites ... 173, **187**
belemnites, extinction of ... 187
Belinurus ... 365
Bell River basin ... 206
Bell River System ... 205–206
Bell, Robert ... 272
Bella Coola, BC ... 139, 291
Belt-Purcell Basin ... 93, 95, 249, 355, 375
beluga whale ... 348
Bennett Lake, BC & YK ... 247
Benton, NL ... 288
bentonite ... **290**
bentonite clay ... 342, 345
bentonite deposits ... 291
bentonite, uses of ... 291
benzene ... **265**, 345
benzene rings ... 347
Bering Sea ... 211, 238, 363
Bering Strait ... 234
Beringia ... 191, 214, 229, **234**–235, 369
berms ... 330
beryllium isotopes ... 44
Bessborough Hotel, Saskatoon, SK ... 240, 287
BIFs ... **84**
Big Bang ... 2, **74**–75
Big Bar Creek, BC ... 211
Big Bowl, AB ... 160
Big Muddy Badlands, SK ... 38
Big Rock, AB ... 221, 223
Bighorn Creek, AB ... 169
bilaterians ... 56–**57**, 105, 375
Billings, Elkanah ... 100, 104
bioaccessibility ... **353**
biochemical weathering ... 68
bioconcentration ... **347**, 348
biodegradation ... **268**
biogenic gas ... **273**
biogenic methane ... 274, 341
biomagnification ... **348**
biomineralization ... 375
biomineralization, earliest ... 98
biomolecules, earliest ... 82
biosphere ... **65**–**66**, 67–68, 241, 364–365
biostratigraphy ... **40**
biotite ... 9, 15, **19**–20
birches ... 194, 201, 212, 218, 229
bird footprints ... 185
bird physiology ... **67**
birds ... 56, 60, **62**, 64, 163, 185, 189, 205, 367
birds' eggs ... 62
birds, toothed ... 186
Bishop's Falls, NL ... 288
bison ... 234–235, 237
bitumen ... **261**–262, **265**, 268, 272–273
bitumen and Aboriginal people ... 261
bitumen, perceived medicinal properties ... 261
bituminous coal ... 14, **263**
bituminous coal, use of ... 263
bivalves (see also clams) ... 56, **59**, 107, 153, 168, 173, 217
Bizard Island, QC ... 288
black bear ... 218, 235
black granite, as commercial term ... 281

Black Granite, Saint-Nazaire ... 287
Black River Limestones ... 284
Black Sea ... 120, 248
black shales ... 120–121, 170, 248
black shales, as source of mercury ... 348
black-smoker chimneys ... 248
black-smoker fauna ... 247
black smokers ... 92, 247–**248**
Blackfoot ... 264
Blairton-Marmora, ONT ... 259
Blind River, ONT ... 343
blowout ... 269
Blue Beach, NS ... 149, 366, 375
Blue Eyes Stone ... 287, **289**
Blue Mountains, NUN ... 8
Blue Rocks, NS ... 15, 110
Bluefish Caves, YK ... 235, 374
blueschists ... **16**, 157, **165**, 173
Bobastrania ... 164
bolide ... **334**
Bonanza Arc ... 166, 172, 250, 374
Bonanza Creek, YK ... 246, 254
bone bed ... 184
bone deposits ... 238
bone piles ... 238
bones ... 52
Bonnet Plume Basin ... 173
Boothia Highland ... 203
Boothia Peninsula, NUN ... 71, 118
Boothia Uplift ... 118–119, 138, 140
Bordeaux, France ... 66
Borden Basin ... 93–94
boreal forest ... 212, 237, 243, 369
borings ... **54**
Boston, Massachusetts ... 86, 109
Bouctouche, NB ... 314
boudins ... **200**
Boularderie Island, NS ... 152
boulevard Champlain, Québec City, QC ... 295
Bourgeau Thrust Fault ... 46, 179
Bow River, AB ... 46, 283, 306, 332
Bow River Basin, AB ... 306
Bow Valley, AB ... 163, 284
Bowmanville, ONT ... 57
Bowser Basin ... 172–173, 178
Bowser Basin, coal in ... 263
Boyle, Dan ... 347
brachiopods ... 53, 56, 58–**59**, 106–107, 117, 120–121, 133, 135, 137, 147, 158–159, 163, 168
Brachiosaurus ... 168
brain ... 58
branchiopod crustacean ... 107
Branchiostoma ... 61
Brantford, ONT ... 350
Bras d'Or Lake, NS, origin of ... 149
bread ... 19
Break Neck Steps, Québec City, QC ... 294
breaker ... **312**
breakwaters ... 287
breccia ... **9**
breccia pipes ... **250**–**251**
breccias, Sudbury impact structure ... 252
breccia, volcanic ... **8**
Brent impact structure, ONT ... 336–337
Brentwood Bay, BC ... 292
brick buildings ... 279
bricks ... 279–280, 283, **291**–292
Bridal Veil Falls, USA ... 4
Bridgeville, NS ... 20
brilliant lustre ... **21**

brines ... 301
Britain ... 128
Britannia Creek, BC ... 342
Britannia Mine, BC ... 342
British Columbia ... 1, 3–4, 7–8, 11, 16–17, 27–28, 30, 36–37, 49, 54, 61, 68–71, 73, 95, 103, 105–107, 113–114, 122–125, 129, 139, 141, 145–146, 155–157, 161–162, 164–166, 169–170, 172, 176–178, 181, 192, 197–202, 204, 207–211, 219, 221, 223–224, 230–231, 237, 239–244, 247–250, 258–259, 263–264, 270, 274, 276–277, 282–283, 287, 290–292, 302–306, 308–309, 311–312, 314–316, 323–324, 326–332, 337, 339–340, 344, 350, 356, 359–361, 365, 367–368, 370–372, 374
British Columbia, earthquakes ... 323
British Columbia Legislative Building, Victoria, BC ... 287, 290
British Isles ... 356, 370
British Mountains ... 72
British North America ... 130
British North American Exploring Expedition ... 298
British Parliament ... 298
Brock River, NWT ... 111
bromine ... 19
brontotheres ... 194, 206
Brooks, AB ... 10, 184
brown algae ... 126
Bruce County, ONT ... 53
Bruce Peninsula NPC, ONT ... 136–137
Bruce Peninsula, ONT ... 71, 137, 285
Bruce Point, ONT ... 343
bryophytes ... **126**
bryozoa ... 10, 56, **60**, 117, 158–159, 285
Buchans, NL ... 249
buffalo (bison) ... 217
Buffon, Comte de ... 41
building stone finishes ... 279
building stone industry ... 282
building stones ... 278–280, **281**, 282–290, 294–296
Burgess Pass ... 122
Burgess Shale ... 54, 61, 107, 114, 122–125, 365, 375
burial mounds ... 238
buried canyons ... 205
Burin Peninsula, NL ... 44, 106, 326–327
burins ... **235**
Burlington, ONT ... 292
Burntcoat Head, NS ... 37
burrowing animals ... 14
burrowing invertebrates ... 84
burrows ... **54**, 120, 284, 365
butane ... **265**
Butchart Gardens, BC ... 292
Buttle Lake, BC ... 165
Buzzard Coulee, SK ... 333
Bylot Island, NUN ... 71, 97

C
Cabot, John ... 373
Cabot Tower, St. John's, NL ... 286, 373
Cache Creek, BC ... 156, 361
Cache Creek Terrane ... 156–157, 165–166, 172–173, 357, 374
Cadieux Island, QC ... 288
cadmium ... 347–348, 350
cadmium in caribou ... 348
cadmium, sources of ... 347
Cadomin, AB ... 342, 362
Calamites ... 2, 128, 153–154
calcareous algae ... 119, 126

calcareous deposits ... **11**
calcareous mudstone ... 11
calcareous shales ... 11
calcite shells ... 52, 68
calcite skeletons ... 68
calcite ... **9**, 10, 16, **19**–21, 84, 175
calcium carbonate ... 9–11, 19
calcium carbonate precipitation ... 84
calcium-magnesium carbonate ... 10, 19
calcium phosphate microfossils ... 98
calcium sulphate ... 11
calcrete ... 15
Caledonia, NB ... 279
Caledonia, ONT ... 11, 137
Caledonia Highlands, NB ... 148–149
Caledonian Mountains ... 138, 141–142
Caledonian Orogen ... **134**, 139
Calgary, AB ... 90, 113, 118, 146, 148, 178, 180, 268–269, 273, 306, 320, 339
Calgary, AB, fire of 1886 ... 282–283
Calgary, AB, 2013 floods ... 332
Calgary, AB, water supply ... 306
California ... 7, 37, 177
California coast ... 242
Cambria ... 40
Cambrian ... **40**, 46, 56, 60, 64, 100, 101, 105–115, 117–118, 120, 122–123, 125, 136–137, 139, 145, 178–179, 266, 286, 290, 293, 335, 356, 364
Cambrian back-arc basins ... 109
Cambrian continental shelf ... 114
Cambrian Explosion ... 58–59, 67, **105**–106, 108, 123, 365
Cambrian fluvial sedimentary rocks ... 99
Cambrian geology map ... 101
Cambrian life ... **107**
Cambrian magmatic arcs ... 109
Cambrian-Ordovician boundary ... 44
Cambrian paleogeography ... 106
Cambrian Period ... 44, 101
Cambrian quartz-rich sandstones ... 112–114
Cambrian reefs ... 107–108
Cambrian sea floor ... 58
Cambrian trace fossil ... 105, 364
Cambrian trackway ... 107
Cambrian turbidites ... 109, 116
Cambridge Bay (former name), NUN ... 193
Cambridge University ... 122
camels ... 191, 212–214, 229, 234–235, 374
Campbell River, BC ... 165, 264
Campbellton, NB ... 7, 134
Canada Basin ... 174–175
Canada's coastline ... 1, 70, 307–319
Canadian Confederation (see Confederation)
Canadian Cordillera (see also Cordillera)
Canadian Cordillera, physiography ... 70, 71–72, 211
Canadian Museum of Nature ... 52
Canadian National Railway ... 330
Canadian Naturalist and Geologist ... 100
Canadian Pacific Railway ... 51, 122, 252, 264, 273
Canadian Rocky Mountains (see also Rockies and Rocky Mountains) ... 11
Canadian Shield ... 23, 34, 37, **70**–71, 77–83, 87, 92, 95, 112, 119, 136, 146–147, 219, 221, 236, 260, 281–282, 287, 291, 293, 302, 324, 342, 349, 355, 369, 372
Canadian Shield, uranium deposits ... 275
Canadian transcontinental railway ... 259, 264
Canadian Volunteers Monument, Toronto, ONT ... 289

Canadian-American Great Lakes International Multidisciplinary Program on Crustal Evolution 96
CANDU reactors 274
canines (teeth) 189–190
Canmore, AB 14, 46, 147, 180, 263–264, 267–268, 368
Cannonball Sea 205
Canol pipeline 271–272
Canol Project 271–272
Canso Causeway, NS 292
Cap Bon Ami, QC 135
cap carbonate **104**
Cap Diamant Blackstone **294**–296
Cap Diamant promontory, QC 293–294, 296–297
Cap Gaspé, QC 70
Cap Tourmente, QC 297
Cape Blomidon, NS 30
Cape Breton Highlands, NS 148–149
Cape Breton Highlands NPC, NS 109, 314
Cape Breton Island, NS 13, 109, 149, 229, 258, 263, 271, 292, 356
Cape Cod, Massachusetts 110
Cape d'Or, NS 19, 168, 258
Cape Dyer, NUN 205
Cape Enrage, NB 152
Cape Smith Belt, QC 91
Cape Split, NS 168
Cape Tormentine, NB 314
Cape Tryon, PEI 130
Cap-Rouge, QC 295, 297
Cap-Rouge, QC, and Jacques Cartier 258
carbon atoms **41**, 265
carbon-based molecules 82
carbon capture 265
carbon cycle **67**–68, 241, 83–84
carbon dating **42**
carbon dioxide 14, 65–68, 83, 104, 126, 241–244, 256, 333
carbon dioxide, Canada's contribution . . . 344
carbon dioxide exchange 67
carbon dioxide levels in atmosphere . . . 66, **68**, 127, 243, 276, 339
carbon dioxide levels in atmosphere by 2050 243
carbon-dioxide-rich atmosphere 256
carbon dioxide storage underground 344
carbon emissions 344
carbon isotopes **41**, 83
carbon monoxide 333
carbon reservoirs **68**
carbon sequestration **344**
carbon sink **68**
carbon source **68**
carbon 19, 21–22, 68, 256, 265
carbon-12 . . . . **41**, 83
carbon-13 . . . . **41**, 83
carbon-14 . . . . 41–42, 241
carbon-14 dating **42**–43
carbonate banks 108, 115, 117
carbonate banks in Sverdrup Seaway 151
carbonate buildup in Scotian Basin 171
carbonate deposits 104
carbonate mounds 175
carbonate mud 11
carbonate reefs 365
carbonate rocks 68, 281
carbonate rocks, ores in 250
carbonate, gold in association with 256
carbonates **10**–11, 18–19, 68, 247, 355
carbonates and MVT deposits 248

carbonates, as reservoir rocks 267
Carboniferous **40**, 46, 53, 58–59, 61, 67, 128–129, 131–132, 136, 139–140, 145–157, 159, 169, 175, 180, 250, 268–269, 278–279, 285–286, 289, 311, 320, 359, 366, 370
Carboniferous amphibian footprints 366
Carboniferous coal 263
Carboniferous environments 153
Carboniferous forests 127, 153–154
Carboniferous, geological map 132
Carboniferous lake scene 150
Carboniferous paleogeography . . 148, 150, 152
Carboniferous-Permian boundary 128
Carboniferous redbeds 151
Carboniferous rift basins 150
Carboniferous uplands 148
carbonized film 53
carbonized plant material **54**
Cardinal River Mine, AB 342, 362
Caribbean Plate 29
Cariboo, BC 259
Cariboo basalts 208, 374
Cariboo gold rush 259
caribou 4, 234–235, 238–239, 347
Caribou basalt lavas 207–208
Caribou Hills, NWT 205
caribou meat 347
Carleton University, ONT 276
Carmack, George 246
carnallite 147
carnivores (dietary strategy) 54, 59, 61
carnivores (mammal group) 190–191, 205–206, 212
Carolinia 109
Carrara Marble 289
carrier pigeon 64
Carrot River, SK 186
Carswell impact structure, SK 337
Carthew-Alderson Trail, AB 94
Cartier, Jacques 258, 293
cartilaginous skeleton **61**
carvings in stone 280, 283, 287
Casa Loma, Toronto, ONT 216
Cascade Magmatic Arc . . . 198, 207–208, 231, 276, 327
Cascade Mountain, AB 46, 146, 180
Cascade Range 7, 27, 72
Cascade volcanoes 328
Cascadia Subduction Zone 197
Cascadia Subduction Zone . . 27–28, 197–198, 207–208, 311, 323–324, 326, 363
Cassiar, BC 157
Cassiar Mountains 72, 113, 172–173
cassiterite 251
Castle Creek, BC 103
Castle Mountain, AB 113
casts **54**
catastrophe 322
Cathedral Escarpment 113–114, 122, 125
Cathedral of Immaculate Conception, Saint John, NB 290
cats 190–191
Catskill Delta 136
cave 136
cave lions 369
Cedar Lake, MB 187
Celebes Sea, as analogy 135
celestite 175
cell-like microfossils 98
cement (industrial) . . . 242, 280, 283, **292**, 295
cement (natural) 9, 11, 84

cement kilns, as mercury emitters 343
cementation 9
Cenozoic **40**, 62, 172, 174, 176, 179–180, 201, 229, 368
Cenozoic dominance of mammals 62, **64**
Cenozoic Era 195
Cenozoic geological map 195
Cenozoic ocean temperatures 197
centipedes **59**
Central America, hurricanes in 322
central Asia 32
central Europe, flooding in 322
Central Park, New York, bridges in 286
central uplift 335
Centre Block, Parliament Buildings, Ottawa, ONT 290
centrosaur herds 185
Centrosaurus 185
Centrosaurus bone bed 185
cephalopods 56, **59**, 107, 120, 187
ceratopsian dinosaur 185, 368
ceratopsians 185
cereals 129
Ceres 333
cetartiodactyls **191**, 205, 214
chalcopyrite **19**–20, 249–250
Chaleur Bay 140, 143, 366
chalk 49, 55–56, **183**, 374
chalk deposition, Scotian Basin 183
Chalk River, ONT 343
Champlain Monument, Québec City, QC . . 296
Champlain Sea 233, **237**, 293, 297, 312, 330, 374
Chancelloria 125
channels **12**, 18
channel, ancient fluvial 13
channel, ancient deep-sea 13
channel migration 12
channel switching 316
channels, fixing the position of 315
Characodictyon 98
charbroiling meat, as source of PAHs 347
Charlesbourg-Royal 258
Charlevoix, QC 335–336
Charlevoix impact structure, QC 337
Charlie Lake Cave, BC 237
Charlie, Tagish "Dawson" 246
Charnia 105
Charnia wardi 104
Charniodiscus 101, 105
Charny, QC 9, 116, 294
Chasmosaurus 368
Château Frontenac, Québec City, QC 296
Chateau Laurier Hotel, Ottawa, ONT 278
Château-Landon, France 296
Château-Richer, QC 295
Château-Richer Limestone 294
Chazy Limestone 285–286
chemical analysis 21
chemical breakdown 9–10
chemical sedimentary rocks **10**
chemical weathering 68
chemistry of wastes 345
chemosynthesis **68**, 82
Chengjiang, China 122
cherries 201
chert 84–86, 97, 235, 249
chert, Aboriginal use 258
Chesterman Beach, BC 199
Chevré, Paul 296
chewing action 189

Chezzetcook Inlet, NS 308
Chibougamau QC 90, 222
Chicago River 233
Chicago, Illinois 318
Chicoutimi, QC 332
Chicxulub, Mexico 187
Chicxulub crater 187
Chidliak project 19
Chignecto Bay 152
Chilcotin basalts 207–208, 374
Chile 235, 374
Chilkoot Pass 247
Chimney Corner, NS 263
chimney structures 247
chimneys **248**
China 61, 128
China, earthquakes in 322
Chinese golden larch 201
Chinese rubber tree 201
Chinese water pine 194, 201
chlorine 19
chlorite 15–16, 79
chlorophyll 126
chloroplasts **84**, 126
chloroplasts in branches 140
chloroplasts in stems 140
chocolate 202
chondrites **334**
Chondrites 53
chondrules **334**
chordates 41, 56, 58, **60**, 127, 129
chordates, primitive 125
Chortis, location in Rodinia 97
Christchurch, New Zealand 325
chrysotile **352**–353
chrysotile asbestos 352
chrysotile, health risks 353
Chu Chua, BC 200
Chuaria 98
Chubb Building, Saint John, NB 280
Chuckanut, Washington State 200
Chugach Terrane **166**, 199, 209
Chukotka 102, 111, 115, 357
Chukotka, location in Rodinia 97
Churchill, MB 92, 120, 137, 237
Chutes-de-la-Chaudière, QC 9, 116, 294
Cigar Lake Mine, SK 275
cirques **221**
Cirrus Mountain, AB 146
Cisco Oil Field 272
cities, resources for 339
City Market, Saint John, NB 280
clam burrows 54
clam coals **153**
Clam Harbour Beach, NS 220
clams (see also bivalves) . . 52, 56, **59**, 117, 133, 163, 175, 186
Clark, Karl 272–273
Clarke Head, NS 13, 16
Clarkson, Adrienne 260
classes **55**
classification of life **55**
classification of organisms **55**
clastic sedimentary rocks 10
clastic sediments 9
clastics **9**–11
clay **9**–10, 330, 345
clay minerals **10**, 68, 302
clay, uses of 280, 290
clays, source of fine atmospheric particles 351

clay-sized particles 12
claystone **9**
cleavage **15**
cleavage planes 15
Clementsport, NS 282
Clevosaurus 168
cliffs 307, 311
cliffs as sediment sources 313
cliffs in permafrost 309
cliffs, erosion of 50, 313
Clifton, NB 152, 154
climate **240**–244, 367, 369, 374
climate change (see also global change) 224, 240, 242–244, 345
climate warming (see also global warming) 121, 244, 339
climatic cycles 237
climatic stability 240
clinker **292**
Cloud, Preston 77
Cloudina 105–106, 375
Clovis points **235**, 237
clubmosses 127–128, 140, 153–154, 159, 366
cnidarians **56–57**, 106
coal **14**, 51, 68, 130–131, 153, 161, 242, **262**–265, 339, 362, 373
Coal Age 152, 154
coal, as a source of gas 274
coal-bed methane **274**
coal-bed methane, environmental issues 274
coal beds 356
coal, carbon content 263
coal deposits 40, 131, 186, 202, 368
coal deposits, Vancouver Island 178
coal, earliest 262
coal, environmental aspects 265
coal exports 265
coal-fired power plants 265, 340
coal-fired power stations, as source of mercury 348
coal forests 154
coal formation 263
Coal Mine Point, NS 131
coal mining 320–321
coal mining, strip mines 263
coal mining, underground 263
coal, origin of 266
coal, ranks of **263**
coal seams **14**, 128, 131, 344
coal swamps 188
coal tar 347
coalfields 128
Coast Mountains, BC 8, 72, 139, 172–173, 176, 178, 208–211, 219, 231, 241, 287, 290, 371–372, 374
Coast Mountains as source of sediment 198
Coast Mountains, Cretaceous magmatism 177
Coast Mountains, Cretaceous rise of 209
Coast Mountains, granitic rocks in 209
Coast Mountains, magmatism in 198
Coast Mountains, recent rise 209
Coast Mountains, volcanic 198
coastal bluffs 237
coastal dykes 315
coastal erosion 307–308, 310–311
coastal floods 332
coastline, movement of sediment 308
coastline, retreat of 311
coastlines 1, 70, 307–319
coastlines, protection of 319
Cobalt, ONT 258
cobalt-60, uses of 274
Cobequid Fault, NS 145
Cobequid Highlands, NS 148–149
Cobequid-Chedabucto Fault System 145
coccolithophores **56, 183**, 368
coccoliths 55–**56**, 163, 183
Cocos Plate 28–29, 198
cod 143
Codroy, NL 229
Codroy Valley, NL 149
coelacanths 60–**61**, 164
Coelophysis 166, 367
coelurosauravids **164**
coffinite **275**
cold currents **66**
Cold Lake Oil Sands 273
Cold Lake, AB 273
Cold War 259, 275
Coleman, Arthur 216-217, 219, 291
Coleman Mine, ONT 253
colliding tectonic plates 32
Collingwood, ONT 274
Collins, Desmond 122
collision 33
collisions in Cordilleran Orogen 172-173
Colonial Building, St. John's, NL 284
Colwood, BC 199
colour (mineral) **21**
Columbia Icefield 219
Columbia Mountains, BC 72, 113–114, 141, 172–173, 211, 259
Columbia Mountains, igneous rocks 141
Columbia River basalts 208
Columbia River Basin, BC 305
Columbia River valley, BC 330
columnar basalt **7**
columnar jointing in komatiite 79
columnar jointing 7, 355
coma **333**
combustion of fossil fuels, as source of PAHs 347
comets 66, **333**–334, 336
comets, and life 333
common clay **290**
Comox, BC 178
compaction 9
complex impact structures **334**–335
composition of atmosphere **66**
compounds 19
compression 16–17
concentric dykes, Sudbury 252
concentric pattern of crevasses 220
Conception Bay, NL 286
concrete 283, 291, 342
concrete, pebble compositions 291
concrete, shells in 291
condensates **265**
conduction 22
condylarth 205
cones 43, 129
Confederation 131, 259, 285, 373–374
Confederation, role of coal 263
conglomerate **9**, 79, 281
Congo, location in Rodinia 97
conifer needles 186
conifer pollen 129, 218
conifers 127–**128**, 158, 196, 229
conifers, early 128
conodont animal **61**
conodonts 41, 44, 56, **61**, 107, 121, 168
conservation 244
conservation of fresh water 303
Constitution Square, Ottawa, ONT 287
construction materials 292
consumer waste 341
contaminant plumes 49, 339, 341, 345
contaminants 339–340
contaminants in homes 353
contaminants, underground storage 344
contaminated sites 345
contaminated storm water 340
contaminated waters 340
contamination 341–342
contamination in sediments 345
contamination monitoring 341
continental arcs **27**, 33, 36, 169
continental blocks 354
continental collision 375
continental crust 9, **22**–23, 65, 74, 82, 355
continental drift 24–27, 29
continental drift, concepts of 184, 202
continental glaciers 241
continental ice sheets 237, 241
continental lithosphere 27, 29–33, 87, 248
continental magmatic arc (see continental arcs)
continental monsoon conditions 366
continental slope 14, 30–31
continents 22–24, 26, 30, 87, 357, 361
continent-to-continent collisions 87, 203, 355
continent-to-ocean convergence 355
continuous permafrost 224–225, 316
convection 22
convection currents 23, **27**, 29
conventional gas 274
conventional oil and gas 268
conventional petroleum reserves **273**
convergence 33
convergent margins and gold 256
convergent plate boundaries 28, 32–34, 36, 165, 355, 359
Convocation Hall, University of Toronto, ONT 291
Conway Narrows, PEI 315
Cook, Captain James 308
Cooksonia 127
cooling, within rock cycle 5
copper 19, 247, 250, 254, 258, 350
copper carbonate 21
copper cent coin 21
Copper Cliff North Mine, ONT 253
Copper Cliff South Mine, ONT 253
Copper Cliff, ONT 342
copper deposits 252
copper-iron sulphide 19
copper mine, first Canadian 259
copper ore 92, 341
copper, traded 258
copper, uses of 260
Coppermine Arch 119
Coppermine Basalts 95
Coppermine River, NUN 259
copper-nickel-cobalt deposit 95
coprolite **54**
coral atolls 156
corals **56–57**, 117, 120, 133–135, 137–138, 147, 159, 168, 367
corals, scleractinian 163
corals, tabulate 159
Cordaites 127, 154
Cordillera 24, 46, 69, **71**, 103, 112, 139, 156–157, 169, 172, 177–178, 180–181, 200, 236, 239, 303, 370, 372, 374,
Cordillera, as a sediment source 205
Cordillera, Cenozoic basins in 201
Cordillera, Cenozoic crustal thinning 200
Cordillera, Cenozoic normal faults 200–201
Cordillera, Cenozoic strike-slip faults 176, 200
Cordillera, influence on climate 304
Cordillera, oblique convergence 176
Cordillera, potential for geothermal energy 276
Cordillera, volcanoes in 192
Cordillera, westernmost terranes 199
Cordilleran basins, coals in 264
Cordilleran belts of uplift and erosion 173
Cordilleran collisions 173
Cordilleran continent-ocean collision 172
Cordilleran glacier 219
Cordilleran granitic plutons 172
Cordilleran granitic rocks 173
Cordilleran Ice Sheet **219**–221, 223, 230–231, 233–234, 236, 374
Cordilleran metamorphism 173
Cordilleran mountain building 359
Cordilleran mountains 70, 197, 360
Cordilleran mountains, earliest 171–172
Cordilleran Orogen 34, 112, 119, 146, 260, 287, 361, 375
Cordilleran Orogen, origin of 141
Cordilleran passive margin 111
Cordilleran sedimentary basins 173
Cordilleran tectonics 169
Cordilleran terranes 155–156, 360, 374
core holes 27, 46
Core Zone Craton 78, 90, 93
cores **22**, 65, **76**, 259
corn 239
Cornwallis Island, NUN 56, 71, 238
correlation of strata **39–40**
corundum **19**, 21
Coryphodon 194–195
cosmic dust 6, **74**–75, 82
cosmic rays **44**, 241
cosmogenic isotope dating **43–44**
Côte-de-Beaupré, QC 295
cotton 202
coulee 234
Court House, St. John's, NL 288
Couture impact structure, QC 337
Cow Head, NL 14, 115
cracking complex organic molecules 274
Craig Mine, ONT 253
Craigleith Provincial Park, ONT 274
Craigleith Shale Oil Works, ONT 274
Crater Lake, Oregon 328
crater **334**
craters on Moon **77**
craters, volcanic 8
cratonic shelves 88
cratons **78**
crayfish 186
cream milk vetch 350
Credit River, ONT 285
Cree 264
creep **22**
creep, as landslide process **329**
Creighton Mine, ONT 22, 253
creodonts 190, 206
Crescent Terrane 199
Cresswell, Samuel 239
Creston, BC 103

Cretaceous....... 39–40, 45–46, 48–49, 54, 56, 59, 62, 68, 113, 128–129, **162**, 172–188, 190, 199, 203, 214, 269, 283, 320, 360, 362–363, 368
Cretaceous Arctic magmatism...........175
Cretaceous climate......................183
Cretaceous coal.............. 187, 263–264
Cretaceous deltaic deposits 174, 178, 268
Cretaceous deposits in Maritimes........182
Cretaceous dinosaurs......... 184–185, 368
Cretaceous fossils......................368
Cretaceous geological map..............162
Cretaceous granitic intrusions...........287
Cretaceous invertebrates................368
Cretaceous kimberlite............ 256, 258
Cretaceous lava flows in Arctic..........175
Cretaceous lignite......................263
Cretaceous mafic sills..................362
Cretaceous magmatism in Arctic........176
Cretaceous marine strata.......177, 179, 268
Cretaceous non-marine deposits ... 178–179, 182
Cretaceous, oil sands host rocks.........272
Cretaceous, paleogeography........ 176, 182
Cretaceous reservoir rocks..........270–271
Cretaceous river channel................181
Cretaceous shales and selenium.........346
Cretacous-Tertiary boundary
  (see also K-T boundary)......... 187–188
Cretaceous vertebrates.................368
crevasses..........................219–220
crinoids...... 56, **60**, 107, 117, 133, 135, 147, 284, 296
crocidolite............................**352**
crocodile-like reptiles..................168
crocodiles.......... 56, 60, **62**, 163, 185–186
cross-bedded sandstone.................81
cross-bedding......................11, **12**
cross-cutting relationships...............39
cross-sections......................**45**–46
Crowsnest Pass, AB........ 95, 146, 320–321
Crowsnest River, AB....................321
crude oil.........................**265**, 274
crude oil, direct use....................266
crushed limestone......................292
crushed stone................ 280, 291–292
crust....................18, **22**–23, 65, 363
crustaceans......................56, 107, 364
crustal stretching.......................96
Cryogenian.........................101–104
Cryogenian Period......................**101**
crystal shape...........................21
crystal size.............................8
crystal systems.........................21
crystalline rocks.......................**9**, 15
crystallization.........................6, 8
crystallization order, Sudbury...........253
crystals, bubbles in....................335
crystals, fracture surfaces in...........335
cubic system...........................21
Cumberland Batholith...................91
Cumberland County, NS.................263
Cumberland Sub-basin........ 148–149, 152
Curie, Marie and Pierre..................41
current direction.......................11
Current River, ONT......................47
currents......................... 313, 317
Curryville, NB.........................279
Cushing Creek, BC.....................103
Custom House, Saint John, NB..........279
cyanobacteria........67, **84**, 97–98, 117, 126
cycad leaves..........................186

cycads.............. 127, **128**–**129**, 163, 189
cyclicity...............................227
cyclothems.......................**152**–153
cypress........................... 186, 204
Cypress Hills...............205, 363, 368
Cypress Hills, Tertiary fluvial gravels......206
cypress swamps.......................202

D
dacite.............................**7**, 287
Daily Sun, Saint John, NB...............280
Daley's Cove, NL.......................104
Dalhousie Junction, NB ... 126–127, 140–141, 365–366, 375
Dalhousie University, Halifax, NS........290
dams...............277, 303–305, 318, 332, 336
dams, natural.........................331
dancing elephants......................47
Daniel's Harbour, NL............... 21, 250
Darlington, ONT.......................343
Dartmouth, NS.........................50
Darwin, Charles..................... 41, 55, 129
Dashwoods Microcontinent........102, 106, 108–109, 114–116, 134, 355, 357, 360, 375
Davis Strait....................71, 170, 184
dawn redwood cones...................194
dawn redwood foliage..................194
dawn redwoods...2, 128, 194–195, 201, 204, 212, 368
Dawson City, YK...............161, 211, 247
Dawson Creek, BC.............. 161, 211
Dawson, George .. 51, 161–162, 188, 219, 372
Dawson, William.................. 155, 160
daylight, dearth of....................173
de Champlain, Samuel...............4, 293
de Monts, Sieur.......................258
de Roberval, Jean-François de la Rocque...258
Debert, NS................. 238, 374
debris field...........................321
debris-flow deposit.....................89
debris flows............**14**, 115, 328–329
debris slide..........................**329**
decapods.............................186
Decarie Interchange, Montréal, QC.......292
deciduous plants......................217
deciduous trees................. 194, 218
Decodon.............................203
Deep Bay impact structure, SK...........337
deep crust exposed....................96
deep marine sediments.................30
Deep Panuke Platform..................272
Deep River, ONT......................343
deep-sea channel.....................103
deep-sea cores................. 196, 227
Deep Sea Drilling Project................27
deep-sea sediments...................165
deer.........................191, 217, 235
deer-like mammals....................212
deforestation........................242
deformation................. 18, 370
deformation timing.....................45
deformed mafic dykes...................79
deglaciation.......... 222, 232, 236–237
degradation of oil.................272–273
Delphi, Greece........................261
Delta, BC............................315
delta deposits, as source of aggregate....291
delta plains.........................244
delta, eastern Baffin Island, NUN........223
delta, sediment supply................316
deltaic environments...................12
deltas..................... 12–13, 224, **315**

deltas in Scotian Basin..................183
deltas in Sverdrup Basin................173
Demascota G-32 well...................170
Denali Fault...................177, 200, 208
Denali (formerly Mount McKinley), Alaska.............................209
Dendrerpeton..........................155
dendrochronology....................**43**
Dene communities....................347
dense non-aqueous phase liquids
  (see also DNAPLs)....................**345**
dental fluorosis......................**350**
dentary.............................189
dents (in sediment)....................**12**
Denys, Nicholas.......................258
deposition.........................5, 66
desertification.......................243
detrital zircons........... **42**, 78, 95, 97, 375
deuterostomes..................56, **58**, **60**
Devil`s Coulee, AB.....................185
Devil's Elbow Brook, NB................259
Devon, England........................40
Devon Island, NUN ...71, 142, 191, 212–213, 337, 368
Devon Island Ice Cap, NUN..............236
Devonian........**40**, 46, 51, 59, 61, 127–128, 131–150, 157, 159, 169–170, 180, 249–250, 262, 269–270, 279–280, 287–288, 294, 311, 315, 356, 365, 374
Devonian deformation..................142
Devonian deltaic sediments............142
Devonian estuarine sediments..........142
Devonian estuary.....................366
Devonian evaporite deposits...........147
Devonian fluvial sediments............142
Devonian folding, Arctic..............142
Devonian, geological map............132
Devonian granitic batholiths...........145
Devonian granitic plutons.............145
Devonian gum beds...................262
Devonian gypsum deposits............137
Devonian halite deposits..............137
Devonian mafic volcanics..............148
Devonian mass extinction event........**147**
Devonian, oil sands source rocks........272
Devonian ore in granitic rock..........251
Devonian paleogeography......... 142, 148
Devonian plants......................126
Devonian redbeds....................136
Devonian reefs as reservoir rocks ...267, 269, 271
Devonian reefs..........146–147, 268–269
Devonian source rocks.................274
DEW Line............................345
diagenesis.................. **9**, 16, 364
diamond deposits....................258
diamonds......... **19**, 21–22, 248, **256**–258
diapiric salt structures.................**150**
diapirs......................... 175, 268
diapsids 56, 60, **62**, 163, 189, 367
diatoms.............................368
diatreme breccia.....................257
diatremes..........................**257**
Diavik Mine, NWT.....................258
diesel......................... **265**, 274
diesel-engine exhaust as source of PAHs ..347
diets of Aboriginal peoples.............347
differences between eastern
  and western Pacific Ocean 169
digging traces.......................107
dimension stone................ 279, **281**
Dimetrodon..................**62**, 158, 189

dinoflagellates.................**56**, 163, 368
Dinomischus...........................125
dinosaur embryo.......................185
dinosaur footprints....................185
dinosaur habitats.....................184
dinosaur physiology..................**67**
Dinosaur Provincial Park, AB.......10, 52, 54, 181, 184–185
dinosaur footprints, bipedal.............166
dinosaur remains.....................161
dinosaurs..... 51–52, 54, 56, 60, **62**, 64, 128, 163, 168, 184–185, 187–190, 367–368, 374
dinosaurs, earliest in Canada 168
dinosaurs, extinction of 187
diorite.................................9
dip....................................**45**
Dirty Thirties........................299
disaster.............................322
discontinuous permafrost...........224–225
Distant Early Warning (DEW) Line
  radar stations......................345
Distichophyton.......................140
distribution of continents and oceans.....66
disused mine.........................342
divergent plate boundaries............28, 34
DNA..................................55
DNAPLs..............................**345**
dodo.................................64
dogs.............................190–191
Doliodus.............................144
dolomite............. 10, **19**–20, 250, 284
dolostone cliffs.......................337
dolostones................ **10**, 281, 284, 364
dolphins.............................164
domain..............................**55**
Dome Mine, ONT.....................255
domes..............................231
doming..............................31
Don Beds..................... **217**–218, 374
Don Valley Brick Works, Toronto, ONT................ 216–218, 227, 291
Door Jamb Mountain, AB..............179
Dorchester, NB.......................279
Douglas Channel, BC..................327
Downie Slide, BC.....................330
drainage and rockfalls.................330
drainage basins, middle Cenozoic.......206
drainage basins, modern..............206
Drake Passage, opening of 196
Drake Point, NUN.....................271
Drake Point Gas Field..................272
Drake, Edwin........................262
dredge, use in mining placer deposits....254
dredges.............................255
drift................................**219**
drift of North America.................175
drifting.................. 30, 36, **171**, 359
Driftwood Canyon, BC 200–201
drives...............................237
dropstones...................... **103**, 159
droughts................. 298–299, 304, 321
drowned estuaries....................314
drowned landscapes..................311
Drumheller, AB...... 10, 37, 51–52, 147, 164, 181, 185, 187, 214, 264, 368
drumlin fields.......................309
drumlins..........**222**–223, 308–309, 364
dry gas.........................**265**–266
drywall plasterboard...................19
duck-billed (hadrosaur) dinosaur eggs ...185
duck-billed platypus..................189

dugongs . . . 190
Dune de Bouctouche, NB. . . . 314
dunes . . . **11**–12, 307
dung . . . 54
Dunnage Zone . . . **102**
dust-bowl conditions . . . 299
dust storms . . . 321
dwarf birch . . . 229
dwarf willow . . . 229
dwelling burrows . . . **54**
Dyer's Bay, ONT . . . 53
dyke swarms . . . **9**
dyke swarms, radiating . . . 9
dykes . . . **5**, 8–9, 27, 39, 244
dykes, associated with rivers . . . 331–332
dykes, coastal . . . 332
dynamite explosions . . . 47
E. coli . . . 338
Eagle Butte impact structure, AB . . . 337
Eagle Plains Basin . . . 173
ear ossicles . . . 189
early Lake Erie . . . 233
early Lake Ontario . . . 233

E

Earth . . . 22, 65–66, **75**, 77, 241, 375
Earth-Moon distance apart . . . **76**
Earth-Moon system . . . **76**
Earth's age . . . 38, 42
Earth's atmosphere . . . 241
Earth's ecliptic axis . . . **227**–228
Earth's interior . . . 22
Earth's magnetic field . . . 44, 49, **76**
Earth's magnetic poles . . . 43
Earth's orbit . . . 227–228
Earth's origin . . . 6
Earth's rotation . . . 66
Earth's rotational axis . . . **227**–228, 241
Earth's tilt . . . 227
earthquake damage . . . 324
earthquake forecasting . . . **324-325**
earthquake precursor events . . . 325
earthquake prediction . . . **324-325**
earthquake types on the west coast . . . 323
earthquakes . . . 22, 28, 47, 115, 207, 308, 321, **322**, 323–327, 332
earthquakes and building codes . . . 325
earthquakes and building design . . . 324
earthquakes and fire . . . 324
earthquakes and public trust . . . 325
earthquakes and tsunamis . . . 326
earthquakes, as landslide triggers . . . 329
earthquakes at subduction zone . . . 323
earthquakes, causes of . . . 322
earthquakes in Atlantic Provinces . . . 324
earthquakes in Canada . . . 322
earthquakes in eastern Canada . . . 324
earthquakes on the west coast . . . 323
earthy lustre . . . **21**
East Africa, location in Rodinia . . . 97
East African Rift System . . . 30
East Antarctica, location in Rodinia . . . 97
East Brazil, location in Rodinia . . . 97
East Coulee, AB . . . 187
Eastend, SK . . . 185
easterlies . . . 65
eastern Pacific Ocean . . . 169, 242
Eastern Shore, NS . . . 307–308
Eastern Townships, QC . . . 259, 296, 353
Eau-Claire (east) impact structure, QC . . . 337
Eau-Claire (west) impact structure, QC . . . 337
Ebony Black . . . 288
eccentricity cycle . . . **227**–228
echinoderms . . . 56, **58**, 59, **60**, 106–107, 117, 159, 168
echinoids . . . 56
eclogite . . . 96, 157
Economy, NS . . . 166, 313
ecosystem destruction . . . 339
Ecuador . . . 242
Ediacara biota . . . **104**–105, 365, 375
Ediacaran . . . 100, 104, 106, 109, 135, 286, 373
Ediacaran paleogeography . . . 102
Ediacaran Period . . . **101**, 105
Ediacaran quartz-rich sandstones . . . 112–113
Ediacaran sea floor . . . 105
Ediacaran to Silurian magmatic arc rocks . 139
Ediacara-type fossils . . . **104**
Edmonton, AB . . . 90, 118, 146–148, 269, 273, 284, 343
Eiszeit . . . **219**
ejecta . . . 334, **335**, 336
ejecta blanket . . . **334**
ejecta curtain . . . **334**
ejecta layer . . . 188, 335
Ekati Mine, NWT . . . 257–259
El Niño . . . **242**
El Niño, 1997–1998 . . . 242
Elbow impact structure, SK . . . 337
Elbow River, AB . . . 306, 332
electrical properties of rock . . . 47
electricity . . . 265
electricity, generated using coal . . . 262
electromagnetic energy . . . 50
electromagnetic methods . . . **49**
electrons . . . **41**
Elephant Rock, PEI . . . 67
elephants . . . 190, 214
Elie, MB . . . 243
Elizabeth Avenue, St. John's, NL . . . 289
elk . . . 229
Elkview Mine, BC . . . 68
Ellef Ringnes Island, NUN . . . 17, 45, 71, 174, 202, 225, 268
Ellef Ringnes Island, NUN, discovery of . . . 193
Ellesmere highlands . . . 205
Ellesmere Island, NUN . . . 8, 39, 61, 71, 111, 121, 138, 142, 144, 150–151, 158–159, 174, 193, 202–203, 206, 212–213, 220, 222, 236, 239, 300, 304, 356, 365, 370, 374–375
Ellesmere Island, possible Neogene glaciers . . . 212
Ellesmerian Orogeny . . . **142**, 150, 357, 375
Ellicott brothers . . . 4
Elliot Lake, ONT . . . 88–89, 275, 342
elm fruit . . . 368
elms . . . 201, 218
Elonichthys . . . 150
Elu Basin . . . 93
Elwyn Bay, NUN . . . 256
Elzevir Basin . . . 94
Elzevir Basin minerals . . . 94
embryo . . . 58
embryophytes . . . **126**
emery cloth . . . 21
end-Cretaceous mass-extinction event . . . 62, 64, **187**–188
end-Cretaceous mass-extinction event, recovery from . . . 196
end-Cretaceous mass-extinction, selectivity of . . . 187
end moraines . . . **222**
end-Ordovician mass-extinction event . . . 64, **121**, 375
end-Permian mass-extinction event . . 59, 64, **159**, 163, 166, 189
end-Permian mass-extinction event, recovery from . . . 163
end-Triassic mass-extinction event . . . 62, 64, **166**, **168**
energy . . . 261–277
energy spectrum . . . 241
energy supply . . . 130
England . . . 239
England, flooding in . . . 322
Enniskillen Township, ONT . . . 262
Encrinurus deomenos . . . 40
entelodonts . . . **205**–206, 214
Enterprise Mine, BC . . . 258
environmental assessments . . . 339, 345
environmental issues . . . 338–353
Eocene . . . 40, 48, 189, 193–194, **195**, 196–206, 212, 369
Eocene climate . . . 194, 196, 202, 368
Eocene coals . . . 264
Eocene Epoch . . . **195**
Eocene fluvial deposits . . . 129
Eocene lake deposits . . . 129, 201–202, 204, 367–368
Eocene paleogeography . . . 202
eocrinoids . . . 58, **60**
Eoentophysalis . . . 97
eolian deposits . . . **11**
eolian dunes . . . 107
eolian sandstone . . . 168
eonothem . . . **40**
eons . . . **40**
epicentre . . . **323**–324, 326
epidote . . . 16
Epipremnum . . . 212
epochs . . . **40**
Equator . . . 25, 242
Equisetum . . . 154
Eramosa Marble . . . 285
eras . . . **40**
erathem . . . **40**
Erebus . . . 192
Erly Lake, NWT . . . 226
erosion . . . 5, **9**–10, 18, 22, 66, 369
erosion surfaces . . . 12, 18
erratic . . . 284
erratics . . . **219**, 221
eruption cloud . . . 328
eruption column . . . 328
escalier Casse-Cou, Québec City, QC . . . 294
Escherichia coli . . . 338
Escuminac Point, NB . . . 14
Escuminaspis . . . 143
esker, source of gravel . . . 292
eskers . . . **222**, 224, 364
eskers, as source of sand . . . 291
Esterhazy, SK . . . 234
Estevan, SK . . . 263
estuary . . . **317**
estuary, water layering in . . . 317
Etawney-North Knife Moraine . . . 236
ethane . . . **265**
Eubacteria . . . 54, **55**
Eukaryota . . . 54, **55**
eukaryotes . . . **55**–56, 64, 87, 98, 126
eukaryotes, earliest . . . 55, 97
eukaryotic cells . . . 98
Euramerica . . . 132, **133**–134, 136, 139, 141–142, 145–148, 150, 152, 155, 157, 169, 355–357, 360, 375
Euramerica, Avalonian margin . . . 145
Euramerica, western margin . . . 141
Euramerican margin on Ural Ocean . . . 138
Eurasia . . . 162, 191, 203, 219, 234, 359–361
Eurasian Plate . . . 29, 31, 169, 203
Eureka, NUN . . . 39, 150, 174, 193, 203, 304
Eurekan Mountains . . . 202, 203
Eurekan Orogeny . . . **202**, 205, 213, 361, 374
European settlers . . . 238, 282
Europeans . . . 4, 372
Eurynotus . . . 150
eurypterids . . . **57**, 119, 134–135, 159
Eurypterus remipes . . . 57
Eusthenopteron . . . 51, 143–144, 365
Eusthenopteron foordi . . . 144
Eustis Mine, QC . . . 259
Evangeline Beach, NS . . . 319
evaporation . . . 11, 300
evaporite basins . . . 148
evaporite deposits . . . 301, 367
evaporite minerals . . . **19**
evaporite rocks . . . **19**
evaporites . . . **11**, 97, 171, 268
evaporites, as industrial mineral . . . 292
evaporites in Sverdrup Seaway . . . 151
Eve Cone, BC . . . 192
evolution . . . 40, **54**–61, 62, 64, 84, 87, 89, 364–368
evolution, early . . . **82**
excavation stage . . . 334–335
exosphere . . . **65**
Explorer Plate . . . 197, 207
Exselsior (steamship) . . . 246
Exshaw, AB . . . 46, 141, 179
Exshaw Thrust . . . 46
extension . . . 16–17
extinct phyla . . . 123
extinction and volcanicity . . . 168
extinctions (see also mass-extinction events)
extinctions of large mammals . . . 236
extraterrestrial bodies . . . 355
extraterrestrial impact structures . . . 64
extraterrestrial impacts . . . 64
extraterrestrial radiation . . . 76
extraterrestrial water . . . 66
extremophiles . . . **82–83**
extrusive rocks . . . **6**
eyes . . . 58
eyes, evolution of . . . 106
Eyjafjallajökull, Iceland . . . 6, 328, 352

F

Fairbanks, Alaska . . . 271
Fairview Lawn Cemetery, Halifax, NS . . . 8, 282
Fairweather, Henry . . . 278
falls, as type of landslide . . . **329**
families . . . **55**
Farallon Plate . . . 198
farmland, rezoning of . . . 339
Farwell Canyon, BC . . . 371
fault-bound basins . . . 374
fault breccia . . . **17**
fault movement and tsunami generation . . 325
fault rocks . . . **17**
fault zones as conduits for mercury . . . 349
fault zones, as source of mercury . . . 348
faults . . . **16–17**, 28, 45, 48, 363, 369
faults and earthquakes . . . 322
faults, as petroleum traps . . . **267**
faunas associated with deep-sea seeps . . . 175
Favosites . . . 133
feathers . . . 187, 201, 204

feeder dykes **30**–31, 34
feeding burrows 53
feeding traces **54**
feldspar, orthoclase 9
feldspar, plagioclase 9
feldspars 7, 9, **19**, 10, 15, 20, 250, 289, 335
felsenmeer **225**–226
felsic **7**
felsic gneiss 79
felsic magma 8, 31
felsic melt 16
felsic rocks **7**, 9, 82
felsic volcanic rocks **7**, 82
Feltzen South, NS 145
Fenian Raids 289
Fergus, ONT 244
Fermont, QC 89, 365
fern foliage 53
fern fronds 186
fern prairie 188
fern spike **188**
fern spore 129
ferns 127–129, 153–154, 163, 167, 188, 196, 203–204
ferrous-iron-rich oceans 256
Ferryland, NL 286
fibrous habit 352
fibrous magnesium silicate 352
Fidler, Peter 264
Field, BC 114, 118, 122
field geology 47
field mapping 45
fieldstone 281–**282**, 284
filter feeders 61
filtration feeding structure 107, 364
fin structure 144
fingernail 21
fingers 144
fins **61**, 143
fiords 223, **309**, 326
fir 128, 212, 218, 229
fish 51, 56, 60–**61**, 142, 144, 174, 185–186, 201, 205, 212, 237, 239, 244, 365–366, 375
fish bones 217
fish, Devonian 143
fish evolution 138
fish fossils 143
fish habitat 340
fish, primitive 135
fish, Silurian 143
fish stocks 50
Fisher site, ONT 238, 374
fission-track dating **43**
fission-track studies 209
fission tracks **43**
fissures 7
Five Islands, NS 12, 372
Five Islands Provincial Park, NS 61, 167
Flat Island Spit, NL 314
Flathead Range, AB 147
Flathead Valley, BC 200
flatulent ruminants 68
flexible neck 144
flight disruptions 6
Flin Flon, MB 90, 92, 246, 355
Flin Flon Arc 91–92
Flin Flon District 249
flood plains 12, 18
flooding 320
flood-plain management program 332
floods 322, **330**–332
floods, coastal 321
floods, damage from 330, 332
floods, river 321
Florida (terrane) 109
Florissantia 201
Florissantia quilchenensis 129, 204
flow of rivers 304
flower calyx 129
Flower Pot, Graham Island, BC 165
flowering plants (see also flowers) 127, **129**, 367–368
flowerpots 152
flowers (see also flowering plants) 188, 201–202, 204, 367
flowers and insects 58
flowers, origin of **129**
flows, as type of landslide **329–330**
fluid content of rocks 47
fluorbritholite 182
fluoridated water 350
fluoride 347
fluoride and teeth 350
fluoride in drinking water 346
fluoride in groundwater 347
fluoride levels 337
fluorides, natural occurrences of 350
fluorine 19, 350
fluorite 21, 251
flute marks **12**
fluted points **235**
fluvial deposits **9**–10, 167
fluvial deposits, as source of aggregate 291
fluvial environments 12
fly in amber 186
focus of an earthquake **322**
folded strata 14
folding **16**, 36, 45, 374
folding, role in petroleum traps 267
folds **16**–17, 33
Fond du Lac, QC 90
food chain 340
fool's gold 21
Foothills Erratics Train, AB 221, 223
Foothills (see Rocky Mountain Foothills)
footprints 53–**54**, 158
foraminifera **55**, **56**, 163, 227
Ford's Harbour, NL 74
fore-arc basin **36**
foreland basins **36**–37, 115, 178–180, 264
foreshocks **325**
forest fire 188
forest fires, as source of PAHs 347
forest hunters 238
forests, early 128, 143
Forillon NPC, QC 70, 135
formation of Earth 55
formation of elements **74**
Fort Macleod, AB 238
Fort McMurray, AB 186, 273
Fort St. James, BC 157
Fort St. John, BC 211, 237
Fort Wayne Spillway 233
Fortress of Louisbourg, NS, coal for 263
Fortune Bay, NL 326
Fortune Head, NL 44, 106
fossil assemblages 39
Fossil Creek, NUN 57
fossil droppings 54
fossil dung 54
fossil fish 143
fossil forest, Axel Heiberg Island, NUN 193–195, 368
fossil forest, Willow Creek, AB 187
fossil fuels 67–68, 243, 276, 305, 339
fossil fuels as source of arsenic 349
fossil preservation 52, 125
fossil ranges 40
fossilization **52**
fossils 10, 39–40, 44, 51–64, 364, 374–375
fossils of soft-bodied animals 100–101
Foxe Basin 71, 119, 231, 236
fracking **274**
Fractofusus 105
fractured rock 334
fractures 9, 16–17, 219
Fram 194
France 40
François-Xavier Garneau monument, Québec City, QC 8
Frank, AB 264, 320–321
Frank Slide, AB 146, 320–321, 330
Franklin Dyke Swarm 9, 110–111, 375
Franklin Dyke Swarm, associated mafic sill 111
Franklin Expedition 192
Franklin Island, ONT 96, 355
Franklin, John 192
Franklin Mountains 72
Franklinian Mountains **142**, 144, 150, 375
Fraser Delta, BC 244, 315–316, 339
Fraser Fault 17, 177, 200
Fraser Mine, ONT 253
Fraser River, BC 4, 17, 211, 259, 303, 315, 374
Fraser River, evolution of flow 210–211
Fraser Valley, BC 331
free oxygen in atmosphere 88
freezing and thawing 9
Frenchman Valley, SK 54, 185
Frenchman's Bay, ONT 340
Frobisher Bay, as estuary 317
Frobisher, Martin 192
Frobisher, Martin, and minerals 258
frond-like organisms 105
Frondophyllus 105
Front Ranges (see Rocky Mountain Front Ranges)
Frontenac Arch 119
Frood-Stoble Mine, ONT 253
Frost Building North, Toronto, ONT 289
frozen ground 224
fruit 201
fuel burning 347
fumaroles 328
Fundy Basin 166–167, 170, 359, 374
Fundy Geological Museum, Parrsboro, NS 155
Fundy NPC, NB 109
fungi 14, 54–**55**, 117, 126
fur trade 78, 293
fusulinid foraminifera **157**

G

gabbro 8–**9**, 30–31, 281, 288
gabbro gravestone 282
gabbro, Sudbury 253
Gaia **240**
Gakkel Ridge 174–175, 202
galaxies 75
galena **19**–20, 249
galena-rich ores 259
Galileo 334
Galway, Ireland, limestone from 286
gamma radiation 21
gamma ray spectrometers 351
gamma rays 351
Gander River, NL 109
Gander Zone 102
Ganderia 32, 102, 108–110, 115–117, 132–135, 140, 355, 358, 375
Ganderia, magmatic activity 133–134
Ganderian margin of Euramerica 141
Ganges-Brahmaputra Delta 315
garbage 341
garbage dumps 341
Garnet Rock, NB 309
garnets 15–16, **19**–20
Garnish, NL 326
Garson, MB 119–120, 284
Garson Mine, ONT 253
gas 47, 68, 242, 261, **265**–267, 271–273, 373
gas, biogenic **267**
gas field 268
gas, flaring off of 270
gas from Scotian Basin 171
gas hydrates 196, **274**
gas, origin of **266**
gas, piped through bamboo in China 261
gas station spills 345
gas, thermogenic **267**
gas, uses of 273
gaseous planets **22**
gasification **274**
gasoline **265**, 274, 345
Gaspé, QC 136
Gaspé Highlands, QC 148–149
Gaspé Peninsula, QC 70, 135–136, 149, 238, 266, 270,313, 343, 346, 350, 365
Gaspestria genselorum 140
Gastown, Vancouver, BC 283
gastropods (see also snails) 56, **59**, 107, 117, 120, 168, 217, 285
Gatineau, QC 121
Gatineau Hills, QC 94
gemstones 292
genera **55**, 64
Gentilly, QC 343
geochemical analysis 106
geochemical evidence 375
geochemical hazards 347
geological ages 45
geological column **39**, 41
geological maps 39, **45**
geological mapping **45**
geological mapping in Rockies 179
Geological Society of America 86, 161
Geological Survey of Canada 27, 41, 51–52, 100, 122, 131, 161, 178, 193, 219, 252, 268, 272, 347, 351, 373
Geological Survey of Newfoundland 99
geological time 38–44
geological time scale **39–40**, 44
geomagnetic surveys 50
geophones **47**
geophysical exploration 170
Georges Bank Basin 170
Georgia Basin 173, 178
Georgia Basin, coal deposits in 264
Georgian Bay, ONT 18, 22, 93, 96, 136, 238, 355
geosphere **65**, 67–68, 241
geothermal energy **276**–277, 373
geothermal energy and coal mines 276
Gertrude Lawson House, Vancouver, BC 282

Gesner, Abraham............ 131, 261–262
Gesner, Abraham as NB provincial
    geologist 262
giant arthropods 155, 366
giant beaver 217–218, 372
giant bison 229
giant ground sloth.....................191
Giant Mine, Yellowknife, NWT 343
giant short-faced bear 234–**235**
Gilbert, Clothilde 259
gill arches..........................61
Gillam, MB 305
gills 144
Ginkgo 201
ginkgo leaves 186
ginkgos.............................127
glacial advances....................227, 241
glacial boulders 43, 108, 282
glacial cobbles 282
glacial cycles.......................228
glacial deposits............. 103–104, 309
glacial deposits in tropics 104
glacial erosion.......................363
glacial grooves **219**–220
glacial intervals................... 227, 374
Glacial Lake Agassiz ...232–**233**, 236, 331, 374
Glacial Lake Algonquin............ 233, 238
Glacial Lake Barlow-Ojibway 233
Glacial Lake Chicago 233
Glacial Lake Chippewa 233
glacial lake clays.....................341
glacial lake deposits 364, 369
Glacial Lake Hough 233
Glacial Lake Madawaska 358
Glacial Lake Maumee 233
Glacial Lake McConnell 232
Glacial Lake Minong-Haughton.........233
Glacial Lake Ojibway............. 232, **236**
Glacial Lake Stanley..................233
Glacial Lake Whittlesey 233
glacial lakes..........**222**, **224**, 236, 238, 318
glacial map of Canada 219
glacial maximum, last................374
glacial outwash (see also outwash) 371
glacial outwash deposits as source
    of aggregate 291
glacial retreat 23
glacial striations 103, **219**
glacial till 89
glaciation-interglaciation cycles 229
glaciation, last........ 44, 231, 234, 237, 374
glaciations..... 5, 23, 89, 120–121, **227**, 243,
    309–310, 318, 369–370, 374–375
glaciations in Paleozoic 153
Glacier NPC, BC....................114
glaciers...... 5, 35, 70, 89, 121, **219**–223, 230,
    235–236, 242, 244, 300, 302, 310, 364, 374
glaciers, as water supply 303
glass 21–22
glass beads (tektites) 187–**188**
glassy lustre..........................**21**
glaucophane..................... 16, **165**
GLIMPCE project 96
global change........................339
global cooling 121
global reference sections (see also
    stratotype sections) **44**
global warming (see also climate
    warming)..... 68, 121, 241, 243–244, 344
Gloecapsomorpha 266
Glooscap 372
Glossopteris 25

gneiss **15**, 18, 74, 79, 355
gneiss, as building stone 289
gneiss, derivation of 79
gneiss derived from granitic rocks 80
gneiss derived from mafic rocks 80
gneiss derived from sedimentary rocks....80
Goderich, ONT 11, 137, 375
goethite **19**–20
gold............ 19, 246–248, 250, 254–256,
    258–259, 343–344
gold deposits 260
gold mine tailings 350
gold nuggets 259
gold ore 92
gold panning **254**–**255**
gold processing as source of arsenic 349
gold, uses of 260
Golden, BC..........................114
Golden Horseshoe, ONT...............339
golden spikes **44**
gold-mine tailings....................343
gomphothere 214
Gondwana..... **162**, 171, 176, 183, 189–190,
    359–361, 363, 374
Goobies, NL.........................288
Gorda Plate 197, 207
gossan **260**
Gould, Stephen J.....................123
Government House, St. John's, NL 286
Government House, Victoria, BC 290
Gow impact structure, SK 337
GPR (see also ground-penetrating radar)... **49**
GPS (see also global-positioning
    systems).....................29–30, 50
graded bedding.......................**12**
Graham Island, BC 165
grain size 11
grain size in igneous rocks **9**
grain size in sediments.................10
Grand Bank, NL.................. 311, 326
Grand Banks Earthquake 14, 326–327
Grand Banks Earthquake, fatalities 327
Grand Banks of Newfoundland 110, 148,
    170–171, 183, 231, 271, 374
Grand Falls, NB................... 134, 358
Grand Manan Island, NB 32, 110, 168
Grand Pré, NS 318
Grand River, QC 304
Grande Cache, AB 185
Grandview Hills, NWT................205
granite and anorthosite plutons 95
granite as building stone...............287
granite as commercial term 281
granite boulders......................44
granite columns......................280
granite etching techniques.............283
granite lighthouse.....................288
granite quarry 288
granite steeple 288
granite trim 280, 286
granites **8**–**9**, 34, 78, 281, 295
granitic complexes 82
granitic layer, Sudbury 253
granitic magmas **8**, 16, 82
granitic melt 15
granitic pegmatites...................275
granitic plutons 80
granitic rocks **9**, 21
granitic rocks, Sudbury, ONT.......253–254
granular rocks........................**9**
graphite **19**, 21–22

graphite, Aboriginal use 258
graphite deposits.....................349
graptolites 56, **60**, 117–118, 121, 138
grass 129, 206, 369, 374
Grassi Lakes, AB 267
Grassi Lakes Trail, AB................147
grasslands 129, 207, 214, 237
Grasslands NPC, SK 187, 299
grass-like plants......................204
Grassy Lake, AB 186–187
gravel 9–**10**, 280, 291
gravel pit 341
gravels, pre-glacial, as aggregate 291
gravestones 8
graveyards...........................282
gravimeter...........................**49**
gravity....................... **49**, 75
gravity anomalies **49**
gravity surveys 50, 96
grazing animals ... 84, 106, 214, 350, 364, 369
Grease River, SK 372
Great American Bank (reef complex) 108, 114
Great Barrier Reef, Australia 11, 57, 108
Great Bear Lake, NWT...............274
Great Bear River, NWT 301
Great Britain 39
Great Canadian Oil Sands Limited......272
Great Lakes 231, 233, 304, 339–340, 369
Great Lakes, as sediment traps...... 315, 317
Great Lakes, as water regulators 303
Great Lakes basins...................303
Great Lakes, changes in level 318
Great Lakes Lowlands................112
Great Oxidation Event **67**, **88**, 365, 375
Great Slave Lake, NWT 146, 148, 250
Greater India, location in Rodinia 97
green algae 125–126, 135
Green Gardens, NL..................116
Green Head Island, NB 279
Green Point, NL **44**
Green Point, NS 229
greenhouse effect **241**
greenhouse gas emissions 277
greenhouse gases 14, 104, 242, 244, 344
greenhouse world 197
Greenland 71, 78, 83, 134, 141, 162, 184,
    202–203, 213, 227, 230, 238–239, 244,
    355, 361, 370, 374
Greenland, as battering ram........202–203
Greenland-Eurasia Plate...............202
Greenland ice cores..................242
Greenland Ice Sheet 220, 231, 236
Greenland, location in Rodinia..........97
Greenland Plate 184, 203
greenschist **16**, 173
Greenshields Building, Vancouver, BC 283
greenstone belts 79, 80–81, 255
greenstones **16**, **79**
Greenwich Dunes, PEI................315
Grenville belt.......................**34**
Grenville Orogen **34**, 36, 78, 87, 93, 96,
    111–112, 287, 293–294, 296, 369, 375
Grenvillian crustal roots 96
Grenvillian gneiss....................294
Grenvillian Mountains 96–97, 355
Grenvillian Orogeny **87**, 95–97, 112,
    355, 375
Grey Coastal Granite 287
greystone................... 281, 284
Greystone City (Saskatoon, SK) 284
Grinder's disease....................**352**

Grise Fiord, NUN.............150, 174, 203
grizzly bear 235
grooves (in sediment) **12**
Gros Morne NPC, NL 34, 115–116, 358
Grotto Mountain, AB..................46
ground moraine......................**222**
ground-penetrating radar.............**49**
ground shaking **324**
ground sloths 229
ground subsidence 321
groundwater.....9, 16, 54, 65, 244, 300–**301**,
    303–304, 338–339, 341, 345, 349
groundwater contamination 338–339,
    341, 345
groundwater, dissolution of rock 301
groundwater quality.................302
groundwater quality, and impact
    melt rocks 337
groundwater supply..................304
groundwater temperature 304
groundwater use 303
Gulf Islands, BC.......... 178, 199, 305, 314
Gulf Islands landscape 199
Gulf Islands Sandstone............ 283, 287
Gulf Islands water supply.............306
Gulf of Alaska 210
Gulf of Maine 109, 334
Gulf of Mexico......................180
Gulf of St. Lawrence 70, 148–150, 229,
    231, 312–313, 370
Gulf Stream 66, 233
Gully (see The Gully)
Gunflint banded iron formation 85
Gunflint chert 86–87, 97, 364, 375
Gunflint fossils 86, 97
Gunflintia 97
gusher............................ **262**, 270
gut................................ 61
gut contents 122
Guysborough County, NS 350
gymnosperm forests, decimation of......188
gymnosperms 129, 154
gypsum **11**, 16, 19, 21, 97, 175
gypsum cliffs........................149
gypsum, in relation to petroleum 268
gypsum, uses of 280

H

habitat fragmentation 339
Haddington Island, BC 287
Haddington Island Andesite...........287
Hadean...............**40**, 76, **77**–78, 82, 354
Hadean and Archean rocks, map........77
Hadean Eon.........................**76**
hadrosaur babies.....................185
hadrosaur embryos...................185
hadrosaurs.................... 174, **185**
Hagersville, ONT....................137
hagfish 61
Haicheng earthquake, China 325
Haida Gwaii, BC 70, 165–166, 237, 326
Haida Gwaii, BC, and earthquakes.......323
hair 189
hair, source of fine atmospheric particles 351
Haiti, earthquakes in..................322
half life **42**
Halfway Rock Point, ONT............136
halides **19**, 247
Halifax, NS 8, 50, 145, 188, 240, 282, 284,
    290, 343, 350–351
Halifax City Hall, NS.................288
Halifax City Memorial Library, NS 284

Halifax Harbour, NS....50, 239, 272, 311, 340
halite...............**11, 19**, 21, 147, 171, 175
halite deposits.........................149
Hall, James............................131
Halley, Edmond.........................41
Hallucigenia..........................123
halogen elements.......................**19**
Halton Region, ONT....................284
Halton Till..................218–219, 301, 341
Hamilton, ONT..........137, 262, 338, 343
Hampstead, NB.........................279
Hampton, NS...........................317
Hand Hills, AB...................205, 214
Hansen, George........................193
Hantsport, NS....................149, 366
harbours..............................340
hard plate collision..................134
hard water............................**301**
hardened watersheds...................340
hardness...............................**21**
Hare Fiord, NUN..................151, 356
hares.................................190
Harricana Moraine.....................236
Harris, Lawren.........................78
Harris, Mike..........................338
Harvard University, Massachusetts.....86, 122–123
Harvey, the Reverend Mr..........99–100
Haughton impact structure, NUN...191, 212, 337, 368
Hawaii.................26, 33, 35, 242–243
Hawaii hot spot..................**35**, 198
Hawaii, as analogy....................156
Hawkesville Kame, ONT................223
Hay River Embayment..................118
hazards, geological..............320–332
hazel.................................212
head end...............................57
headlands.............................309
heads, evolution of...................106
Head-Smashed-in Buffalo Jump, AB....237–238, 374
Health Canada.........................350
Hearne Craton................78, 87, 90–92
Hearne, Samuel........................259
heart..................................58
heat waves............................321
heavy oil.............................**265**
Hebrew calendar........................38
Hebron Oil Field......................271
hedgehogs........................190–191
helium........................43, 65, 74
Hell`s Half Acre.................269–270
hematite..........**19**, 21, 67, 84, 255–256
hematite, Aboriginal use..............258
hemichordates..........................56
hemlock..........................212, 218
Henderson Lake, BC....................304
Henderson, Robert.....................246
Hennepin, Louis.........................4
herbaceous clubmosses............127, 142
herbivore..............................54
herds..................................54
Herpetogaster.......................123
Hess, Harry........................27–28
hexagonal system.......................**21**
hexane...............................**265**
Hibernia Oil Field...............183, 271
hickory....................212, 218, 229
High River, AB........................332
high-grade metamorphic rocks...**15**–16, 172

Highland Valley, BC...................259
Highland Valley Copper Mine, BC......250
Highvale, AB..........................264
Highway 3, BC...................103, 328
Highway 101, NS.......................149
Highway 401, ONT......................340
Highwood River, AB...................332
Hillsborough, NB.......................54
Hillsborough Bay, PEI.................271
Hillsborough mastodon..................54
Himalayas......22, 24, 32, 96, 163, 242, 355
hippos................................191
Hoggar, location in Rodinia............97
holding ponds.........................342
Holleford impact structure, ONT.......337
Holman, NWT..................150, 174, 203
Holmes, Arthur................27–29, 41–42
Holocene.................211, 236–239, 339
Holocene Epoch......................**195**
Holyrood, NL..........................288
Homo sapiens...................55, 234
homospores...........................**128**
hoodoos..........................10, 185
hoofed mammal.........................129
Hope, BC......................17, 200, 207
Hopedale, NL...........................79
Hopedale Basin........................170
Hopewell Cape, NB.....................152
horizons (soil).......................**14**
horn coral.............................57
hornblende.....................**19**–20, 79
Hornby Bay Basin.................93, 275
Horne Mine............................249
horned dinosaur.......................185
horns................................**221**
hornworts.............................126
Horsefly, BC.....................200–201
horses........190–191, 206, 214–215, 229, 234–235
horses, migrations....................206
horseshoe crabs.......................119
Horseshoe Falls, ONT.................4–5
horsetail trees.......................159
horsetails 2, 127–128, 142, 153–154, 167, 189
Horton House, Acadia University, Wolfville, NS....................290
hot spot traces......................**35**
hot spot volcanic activity.............35
hot spots............9, 34–**35**, 172, 182, 208
hot springs...........................328
Hot Springs Cove, BC..................326
Hotel Vancouver, Vancouver, BC.......287
House of Assembly, Québec City, QC.....8
household contaminants................353
Howe Sound, BC........................342
Hubble Space Telescope.............75–76
Hudson Bay....10, 23, 70–71, 78, 92, 97, 120, 205–206, 236–237, 244, 303, 312, 334,
Hudson Bay, as analogy...............118
Hudson Bay Basin..............118–119, 137
Hudson Bay Company...................264
Hudson Bay Lowlands......69, 182, 229, 369
Hudson Bay Mountain, BC..............302
Hudson Bay Platform..................112
Hudson Strait....................71, 236
Hudson, Henry.........................192
humans...........55, 57, 191, 235, 237, 240
Humber Zone..........................**102**
humus.................................**14**
Hungry Hollow, ON......................10
Hunter, Vern "Dry Hole"...............270

hunter-gatherers.....................235
hunting tools........................237
Huntsville, ONT......................349
Huron................................293
Huronian metasediments...............254
Huronian passive margin..............375
Huronian redbeds......................89
Huronian succession......78, **88**–89, 275, 354
Huronian tillites.....................89
Huronispora.........................97
Hurricane Edna......................308
Hurricane Hazel.....................331
Hurricane Juan......................243
hurricanes................243, 308, 322
Hutton, James........................18
Huxley, Thomas......................160
hydro...............................305
hydrocarbon traps...................**267**
hydrocarbons....................**265**, 347
hydroelectricity........244, 261, **276–277**, 304–305
hydrogen.........65, 74, 83, 265, 274, 299
hydrogen atoms......................265
hydrogen sulphide...........83, 120, 268
hydrological cycle...............**66, 300**
hydrophone streamer..................47
hydrophones.........................**47**
Hydro-Québec........................304
hydrosphere.........**65–66**, 67–68, 241, 364
hydrothermal fluids........**16**, 175, 247–249, 254, 256, 275, 300
hydrothermal fluids and gold........255
hydrothermal fluids, uranium-rich...275
hydrothermal plume..................248
hydrothermal vents.............83, **248**
hyenas..............................191
Hylonomus lyelli..............61, 155
Hylopus............................53
Hyopsodus.........................205
Hypacrosaurus.....................185
Hypsithermal interval...........**237**, 374
Hypsognathus.................166–167
hyraxes.............................190
Iapetan terranes....................139
Iapetognathus fluctivagus..........44

I
Iapetus Ocean......32–33, 99, 102, 108–110, 114–116, 133, 249, 294–295, 324, 357, 375
Iberia..............109, 171, 183, 357, 374
Ice Age....211, 216–218, **219**, 220–239, 242, 308, 364, 369, 374
ice ages..........................**219**, 367
ice caps..........70, 118, 120, 230, 236, 374
ice cores...........................227
ice fields......................69, **219**
ice floes...........................219
ice jams............................331
ice lenses..........................225
ice lobes......................231, 236
ice scours..........................231
ice sheets....23, 45, 147, 211, **220**–222, 228, 230, 236–237, 364, 369–370
ice sheets on Protogondwana.........149
ice shelves.........................103
ice storm...........................322
ice thickness........................49
ice wedges..................225–226, 310
icebergs.......................103, 237
Icefield Ranges.....................209
icehouse world......................197
Iceland.........6, 26–27, 31, 35, 70, 352

Iceland, analogy for Olympic Terrane.....199
Iceland, geothermal energy in...........276
Icemaker Glacier, BC....................241
ice-plucking...........................220
ice-wedge casts........................225
ice-wedge polygons.....................225
ichthyosaurs...60, **62**, 163–164, 173, 189, 367
Idaho..................................231
igloo..................................239
igneous rocks.....................**5–6**, 7–9
igneous rocks and water................301
igneous rocks, as building stones...287–289
igneous rocks, colour....................7
iguanas................................164
Igwisi Hills, Tanzania, kimberlites....257
Île d'Orléans, QC.................294, 297
Île Rouleau impact structure, QC......337
Ilgachuz Range, BC....................208
Illinoian Glaciation.................**227**
illite.................................290
ilmenite...............................77
imbricated thrust sheets...............92
impact breccia............252, 334–335, 337
impact craters..........................77
impact-generated dust plume...........188
impact melt......................334–335
impact, speed and angle...............334
impact stage..........................335
impact structure, evidence of.........335
impact structures.....77, 252–254, 333–**334**, 335–337
impact structures in Canada...252–254, 337
impact vapours........................334
impacts in ocean.......................64
impacts, extraterrestrial...252–254, 333–337
impacts in Canada......................64
impacts on land........................64
Imperial Oil..........................269
incisors..............................190
India.........................25, 32, 359
Indian Head, SK.......................281
Indian Ocean...............32, 196, 244
Indian River, YK.......................255
Indiana Limestone.....................278
indium, uses of.......................260
Indo-Australian Plate..................29
Indonesia...................37, 135, 241
industrial effluents, as source of PAHs....347
industrial emissions...................302
industrial minerals.................**280–281**
Industrial Revolution..........243, 262–263
industrial waste......................341
infrastructure........................339
Inglefield Uplift.................119, 138
Ingonish Beach, NS....................314
inner core.........................**22–23**
Innuitian Ice Sheet........230–**231**, 233, 236
Innuitian Mountains....................**70**
Innuitian Orogen.................112, 119
insect life...........................367
insect pollination....................129
insects..........56, **58**, 67, 201, 204, 217, 367
insects in amber......................187
Insular Mountains......................72
interaction of magma and salty brines...175
interceptor wells.....................341
interglacial climate..................229
interglacial intervals...........227, 364
interglaciation, last (see also Sangamon).................191, 310, 374
interglaciations.............**227**, 243, 369

Intergovernmental Panel on Climate
  Change............................244
Interior Plateaus.................72, 209
interlobate moraines.............**222**, 236
intermediate grain size.................**9**
intermediate magma..................8, 31
intermediate rocks.....................**9**
intermediate volcanic rocks............**7**
internal nostrils....................144
International Boundary Commission..160–161
International Mining and Manufacturing
  Company.......................261–262
intertidal zone......................12
intrusive igneous rocks................**8**
intrusive rocks..............5, **8**, 39, 182
Inuit.........................238–239, 374
Inuit sculptors......................97
Inuit tradition.....................308
Inverness, NS........................53
invertebrate skeletons................11
invertebrates...**52**, 55–**56**, **58**, 186, 365–366,
  368
Investigator.........................239
iodine..............................19
Iqaluit, NUN....................90, 258
Iqalukttuttiaq, NUN.................193
iridium...................187-188, 250, 335
iridium anomalies...................336
iron......7, 21, 74, 76, 247, 254, 258, 334, 364
iron alloys.....................22, 334
iron carbonate......................19
Iron Catastrophe................**76**, 375
iron deposits......................**256**
iron, extraction by Vikings.........258
iron filings........................25
iron formation.....................248
iron in oceans.....................**84**
iron-loving bacteria................97
iron meteorite.................333–**334**
iron minerals......................84
iron, mining and smelting..........259
iron ore........................85, 365
iron ore deposits..................89
iron oxides..............10, **19**, 248, 366
iron sulphide..................**19**, 249
ironstone.........................290
Iroquois.......................4, 293
Iroquois Sand.................218–**219**
irrigation.....................304–305
island arc chain off western Pangea..166
island arcs (see also island magmatic
  arcs)................**27**, 33, 36, 91, 155, 359
Island Intrusions..................166
island magmatic arcs (see also island arcs) 34
Island Stones......................285
islands............................309
isostasy...................**23**, 236, 297
isostatic rebound........237–238, 311–312
Isotelus..........................57
Isotelus rex......................119
isotopes.....................**41**–42, 241
Isthmus of Panama............214, 364, 374
Isthmus of Panama, impact on climate...214
Itcha Range, BC....................208
Ivesheadia lobata.................104

J
jack pine.........................229
Jackpine Mine, AB.................273
Jacksonville, ONT.................259
jack-up rig.......................272

Jacques Cartier and fool's gold.....258
Jacques Cartier and minerals........258
Jacquet River, NB.........13, 133, 359
James Bay, QC...............79, 303–304
Japan....................28, 323, 326
Japan, as analogy...................141
Japan, earthquakes in...........322–323
Japanese earthquake and tsunami
  of 2010........................326
Japanese Island Arc..............28, 169
jasper......................67, 84, 255
Jasper, AB..................103, 146, 211
Jasper NPC, AB............180, 220–221, 303
jaw bones........................189
jawless fish.....................60–**61**
jaws..........................**61**, 143
Jeanne d'Arc Basin...170–171, 183, 205, 271
Jeffrey Mine, QC....................20
jellyfish..................**56**, 119, 147
Jervis Inlet, BC................290, 309
Joggins, NS....18, 54, 61, 131, 152, 154–155,
  366, 375
Joggins "Fossil Cliffs"............152
jointed limbs.......................58
joints...........................**16**–17
Jones Sound, NUN..................213
Jones Sound rift valley............203
Jonquière Dam, QC.................332
Joutel, QC.........................84
Juan de Fuca Plate......28–29, 37, 197–198,
  207, 209, 311–312, 323–324, 326
Juan de Fuca Plate, ancestral......207
Juan de Fuca Ridge.......49, 198, 247–248
Juan de Fuca Strait................340
jumps............................**238**
Jupiter........................22, 333
Jura Mountains.....................40
Jurassic.....39–40, 46, 48, 59, 61–62, 68, 73,
  128, 151, **162**, 165–166, 168–176, 178,
  180–181, 199, 201, 359–360, 363, 367–368
Jurassic carbonate reef............170
Jurassic coal.....................263
Jurassic deltaic sediments.........172
Jurassic diversification of dinosaurs..62
Jurassic drifting.................359
Jurassic, geological map..........162
Jurassic marine strata............179
Jurassic non-marine strata........179
Jurassic paleogeography...........171
Jurassic reservoir rocks..........271

K
K/T boundary (see also Cretaceous-
  Tertiary boundary)..............**187**
K/T boundary, claystone at........187
K/T boundary localities...........187
K/T boundary, spores and pollen at..187
Kakabeka Falls, ON................245
Kalahari, location in Rodinia......97
Kamchatka Arc.....................166
Kamenka stone quarry, AB......163, 284
kames........................**222**, 224
Kaminak Lake, NUN.................348
Kaministiquia River delta, ONT....319
Kamloops, BC..............170, 211, 223
Kamloops Lake, BC.................165
Kamouraska, QC....................317
Kananaskis Country, AB......24, 160, 371
Kanata, ONT.......................343
kangaroos.........................190
Kanguk Peninsula, NUN.............175

Kansan Glaciation..................**227**
Kansas............................157
kaolin clay.................**187**, **290–291**
kaolin, uses of...................291
kapok.............................202
Karelian Craton....................87
Karmutsen Volcanics...........165, 374
Kaskawulsh Glacier, YK.............69
Kay Point, YK.....................310
kayaks............................238
Kazakhstan........................162
Keele Arch........................119
Keels, NL..........................99
Keewatin, NWT................231, 236
Keewatin ice centre...............231
Kejimkujik NPC Seaside, NS....39, 145, 315
Kelly's Island, NL................286
Kelowna, BC.......................240
Keremeos, BC......................328
kerogen..........................**266**
kerogen types....................266
kerosene......................261–262
kettle and kame topography...222, 224
kettle holes...............52, **222**–224
kettle lakes......................222
Kettle Rapids Generating Station, MB..305
Keweenawan Rift...................96
Key Lake Mine, SK.................343
Kicking Horse Rim...........114, 118
Kidd Creek, ONT...................259
Kidd Creek Mine, ONT..............249
kidney damage....................347
Kilauea, Hawaii...............35, 82
Kildare Capes, PEI................158
Killarney Provincial Park, ONT.....88
Kimberley, BC.................95, 249
kimberlite diatreme...............**256**
kimberlite magmas................257
kimberlite pipes...........19, 150, **257**–258
kimberlite volcanoes.............257
kimberlites..............248, **256-257**, 258
kimberlites, explosive eruptions..257
King Christian Island, NUN, discovery of..193
King Louis XIV of France.........258
King Street, Saint John, NB......278
kingdoms.........................**55**
Kingston, ONT...94, 107, 113, 284, 343, 365
Kipling, Rudyard.................274
Kirkland Lake, ONT................81
kitchen-counter tops..............94
Kitimat, BC..................291, 327
Klondike, YK.......246–247, 254–255, 259
Klondike Gold Rush........161, 246, 259
Klondike, lack of glaciation.....254
Klondike River, YK...............255
Kluane Lake, YK...................49
koalas...........................190
Kobe earthquake, Japan...........323
Kola Peninsula, Russia............22
komatiite dykes..................250
komatiite lavas...............79, 250
komatiite magmas..................**79**
komatiite sills..................250
komatiites..........9, **78–79**, 250, 355
Kootenay Basin...................179
Kootenay Lake, BC................200
Kootenay NPC.....................277
Kootenays, BC......................3
Kugluktuk, NUN....................95
Kuiper Belt......................333

Kula Plate.......................198
Kwajalein Atoll, as analogy......156
Kwikhpak basin...................206
kyanite.......................**15**–16

L
L'Anse aux Meadows, NL......239–240, 258
L'Anse-Amour, NL..............238, 374
L'Islet-sur-Mer, QC...........13–14, 109
La Baie, QC......................332
La Grandelle, QC.................116
La Mauricie NPC, QC..............373
La Moinerie impact structure, QC...337
La Ronge, SK......................92
La Ronge-Lynn Lake Arc............91
La Salle Expedition................4
LaBine, Gilbert..................274
Labrador........20, 34–35, 70, 74, 78–80, 82,
  88–89, 91, 93, 95, 106, 108, 139, 142, 155–
  156, 225–226, 231, 238, 250, 256, 287,
  289–290, 309, 312, 354–355, 365, 370, 375
Labrador-Baffin Ocean, potential..203
Labrador-Baffin Seaway......**184**, 202, 324,
  370, 374
Labrador-Baffin Seaway, rift basin
  precursors of..................184
Labrador-Baffin Seaway, sea-floor
  spreading in...................203
Labrador-Baffin Seaway, sedimentary
  basins in......................203
Labrador-Baffin Seaway, volcanism in....202
Labrador City, NL.................89
Labrador Current..................66
Labrador Iron Ranges..............78
Labrador Sea......35, 170, 176, 184, 203,
  205, 360
Labrador Shelf...................271
Labrador Shelf, Cretaceous sediments....183
Labrador Trough.......89, 93, 256, 354, 375
labradorite.............**20**, 95, 289
Lac Beauport, QC.................297
Lac de Gras, NWT.................258
Lac des Arcs Thrust...............46
Lac des Deux Montagnes, QC......340
Lac Guillaume-Delisle, QC.........89
Lac Mégantic, QC.................238
Lac Saint-Jean, QC........108, 119, 331
Lac Saint-Louis, QC..............340
lacewing.........................204
lactation........................189
lacustrine deposits................**9**
lacustrine environments...........12
lacustrine limestone.............168
lagoon......................315, 340
lahar........................328–**329**
LaHave Platform, as analogy......113
Lake Ainslie, NS.................270
Lake Ainslie, NS, oil well........271
Lake Bouchard, QC................373
Lake Erie....5, 71, 115, 136, 233, 262, 318, 344
Lake Huron......22, 71, 136, 233, 344
Lake Louise, AB........70, 103, 113–114
Lake Louise ski hill, AB.........356
Lake Manitoba, MB...........337, 369
Lake Michigan...............136, 233
Lake Minnewanka, AB..............46
Lake Minto, QC....................81
Lake Nipissing, ONT..............108
Lake Ontario....5, 71, 108, 115, 136, 218–219,
  233, 301, 338–340
Lake Ontario, as source of drinking water..340
Lake Ontario, pollution in.......340

lake shorelines ... 319
Lake Superior ... 47, 85, 96, 136, 233, 258, 319
Lake Timiskaming ... 108, 119
lake trout ... 349
Lake Winnipeg, MB ... 319, 337
Lake Winnipeg, tilting of ... 319
Lake Winnipegosis, MB ... 146, 369
lakes (see also glacial lakes) ... 300, 302, 319
lakes, changes in level ... 318
lakes, in water cycle ... 302
lakes, non-renewable water supply ... 302
lakes, renewable water supply ... 302
Lambe, Lawrence ... 52
Lambeosaurus ... 52
lamp shells ... **59**
lampreys ... 60
Lancaster rift valley ... 203
Lancaster Sound, NUN ... 71, 202, 213, 231, 239
lancelets ... 56, **60-61**, 125
land animals ... 375
land animals, early ... 140–141
land bridges ... 25
land link ... 374
land plants ... 55, 126, 128, 140, 375
land plants, early ... 138, 140
land routes between North America and Eurasia ... 195
land snails ... 366
landfill liner ... 341
landfills ... 340–**341**
landing on the Moon ... 259
landslides ... 242, 296, 310, 321–322, 324, 328, **329**, 330, 332
landslides as tsunami trigger ... 325–326
landslides, damage caused by ... 329–330
landslides, velocity ... 329
larches ... 194, 212
laryngeal cancer ... 353
late-Devonian mass-extinction event ... 64, **147**
lateral moraines ... **222**, 223
latitudinal temperature gradient ... 120
Latvius ... 150
Laurasia ... **162**, 171–172, 175–176, 182–183, 189–190, 195, 203, 359–361, 374
Laurasia, western margin ... 172, 177
Laurasia, North American part ... 171
Laurasian Plate ... 179, 203
laurasiatheres ... 190–**191**, 235
laurasiatheres, origin in Laurasia ... 191
Laurentia ... 32, **101–102**, 103, 106, 108–110, 112, 115, 117–118, 120–121, 132–134, 139, 171, 260, 293–294, 355–357, 360, 375
Laurentia, location in Rodinia ... 97
Laurentia, paleoeastern margin ... 110, 138
Laurentia, paleonorthern margin ... 113, 141
Laurentia, paleosouthern margin ... 108–109, 113–114, 116
Laurentia, paleowestern margin ... 112
Laurentian basement ... 112
Laurentian continental margin ... 115
Laurentian continental slope ... 116
Laurentian crust ... 355
Laurentian Highlands, QC ... 34, 93–94, 108
Laurentian lithosphere ... 112, 114
Laurentian margin ... 109, 115
Laurentide glacier ... 219
Laurentide Ice Sheet ... **219**–222, **230**–231, 233–234, 236, 282, 297, 363, 374
lava ... **6**
lava dome ... 328
lava dome collapse ... 328
lava flows ... **6**, 8, 327–328

Lava Lake, BC ... 208
lava tube ... 168
Lawn, NL ... 326
Lawrencetown (near Halifax), NS ... 10
leachate ... **341**
leaching in hot water ... 247
lead ... 19, 247, 249–250, 254, 258
lead-206 ... 41–42
lead-207 ... 42
lead and health ... 353
lead concentrations ... 353
lead ores ... 92, 95, 142, 148
lead sulphide ... 19
leaf and needle fragments, source of fine atmospheric particles ... 351
leaves ... 52, **127**, 129, 154, 201, 366
leaves, origin of ... **127**
Leclerc, George-Louis, Comte de Buffon ... 41
Leda Clay ... **237**, 330
Leduc AB ... 269, 365
Leduc No. 1 well ... 269–270
Leduc No. 2 well ... 270
Leduc Oil Field, AB ... 147, 267, 269
Leduc reefs, AB ... 147–148
left-lateral strike-slip fault ... **17**
Lefty's Falls, SK ... 372
legs, evolution of ... 106
Lemieux Landslide, ONT ... 330
lemurs ... 191
length of Earth days ... **76**
Lepidodendron ... 127–**128**, 154
Leptaena ... 53
Letellier, MB ... 372
Lethbridge, AB ... 283
lichens ... 14, 347
lichens, absorption of cadmium ... 347
LIDAR ... **49–50**
LIDAR surveys ... 50
life, oldest indications of ... 53
life's origins ... **82**
Light Detection and Ranging ... **49**
light non-aqueous phase liquids ... **345**
light oil ... **265**
lignite ... 14, 263
lignite, use of ... 263
Lillooet, BC ... 200, 211
Lillooet Glacier, BC ... 7
Lillooet River valley, BC ... 331
limb bones ... **61**
limbs ... **61**
lime kilns ... 279
limestone breccias ... 115
Limestone City (Kingston) ... 285
Limestone Point, MB ... 319
limestone, as commercial term ... 281
limestone, for cement making ... 280, 292
limestone, from France ... 296
limestones ... 5, **10–11**, 19, 281
limonite, Aboriginal use ... 258
linden ... 212, 218
Lingula ... 107
Lingulella ... 107
lions ... 229, 234
liquefaction of sediment ... 10, 324–325, 330
liquid iron ... 22
liquid petroleum, origin of ... **266**
Lismore, NS ... 234
lithium ... 74
LITHOPROBE Project ... **49**, 82, 92
LITHOPROBE seismic sections ... 80, 91
LITHOPROBE Trans-Canada Transect ... 47

lithosphere ... **22–24**, 27–28, 30, 33, 36, 49, 66, 231, 236–237
lithosphere bulging ... 34
lithospheric plates ... **29**
Little Burin Harbour, NL ... 326
Little Cornwallis Island, NUN ... 142, 250
Little Dal ... 375
Little Dal assemblage ... 98
Little Ice Age ... **239**, 374
Little Mount Mégantic District, QC ... 296
Liverpool, NS ... 145
liverwort spore ... 129
liverworts ... 126
living fossils ... 107, 128
lizards ... 56, 60, **62**, 185
Lloydminster Embayment ... 118
LNAPLs ... **345**
lobe-finned fish ... 60–**61**, **143**–144, 150, 365, 375
lobe-finned fish, limb bones ... 143
lobopods ... **123**
lobsters ... 56, 58
locks ... 318
lode gold ... **255**
Logan, William ... 76, 100, 130–131, 209, 219, 294, 372
Logan's Line ... **114**, **294**, 356
Lomonosov Ridge ... 174–175, 202, 203
Londonderry, NS ... 259
Lone Rock, SK ... 333
Long Beach, BC ... 308
Long Island, NS ... 362
longshore currents ... **314**
longshore drift ... **313**
long-wave infrared radiation ... 241
Lord Kelvin ... 41
Lord Selkirk Provincial Park, PEI ... 159
Los Angeles, USA ... 28
Lougheed Island, NUN ... 272
Louisbourg, NS ... 109
Lovelock, James ... 240
Low, Albert ... 89
Lower Canada ... 130
Lower Darnley, PEI ... 307
Lower Fort Garry, MB ... 284
lower mantle ... 22
Lower Town, Québec City, QC ... 294, 297
Lower Town fire, 1682, Québec City, QC ... 294
low-grade metamorphic rock ... **15**
low-grade metamorphism ... 16
lubricating oil ... **265**
Lulu Island, BC ... 316
Lunenburg, NS ... 110, 145
Lunenburg County, NS ... 349
lung cancer ... 351, 353
lungfish ... 60–**61**, 143–144
lungs ... **61**, 144
lustre ... 21
Lyell, Charles ... 5, 131, 154–155, 160, 219
Lynn Lake, MB ... 92
Lystrosaurus ... 25
Lytton, BC ... 4, 17, 330

M

MacDonald Farmhouse, Bartibog, NB ... 285, 286
Macdonald, John A. ... 131
Mackenzie, Alexander ... 272
Mackenzie Delta, NWT ... 175, 205, 212, 225, 272, 274, 316, 317
Mackenzie Dyke Swarm ... 9, 95

Mackenzie Mountains, NWT ... 72, 85, 96, 104, 324
Mackenzie River, NWT ... 205, 210, 301, 315, 347
Mackenzie River basin ... 206
Mackenzie Valley, NWT ... 230, 272
Macmillan Pass, YK ... 348
Macoun, John ... 299
Mactaquac Dam, NB ... 261
Madagascar ... 33
Madagascar, location in Rodinia ... 97
Madawaska Highlands, ONT ... 108
Madoc, ONT ... 76
Madoc meteorite ... 76
mafic ... **7**
mafic dyke ... 356
mafic dyke, diatreme related ... 257
mafic dyke swarms ... 87
mafic magma ... 27, 30–31
mafic magmatism ... 375
mafic melt ... 16, 31
mafic minerals ... 10
mafic rocks ... **9**, 22
mafic rocks, ores in ... 250
mafic sills ... 8, 92, 95, 362
mafic volcanic rocks ... **7**
mafic volcanic rocks, ore in ... 252
mafic volcanism ... 96
Magdalen Island, QC ... 149–150, 157–158, 314, 354, 356
magma ... 5, **6**, 8–9, 15, 31, 33, 328
magma, felsic ... 8
magma, granitic ... 8
magma, intermediate ... 8
magma ocean ... **76–77**, 82, 354
magma, ores formed in ... 250
magmatic arcs ... **27**–29, 31, 33, 36, 81, 91, 93, 247–248, 294, 355, 374
magmatic arcs, porphyry deposits ... 250
magmatic fluids, gold from ... 255
magmatic fluids, late-stage ... **250**
magmatism ... 370, 374–375
magnesite, as industrial mineral ... 292
magnesium ... 7, 250
magnetic anomalies ... 26, **49**, 197–198
magnetic field ... 22, 25, 27, 30
magnetic inclinations ... **25**–26
magnetic intensity ... 27
magnetic maps ... 49
magnetic methods ... **49**
magnetic minerals ... 49
magnetic north pole ... 26
magnetic poles ... 25–26
magnetic properties of rock ... 47
magnetic reversals ... **27**, 43
magnetic stripes ... **27**–28, 30
magnetic surveys ... **49**, 96
magnetic timescale ... 227
magnetite ... **19**, 25, 67, 84, 255, 266
magnetochronology ... **43**
magnetometers ... 27, 49
Main Channel, Fraser Delta, BC ... 316
Main Ranges, Rocky Mountains (see Rocky Mountain Main Ranges)
Maison Cureux, Québec City, QC ... 294–295
Maison Parent, Québec City, QC ... 296
Maitland Creek, BC ... 73
maize ... 239
major faults and gold ... 256
major volcanic episodes ... 64
Makkovik Orogen ... 90

malachite..................................**21**
Malaspina Marble.....................290
Maldives..................................244
Maligne Lake, AB......................303
Mallik, NWT..............................274
mallow family..........................202
Malpeque Bay, PEI....................314
mammal evolution.....................62
mammal footprints..................185
mammal-like reptiles....56, 60, 62, 159, 163, 166–167, **189**
mammal migrations..........195, 205, 214
mammals...54, 56, 60, **62**, 67, 185, **189**–191, 205, 217, 367–368, 374
mammals, blood circulation in..........189
mammals, bone fragments.................201
mammals, continental nurseries.........189
mammals, Gondwanan nursery............189
mammals, hearing in.......................189
mammals, large.....................235, 374
mammals, large, extinction of......235–236
mammals, Laurasian nursery.............190
mammals, live birth in...................189
mammals, origin of.......................189
mammals, parental care in..............189
mammoths......4, 190–191, 229, 234–235, 236, 369, 374
manatees..................................190
manganese........................21, 259
manganokomyakovite...................182
Manicouagan, QC...........89, 96, 119
Manicouagan impact structure, QC...336–337
Manicouagan Reservoir, QC............336
Manikewan Ocean..90, **91**–92, 249, 355, 375
Manitoba.....1, 10, 20, 78, 91–92, 118–120, 126, 137, 146, 159, 161, 180, 182, 186–187, 203, 214, 223–224, 226, 229–231, 233, 235–237, 243, 246, 249–250, 270, 281–282, 284, 291, 299, 305–306, 312, 319–320, 331–332, 337, 350, 355, 368–369, 372
Manitoba Escarpment, MB.............369
Manitoba, flooding in..................322
Manitoba Museum (see The Manitoba Museum)
Manitoba, petroleum in................270
Manitoulin Island, ONT........137, 285, 292
mantle........**22**–23, 27–31, 65–67, 76, 169, 323, 370
mantle plumes.......................**35**, 87
manto deposits..................**250**, 251
Manuels River, NL......................107
Maple Beds.............................218
Maple Creek impact structure, SK......337
maples....................201, 217–218
marble..............................**15**–16, 281
marble, as building stone..............289
marble, as commercial term...........281
Marble Canyon, BC......................157
marble from Italy......................283
marble from Vermont....................283
March fly................................367
Maria, QC......................346–347, 350
Marian Belle Wolfe (schooner)........326
Mariana Islands......................27, 31
marine biota and contamination......342
marine channels........................108
marine deposits..........................**9**
marine environments....................12
marine micro-organisms.................67
marine plankton and oil...............266
marine reptiles..................186, 367
Marinoan glaciation................103–104

Maritimes Basin..129, **148**–149, 152, 270, 356
Maritimes colonies....................131
Maritimes Cretaceous deposits........182
Market Square, Saint John, NB........278
marl......................................**11**
marl, as industrial mineral..........292
Mars..................22, 82, 333, 366
Marshall Islands, as analogy.........156
marshes..........................**315**, 324
marsupials........56, 63, **189**–190, 205, 374
marsupials, birth in..................190
marsupials in Australia...............190
marsupials in South America..........190
marsupials, Laurasian origin of......190
marsupials, pouches in................190
Martello towers, Québec City, QC.....295
Mary Rose................................50
Marys Point, NB........................279
Marystown, NL..........................326
Mason, Skookum Jim....................246
mass-extinction event, end-Cretaceous.....62, 64, **187–188**, 196
mass-extinction event, end-Ordovician...64, **121**, 375
mass-extinction event, end-Permian...59, 64, **159**, 163, 166, 189
mass-extinction event, end-Triassic...62, 64, **166**, 168
mass-extinction event, late-Devonian..............64, **147**
mass-extinction events.....60, **64**, 159, 121, 274, 365–366, 374–375
mass extinction, survivors............159
Massey Hall, Toronto..................216
mastodons 4, 63, 190, 214, 233, 235-236, 369, 374
Matachewan dyke swarm.................87
Matagami, QC..........................236
matrix......................................8
matryoshka dolls........................65
Matterhorn, Switzerland..............221
Matthews, Drummond......................27
mature zone...........................**266**
Maui, Hawaii............................35
Mauna Loa, Hawaii................242–243
McAbee, BC.......................200–201
McAdam, NB............................292
McArthur River Mine, SK..............343
McAvity, James........................280
McClean Lake Mine, SK...........276, 343
McClure, Robert........................239
McConnell, Richard..........122, 161, 178
McConnell Thrust Fault.........46, 179–180
McGill Faculty Club, Montréal, QC....291
McGill University....................155
McKinley Bay, NUN....................300
Meadow Lake Escarpment.......138, 146
mean annual temperatures.............230
medial moraines...................**222**–223
Medicine Hat, AB....229, 273–274, 299, 332
Medieval Warm Period.......238, **239**–241
Mediterranean Sea, drying up of......196
medium oil............................**265**
Medullosa.............................129
medusas............................**56**–57
Megalonyx.............................**229**
Meganeura.............................**154**
megaquakes......................**323**, 324
megaspores...........................**128**
Meguma...102, 108–110, 115, 133–134, 145, 355, 357, 360, 375
Meguma margin of Euramerica.........148

Meguma metasandstone................290
Meguma Terrain.........**110**, 118, 145, 290
Meguma Zone..........................102
mélanges....................**156**, 165, 178
Melmerby Beach, NS..............314–315
melt..22
melt layer............................335
melting rocks..........................8–9
melting temperatures...................31
meltwater..........**219**, 224, 233, 303, 364
meltwater streams.................**222**–223
Melville Island, NUN......71, 111, 138, 142, 151, 213
Melville Peninsula, NUN............71, 91
Melville Sound, NUN..................239
Memorial University of Newfoundland...100
Mendeleev Ridge......................175
Mendham, SK.........................281
mercury...........22, 347–348, 349–350
mercury, concentrations in lakes....349
mercury contamination...............339
mercury in food chain................348
mercury in the environment..........348
mercury levels in fish...............349
mercury, national consumption guideline........................348
mercury, uses of......................348
mercury vapour.......................348
mercury vapour emissions, measurement of..................348
Mercy Bay, Banks Island, NWT........239
Merritt, BC......................202, 264
Merychippus......................214-215
Merycodus........................214–215
Mesabi Range, Minnesota..............85
Mesohippus...........................206
Mesoproterozoic.....**40**, 64, **86**–87, 93–97, 111, 249, 337, 355
Mesoproterozoic anorthosite.........289
Mesoproterozoic geology map...........87
Mesoproterozoic granite from Finland...287
Mesoproterozoic lava flows............94
Mesoproterozoic life..................98
Mesoproterozoic passive margin........95
Mesoproterozoic redbeds...............96
Mesoproterozoic rift basin............95
Mesoproterozoic river systems.........95
Mesoproterozoic tidal flats...........97
Mesoproterozoic tidal mudflats........94
Mesosaurus............................25
mesosphere............................65
mesothelioma........................**353**
Mesozoic............**40**, 62, 162, 367–368
Mesozoic Era.........................367
Mesozoic-Cenozoic sedimentary basins, offshore eastern Canada............170
Meta-Incognita Craton.................90
metallic elements.....................19
metal sulphide ores..............248, 342
metal sulphides.............175, 248, 349
metal-bearing brines.................142
metallic lustre........................**21**
metalloids...........................**347**
metal-rich fluids................249–250
metals........................**247**, 373
metals in food chain.................348
metals, modern use of................260
metamorphic rocks......**5**, 9, **15**–16, 18, 173
metamorphic rocks and water.........301
metamorphic rocks, as building stone............................289–290

metamorphism...................**15**–16
metamorphosed mafic dykes..........355
metasandstone........................**15**
metasedimentary rocks................81
metavolcanic rock....................**15**
metazoan evolution..................106
Metazoan reefs......................115
metazoans....................**56**, 104, 117
Metchosin, BC........................199
meteorite impacts...............77, 159
meteorites......22, 42, **75**–76, 82, **333–334**, 339, 375
meteoroids.....................**333–334**
meteors..............................333
methane......66, 68, 196, 241, 244, **265**, 267, 274, 333,
methane-bearing fluids..............256
methane, biogenic...................266
methane gas escape trenches..........50
methane releases.....................68
methane, thermogenic................266
methanogens........................**273**
methylmercury...................**347–348**
methylmercury, as toxin.............348
methylmercury, in fish and sea mammals.........................348
Métis communities...................347
Metoposaurus...................166, 167
Mexico...............................239
Mi'kmaw legend......................372
mica, as industrial mineral........292
mica, biotite..........................9
mica, muscovite........................9
micas..................15, 16, **19**, 21
Michigan........................137, 375
Michigan Basin...36, 118, 119, 136–137, 375
microbes...**55**, 66, 83, 106, 135, 138, 364, 375
microbial films.......................**84**
microbial mats......................**84**
microcontinents....**32**, **33**–34, 108, 355–357, 361
microcontinents in Ural Ocean.......111
microdiamonds......................**188**
microfossils...........39, 41, **56**, 84, 235, 375
microfossils, early..................86
micro-hydro facilities..............276
microscopic marine organisms........266
microscopic plankton................163
microscopic plant debris............117
microspores........................**128**
microstructures.....................375
Mid-Atlantic Ridge....**26**–27, 29, 31, 36, 70, 174–175, 248, 324
midbody trace.......................107
Mid-Continent Arch..................119
Midcontinent Rift.........78, 93, **96**, 375
Midcontinent Rift, mafic sill.........96
middens..............................341
Middle Head, NS......................109
migmatites......................**15**–16, 79
Miguasha, QC......54, 61, 128, 136, 143–145, 152, 365–366, 375
Miguasha National Park, QC.....51, 143–144
Milankovitch cycles.......**227**, 228, 241, 369
Milankovitch, Milutin...............227
Milford, NS.........................236
Milky Way Galaxy................22, **75**–76
millipedes.............59, 140, 141, 366
Minas Basin, NS.....................37
Minas Passage, NS...................277
Minas Sub-basin, NS.............148–149
mine tailings, as source of mercury.....348

mine waste dumps 341
mineral aggregate (see also aggregates) 280, **291**
mineral deposits 246–260, 369
mineral industry in Canada 259
mineral properties **21**
mineral-rich hydrothermal systems 92
mineralization **52**
minerals (see also mineral deposits) . . 5, **19**–21
minerals, metal-bearing 247
Mingan Archipelago, QC 115, 116
mining 246–260
mining and the environment 341
mining on Canadian Shield 78
mining waste 342
Minto, NB 128
Minto Arch 119
Minto, NB, early coal mining 263
Minto Sub-basin, NB 149, 152
Miocene 40, 48, **195**, 196, 198, 205–207, 209–215, 287, 369
Miocene Arctic vegetation 212
Miocene climate 196
Miocene Epoch **195**
Miocene lacustrine deposits 212
Miocene landscape 214
Miocene paleogeography 207
Miocene-Pliocene climate deterioration . . 212
Miramichi Highlands, NB 148–149, 231
Miscou Island, NB 314
Misener family 307
Misener, Harris 307–309, 313, 319
Miseners Island, NS 307–308, 319
misfit streams **234**
Misra, Shiva Balak 100–101
Missisquoi Marble 290
Mississippi River 233
Mississippi Valley 249
Mississippi Valley Type deposits 248–**249**
Mississippian **41**
Mistaken Point, NL 100–101, 104–106, 109, 365, 375
Mistaken Point ecosystem 105
Mistastin impact structure, NL 337
moa 64
modification stage 334–335
Mohs, Frederic 21
Mohs Scale **21**
Moira River, ONT 349
molars 189
molds 53, **54**
molecular evidence for eukaryotes 97
molecular structure 21
molecular studies **55**, 60, 62, 82, 123, 189–190
moles 190–191
mollusks . . . 39, 56, 58–**59**, 106–107, 117, 163, 168, 217, 237, 250, 270, 367
molybdenite 250
molybdenum 250
Moncton, NB 270
Moncton Sub-basin, NB 149, 270
monitor lizards 186
monkey-puzzle tree 128
monkeys, New World 191
monkeys, Old World 191
monotreme jaw bones 190
monotreme molars 190
monotremes 56, 62, **189**–190,
monotremes, fossils of 189
monotremes, lack of nipples 189
monotremes, single posterior hole 189
monsoonal flooding 157
Mont Bélair, QC 297
Mont Brome, QC 183
Mont-Laurier, QC 94
Mont Mégantic, QC 182–183
Mont Rougemont, QC 183
Mont Saint-Bruno, QC 183
Mont Saint-Grégoire, QC 183
Mont Saint-Hilaire, QC 182–184
Mont Shefford, QC 182–183
Mont Yamaska, QC 183
Montagnais impact structure, QC . . . 336–337
Montana 221, 231
Montauban-les-Mines back-arc basin 94
Monte Verde, Chile 235, 374
Monteregian Hills, QC 35, 182–184
Monteregian intrusions, QC 288
monteregionite 182
Montmagny, QC 297
Montmartre Basilica, Paris, France 296
Montmorency Falls, QC 294
Montmorency Park, Québec City, QC 296
Montréal, QC 108, 130–131, 182, 259, 285, 288, 292, 324, 339–340, 343
Montréal QC, fire of 1852 282
Montréal Island, QC 288
Mont-Tremblant, QC 96
monuments 282
Moon **76**–77, 82, 334, 375
Moon formation **76**
Moon, recession from Earth **77**
Moon rock thin section 77
Moon rocks 55, **76**
Moon rocks, age of **76**
Moon rocks, oldest 55
Moon's surface **77**
moose meat 347
Moose River Basin, ON 118–119, 182–183
Mooseland, NS 259
Moraine Lake, AB 114
moraine systems 236
moraines 222–223, 236
moratorium on west coast petroleum exploration 199, 270
Moravian mission, Hebron, NL 290
Morden, MB 186
Morley, Lawrence 27, 372
Morocco 25
Morrin Bridge, AB 185
Morris, MB 331
mosasaurs **62, 186**
mosasaurs, extinction of 187
mosses 126
Mount Assiniboine, AB and BC 221
Mount Baker, Washington 207, 327
Mount Brazeau, AB 303
Mount Burgess, BC 122
Mount Cayley, BC 207, 374
Mount Close, YK 72
Mount Costigan, AB 46
Mount Edith Cavell, AB 221
Mount Edziza, BC 192, 208
Mount Everest 35, 163
Mount Field, BC 113, 122
Mount Fuji, Japan 8
Mount Garibaldi, BC . . . 7, 207, 231, 327–328, 374
Mount Gifford, NWT 317
Mount Hood, Oregon 327
Mount Indefatigable, AB 160
Mount Lassen, California 327
Mount Laurie, AB 179
Mount Lister, NL 95
Mount Logan, YK 30, 69, 209, 328, 364
Mount Logan, geology of 209
Mount McKinley, Alaska (now Denali) 209
Mount Meager, BC 207, 327–328, 374
Mount Meager, BC, and geothermal energy 276
Mount Meager, BC, landslide on 329
Mount Pelée, Martinique 328–329
Mount Pleasant deposit, NB 251
Mount Ranier, Washington 327
Mount Robson, BC 210
Mount Royal, QC 182–183
Mount Rundle, AB 46, 146
Mount Shasta, California 327
Mount Slipper, YK 98
Mount St. Elias, YK 69
Mount St. Helens, USA 327–329
Mount Stephen, BC 122
Mount Tambora, Indonesia, 1816 faminerelated to eruption 241
Mount Tambora, Indonesia, eruption of 1815 241
Mount Vesuvius, Italy 328
Mount Waddington, BC 209–210
mountain belts **32**–34, 43, 354–355
mountain building 68, 241
mountain building and climate 196
mountain ranges 18
mountain roots 22
mountain sheep 235
mountains in Cordilleran Orogen 169
mouth, evolution of 106
Mucrospirifer arkonensis 59
mud and boulder flows 168
mud flats **318**
mud slides 125
mudcracks **12**–13
mudstone **9**, 12, 15
mudstones, uses of 290
Muir, ONT 226
mulberry 212
multibeam **50**
multibeam bathymetric sounding **50**
multicellular filaments 98
multicellular organisms 104
multituberculate molars 190
multituberculates 56, 62, **189**–190, 205–206
multituberculates, birth in 190
multituberculates, life in trees 190
Murdochville, QC 343
Murray, Alexander 99, 131, 252, 372
Murray Mine, ONT 252
muscle power 58
muscovite mica 9, 15, **19**
Mushbowl Hill, BC 208
musk oxen 234
Muskaboo Creek, BC 16
muskeg 225
Muskox Intrusion, NUN 250
Muskwa Basin 93, 95
mussels 56, **59**
Musselwhite Mine, ONT 255
MVT deposits (see also Mississippi Valley Type deposits) . . . 248–**249**, 250, 256, 260
myotomes 125
Myra Falls-Buttle Lake District, BC 249
myriapods 56, 155

N

Nachvak Fiord, NL 309
Nain, NL 20, 74, 80, 90, 95, 289
Namacalathus 105, 106, 375
Nanaimo, BC 178, 264
Nanaimo Harbour, BC 283
Nanisivik, NUN 250
Nanisivik Mine, NUN 20, 249
Nanticoke Generating Station, ONT 344
Nares Strait 71, 184, 202, 203
nascent North Atlantic 359
Nass River 208
National Building Code 325
National Museum of Canada (now Canadian Museum of Nature) 52
native copper **19**, 95, 247, 258–259
native gold **19**
native sulphur 175
natural disasters, public awareness 332
natural hazards **321**
natural hazards, cost of 322
Natural Resources Building, St. John's, NL 289
natural water-storage reservoirs, as water regulators 304
nautical charts 50
nautiloids 53, 56, **59**, 117, 120, 147, 284
navigable waterways 340
Nazca Plate 29, 198
Nebraskan Glaciation **227**
nebulae 76
Nechako Basin 210, 211
Nechako drainage 211
Nectocaris 123
negative feedback processes **240**
Nelson River, MB 319
Neoacadian Orogeny . 110, **145**, 290, 357, 375
Neocalamites 167
Neogene . . . 40, **195**–196, 198, 204, 206–207, 209–215
Neogene, Arctic climate 212
Neogene climate 212
Neogene cooling trend 196
Neogene Period **195**
Neogene volcanism 207
Neoproterozoic 40, 58, 85, **86**, 87, 96–98, 101–106, 109–112, 114, 117, 202, 296, 359–360, 365
Neoproterozoic coastal habitats 98
Neoproterozoic deep-sea deposits 356
Neoproterozoic Era 101
Neoproterozoic geology map 101
Neoproterozoic glaciations 64
Neoproterozoic life 98
Neoproterozoic marble 279, 290
Neoproterozoic metazoans 101
nepheline syenite, as industrial mineral . . . 292
Nepisiguit River, NB 116
Neptune 22, 333
neutrons **41**
New Brunswick 7–8, 11, 13–15, 30, 32, 54, 58–59, 102, 105, 108–110, 116, 126–128, 132–134, 140–141, 144, 148–150, 152–154, 157, 166, 168, 182, 231, 233, 249, 251, 259, 261–263, 270, 274, 278–280, 282, 285–286, 288, 290, 292, 305, 309, 314, 317–318, 324, 329, 332, 343, 350, 355–356, 358–359, 364–366, 370
New England Seamounts 182–183
New Guinea 189
New Guinea, as analogy 135
New London, PEI 158–159
New Orleans, hurricane in 322

New Quebec impact structure, QC 336
New Quebec Orogen 90, 93
New York Brownstone Buildings 286
New York City, New York, USA 296
New York City, hurricane in 322
New York State 136
New Zealand 325
New Zealand as analogy 108
Newcastle Island, BC 283
Newfoundland 14, 21, 33–34, 44, 70, 99–102, 104–106, 107–110, 114–116, 133–136, 141, 148–149, 218, 220, 229, 231, 233, 236, 239–240, 249–250, 258, 266, 270, 282, 284, 286, 288–290, 305, 309, 311, 314, 324, 326–327, 340, 341, 343, 355–356, 358, 365, 370, 373
Newfoundland Railway 288
Newmarket Till 218, 301
Newport Landing, NS 7
Niagara, ONT 137, 284
Niagara Escarpment ONT 5, 36, 71, **136–137**, 138, 284, 369
Niagara Escarpment, Silurian cap rock 71, 137
Niagara Falls, ONT 4, 5, 136–137
Niagara Gorge, ONT 4–5
Niagara Gorge, Whirlpool section 284
Niagara River, ONT 5, 339
Niagara River sediment plume 71
Niagara Whirlpool, ONT 5
Nicholson impact structure, NUN 337
nickel 76, 250, 334
nickel alloys 334
nickel-copper-PGE ore deposits 252–253, 260
Nickel Rim South Mine, ONT 253
nickel sulphides 252
nickel, uses of 260
Nicola Lake, BC 223
nineteenth century 241, 372
Nipigon, ONT 96
Nisga'a oral tradition 327
Nisga`a people 208
nitric acid 343
nitrogen 66, 265
nitrogen-14 42, 241
nitrogen-16 41
nitrogen oxides in atmosphere 353
non-renewable resource 276
Noranda, QC 259
Norfolk Island pine 158
norite 9, **253**
norite sublayer, Sudbury 253–254
normal faults **17**, 201
normal polarity 27, 30, 227
Norman Wells, NWT 271, 272
Norman Wells Oil Field, NWT 271
North America 28, 37, 162, **195**–196, 203, 207, 214, 219, 227, 230, 232–238, 241–242, 355–356, 359–361, 363–364, 369, 374–375
North America, direction of movement 176, 200
North America-Eurasia land link (via Greenland) 195, 374
North American Continental Platform 112
North American Great Plains 237
North American part of Laurasia 363
North American Plate 28–31, 37, 177, 195, 197, 203, 207–209, 312, 323–324, 326
North American Plate, movement of 197
North American Plate, western margin 198
North American-Eurasian plate boundary 195, 363
North Atlantic borderlands 109
North Atlantic Craton 74, 78–80, 88, 90, 93
North Atlantic Ocean 108, 195, 203, 237, 359
North Cape, PEI 277
North China, location in Rodinia 97
North Dakota 146, 235
North Howser Tower, BC 3
north magnetic pole 27
North Mountain Basalt 167–168, 362, 374
North Saskatchewan River 291, 303
North Slope of Alaska 174
northern ecosystems 343
northern Europe 233
Northern Hemisphere 219, 228, 230, 364, 374–375
Northern Hemisphere biomes 369
Northern Hemisphere ice sheets 214
Northern Rocky Mountain Trench 200
Northumberland Shore, NS 158
Northumberland Strait 158, 234, 314
northwest passage 192, 239
Northwest Passage 71, 202
Northwest Territories 9, 17, 69–70, 72, 77–78, 80–82, 85, 94, 96–98, 104, 110–111, 137, 138, 142, 146–148, 174–175, 205, 212, 224–226, 230, 237, 239, 243–244, 250, 256–258, 264, 271–272, 274, 301, 315–317, 324, 335, 347, 349, 364
Norway 139
Norwegian Sea 203
Nose Hills, AB 205
notochord **60**, 125
Notre Dame Bay, NL 341
Notre-Dame-du-Portage, QC 108
Nova Craton 90
Nova Scotia 1, 7–8, 10–13, 15–16, 18–21, 25, 28, 30, 36–37, 39, 43–44, 50, 53–54, 60–61, 62, 102, 108–110, 118, 127–128, 131, 133–134, 141, 145, 148–150, 152–155, 157–158, 160, 166–168, 171, 182–183, 188, 220, 229, 231, 233–234, 236, 238–240, 243–244, 258–259, 261–263, 265, 270–272, 274, 276–277, 282, 284–286, 288, 290–292, 298, 305, 307–309, 311, 313–315, 317–319, 324, 332, 340, 343, 349, 350–351, 355–356, 359, 362, 365–367, 370, 372 374
Nova Scotia, colony of 131
Nova Scotia, shape of 145
Nubia, location in Rodinia 97
nuclear bombs 275
nuclear decay **41**
nuclear fission **274**
nuclear fuel rods, storage of 342
nuclear fusion reactions in stars **74**
nuclear power **274**–276
nuclear power plants 274, 305, 340
nuclear reactors 342
nuclear-related localities in Canada 343
nuclear waste 343
nuclear waste, underground storage 343
nucleus (of a cell) 55
nuggets (of gold) 254
Nuna **87**, 90–93, 95–96, 254, 275, 354–355, 375
nunataks **231**
Nunavut 2, 8, 9, 15, 17, 20, 34–35, 39, 45, 56, 61, 70–71, 78–79, 82, 91–92, 94–95, 97–98, 110–111, 118, 121, 127, 129, 138, 140–142, 144, 150–151, 158–159, 173–175, 191, 192–195, 202, 205–206, 212–213, 216, 220, 222–223, 225, 230–231, 236, 238–239, 243, 249–250, 256, 258–260, 264, 267–268, 271–272, 275, 300, 304–305, 308–309, 317, 324, 337, 345, 348, 355–356, 362, 364–365, 368–370, 375
Nut Cove, NL 290
Nuvvuagittuq Greenstone 78

O

O horizon (soil) **14**
Oak Ridges Moraine, ONT 222, 301, 341
Oak Ridges Moraine sediments 301
oaks 127, 194, 218, 229, 217
Oaxaquia, location in Rodinia 97
oblique collision 360
obliquity cycle **227–228**
obsidian **6**
obsidian, mining of 258
ocean basin closing 37
ocean basin opening 37
ocean basins 31, 66
ocean chemistry 159
ocean circulation 66, 120–121, 214, 242
ocean currents 66, 120
ocean ecology 163
ocean floors 7, 22–23, 27–28, 32–33, 55, 69–70
ocean-ridge system **26**
ocean trenches **26**
oceanic crust 9, **22**–23, 30, 64–65, 354, 359
oceanic lithosphere 27–29, 31, 33, 87, 91, 169, 248
oceanic plate 32
oceanic plateaus 33, 91, 156, 166
oceanic ridges **27**, 31
oceanic trenches **27**–28, 33, 169
oceans 357, 361, 365
oceans, early **82**
oceans, salinity of 300
octopuses 56, 59
Odontogriphus 123
offshore eastern Canada 48–49, 55, 170–171, 176, 182–184, 202–203, 231, 267, 271–272, 324, 327, 336, 360, 368, 370, 374
offshore oil and gas wells 170
Ogilvie Mountains 72
oil 47, 68, 142, 242, 261, **265**–267, 272–273, 373
oil and gas 175, 261–262, 265–274
oil and gas exploration 171
oil and gas exploration in Rockies 179
oil and gas reservoirs 344, 355
oil and gas, conventional **268**
Oil City, AB 269
oil exploration renewed in Quebec 270
oil exploration wells 36
oil exploration 269
oil field 268
oil fields on Grand Banks 183
oil, natural seeps 266
oil production 269
oil production, conventional 306
oil production, unconventional 306
oil refinery 262
oil refining 266, 271
oil reserves in Canada 273
oil sands 265, **272**, 306, 344, 373
oil sands and elevated arsenic levels 349
oil sands mining 342
oil sands, processing 272–273
oil sands, vertebrate remains in 186
oil shale **274**
Oil Springs, ONT 120, 262, 274
oil well, first commercial in North America 262
Oka, QC 182–183
Okanagan Falls, BC 201
Okanagan Valley, BC 200, 305
Okanagan Valley Fault 201
Oklahoma Museum of Natural History 105
Okotoks, AB 223
Okotoks Erratic 221
Old Crow, YK 236
Old Red Sandstone 139
Oldman River, AB 306
Oldman River, AB, coal deposits 264
Oldman River Dam, AB 304
Oligocene 40, 129, **195**, 196, 198–199, 203–204, 206, 369
Oligocene climate 196
Oligocene Epoch **195**
Oligosphaeridium 368
Oliver, BC 305
olivine 7, 9, **19**, 79, 334
Olympia, Washington 323
Olympic Terrane 198–199, 374
Olympic Terrane, accretion of 198-199
Omineca Mountains 72, 113, 172–173
Ontario 3–5, 8, 10–11, 16, 18–20, 22, 34, 36, 41, 43, 47, 52–53, 57, 59, 67, 70–71, 76, 78, 80–81, 83–86, 88–89, 93–94, 96–97, 107–108, 112–113, 115–116, 118–120, 130–131, 136–137, 182, 216–219, 222–223, 225–227, 229–231, 233, 236–239, 244–245, 248–249, 252–256, 258–259, 262, 270, 274–276, 278, 282–285, 287, 289–292, 301–303, 305, 312, 318–319, 322, 324, 329–331, 334–336, 338–345, 348–350, 353–355, 358, 364–365, 369, 373–375
Ontario, ice storm in 322
Ontario Legislative Building, Toronto, ONT 284–285
Opabinia 123, 365
open-hearth process, converting iron to steel 259
open-pit coal mine 68, 264
Operation Franklin 193
ophiolite suites **33**–34, 91, 115, 199
ophiolites **33**–34
opossums 190, 205, 214
opossums, migration to North America 190
oral hygiene 346
Orangeville River, ONT 285
orders **55**
Ordovician 40–41, 44, 53, 56–57, 60–61, **101**, 108, 110–121, 126, 133–134, 136–137, 139, 142, 145, 217, 250, 281, 284–286, 290, 294, 365
Ordovician evaporites, Arctic 111
Ordovician foreland basins 120
Ordovician geology map 101
Ordovician glaciation 135
Ordovician isolated islands 119
Ordovician lagoon environments 119
Ordovician life 117
Ordovician marine life 117
Ordovician non-marine deposits 118
Ordovician oil shales 274
Ordovician paleogeography 117
Ordovician Period **101**
Ordovician shoreline 120
Ordovician-Silurian boundary 121
Ordovician-Silurian quartzite 290
Ordovician tidal flat environments 119
Ordovician-Silurian boundary 121

ore bodies 49, 78, **247**, 249, 251
ore deposits.................. **247**, 252, 260
ore deposits in Canada.................. 260
ore-forming fluids..................... 250
Oregon.............................. 207
oreodonts **214**
ores...................... 21, **247**, 250, 341
ores as source of arsenic 349
organic contaminants 348
organic debris burial.................... 67
organic matter, conversion to petroleum.. 266
organic-rich muds, ... 248, 365
organic-rich muds in Sverdrup Basin..... 173
organic-rich shales as source rocks....... 148
Orient Point, AB 46
origin of life............................ **82**
Orion Nebula............................ 76
ornamental stone **281**
orogen **34**
orogenic belts........................ **34**, 91
orogenic gold deposits 248, **256**
orogenic interval 361
orogenies 357, 361
orogeny **34**
Orphan Basin 170
Ortelius, Abraham 24
orthoclase feldspar 9, **19**, 21
Osgoode Hall, Toronto, ONT............. 216
Oshawa, ONT 339
Oshika Peninsula, Japan................. 326
osmium 250
ostracods 58, **107**
Ottawa, ONT 2, 41, 52, 94, 108, 233, 276, 278, 284, 353, 355, 358
Ottawa-Bonnechere Graben............. 108
Ottawa Citizen.......................... 100
Ottawa River............ 108, 121, 233, 236
Ottawa Valley................. 237, 312, 330
Otto Fiord, NUN 158
outer core 22–**23**
outwash.......................... **222**–223
overburden **216**
overmature zone 266
overriding Pangean Plate 169
ovule.................................. **128**
Owen Sound, ONT...................... 137
oxic conditions..................... 88, 266
oxic layer 248
oxidation 248
oxides **19**, 247
oxygen 19, 65–67, 265, 299, 365
oxygen-16 243
oxygen-18 243
oxygen as waste product............ **84**, **126**
oxygen-consuming organisms 66
oxygen-free oceans..................... **84**
oxygen in atmosphere 87–89, **126**
oxygen isotopes.............. 196, 227, 243
oxygen levels 64, 66–**67**, 106, 155, 159, 163, 256, 265, 366
oxygen-rich atmosphere 58
oxygenated surface waters............... **84**
oxygenation of oceans.................. 120
ozone layer **65**

P

pachycephalosaurs **185**
Pachygenelus.......................... 168
Pacific coast........................... 308
Pacific continental shelf................ 231
Pacific crust........................... 360
Pacific lithosphere, subduction under North America 198
Pacific margins 169
Pacific Ocean.... 26–28, 31–33, 36, 162, 169, 171, 173–174, 192, 196–197, 208–209, 214, 234, 242, 244, 248, 304, 326, 356, 359, 360–361, 363
Pacific Ocean floor....................... 32
Pacific Ocean, sediment shed onto....... 178
Pacific Plate...28–29, 35, 166, 177, 197–198, 207, 209, 323, 364
Pacific Plate, subduction of............. 208
Pacific Rim NPRC, BC........... 178, 199, 308
Pacific Rim Terrane 178, 198–199, 374
Pacific Rim Terrane, accretion of......... 198
pack ice 239
PAHs (see also polycyclic aromatic hydrocarbons) **347**
Palatine Building, Saint John, NB......... 280
Paleo-Eskimo culture 374
Paleo-Eskimos **238**
Paleo-Indians 4
Paleocene 40, 175, 180, 189, **195**–199, 202–203, 205, 283, 362, 368–369
Paleocene climate...................... 196
Paleocene Epoch **195**
Paleocene landscapes 196
Paleocene paleogeography 196
Paleocene tectonics 320
Paleocene, kimberlite.................. 258
Paleocene-Eocene Thermal Maximum... 197, 274, 374
paleoenvironment....................... **11**
paleoequator 355
Paleogene....... 40, 193,**193**, 194–206, 212
Paleogene Period...................... **195**
paleogeographic reconstructions, global 102, 106, 117, 133, 142, 152, 163, 171, 176, 182, 196, 202, 207, 214, 359, 372–373
paleogeographic reconstructions, North America ... 102, 106, 117, 133, 142, 152, 163, 171, 176, 182, 196, 202, 207, 214, 360
paleomagnetic data 103, 177
paleomagnetic evidence................ 194
paleomagnetism **25**, 32
paleontology 51, **52**, 53–64
paleoplacers **255**
paleopoles 26
Paleoproterozoic **40**, 67, **86**–94, 120, 249, 254, 256, 274–275, 349, 354–355, 365
Paleoproterozoic atmosphere 88
Paleoproterozoic geology map 87
Paleoproterozoic mafic dykes 88
Paleoproterozoic oceans 88
Paleoproterozoic orogenic belts 78
Paleoproterozoic passive margins......... 89
Paleoproterozoic pillow lavas........... 355
Paleoproterozoic raindrop prints.......... 94
paleosol, near Parrsboro, NS 366
paleosols **15**, 366
paleotalus **168**
Paleotethys **133**, 157, 162
Paleozoic **40**, 355–356, 359, 365
Paleozoic Era............... **40**, 44, 101, 367
palladium............................. 250
palladium, uses of 260
Palliser, John..................... 298–299
Palliser's Triangle................. 298–299
palm trees in Arctic.................... 196
palsas **225**
Panarctic Oils 271
Pangea..........24, 37, **132**–134, 152, 155, 157–158, 162, 164, 171–173, 175, 189, 294, 354, 356–357, 359, 361–363, 366, 374–375
Pangean continental margin ...165, 284, 360
Pangnirtung Fiord, NUN 309
pangolins............................. 191
Panthalassa... 36, 97, 107, 133, **133**, 139, 155, 157, 162, 165–166, 171, 173, 357, 359, 361, 375
pantodonts **206**
Paradoxides davidis..................... 107
Parana, location in Rodinia 97
Parliament Buildings, Ottawa, ONT...... 119, 284, 355
Parrsboro, NS 11–13, 16, 18 25, 53, 62, 155, 168, 262, 277, 365–366
Parry Sound, ONT.................... 18, 96
Parsons, Mr............................ 270
Parson`s Pond, NL................ 266, 270
partial melting 8, 79
particulate matter 347
Paskapoo Sandstone.......... 283–284, 291
passive continental margins............ 356
passive margin basin 36
passive-margin strata.................... 89
passive margins......**31**, 36, 87, 91, 113, 248
Patterson, Clair.......................... 42
Peace River, AB............... 210, 273, 333
Peace River Arch 118–119, 137, 146, 148
Peace River Oil Sands.................. 273
Peace Tower, Parliament Buildings, Ottawa, ONT 290
Pearya.....111, 115, 138–142, 150, 356–357, 375
peat................12, 14, 68, 155, 212, **263**
peat as fuel 263
peat as soil conditioner 263
peat as water filter..................... 303
peccary............................... 191
pectoral fins 143–144
pectoral girdles 143
Peel Channel, NWT 317
Peel Region, ONT....................... 284
Peel River basin, ancestral.............. 206
Peel Trough........................... 173
Peggys Cove, NS 28, 43, 145, 375
Peggys Cove Lighthouse, NS 8
pegmatites **275**
pelvic fins............................. 143
pelvic girdles.......................... 143
Pemberton, BC......................... 331
Pembina Oil Field....................... 270
Pembroke, ONT 237
Pennsylvania.......................... 128
Pennsylvanian **41**
Penny Ice Cap, NUN 236
Penokean Orogen 90
Penticton, BC 323
people 237–239, 374
peridotite **9**, 22, 30–31
perihelion **228**
period boundaries....................... 44
periods **39**–**40**, 44
perissodactyls..................... **191**, 205
permafrost ..**224**–225, 239, 244, 274, 309–310
permafrost and landslides.............. 330
permafrost zone....................... 224
permeability 47, **267**, 301, 306
Permian ... 40, 46, 62, 67, 131, **132**, 133, 148, 155–159, 163, 166, 189, 222, 285, 307, 356, 366–367, 370
Permian climates 158
Permian coal seams 159
Permian fluvial deposits................. 130
Permian, geological map................ 132
Permian paleogeography 148, 150
Permian Period........................ **132**
Permian redbeds 157–158
Permian sand dunes.................... 158
Permian sea floor....................... 165
Permian seasonal ice cover............. 159
Permian-Triassic transition 159
permineralization **52**–**54**
Persian Gulf, as analogy................. 111
persistent organic pollutants (see also POPs)...................... **347**
Peru............................. 157, 242
Peterborough, ONT..................... 345
Petites, NL 288
Petites Granite 288
petrification............................ **54**
petrified fossils 16
petrified tree stump 53
petroleum **261**, **265**, 268, 272, 339, 365
petroleum and plastic 265
petroleum entrapment 268
petroleum exploration and production... 146
petroleum generation 266
petroleum industry........... 261–262, 269
petroleum migration 266, **267**–268, 272
petroleum, origin of **266**
petroleum reservoirs.................... 147
petroleum resources of Western Canada .. 363
petroleum source rock.................. 120
petroleum traps **267**
petroleum, uses of 265
Peyto Lake, AB 13
PGEs (see also platinum-group elements)...................... **250**, 252
Phanerozoic **40**, 64, 112, 369
Phanerozoic Eon 44
Phanerozoic orogens 260
phase changes **22**
phenocrysts.............................. **8**
Philip Edward Island, ONT................ 93
Philippine Plate 29
Philippines, hurricanes in................ 322
Philipsburg, QC......................... 290
phosphate deposits 94
phosphate, as industrial mineral......... 292
photic zone 266
photosynthesis........... 56, 67–**68**, 83–**84**, 126–127, 155, 188, 375
photovoltaic cells....................... 276
phyla **55**
phyla, extinct 123
Phyllanthus............................ 212
phyllite **15**
Phyllopod Bed 122
physiography of Canada 69
physiography of Cordillera 72
phytosaur............................. 367
Pickering, ONT 343
Pictou, NS 155, 160
Pictou County, NS 263
Pigeon Mountain, AB 46
pigs 191
Pikaia 61, 124, **125**
pikas 214
pillow lava **7**, 30–31, 81, 116, 355
Pilot impact structure, NWT 337
Pincher Creek, AB................. 276, 304
Pine Point, NWT.............. 148, 250, 259

pine....128, 204, 212, 218
pingos....**225**–226
Pingualuit impact structure, QC....336–337
Pink Mountain, BC....181
pinnipeds....**212**
pitchblende....**274**
Place Royale, Québec City, QC....295
placenta....62, **190**
placental mammals....56, 62, **189**–190, 205, 374
placentals, migration to South America...190
placentals, origin of....190
Placentia, NL....311
Placentia Bay, NL....326
placer....**254**
placer deposits....**88**
placer gold....246
plagioclase feldspar....**19**, 77, 287
plagioclase feldspar, calcium-rich....9
planetary discs....76
planetary ring....76
planetesimal bombardment....**76**
planetesimals....**75**, 77
planets....**75**
plankton....55, 367
planktonic foraminifera....368
planning....345
plant adaptation to open air....**126**
plant cells....126
plant evolution....**126–129**
plant fossils....126–129, 140
plant fossils and past climates....197
plant stems....186
plant traces....**54**
plant transport system....126
plants in weathering....9
plants....14, 54–**55**, 62, 126–129, 366, 368
plants, and the water cycle....300
plants, earliest....**126**–127
plate boundaries....28, 247, 363–364
plate collision....33–34, 36, 169
plate convergence....34, 36, 169, 363
plate movements....29
plate-tectonic revolution....373
plate tectonics....18, 23–28, **29**–37, 66–67, 82, 87, 354–364
plate tectonics and mammal evolution...189
Platform....**112**, 260
Platform deposits....112, 118
platinum....247, 250
platinum-group elements (see also PGEs)....**250**, 254
platinum, uses of....260
Pleistocene....**195**, 216–236
Pleistocene Epoch....**195**
plesiosaurs....163, 173, **186**
plesiosaurs, extinction of....187
Pliocene....40, 48, 73, **195**, 209–214
Pliocene climate....209, 212
Pliocene Epoch....**195**
Pliocene paleogeography....214
Pliocene-Pleistocene transition....229
pliosaurs....**186**
Plourdosteus....143
plucking....**220**
plumbing systems....222
plums....201
Pluto....333
plutonic rocks, as building stone....287
plutons....5, **8**–9
plutons in Grenville Orogen....94
pneumonoultramicroscopicsilicovolcanoconiosis....352
Point Amour, NL....106
point bar....**12**
Point La Nim, NB....140
Point Lepreau, NB....343
Point Pelee NPC, ONT....318
Pointe-aux-Trembles Limestone....294, 296
polar regions....374
polar wandering....**26**
polar-wandering curve....**26**
Polaris Mine, NUN....142, 250
polarized light....21
polished surfaces....89
Pollard, William....274
pollen...52, 56, **128**, 129, 154, 206, 212, 217, 229, 235, 301, 351
pollen and spores....188
pollen and spores, source of fine atmospheric particles....351
pollution....301
polonium-218....351
polycyclic aromatic hydrocarbons (see also PAHs)....**347**
polygonal patterned ground....225
polygonal patterned ground, fossil....226
polymorphs....**19**
polyps....**56–57**
Pompeii, Italy....328
Pond Inlet, NUN....90
Pond, Peter....272
Pont Rouge, QC....15
POPs (see also persistent organic pollutants)....**347**
population migration....339
porcelain....21
Porcupine River, YK....236
Porcupine River basin, ancestral....206
Porcupine-Timmins, ONT....259
porosity....47, **267**, 301, 306
porous rock, filtration of microbes....302
porphyritic granitic rocks, gold in....255
porphyritic textures....**250**
porphyry....**8**, 251
porphyry deposits....**250**, 260
porphyry ores....248
porphyry-type gold deposits....255
Port Alberni, BC....326
Port au Bras, NL....326
Port au Port Peninsula, NL....270
Port aux Basques, NL....288
Port-Daniel, QC....136
Port Greville, NS....145
Port Hope, ONT....343
Port Morien, NS....152
Port Radium, NWT....274
positive feedback processes....**240**
Post Office Building, Saint John, NB....279
postglacial beaches....237
post-impact deposits, Sudbury....253
potash....**11**, 355
potash deposits....11, 147–149
potash exploration and production....146
potash ore....147
potash, uses of....280
potassium....19, 22, 250
potassium chloride....147
potassium isotopes....42
Potter's rot....**352**
Pottery Road Beds....218
Powassen, ON....8
power generation....339
power generation, role of water....305
power plants, coal-burning....343
Prairies....70, 161, 224, 230–231, 233–236, 243, 299, 368–369, 372, 374
Precambrian....**40**, 136, 146, 229
Precambrian basement....118
Precambrian-Cambrian boundary....44
Precambrian crystalline rocks....369
Precambrian fossils....**86**, 100–101
Precambrian life....85, 364
Precambrian microfossils....86
Precambrian rocks....112, 355
precession cycle....**228**
precious metals....**247**
precipitates from micro-organisms....11
precipitation....10–11, 16, 240, 304
predatory animals....106
preserved soil (see paleosol)
Presqu'ile Barrier....146–148
Presqu'ile Barrier reef system....250
Presqu'île impact structure, QC....337
pressure....16
Price Building, Québec City, QC....296
primates....190–**191**, 205
primates, binocular colour vision in....191
primates, fingers and fingernails....191
primates, opposable thumb....191
Primocandelabrum....105
Prince Albert impact structure, NWT....335, 337
Prince Edward Island....67, 130, 157–159, 189, 233, 271, 276–277, 282, 285–286, 304–305, 307, 314–315, 324, 343, 350, 355–356, 366, 370
Prince Edward Island NPC, PEI....315
Prince George, BC....210–211
Prince Leopold Island, NUN....138
Prince of Wales Island, Alaska....71, 139
Prince of Wales Martello Tower, Halifax, NS....290
Prince Patrick Island, NWT....17, 71
Prince Rupert, BC....90, 197, 208
Prince Street, Truro, NS....291
Prince William Sound, Alaska....326
Prince William Street, Saint John, NB..278–280
Princess Margaret Range, NUN....205
Princess Street, Saint John, NB....279
Princeton, BC....198, 200, 264
Principle of Faunal Succession....**39**
Principle of Superposition....38, **39**
prokaryote microfossils....364
prokaryotes...**55**, 64, 67, 82, 86–87, 97, 104, 126, 365
pronghorns....214–215, 229
propane....**265**
prosauropods....62, 367, **168**
Prospect, NS....145
Proterozoic....40, **86–87**, 179, 364
Proterozoic Eon....**86**
Proterozoic life....97
Proterozoic microfossils....97
Protichnites....107
protists....55–**56**, 98, 183
protocontinents....**81**, 87, 354, 375
protofeathers in amber....187
Protogondwana..102, **109**–110, 115, 120, 132, 134, 142, 145, 147–148, 152, 357, 360, 375
Protogondwana, glaciation....157
protons....**41**, 44
Protopanthalassa....102, 110, 113–115, 133, 141, 145, 169, 356–357, 359
protostomes....56, **58**, 59
Protosuchus....**168**
Province House, Charlottetown, PEI....285
Province House, Halifax, NS....285
Psammichnites gigas....105, 364
Pseudocraspedites....39
Psilophyton....126, 140
pterosaurs....56, 60, **62**, 163, 185, 367
pterosaurs, extinction of....187
Puget Sound....323
Pugsley Building, Saint John, NB....279
Pugwash, NS....149
Puijila darwini....212
pull-apart basin....36
pulp-grinding wheels....283
pumice, as industrial mineral....292
Purcell Thrust....114
Purcell Trench, BC....200
Purcell's Cove, NS....288, 290
Purtuniq Ophiolite, QC....91
pyrite....**19**–21, 88, 249
pyrite, Aboriginal use....258
pyrite, gold in association with....256
pyroclastic cloud....352
pyroclastic flows....**8**, 328
pyroclastic rocks....**6**
pyroclastics....**6**, 8
pyroxene....7, 9, 16, **19**, 77, 79, 287, 334

Q

quartz....9–10, 15, **19**–21, 250–251, 255, 289, 335
quartz, as industrial mineral....292
quartz crystal, bubble trains in....336
quartz crystal, fractures in....336
quartz, gold in association with....256
quartz, uses of....280
quartzite....**15**
quartz-pyrrhotite-rich replacement....255
Quaternary...40, 52, 59, **195**–196, 207–208, 216–239, 369
Quaternary Ice Age....196
Quaternary Period....**195**
Quaternary volcanism....207
Quatsino Limestone....291
Quebec....9, 13–15, 20, 33–35, 51, 61, 70, 79–82, 84, 88–89, 91, 93–94, 96, 102, 108, 109, 114–116, 119, 121, 128, 130, 135–136, 140, 143–145, 150, 157, 182–184, 222, 225–226, 231, 236, 238–238, 244, 249–250, 255–256, 258–259, 266, 270, 274, 277, 281–283, 285–288, 290–297, 302–305, 312–315, 317–318, 322, 324, 329–332, 335–336, 339–341, 343, 346–347, 350, 352, 353–356, 365–366, 369, 370, 373–374
Québec Bridge, QC....296
Québec Citadel, Québec City, QC....295
Québec City, QC....8, 108, 114, 270, 287, 293–297, 346, 356
Québec City Lower Town fire of 1682....282
Quebec, ice storm in....322
Quebec Parliament Building, Québec City, QC....295
Québec promontory (see also Cap Diamant)....293
Queen Charlotte Basin....173, 178, 199
Queen Charlotte Basin, oil exploration in .270
Queen Charlotte Basin, petroleum exploration....200
Queen Charlotte Sound, BC....199
Queen Charlotte-Fairweather Fault....28, 197–198, 207–208, 363

Queen Charlotte-Fairweather Fault, earthquakes on 323
Queen Elizabeth Islands 71
Queen Elizabeth Park, Vancouver, BC 198
Queens University, Kingston, ONT 285
Queensland, Australia, flooding in 322
Queenston Limestone 284–285, 289, 296
Quesnel, BC 202, 210
Quesnellia 155, 156–157, 165–166, 172–173, 250, 357, 360–361, 374
Quesnellia, granitic rocks 170
Quesnellia, island arcs on 169
Quesnellia, magmatic arc 170
Quesnellian magmatic arcs 169
Quesnellia-Stikinia 152, 360
Quichena, BC 129, 200–201, 204, 367–368
Quinn Point, NB 133, 359
Quinsam Mine, BC 264

R

R horizon (soil) **14**
Rabbit Lake Mine, SK 343
rabbits 190, 205, 212
radar **49**, 50
radial dykes filled with norite 253
radial dykes, Sudbury 252
radial pattern, Baffin Island, NUN 220
radiating dyke swarms 9
radiation 228
Radio Detection and Ranging **49**
radio waves **49**
radioactive decay 29, 42–43, 76, 351
radioactive isotopes, use of 342
radioactive properties of rock 47
radioactive waste 342
radioactive waste, classification of **342**
radioactivity **41**, 342
radiocarbon dates 43, 235
radiocarbon dating **42**, 219, 324
radiolaria 55, **56**, 138, 163, 165
radiolarian chert 156
radiometric dating of meteorites 75
radiometric dating .. 32, 40, **41–42**, 43, 78, 241
radium 351
Radium Hot Springs, BC 277
radium, use of 275
radon 347, **351**
radon-222 351
radon gas, accumulation of 351
radon in homes 351
radon levels in homes 351
radon, migration of 351
radon, risks of 351
radon, sources of 351
radon testing 351
Rae Craton 78–79, 90–91, 355
Rae-Slave-Hearne-Meta-Incognita supercraton 91
rafts of granitic material 354
Raglan Mine, QC 250
Railway and Coastal Museum, St. John's, NL 288
rain 9, 65
rain shadow 306
Rainbow Range, BC 208
raindrop prints 53, 94
rainforests, equatorial 128
rainwater 301
raised beach ridges 313
raised beaches ... 229, 231, **237**, 308, 311–312
raised deltas 312
rangeomorphs **104**, 105

raspberry 201
rate of plate advance 169
rate of trench retreat 169
rauisuchids 167
Ray Lake, ONT 47
ray-finned fish 61, **143**, 150, 164
Raymond Quarry 122
Raymond, Percy 122
rays (fish) 60
RCAF photography program 193
reactive mineral surfaces 82
receptaculitid **119**, 126
reclamation 292, 342–343
recycling 341
red alga 98, 125–126
Red Deer, AB 182
Red Deer River, AB 51, 187
Red Deer River badlands, AB 52
Red Deer Valley 182, 185
Red Harbour, NL 326
Red Mountain, NWT 317
Red River, MB 161, 306, 331, 372
Red River flood 331
Red River Floodway, MB 320, 331–332
Red River Valley, MB 320
Red Rock Canyon, AB 94
red sandstones **88**
Red Sea 31
Red Sea (Manitoba floods) 331
redbeds 11, 30, 37, 67, **88–89**, 94, 166–167, 171, 354
Redwater Oil Field 147
Redwater reef, AB 148
redwoods (see also dawn redwoods) 126
reef builders 163
reef complexes 146–147
reef organisms 168
reefs as petroleum reservoir rocks 148
reefs in Sverdrup Seaway 151
reefs 11, 56–57, 67, 84, 108, 117, 121, 135–138, 147–148, 355
refining 247
Reflect Blue **289**
Regina, SK 118, 146, 148
regression **30**, 118
reindeer 238
relative ages 39–**40**
relative sea-level change **311**
relict permafrost features 225
relief **69**
remanent magnetism **25**, 27
remediation 345
remote sensing 46, **49**
remote-sensing techniques 22
renewable energy 244, **276–277**
renewable resources 277
renewable water supply **302-303**
renewable water supply, per capita 303
replacement **54**
reptile, earliest 1, 61, 155, 366
reptile, largest-known marine 164
reptile remains 164
reptile skulls 62
reptiles 54, 56, 60–**61**, 62, 128, 163, 189, 205, 367, 375
Republic, Washington State 200
reservoir rocks **267–268**
reservoirs 244, 305
residential dirt 353
residual iron ore 255
Resolute, NUN 150, 174, 203, 238

respiration 67–**68**, 84
Restigouche River 358
resting traces **54**
retaining walls 330
retreat of oceanic trenches 169
Revelstoke, BC 1, 113, 118, 141, 201, 330
Revelstoke Dam, BC 330
reverse faults **17**, 28, 141
reverse magnetic polarity **27**, 30, 227
Rhabdinopora flabelliforme 118
Rhadinichthys 150
Rheic Ocean 28, **110**, 115, 134, 145, 148, 357, 375
rhinoceros-like mammals 212
rhinos 190–191
rhodium 250
rhyolite **7**, 9
rhyolite, banded 7
rhythmic bedding **44**
rhythmic couplets **224**
rias **309**
rib cage 144
Richmond, BC 315, 339
Richmond, QC 259
Rideau Canal, ONT 284
Rideau River, ONT 93
Ridgemount, ONT 57
Rielaspis elegantula 40
rift basins **30**, 36, 81, 88, 166, 171, 374
rift sequence 355
rift valleys **30**
rifting **30**, 35, 171, 359, 375
rift-related normal faults 141
rift-shoulder highlands **34**, 166, 370
rift-shoulder mountains **34–35**, 213
rifts 87
rifts associates with Iapetus Ocean 324
right-lateral strike-slip faults **17**, 200
ring dyke 331
Ring Mountain, BC 27, 230
Ring of Fire **27**, 169
Rio de la Plata, location in Rodinia 97
ripples **11**–12
riprap **319**
river bank erosion 12
river bend 12
river capture **211**
river channels, dredging of 315
river current 12–13
river discharge, global 302-303
river dykes 315
river ecosystems 332
river flow in Canada 303
river runoff 303
river terraces 211
River Thames, England 239
rivers 300
rivers, as sediment supply 315
Rivière aux Sables, QC 332
Rivière Gilbert, QC 259
Rivière-à-Pierre, QC 287
Rivière-à-Pierre Granite 296
RNA 55
road salt 339
Rocanville, SK 19
Roche à Perdrix, AB 146
Roche Miette, AB 146
roche moutonné **220**
rock cycle 4–**5**, 6–18, 66
rock flour **17**
Rock Mountain Front Ranges 179

rock salt **11**, 19
rock strength 22
rock type, impact on coastal erosion 313
Rockall, location in Rodinia 97
rockfall nets 330
rockfalls 296–297, 321, **329**
rockfalls along highways 330
Rockglen, SK 63
rocks, oldest terrestrial 1, 55, 77–78
Rockies (see also Rocky Mountains) ... 24, 160, 188, 320
Rocky Mountain Foothills 113, 148, 164, 179–180, 231, 367
Rocky Mountain Foothills, coal in 264
Rocky Mountain Front Ranges 113, 141, 146, 164, 179, 180, 268
Rocky Mountain Main Ranges . 113–114, 171
Rocky Mountains (see also Rockies) 11, 32–33, 36–37, 46, 70–72, 95, 114, 122, 137, 145–146, 161, 172, 178, 180, 210–211, 220–221, 276, 291, 355, 371–372, 374–375
Rocky Mountains thrust-and-fold belt 178–180, 264, 284
Rocky Mountains, thrust-faulting in 363
Rocky Mountains, volcanic ash in 170
rocky planets **22**, 375
rodents 190–191, 206, 214
rodents, origin in Laurasia 191
Rodinia 36, **87**, 95, 97, 101–102, 109, 114, 141, 293, 355, 357, 375
Rodinia break up 110
root bundles 140
root casts 12
roots (plants) 14, 126
Rose Blanche Granite 288
Rose Blanche Lighthouse, NL 288
Rose Blanche Pluton, NL 133
rose family 201
rose quartz **21**
Ross Farm, ONT 244
Ross Ice Shelf, Antarctica 103
rotational axis 241
rotational poles 25
Rouyn-Noranda, QC 79, 81
Rove Shale 349
Rove Shale, emissions of mercury 349
Royal Botanical Gardens, Burlington, ONT 292
Royal Ontario Museum, Toronto, ONT ... 122, 284
Royal Tyrrell Museum of Palaeontology, AB 51–52, 147, 164, 184–185, 368
Rubble Creek, BC 330
ruby **19**
rue Champlain, Québec City, QC 297
rue d'Auteuil, Québec City, QC 295
rue des Remparts, Québec City, QC 296
rue Saint-Louis, Québec City, QC 294–295
rue Saint-Pierre, Québec City, QC 296
rugose corals 159
Ruhr Valley, Germany 128
Runcorn, Keith 25
Rundle Rock 284
Rundle Thrust 46
Rundlestone 163, 284
runoff 300, 340
Russia 139, 355–356
ruthenium 250
Rutherford, Ernest 41

S

Sable Delta 171, 183
Sable Island, NS 49, 171, 271
Saddle Peak, AB 46
Saglek Basin 170
Saglek Fiord, NL.....................80, 88
Saguenay, QC 331
Saguenay flood of 1996.................332
Saguenay Graben, QC...................108
Sahara, location in Rodinia 97
saiga antelopes 234
sail-backed mammal-like reptiles........158
Saint-Flavien, QC 270
Saint-Jean Gate, Quebec City, QC 295
Saint John, NB...........58–59, 66, 278–280
Saint John County Courthouse, NB.......279
Saint John, NB, fire of 1877.... 278–279, 282
Saint John River, NB 11, 134, 261, 358
Saint John River watershed.............358
Saint-Louis Gate, Quebec City, QC........295
Saint-Marc-des Carrieres, QC 295
Saint-Marc-des-Carrieres Limestone 94–296
Saint Martin impact structure, MB 337
Saint Pierre, Martinique 328–329
Saint-Sebastien Granite.................296
Saint-Simeon, QC.......................313
Sainte-Anne-de-Beaupré, QC............296
Sainte-Hélène Island, Montréal, QC 288
saline ground water 344
Salinic Orogeny **133**–134, 357–358, 375
salinity 66, 233
salmon 143
Salmon River Gold District, NS 20
salps 56, **60**
salt...............................300–301
salt canopies **48**
salt deposits 11
salt diapirs **150**–151
salt domes......................... 268, 344
salt marshes 311, **318**
salt marshes, ecosystems 318
salt structures...................... 48, 267
salt water...............................11
Sambro, NS 243
San Andreas Fault 28–29, 145, 177, 197–198, 208, 323, 363
San Francisco, California 28, 197, 208, 246, 283
San Francisco Earthquake 28
sand..........................9–**10**, 12, 291
sand dollars...........................56, **60**
sand dunes 315
sand for golf courses....................182
sand volcanoes.........................**324**
sandbars, as source of sand 291
sandstone **9**–10, 12, 281
sandstone carvings.....................280
sandstone gravestone 282
sandstone trim...................279–280
sandstone, as commercial term..........281
sandy beach ridges.....................224
sandy beaches 307
Sangamon interglaciation **........ 229**, 374
sapphire **19**
Sarnia, ONT 262
Sask Craton 90, 92
Saskatchewan.... 1, 11, 19, 38, 54, 63, 70, 78, 91–92, 94–95, 107, 118, 138, 146–147, 159, 161, 180, 182, 186–188, 205–206, 214–215, 222, 224, 230, 233–236, 240, 243, 249, 263–265, 270, 275–276, 281–282, 284, 287–288, 291, 298–299, 306, 329, 333, 337, 344, 350, 355, 363, 368–369, 372
Saskatchewan Glacier...................220
Saskatchewan, petroleum in 270
Saskatchewan, potash deposits..........355
Saskatoon, SK...... 1, 90, 118, 146, 148, 240, 282, 284, 299, 343
Saskatoon berry........................201
satellite measurements 323
satellites................................29
satin spar gypsum 20
Saturn..................................22
Saturna Island, BC 327
Sault Ste. Marie, ONT 88
sauropods 168
Sawdonia..............................140
Saxby Gale........................ 318, 332
Saxby, Stephen.........................332
scale-like microfossils....................98
scallops..............................56, **59**
Scandinavia.... 134, 141, 202, 203, 355–356, 370, 374
Scarborough Beds**....................218**
Scarborough Bluffs, ONT 218, 227
scarps 369
Scatopsia...............................186
Scaumenacia.......................143–144
Schefferville, QC................... 89, 255
Schei, Per..............................193
schist............................. **15**, 111
Schreiber, ONT 85
scientific drilling.........................27
scleractinian corals 163
scorpions 141, 143, 366
Scotia Plate 29
Scotian Basin..... 36, 170–171, 205, 267, 271
Scotian Basin, as analogy................113
Scotian Basin deltaic deposits 171
Scotland 78, 134
Scottish Enlightenment..................18
Scotty 185
scratches in sediment....................**12**
scratching traces 107
sea anemones...........................**56**
sea cows...............................190
sea floor 50
sea ice........................239, 244, 313
sea ice, damage caused by 313
sea-level and ice sheets.................152
sea-level change 237, 308–**310**, 311
sea–level change, in future.............310
sea-level change, tectonic factors...311–312
sea-level fall **30**, 118, 120–121, 236
sea-level fall and ice caps 136
sea-level fall, distribution in Canada......311
sea-level rise.... **30**, 118, 121, 135–136, 229, 237, 239, 244, 307, 312, 319
sea-level rise and ice caps...............136
sea-level rise, distribution in Canada 311
sea lilies 56, **60**
sea lions 212
Sea of Japan as analogy............ 141, 169
Sea of Japan as back-arc basin...........169
sea scorpions 119, 135
sea squirts 56, **60**
sea stacks..............................152
Sea To Sky Highway, BC.................330
sea turtles 186
sea urchins...........................56, **60**
seabed mapping 311
sea-floor spreading ...**27**, 30, 87, 162, 374–375
sea-floor spreading in Atlantic. 171, 182–183
sea-floor spreading rates................118
seahorses..............................143
seals (animals) 191, 212, 237, 239, 348
seals (oil and gas) **267**
seamounts.............................156
seasonal cycles 227
seasonal temperature fluctuations.......240
seasonality.............................228
sea-surface temperature anomalies......242
seawater circulation 147
seaweeds..............................126
Second Norwegian Expedition...........192
Second Relief Mine, Salmo, BC...........343
SEDEX deposits (see also sedimentary-exhalative deposits) ...248–**249**, 256, 260
sediment load...................... 12–13
sediment movement at coastline........312
sediment plume.........................10
sediment storage in sea.................310
sediment transportation 12
sedimentary basins.............**35**–36, 344
sedimentary basins, groundwater in 301
sedimentary environments...............12
sedimentary rocks..............**5**, **9**–15, 39
sedimentary rocks and water............301
sedimentary rocks as building stones.........................283–286
sedimentary structures **11**
sedimentary textures.................11, 12
sedimentary-exhalative deposits (see also SEDEX deposits) 248, **249**
sediments brought to coast by glaciers...310
sediments brought to coast by rivers.....310
sediments **5**, 9–16, 45, 310
seed ferns 127, **129**, 154
seed plants 128
seeds....................**128**–129, 154, 201
seeds, origin of.........................127
segments, evolution of 106
seismic exploration in Rockies 179
seismic hazard and risk maps............324
seismic reflection........................**47**
seismic refraction........................**47**
seismic retrofitting 325
seismic sections.................... 47–48
seismic survey, 3D 47
seismic surveys.....................36, 49
seismic technology **49**
seismic tomography.....................31
seismic trucks 47
seismic waves..................22, **47**, **324**
seismographs 327
seismologists 325
seismology **47**, **49**
seismometers...........................22
selenite.................................20
selenium 346–347. 350
selenium, export of.....................350
selenium in the food chain 346
selenium poisoning 350
selenium, sources of....................350
selenium toxicity 350
selenium, uses of.......................350
Selwyn Basin...........................249
Selwyn Mountains, NWT and YK..........72
semi-submersible rig 272
Sept-Îles, QC 93, 324, 336
sequestering of carbon dioxide..........**344**
serandite 184
serpentine group asbestos **352**
serpentinite............................281
Service, Robert.........................246
serviceberry............................201
sessile organisms.......................**105**
Seven Days Work Cliff, NB 168
Severn Arch..................119, 137, 146
sewage treatment plants................340
sexual reproduction 1, 98, 375
shale gas **274**
shale gas, environmental issues 274
Shale Lake, NWT........................104
shale oil **274**
shales 5, 8, **9**–10
shales, uses of..................... 280, 290
Shark Bay, Australia.................. 83–84
shark, earliest-known 144
sharks 60, 186
sharks' teeth 186
Shastasaurus................. 164–165, 367
shatter cones 254, **335**
Shawinigan, QC 93
sheep 191
shelled eggs 128
shelled invertebrates 186
shells.......................... 11, 52, 106
Shepody Bay, NB 318
Sherbrooke, QC 287, 290
Sherbrooke Lake, NS.............. 298, 349
Sherman Mine, ONT 84
shield volcanoes................. **7**, 35, 208
Shoal Lake, MB.........................306
shock metamorphism **335**
shock wave 334–335
shock waves from supernovae............75
shocked quartz................... 187, **188**
shooting star...........................**333**
shoreline environments..................12
short-faced bear..................234–235
shrews 190–191, 212
SHRIMP II ion microprobe 41
shrimp-like crustaceans............ 135, 364
shrimps.................................56
Shubenacadie H-100 well 48
Shuldham Island, NL.....................80
Siberia......... 115, 173, 195, 230, 234–235, 238–239, 363, 374–375
Siberia (continent)............102, 132, 139
Siberia, location in Rodinia 97
Siberian Continental Shelf.......... 173, 175
Siberian Traps volcanism........... 159, 375
Sibley Basin...........................93, 96
siderite **19**, 84
side-scan sonar **50**
Sidney Spit, BC 314
Siemens, Friedrich......................259
Sigillaria.................127–**128**, 153–154
Signal Hill, St. John's, NL 286, 373
Sikanni Chief River, BC 164
silica **7**, 9, 19, 53, 250
silica rich magmas.......................82
silicates.................................**19**
siliceous iron carbonate.................255
silicon 7, 19, 76
silicon dioxide......................... 7,19
silicosis **264**, **352**
Sillery, QC.............................295
Sillery Sandstone 296
Sillery-Cap-Rouge Sandstone.......294–296
sillimanite 15–16, 18
sills......................................9
sills, mafic 8

silt 9, **10**
siltstone **9**
Silures 40
Silurian **40**, 53, 56–57, 60–61, 113, 121, 127, 131–138, 140, 143, 229, 256, 280, 284–285, 288, 288–289, 365
Silurian carbonate cap rock 71, 137
Silurian carbonate platform 135
Silurian, geological map 132
Silurian paleogeography 133
Silurian Period **132**
Silurian sea-floor scene 135
silver 19, 247, 249, 254, 258–259
silver, Aboriginal mining of 258
Silverwater Marble 285
simple impact structure **334**
single-celled eukaryotes 104
sink holes 225
sinters **250**, 251
Skagway, Alaska 247
skarn 248, **251**
skarn deposits **250**
skarn gold deposits 255
Skeena Mountains, BC 16, 72–73, 172
skeletal fluorosis 346, **350**
skeletons 52, 105
Skihist Provincial Park, BC 4
skin, source of fine atmospheric particles 351
skull structure 144
slate **15, 281**
slate, as building stone 289–290
slate, as commercial term 281
Slate Islands impact structure, ONT 337
slate, uses of 290
Slave Craton . . .78–81, 87, 90–92, 112, 256, 354
Slave Craton, geological map 80
Slide Mountain Basin 141, 155, 169, 357, 361, 375
Slide Mountain Terrane . . . 141, 156, 157, 173
slides, as type of landslide **329**
Sliding Mountain, BC .141
sloths 105, 190
slugs 56, **59**, 105
sluice box 255
slump **329**
small shelly fauna **106**, 375
smallmouth bass 349
smectite clay **290**
smelter stacks, as source of mercury 348
smelters, as mercury emitters 343
smelting as source of arsenic 349
Smith Creek, NUN 174
Smith, William 39
Smithers, BC 198, 201, 211, 302
Smithsonian Institution 122
smog 339, 353
Smoking Hills, NWT 205
smoking tobacco, as source of PAHs 347
smoky quartz 20, **21**
snails (see also gastropods) . . . 52–53, 56, **59**, 117, 135, 186, 366
snakes 60, **62**, 186
Snap Lake Mine, NWT 258
snow 65
snow avalanches 321
Snow Lake, MB 92
snow line 239
snow melt 304
Snowball Earth and volcanoes 104
Snowball Earth hypothesis **104**

Snowball Earth 256, 375
Snowbird Orogen 90–91
soapstone, as industrial mineral 292
sodalite 94
Soddy, Frederick 41
sodium 19
sodium arsenite 350
sodium chloride 11
soft-bodied animals .122
soft-bodied animals, fossils of 104
soft plate collision 134
soil associated with kimberlite 256
soil maps 45
soil on glacial deposits 308
soil profile 14
soils 12, **14**, 68, 117, 126, 141, 348, 364, 366, 369
soils and selenium 350
solar activity 241
solar energy 228, 373
solar energy from Sun 276
solar heating 66
solar output 241
solar output changes 241
solar power **276**
solar radiation 227, 230, 240–241, 300
Solar System 22, 42, **75**–77, 333, 375
Solar System, age of 42
solar systems, origin of 76
solar winds 22, 333
solid organic waste .340
Solitude Range, BC 114
Somerset Island, NUN 71, 98, 138, 213, 256, 364
sonar **50**
Sooke, BC 199
Sooke Potholes Regional Park, BC 199
sour gas **268**
source rocks **266**–267, 268, 273
Sous-le-Capin, Québec City, QC 294
Sous-le-Fort, Québec City, QC 296
South Africa 83, 159
South America 24–25, 29, 234–235, 242, 359, 363
South American Plate .33
South Atlantic Ocean 363
South China (ancient continent) 102
South China, location in Rodinia 97
South Mountain Batholith, NS 8, 21, 145, 220, 288, 290, 351
South Nation River, ONT 330
South Pole 109, 242
South Saskatchewan River 1, 229, 284, 329, 332
South Shore, St. Lawrence River, QC 294
Southampton Island, NUN 57, 71
southeastern Asia, as analogy 135
Southern Oscillation **242**
Southern Rocky Mountain Trench . . . 113–114, 200, 210
Southern Rocky Mountain Trench, BC 303
Southminster United Church, Lethbridge, AB 283
space 44, 76
Sparwood, BC 68, 264
species **55**
species loss 339
sphalerite **19**, 249–251
Sphenophyllum 154
spherical microstructures 83
spicules 56
spiders 56, **58**, 143

spillways 233
Spinifex 79
spinifex texture **79**
spits 308, **314**, 319
sponges **56**, 58, 106, 117, 120, 122–123, 125, 135, 158–159, 168, 170,
sporangia **127**, 140–14
spore-bearing plants **128**
spores 117, 126, **128**–129, 140, 154, 375
spreading centres 9
spreading ridge volumes 118
spreading ridges . . . **27**–29, 31, 34, 91, 247–248
Springhill, NS 276
Springhill, NS, disasters 263
Springhill, NS tragedies 274
springs 304
Springwater, SK 333
spruce 128, 212, 218, 229
spruce cone 129
Squally Point, NS 311
Squamish, BC 177, 207, 342
squids 56, **59**
St George, NB 280
St George's Bay, NL 314
St. Anthony Basin 149
St. Croix River, NS 149
St. David's Buried Valley, ONT 5
St. Elias Mountains 30, 69, 72, 139, 208–209, 324, 364, 371–372, 374
St. George, NB 279
St. George Batholith, NB 133
St. George Granite 279, 288
St. Jean Baptiste, Manitoba 372
St. John's Church, Saint John, NB 279
St. John's, NL 99, 101, 135, 270–271, 284, 286, 288, 340, 373
St. John's Railway Station, NL 288
St. John's, NL, early fortifications 286
St. John's, NL, fire of 1892 282
St. Lawrence Estuary 244, 297, 315, 317
St. Lawrence Graben 108
St. Lawrence Lowlands 69, 112, 182, 237, 297, 302, 330, 369–370, 374
St. Lawrence North Shore 115
St. Lawrence Platform 293–295
St. Lawrence River valley 330
St. Lawrence River 14, 93, 108–109, 287, 293–294, 303, 315, 324, 339, 340
St. Lawrence Seaway 304, 318
St. Lawrence South Shore, QC 108–109
St. Lawrence Valley 237, 297
St. Martins, NB 15, 105
St. Mary's Basilica, Halifax, NS 288
St. Marys Bay, NS 258
stampeders 247
Stanstead, QC 287
Stanstead Grey Granite . . . 281, 287–289, 296
starfish 56, 59, **60**
stars 76
Statue of Liberty, New York 296
statues 282, 289
staurolite 15–16
steel 259, 283
steel knife 21
steelworks, as mercury emitters 343
Steen River impact structure, AB 337
Steenson, Nicolaus 39
Steep Rock Lake, ONT 47, 83–84
Stellarton Sub-basin, NS 149
stems 140
Steno 39

Stephan's Quintet 75
Stephenville, NL 314
Sterling Mine, NS 344
Sternberg, Charles H. .52
Sternberg, Charles M. 52
Sternberg, George 52
Sternberg, Levi 52
Stewamus Chief, BC .177
Stewart Museum, Montréal, QC 288
sticklebacks 143
Stigmaria **154**
Stikine Plateaus 72
Stikine Volcanic Belt 73, 192, 208, 231, 327
Stikinia 155–157, 165–166, 172–173, 250, 357, 360–361, 374
Stikinia, island arcs on 169
stockwork **250**–251
Stokes Point, YK 310
stomach **61**
stomata **126**
Stone Church, Saint John, NB 279
stone **281**
stone tools 238
stone carvings 280, 283, 287
stone, local 282
stone, on Prairies 281
stones, glacial transport 281
Stoney Creek Oil and Gas Field, NB 270
stony-iron meteorite 333, **334**
stony meteorite 333–**334**
Stony Rapids, SK 372
stoping 8
Storm Hills, YK 205
storm ridges **312**
storm surges 332
storm water runoff 340
storm waves 313
storms 318, 321
storms, effect on coastline 312–313
Strait of Belle Isle 238
Strait of Canso, NS 292
Strait of Georgia, BC 219, 291, 323
Strand Fiord, NUN 151
strata 9, 16, 39
Strathcona Fiord, NUN 212
Strathcona Mine, ONT 254
stratified oceans 120, 248
stratigraphic traps 267–**268**
stratigraphy 9
stratosphere 65
stratotype sections 44
stratovolcanoes **8**, 207, 327–328
streak 21
streak plate 21
stream discharge 244
stream ecosystems 301
streamer **47**
Streptelasma 57
striations (glacial) 89, **220**–221
strike-slip fault 17
string bogs **225**
string bogs, James Bay Lowlands, near Quebec-Ontario border 226
strip mining 263
stromatolite-like features 84
stromatolite-like structures 375
stromatolites **83**–**84**, 88, 97, 121, 138, 364
stromatolites, earliest 55
stromatolitic reefs 98
stromatoporoid sponges 117
stromatoporoid-coral reef 147

stromatoporoids **56**, 117, 119, 136, 138, 147, 268
Struan, SK. 70
structural traps 267–**268**
sturgeon. 164
Sturtian glaciation 103–104
subaerial trackways. 113
sub-bituminous coal. 14, **263**
sub-bituminous coal, use of 263
subducting oceanic lithosphere 169
subducting plates 32
subduction **28**, 33, 67, 363
subduction-dominated plate boundary ... 374
subduction zone, Cordilleran inception ... 141
subduction zone in Ural Ocean 111
subduction zones ... 9, 28–29, 31, 33, 36, 66, 82, 91, 169, 364
subduction zones, deformation at 165
subduction zones, metamorphism at 165
submarine canyon. 356
submarine channel 12
submarine fans. **12**, 138
submarine landslides 156, 165, 316, 327
submarine volcano 165
subperiods. 41
subsidence 5, 18
subsidence after earthquake 324
subsoil 14
subsurface sections. 39
Sudbury, ONT 22, 88, 90, 108, 252–253, 259, 341–342, 355
Sudbury Basin. 253–254
Sudbury Igneous Complex **253**
Sudbury impact event 375
Sudbury impact structure, ONT 252–254, 334, 336–337, 355
Sudbury impact structure, tectonic compression of. 252
Sue C open uranium pit, SK. 276
Suess, Eduard 66
Suffield Experimental Proving Ground, AB 345
Sugarloaf Mountain, NB. 134
sugars 126
Sullivan, BC. 249, 259
Sullivan Mine, BC. 95, 249
sulphates **19**, 247
sulphide minerals 249–250, 342, 349
sulphide ore minerals. 251
sulphide ores 254
sulphide particles. 248
sulphide-rich liquid. 250
sulphide-rich metamorphic rocks 252
sulphide-rich sedimentary rocks 252
sulphides **19**, 247–248
sulphur 19, 66, 83, 265
sulphur, as industrial mineral 292
sulphur, as resource associated with petroleum 268
sulphur dioxide 343
sulphur dioxide in atmosphere 353
Sulphur Mountain, AB 46
Sulphur Mountain Thrust 46
sulphur-rich fluids 249
sulphuric acid 342–343
Sun 44, 75, 228, 240–241, 288, 333
Sun Life Building, Montréal, QC. 288
Suncor Energy Centre, Calgary, AB ... 287
Sundance Canyon, AB 179
sunlight 126, 128
Sunnybrook Drift. **218**
Sunwapta Pass, AB 146

supercontinent cycle **37**
supercontinents 24, **36**–37, 354
superheated water 31
Superior Craton 78–81, 83, 87–93, 97, 112 226, 249, 253, 255–256, 336, 354–355, 365, 375
Superior Craton, geological map 81
supernova events **74**
superocean **36**
surface runoff 339
surface veneer 10
surficial deposits **45**
Sussex, NB. 149, 182
Sustut Basin, BC 173, 178
Svalbard, Norway. 111
Sverdrup Basin ... 159, **173**–175, 193, 267, 375
Sverdrup, Otto 192
Sverdrup Seaway 150–**151**, 158, 375
Sverdrup Seaway, deepening of 151
Swallowtail Head, NB 110
Swallowtail Light, NB 32
swamp cypress 204, 212
Swan Hills, AB. 205
Swan Hills Complex, AB 147–148
swans 212
swash **312**
Sweetgrass Arch 119, 137, 146
Sweetgrass Hills, Montana 205
sweetgum 229
swell waves **312**
Swift Current, SK 205
swifts. 204
Swiss Alps 239
Switzerland 40, 219
sycamore 194
Sydney, NS. 53, 349
Sydney Harbour, NS 263
Sydney Street, Saint John, NB. 279
Sydney Sub-basin 149, 152
sylvinite 147
sylvite 147
symbiosis 126
symmetrical ripples/dunes **11**
synapsids 56, 60, **62**, 163, 189
synclines **16**–17, 46
Syncrude 272
system **40**

T
T. rex Discovery Centre, SK 185
table salt. 19
tabulate corals 159
Taconic metamorphism 290
Taconic Mountains 116, 118
Taconic Orogeny **114**–116, 294–295, 357–358, 375
Taconic Seaway **108**, 114–115, 357
Taft Creek, BC 172
Tagish Lake meteorite 334
tail (of a comet) **333**
tail trace 107
tailings **341**–342
tailings, discharged into sea 341–342
tailings ponds 291
tailings, treatment of 342
talc. 21
talc, as industrial mineral 292
Talmudic prophecy 38
Taltson-Thelon Orogen 90–91
talus cone **329**
Tangshan earthquake, China 325
Tantramar Marshes 318

tapirs 191, 206
Tarim, location in Rodinia 97
tarsiers 191
Tasiuyak Arc, NL 91
Tathlina Arch. 119, 137, 146
Tawuia 98
Taylor, Frank 24
TCU Financial Group Building, Saskatoon, SK 284
tea family 204
tectonic plates **23**, 28, 29, 31, 37, 77, 321–322
tectonic processes 66
teeth **61**, 143, 189, 236, 368
teeth, differentiation of 189
teeth, evolution of 106
tektites **188**, 335
teleosts 164
telephone cables 26
Temagami, ONT 67, 84
temperatures, global 375
Ten Mile Pond, Gros Morne NPC, NL 220
Terminonaris 186
Terra Nova Oil Field 271
Terrace, BC. 291, 327
terranes **33**, 357, 361
terranes across North Atlantic 109
terrasse Dufferin, Québec City, QC ... 296–297
terrestrial ecosystems 159
terrestrial invertebrates 366
terrestrial organisms **52**
terrestrial planets **22**
terrestrial plant communities 366
terrestrial spheres 65
Terror 192
Tertiary 40, 129, 187, 192–194, **195**, 196–215, 363
Tetagouche-Exploits Basin 116, 133–134, 357, 375
Tête Jaune Cache, BC 103
tethered pizza disc 104
Tethys **32**, **162**, 171, 359, 361, 374
tetrahedral structure. **19**, 21
tetrahedron **19**
tetrapod ancestors 365
tetrapod community, earliest 149
tetrapod footprints 53
tetrapod limbs 144
tetrapod tracks 149
tetrapods 56, **61**, 143–144, 155, 159, 375
Texada Island, BC 290–291
Texas 235, 374
thalattosaurs **164**
Thayer-Lindsley Mine, ONT 253
The Ancient Wall, AB 180
The Barrier, BC 330
The Big Dry 242
The Chief, BC 177
The Cremation of Sam McGee 246
The Gully, offshore NS 205
The Manitoba Museum 117
The Map That Changed the World 39
The Origin of Continents and Oceans 24
The Rooms, St. John's, NL 288
The Topsails, NL 288
Thelon Basin 93–95, 275
Thelon-Taltson Orogen 118
thermal convection 22, 66
thermo-clock dating 43
thermogenic gas 273
thermokarst landscape **225**–226

thermokarst, Wapusk NPC, MB. 226
thermosphere. **65**
theropods 174, **185**
Thetford, QC 20
Thetford Mines ophiolite, QC 115
thin section **21**, 77
Thingvellir, Iceland 31
Thompson, MB. 92, 250, 355
Thompson Belt, MB. 91
Thompson Glacier, Axel Heiberg Island, NUN. 362
Thompson River, BC 4
Thomson, Tom 78
Thomson, William 41
thorium 22, 351
thorium-292 43
Thorncliffe Beds 218
Three Valley Gap, BC 201
thrust-and-fold belt of the Rocky Mountains 177, 360, 362–363
thrust-and-fold belts 32–**33**
thrust faults **17**, 33,36, 46, 138, 294, 374
thrust sheets 32, 294
Thule Basin 93
Thule Inuit 238–239
Thule villages 239
Thunder Bay, ONT . 20, 96, 233, 319, 348–349
Tibet 32
Tibetan Plateau 32, 242
tidal channels 315
tidal currents 12, 310, 317
tidal cycles 76
tidal deposits 181
tidal flats 12, **315**
tidal gauges 327
tidal marshes 316, 318, 324
tidal power **276**–277
tidal power potential in British Columbia .. 277
tidal wetlands 244
tide-gauge records 237
tides 313, **318**
tides and resonance 318
tides, causes **317**
Tignish, PEI 67
Tijuana, Mexico 28
Tikkoatokak Bay, NL 95
Tiktaalik 61, 144, 365, 375
Tiktaalik roseae 144
till **217**, **221**–222, 227, 229, 341, 348, 363–364, 369
till deposits 219
tillites 8, **89**, 103, 120, 375
Tilt Cove Mine, NL 341
tilt to Earth's axis **76**
tilted strata 13
time (see geological time)
Timiskaming Graben. 108
Timmins, ONT. 19, 81, 249, 256
Tingin Fiord, NUN 309
Tintina Fault 177
Tintina Trench 200
Titanic 282, 296
Titanic gravestones 8
titanium 21
Titusville, Pennsylvania 262
Tofino, BC. 324, 326
Tofino Basin. 173, 199
Tofino Basin, oil exploration in. 270
Tofino Basin, petroleum exploration 200
Tolman Bridge, AB 185
tombolo **314**

Tombstone Pluton.......................177
Tombstone Range, YK 177
Tonga...................................31
Tonga-Kermadoc-New Zealand region...109
tool marks (in sediment) **12**
toothed birds 186
toothpaste..............................19
topaz...................................21
topographic maps........................70
topography.............................369
topography, early 82
topography of Canada **69**
topple, as type of landslide.............**329**
Torbay, NL.............................286
tornadoes..........................243–244
Torngat Mountains, NL 35, 70
Torngat Orogen 90, 93, 95
Toronto, ONT 71, 218, 222, 233, 259, 284, 318, 331, 339–340, 343, 374
Toronto, ONT, fire of 1904...............282
Toronto, City of, ONT 301
tourmaline..........................21, 94
toxin solubility 353
trace fossils **54**, 58, 105–106, 364, 375
traces of blood vessels 144
tracheophytes.................126–**127**, 375
tracks..................................**54**
trackways........**54**, 107, 154–155, 159, 185
trackways on land 107
trade winds 242
trails..............................**54**, 365
transatlantic cables 14
Trans-Canada Highway 88, 90–91, 96, 113–114, 118, 141, 165, 178–181, 201
transcontinental railway 282
transform faults **28**, 31, 36, 363–364
transgression **30**, 118–119
Trans-Hudson mountains, erosion of......94
Trans-Hudson Orogen 92, 112, 120, 246, 249, 355, 375
Trans-Hudson Orogeny 91–92, 355
transitional oceans 256
transpiration..........................**300**
transportation of sediment..............5, 66
tree line 238
tree of life..............................**55**
tree-ring counts.....................42–43
tree rings 187, 241
tree roots 9
trees 127, 217, 366, 369, 375
trees, early 128, 142–143, 262
trees, fossil 193–195
Treherne, MB 186
trenches............**26**, 33, 36, 248, 323, 373
Trenton, ONT..........................112
Trenton Limestone 285, 291
Treptichnus.............................106
Treptichnus worm burrow 44
Triassic 40, **41**, 46, 67, 156–157, 159, 162–171, 173–175, 189–190, 250, 284, 359, 366–367
Triassic eolian deposits..................168
Triassic flash flood 168
Triassic fluvial deposits.................168
Triassic fossils 164
Triassic, geological map.................162
Triassic-Jurassic boundary..............168
Triassic-Jurassic paleogeography of Arctic 174
Triassic lacustrine deposits 168
Triassic magmatism 362
Triassic marine sediments..............174

Triassic paleogeography 163
Triassic redbeds 151
Triassic rifting 359
Triassic sand dunes 168
Triassic source rocks 274
Triceratops 185
trigonal system........................**21**
Trilobite Beds 122
trilobites... 40, 56–57, **58**–59, 106–107, 117, 119, 121–122, 159, 365, 375
trimline...............................241
Trinacromerum kirki...................186
Trinity Royal Preservation Area, Saint John, NB 280
Tripp, Charles 261–262
Trois Rivières, QC 94, 270
tropical forests 366
tropics.................................65
troposphere **65**–66
Trout River, NL 311
Trout River Pond, NL..................358
Truro, NS 291
Truro police station, NS 291
Tseax Cone, BC 208, 327
Tseax eruption 208
tsunami caused by Grand Banks Earthquake 326
tsunami-deposited sediment............324
tsunami risk, maps of 327
tsunami warnings 327
tsunamis 308, 321–322, **325**–327, 332
tuatara 168
tube worms...........................247
tufa 175
Tuktut Nogait NPC, NWT 111, 226
tulip tree 212
Tulita Spring, NWT....................301
Tumbo Island, BC......................177
tundra................................369
tundra biomes 237
tundra environment 218
tundra shoreline......................310
tundra vegetation 229, 235, 243
turbidites...........12–**14**, 81, 95, 103, 326
turbidites, in Sudbury Basin 253
turbidity currents............ 12, **14**, 205, **330**
Turner Valley, AB 269–270
Turner Valley, Royalite No. 4 well, AB.....269
Turtle Mountain, AB 146, 320–321, 330
turtles 56, 60, **62**, 185
Tutankhamun.........................247
tuyas 27, 230–**231**
Tuzo Wilson, see Wilson, John Tuzo
Tyaughton-Methow Basin...............173
Tyler, Stanley......................85–86, 97
Tyndall Stone 119–120, 126, 281–282, 284, 355, 375
typhoons 243
Tyrannosaurus rex 51, 185
Tyrrell Sea 232–233, **236**
Tyrrell, Joseph..........51–52, 219, 264, 372

U
U-shaped valleys 220–221
Ucluelet, BC 326
ultramafic magmas......................30
ultramafic rocks **7**, **9**, 22, 34
ultramafic rocks, ores in.................250
ultramafic volcanic rocks 7
ultraviolet rays 65
umiaks 238
unaltered preservation..................**52**

unconformity **18**, 150, 268, 294
unconventional petroleum reserves......**273**
undersea cable breaks, Grand Banks Earthquake 326–327
UNESCO World Heritage Sites 51, 54, 61, 122, 143, 152, 178, 184, 238, 293, 371
Ungava Bay, QC 89, 277
Ungava Bay, as estuary.................317
Ungava Peninsula, QC 336
United Nations Convention for the Law of the Sea (UNCLOS) 174
Universe, origin of **74**
University of Alberta, Edmonton, AB 291
University of Calgary, AB 333
University of Regina, SK................276
University of Saskatchewan, Saskatoon, SK.....................284
University of Saskatchewan Observatory, Saskatoon, SK........281
University of Toronto 216
University of Wisconsin, USA 85
untreated sewage 340
Unuk River valley, lava flow.............327
uplift 5, 18, 22, 33, 35, 43
uplifted rim 334
Upper Canada....................130–131
upper mantle 22
Upper Town, Québec City, QC 294
upwellings....................... 68, 370
Ural Mountains...................355–356
Ural Ocean passive margin 111
Ural Ocean.........102, 110–111, 115, 132, 138–140, 155, 356–357, 375
Ural Ocean, microcontinents in 138–139
uraninite..................88, **274**–275
uranium... 22, 43, 88, **274**–276, 342–343, 351
uranium-235...................... 42, 274
uranium-238............. 41–43, 274, 351
uranium, as an export...................275
uranium, Canadian reserves 275
Uranium City, SK 337
uranium deposits................. 95, 260
uranium isotopes......................274
uranium mineralization in Athabasca Basin 275
uranium mining.......................337
uranium ore 275, 342, 355
uranium-rich layers.....................88
uranium, sources of....................351
Uranus 22
urban air pollution, sources of 353
urban development 339–340
urbanization 339
Ussher, James 38

V
Val d'Or, QC...........................81
Val d'Or mining camp, QC..............256
valley glaciers 69, **219**, 222–223
valleys, pre-glacial.....................221
Vancouver, BC.......8, 28, 107, 198, 207, 208, 210–211, 240, 259, 276, 283, 287, 291, 316, 323, 326–327, 339
Vancouver, BC, fire of 1886..............282
Vancouver, Greater, BC.................339
Vancouver Island, BC 20, 37, 49, 70, 166, 172, 178, 198–199, 207, 248–250, 259, 264, 287, 291–292, 304, 308, 311–312, 323–324, 326, 340
Vancouver Island, BC, earthquakes.......323
Vancouver Winter Olympics, 2010 342
Varley, Frederick.......................78
varves **224**

varves, Lac Témiscamingue, QC..........225
vascular cambium **127**
vascular plants 127, 365, 375
vascular plants, early...................140
vascular system (plants) ... 126–**127**, 365–366
veins (mineral) **16**, 250–251
velvet worms 123
vents................................7–8
Venture Gas Field......................271
Venus 22
vertebrate evolution 143
vertebrate fauna 367
vertebrates 52, 55–56, **58**, **60**–63, 368
vibrating trucks 47
Victor Mine, ONT 258
Victoria, BC .199, 259, 283, 287, 292, 323, 340
Victoria, Greater, BC 340
Victoria Island, NUN and NWT 71, 95, 213, 335
Vieux séminaire de Saint-Sulpice, Montréal, QC.....................285
Viewfield impact structure, SK...........337
Viking settlements.....................239
Vikings 239, 240, 258
Vine, Fred...........................27
Virgiana..............................137
viruses, source of fine atmospheric particles................351
viscosity 7
Viscount Melville Sound 71
vitreous lustre.........................**21**
VMS deposits (see also volcanogenic massive sulphide deposits).......... 246, **248**–**249**, 254, 256, 260
Voisey Bay, NL. 95, 250, 259
volcanic ash **6**, 8, 169–170, 290, 328, 351–352
volcanic ash, as health hazard 352
volcanic ash, damage from.............328
volcanic bombs **6**, 8, 328
volcanic breccia....................**8**, 165
volcanic breccia, as building stone 288
volcanic eruptions 64, 68, 208, 241, 321–322, 327–329, 332
volcanic eruption as tsunami trigger 325
volcanic eruptions as source of arsenic ...349
volcanic eruptions, as source of PAHs 347
volcanic gases.......................8, 82
volcanic glass 6
volcanic necks......................**8**, 73
volcanic outpourings 374
volcanic particles......................241
volcanic rocks 5, **6**–9
volcanic rocks, as building stone.........287
volcanic rocks, felsic 7
volcanic rocks, intermediate.............7
volcanic rocks, mafic...................7
volcanic rocks, ultramafic 7
volcanic vents........................22
volcanism............................375
volcanism in Canada..................327
volcano, dormant **320**
volcanoes............6, 8, 27, 35, 66–67, 231
volcanoes, as landslide triggers..........329
volcanoes, as source of mercury 348
volcanoes, cone-shaped 7–8
volcanoes, gold associated with 255
volcanoes, source of fine atmospheric particles................351
volcanogenic massive sulphide deposits (see also VMS deposits) **248**
Vosges granite 296
voyageurs 78

W

Wabamun Lake, AB 264
Wabush, NL 89, 365
Wadati, Kiyoo 28
Walchia 158
Walcott, Charles 122, 125
Walcott Quarry 122
Wales 40
Walkerton, ONT 338
Walkerton Commission 338
Walkerton tragedy 338–339
walking traces 107
Wallace Sandstone 285–286
walnut 194, 212
walnut seeds 194
walruses 191, 212, 237
Wanapitei impact structure, ONT 337
Wapiti Lake, BC 164
Wapta Mountain, BC 122
Wapusk NPC, MB 10, 226, 313
War of 1812 284
warm-bloodedness 189
warm currents **66**
warming trends 374
warmth-loving plankton 196
Warner, AB 185
Washington (State) 37, 207, 231, 327
Wasson Bluff, NS 11, 168
waste rock **341–342**
wastewater disposal 339
wastewater treatment plant 340
water bodies, contaminants in 340
water 66, 245, 298–**299**, 300–306
water, agricultural use 305
water, and Earth's position
    in solar system 300
water, Canada's resources 302
water, commercial and industrial use . 305
water cycle **300**, 302
water cycle, in budgetary terms 302
water, domestic use 306
water gauging stations 304
water lily 204
water, locked in ice sheets 310
water management 304
water molecule 299
water, municipal use 305
water, natural reservoirs **303**
water on early Earth 78, **82**
water pollination 129
water, properties of 299
water quality 340, 345
water quality testing 349
water recharge **301**
water reservoirs 302
water, residential use 305
water runoff 303
water seeps 304
Water Street, St. John's, NL 288
Water Street, Vancouver, BC 283
water supply problems 305
water table 49, **301**, 303, 304, 345
water, use in manufacturing 305
water, use in mining 305
water, use in petroleum industry .. 306
water vapour 65–66
water wells 347
water willow 204
Waterloo, ONT 43, 52, 223, 341
watersheds 244, 339
watersheds, hardening of **339**
waterstrider 204
Waterton Lakes, AB 161
Waterton Lakes NPC, AB 94, 269
Watson Lake, YK 20
wattle and daub 281
wave action 12
wave-cut terraces 224
waves 11–12, 310–311, **312**, 313–314, 317
waves, breaking of 312
waves, cause of 312
Wawa, ONT 20
weasels 191
weather **240**
weather and radar stations 193
weathering **9**–10, 14, 66–67, 347–348
weathering as source of arsenic ... 349
Wegener, Alfred 24–26, 29
well logs **47**
Welland Canal, ONT 284
Wellington Arch 119
wells **45**, 304–341
Wells Gray Provincial Park, BC 208
Wernecke Basin 18, 93, 95
Wernecke Mountains, YK 18
West Africa, location in Rodinia 97
West Alberta Ridge 137, 146–147
West Bay, NS 12
West Georgia Street, Vancouver, BC . 287
West Hawk impact structure, MB ... 337
West Indies 214
westerlies 65
Western Canada Sedimentary Basin ...**180**, 269
Western Interior Basin 37, 173, **180**–181, 185–187, 205, 363, 368, 374
Western Interior Plains 36, **70**, 112, 137, 145–146, 180, 182, 184–185, 203, 205, 222, 235, 237, 369
Western Interior Plains,
    Cenozoic deposits 203, 205
Western Interior Plains, Cenozoic
    landscape of 206
Western Interior Plains, Neogene faunas ...214
Western Interior Seaway **173**, 180–182, 184–186, 187, 374
Western Newfoundland Highlands, NL . 149
western Pacific Ocean 169, 359
western Pacific Ocean, as analogy . 155
Western Platform **112**–113, 118–119, 146, 148–149, 152, 180, 196, 363
Westham Island, BC 316
Westray tragedy 263, 274
westward drift of North America .. 172
wet gases **265**–266
wetland ecosystems 301
wetland environments 303
wetlands 14, 262–263, 302, 340
wetlands as habitat 315
Weyburn Oil Field, SK 344
whale hunting 239
whale oil 262
whalebones 239
whales 191, 237–238
Whirlpool, ONT (see Niagara Whirlpool)
Whirlpool Sandstone 284–285
Whistler, BC 230, 330
White Mountains intrusions,
    New Hampshire 182–183
White Man Gap, AB 268
White Mountain Pass 247
white pine 218, 229
White River Ash 208
White Rose Oil Field 271
white spruce 218
Whitea 164
Whitehorse Trough 173
Whitehorse, YK 21, 197, 271
Whitney Block, Toronto, ONT 285
Whittington, Harry 122–123
wildfire 321
wildfires as source of arsenic 349
Williams, James 262
Williams Lake, BC 210–211
Williston Basin 118–119, 146
Willow Creek, AB 187
willows 193, 212
Wilson Cycle **37**
Wilson, John Tuzo 28, 35, 37, 372
Winchester, Simon 39
wind dispersal 128
wind patterns 65
wind pollination 129
wind power 276, 343
Windermere rift basin 103
Windermere Seaway **103**–104, 110, 356–357, 375
Windermere strata 103, 114
Windsor, NS 149
Windsor, ONT 137, 375
Windsor Sea 149, 152
Windsor Sea evaporates 149
Windsor Station, Montréal, QC ... 285–286
Windsor succession 150
Winnipeg, MB ... 20, 90, 117, 146, 231, 247, 284, 306, 319–320, 331–332, 343
Winnipeg, MB, water supply 306
Wisconsin Arch 119
Wisconsinan Glaciation **227**
Witchekat Creek, MB 10
Witty`s Lagoon Regional Park, BC . 199
Witwatersrand, South Africa 255
Wiwaxia **123**–125
Wolfville, NS 118, 290
Wollaston Lake, SK 276
wolves 234
wombats 190
Wonderful Life 123
wood 52, 366
wood bison 237
Wood Buffalo NPC, AB and NWT ... 237
wood, evolution of 128
Wood Mountain, SK 205, 214–215
Wood Mountain Creek, SK 187
wood, origin of **127**
wood preservatives as source of arsenic ... 349
woodland savannah 206
woodlands 214
Woodstock, ONT 226, 259
Woody Cove NL 229
woolly mammoth 64
Wopmay Orogen 90–91
World War I 259
World War II 193, 259, 271, 275, 313
worm burrow 44, 106
worms 14, 53, 56, **58**, 106, 122, 175
Wrangellia 152, 155–156, 165–166, 172, 173, 176, 178, 198, 249–250, 357, 360–361, 374
Wrangellian margin of North America ... 199
Wrangellian volcanic rocks 165
wrist bones 144
Writing-on-Stone Provincial Park, AB ... 185
Wyoming Craton 90

X

xenarthrans **190–191**
xenarthrans, origins in South America ... 190
xenoliths **8**
Xiphactinus 186

Y

Yabeina 157
Yakutat Terrane 197, 199, 208–209, 324
Yamnuska, AB 46, 179–180
Yarmouth, NS 351
Yellowhead Highway 103
Yellowknife, NWT 78, 80–81, 90, 146, 256, 349
Yellowknife Fault 80
Yellowstone hot spot 198, 208
Yellowstone, USA 35
Yoho NPC, BC 54, 122
York Till **217**–218
Younger Dryas **233**–234, 374
Yucatán Peninsula, Mexico 64, 187
Yukon 18, 20–21, 49, 69, 95–98, 110, 161–162, 165–166, 172, 174, 176–177, 198, 208–211, 229–231, 234–236, 243–244, 246–247, 249, 254–255, 259, 264, 272, 276, 305, 310, 316, 324, 327, 347–348, 364, 370, 372, 374–375
Yukon, crustal stretching 174
Yukon Plateaus 72, 211, 231
Yukon River 247, 374
Yukon River basin 206
Yukon River basin, ancestral 206
Yukon River, evolution of flow 210–211
Yukon-Tanana Terrane ... 155–156, 166, 173, 357, 361

Z

Zahursky Point, SK 187
Zeballos, BC 326
zeolite, as industrial mineral 292
zinc 19, 247, 249–250, 254, 258, 350
zinc-lead ore 250
zinc ores 92, 95, 142, 148
zinc sulphide 19
zinc, uses of 260
zircon crystals 42
zircons 42, 97
zone of low pressure 65
Zosterophyllum 140
Zygolophodon 63